Science and Fiction

For further volumes:
http://www.springer.com/series/11657

Science and Fiction – A Springer Series

This collection of entertaining and thought-provoking books will appeal equally to science buffs, scientists and science-fiction fans. It was born out of the recognition that scientific discovery and the creation of plausible fictional scenarios are often two sides of the same coin. Each relies on an understanding of the way the world works, coupled with the imaginative ability to invent new or alternative explanations - and even other worlds. Authored by practicing scientists as well as writers of hard science fiction, these books explore and exploit the borderlands between accepted science and its fictional counterpart. Uncovering mutual influences, promoting fruitful interaction, narrating and analyzing fictional scenarios, together they serve as a reaction vessel for inspired new ideas in science, technology, and beyond.

Whether fiction, fact, or forever undecidable: the Springer Series "Science and Fiction" intends to go where no one has gone before!

Its largely non-technical books take several different approaches. Journey with their authors as they

- Indulge in science speculation – describing intriguing, plausible yet unproven ideas;
- Exploit science fiction for educational purposes and as a means of promoting critical thinking;
- Explore the interplay of science and science fiction – throughout the history of the genre and looking ahead;
- Delve into related topics including, but not limited to: science as a creative process, the limits of science, interplay of literature and knowledge;
- Tell fictional short stories built around well-defined scientific ideas, with a supplement summarizing the science underlying the plot.

Readers can look forward to a broad range of topics, as intriguing as they are important. Here just a few by way of illustration:

- Time travel, superluminal travel, wormholes, teleportation
- Extraterrestrial intelligence and alien civilizations
- Artificial intelligence, planetary brains, the universe as a computer, simulated worlds
- Non-anthropocentric viewpoints
- Synthetic biology, genetic engineering, developing nanotechnologies
- Eco/infrastructure/meteorite-impact disaster scenarios
- Future scenarios, transhumanism, posthumanism, intelligence explosion
- Virtual worlds, cyberspace dramas
- Consciousness and mind manipulation

H. G. Stratmann

Using Medicine in Science Fiction

The SF Writer's Guide to Human Biology

 Springer

H. G. Stratmann
Springfield
Missouri
USA

ISSN 2197-1188　　　　ISSN 2197-1196 (electronic)
ISBN 978-3-319-16014-6　　ISBN 978-3-319-16015-3 (eBook)
DOI 10.1007/978-3-319-16015-3

Library of Congress Control Number: 2015940326

Springer Cham Heidelberg New York Dordrecht London

Springer International Publishing is part of Springer Science+Business Media (www.springer.com)

To my wife, Dr. Maryellen Amato, and our sons, Henry III and Joseph. You have enriched my life beyond measure.

Preface

Over the past two decades I have given talks and been on panels dealing with space medicine, suspended animation, medical nanotechnology, and similar topics at science fiction conventions. Audience members have often asked me where they could read more about these subjects that blend medicine with science fiction. Until now I could only tell them that information is scattered over numerous books, research papers, and other sources, with many of them written at a technical level not geared for a general readership. Some of those individuals have urged me to write a book covering those topics.

Using Medicine in Science Fiction is my response to that request. Its purpose is to describe how the human body actually works, the myriad ways that space travel, radiation exposure, and other factors affect it, and the biological challenges to modifying and "improving" it. Science fiction writers can use this information to make their depictions of human biology more accurate and plausible. Readers will gain a better understanding of the medical science underlying such commonly used plot elements as life extension, cloning, genetic engineering, and bionics.

The subjects I cover in this book are complex and vast. No single work can comprehensively include what is currently known about them, and new research will further expand our knowledge and capabilities. Instead each chapter focuses on presenting core concepts about human biology and medical care, summarizing the current status of individual topics (e.g. stem cells and organ transplantation), and describing what would need to be done to turn ideas now solely the provenance of science fiction into reality.

Science fiction does not need to be ultrarealistic in its portrayals of medical issues to be effective, enjoyable, and thought-provoking as fiction. Sometimes a little "blurring" of medical detail is, in fact, justified by the greater goal of making a story, television show, or movie succeed as a whole. A science fiction work can still triumph as an exercise in imagination, entertainment, and/or exploration of the human condition even if the medicine in it stretches plausibility to or beyond its breaking point.

But there is a danger with putting the priorities of fiction over those of science. A work depicting suspended animation, advanced genetic engineering, or other medical elements well beyond current capabilities may give those unfamiliar with real medicine an erroneous impression of how "easy," plausible, or even possible they are. This can do a disservice by inadvertently "teaching" its audience something that might be dubious at best or perhaps simply wrong.

While I hope you find the information in this book interesting and enlightening in its own right, it is also meant to serve another very important purpose. It will give you a better appreciation for how much of a difference there is between current medical science and that shown in science fiction. Sometimes the gap between them is small and might be bridged in the near future. In other cases there is a huge chasm that will require overcoming tremendous challenges, with no guarantee that they can be actually be done at all or at least used in a practical way. And it will also help you identify what medicine is "impossible" so that you can, if you choose, decide to suspend your disbelief to enjoy a science fiction work rather than learn something that is actually not correct.

In medicine as in life, it is best for each of us to know what is actually true. The second best thing, as Socrates knew, is to know that we really do not know something with certainty, and to try our best to understand and learn it. By far the worst situation is to "know" something is true when it really is not, or is at least less certain to be right or more complex than we think. Similarly, medical science must also continuously reevaluate what it "knows" based on new discoveries, facts, and concepts.

This book gives a snapshot of where many medical concepts used in science fiction are right now in their development. However, even the most informed speculation on when or whether they will eventually become real has an unavoidably wide margin of uncertainty. Fortunately, whatever path the future of medicine does take, we have science fiction to help imagine what it might be.

Acknowledgments

The best thing about becoming a science fiction writer is that it has brought me into contact with so many fine, accomplished individuals in the field. This book would not have been possible without their inspiration, example, and help.

Two people in particular deserve special thanks. The first, Gerald (G. David) Nordley, is an astronautical engineer and creator of superb, scientifically accurate science fiction. Gerald offered me the opportunity to be his coauthor on a futuristic story with a medical theme. His proposal made me wonder for the first time whether I had the ability to go from just reading science fiction to actually writing it. Our collaboration (and my first sale!), "Tin Angel," was published in the July 1994 issue of *Analog Science Fiction and Fact*. Everything I have accomplished in the field since then I ultimately owe to Gerald, for which he has my undying gratitude.

The second person is Dr. Stanley Schmidt, who recently retired as editor of *Analog* after over three decades in that position. Encouraged by the success of "Tin Angel," I began submitting my own stories to *Analog* in the mid-1990s. Despite the many failings and weaknesses of those early efforts, Stan patiently encouraged me to keep trying and suggested ways to improve my writing. The fact that I was eventually able to meet the high standard of quality he set at *Analog* and sell so many stories and science fact articles to him is a testament to his incomparable abilities as an editor and mentor. Anyone who has ever met and worked with Stan cannot help but be impressed by his consummate professionalism, erudition, and friendliness. I feel privileged to have known and worked with him.

Writing for *Analog* has also had the delightful fringe benefit of enabling me to meet and form friendships with other contributors to that magazine. A common characteristic among them is that they combine successful careers in science-related fields with great skill as writers. Among examples too numerous to mention here I will single out Edward M. Lerner, Nick Kanas, and Brad Aiken. Ed, with a strong background in physics and computer science, is equally adept at writing top-quality short fiction, novels, and nonfiction. Like me, Nick and Brad have MD degrees and incorporate their own extensive medical experience into their works.

Next I would like to extend thanks and appreciation to my fellow members of the SIGMA science fiction think tank. Its membership includes the highest ranks of creative and productive science fiction writers, and it is an honor for me to be listed among the company of so many individuals whose writing accomplishments far exceed my own. Many of them responded to my request for what they considered good examples of medicine in science fiction and gave me excellent suggestions. Besides G. David Nordley, Stan Schmidt, and Ed Lerner, listing them in alphabetical order I am grateful for the help provided by Arlan Andrews, Sr. (the founder of SIGMA); Chris Christopher; Alan Dean Foster; Nancy Kress; Geoffrey A. Landis; Larry Niven; Jerry Pournelle; Allen Steele; and Bruce Sterling.

I am also grateful to Dr. Christian Caron, publishing editor for Springer's "Science and Fiction" series, for reviewing this book and providing me with helpful suggestions regarding it.

Finally, I wish to thank Dr. Robert Mayanovic, a professor and research physicist in the Department of Physics, Astronomy, and Materials Science at Missouri State University, for reviewing some of the calculations involving physics-related subjects I have included in this book. I am grateful to him for reassuring me that I correctly learned what he taught so well in the physics courses I have taken with him.

Contents

1 How the Human Body Works: From Quarks to Cells 1

  1.1 The Stuff of Life. 2
  1.2 What Is Life? 6
  1.3 The Chemistry of Life 8
  1.4 Cells and Their Contents. 13
    1.4.1 Cytoplasm. 14
    1.4.2 The Nucleus 15
  1.5 Human Anatomy. 18
    1.5.1 Tissues. 19
    1.5.2 Organs and Organ Systems. 20
  1.6 The Bottom Line 34
  References 36

2 Hurting and Healing Characters 41

  2.1 Common Illnesses 43
    2.1.1 Common Causes of Death. 43
    2.1.2 Cardiovascular Disease 44
    2.1.3 Infections 47
    2.1.4 Cancer. 51
    2.1.5 Miscellaneous Health Issues 53
  2.2 Common Injuries 54
    2.2.1 Penetrating Injuries. 55
    2.2.2 Blunt Trauma 69
    2.2.3 Burns. 71
    2.2.4 Explosions. 73
    2.2.5 Poisons and Toxins. 75
  2.3 Delivering Medical Care 76
  2.4 "He's Dead, Jim." 82
  2.5 The Bottom Line 85
  References 86

3 Space Is a Dangerous Place 89

  3.1 The Weakest Link 91
  3.2 Atmosphere and Pressure. 92
  3.3 Temperature, Toxins, and Trauma 99
  3.4 Psychological Stress. 105

3.5   Circadian Rhythms and Sleep 107
3.6   Risks of Meteoroids and Space Debris 108
3.7   Acceleration and Deceleration 113
3.8   The Bottom Line 114
References 117

4   **Microgravity and the Human Body** 121
4.1   Space Adaptation Syndrome 122
4.2   Cardiovascular Effects 125
4.3   Hematological Effects 127
4.4   Musculoskeletal Effects 128
4.5   Effects on Vision 132
4.6   Gastrointestinal Effects 133
4.7   Genitourinary and Endocrine Effects 134
4.8   Pulmonary Effects 135
4.9   Dealing with Microgravity 135
4.10 The Bottom Line 145
References 146

5   **Space Medicine: Paging Dr. McCoy** 151
5.1   Medical Care in Space 152
5.2   Preventive Care 160
5.3   Exercise and Nutrition 165
5.4   Infectious Diseases 166
5.5   First Aid and Surgery in Space 169
5.6   The Bottom Line 178
References 184

6   **Danger! Radiation!** 187
6.1   What is Radiation? 188
6.2   Biological Effects of Radiation Exposure 190
6.3   Radiation Dosage and the Human Body 194
6.4   Realistic Settings for Radiation in Science Fiction 197
6.5   Radiation in Space 200
6.6   Protecting Against Radiation 206
6.7   The Bottom Line 207
References 208

7   **Suspended Animation: Putting Characters on Ice** 211
7.1   Suspended Animation and Hibernation in Animals 212
7.2   Suspended Animation and Humans 214
7.3   How the Human Body Responds to Cold 217
7.4   Therapeutic Hypothermia 221
7.5   Cells and Cold 230
7.6   Making Your Characters Chill Out 232
7.7   Prospects for Reviving the Frozen 236

7.8   Methods Equivalent to Suspended Animation   239
7.9   The Bottom Line   243
References   244

8   **Telepathy, Using the Force, and Other Paranormal Abilities**   249

8.1   Telepathy   250
8.2   Telekinesis   259
8.3   Precognition   264
8.4   Extrasensory Perception   266
8.5   Teleportation   270
8.6   The Bottom Line   276
References   278

9   **The Biology of Immortality**   281

9.1   The Dying of the Light   283
9.2   Longevity and Genes   287
9.3   Mechanisms and Theories of Aging   292
9.4   The Quest for Eternal Youth   297
9.5   The Hayflick and Other Limits   302
9.6   The Future of Aging   308
9.7   The Bottom Line   310
References   314

10   **Sex in Science Fiction**   321

10.1   Why Does Sex Exist?   322
10.2   Technology and Sex   328
10.3   Cloning   332
10.4   Sex in Space   335
10.5   Pregnancy in Space   342
10.6   Space Kids   349
10.7   Sex and Aliens   351
10.8   The Bottom Line   355
References   356

11   **The Promises and Perils of Medical Nanotechnology**   361

11.1   Nanotechnology and the Human Body   363
11.2   Medical Nanotechnology: The Dream   364
11.3   Medical Nanotechnology: The Reality   367
11.4   Risks of Medical Nanotechnology   380
11.5   The Bottom Line   383
References   385

12   **Genetic Engineering: Tinkering with the Human Body**   389

12.1   Genes, Chromosomes, and Nature   391
12.2   Potential Applications of Gene Therapy
       and Genetic Modification   397

12.3   Current Status of Gene Therapy and Genetic Modification   402
12.4   Genomic Medicine   409
12.5   Repairing and Enhancing the Human Body   413
12.6   Dangers of Genetic Engineering   420
12.7   The Bottom Line   423
References   424

13   **Stem Cells and Organ Transplantation: Resetting Our Biological Clocks**   429
13.1   Types of Stem Cells   430
13.2   Reprogramming Cells   439
13.3   Risks of Stem Cells   444
13.4   Organ Transplantation   449
13.5   The Bottom Line   458
References   461

14   **Bionics: Creating the Twenty-Four Million Dollar Man or Woman**   467
14.1   Blending Machine and Flesh in Science Fiction   468
14.2   How to Make a Cyborg   469
14.3   Bionic Limbs   470
14.4   The Bionic Eye   477
14.5   The Bionic Ear   482
14.6   Other "Bionic" Body Parts   484
14.7   Connecting Brains to Computers and Machines   487
14.8   Uploading Our Minds   492
14.9   The Bottom Line   499
References   502

15   **Summing Up: Using Medicine in Science Fiction**   507
15.1   "As You Know, Bob..."   508
15.2   Too Much or Too Little Medicine?   510
15.3   Say or Suggest?   513
15.4   The Limits of Extrapolation   514
15.5   What Price Progress?   524
15.6   The Bottom Line   527

Appendix   529

Index   537

1

How the Human Body Works: From Quarks to Cells

What a piece of work is a man! How noble in reason! How infinite in faculty! In form, in moving, how express and admirable!

William Shakespeare

Hamlet

Man is the measure of all things.

Protagoras

Science fiction is created by and for human beings. By definition it sets characters in times, places, and situations that subtly or overtly involve scientific principles. A science fiction work might take place in Earth's future, such as with the *Star Trek* universe. It may occur on worlds other than our own, as near as the Mars of Ray Bradbury's *The Martian Chronicles* stories or as remote as the intergalactic regions of E. E. "Doc" Smith's classic "Lensman" series. Or it could present science-based ideas and challenges in a contemporary (at least for a particular time) setting, like in early television series such as *Science Fiction Theatre* (1955–1957) or *The Outer Limits* (1963–1965). As a genre science fiction helps expands our conception of what reality is and humanity's place in the universe.

But however exotic the location or (perhaps literally) alien the individuals and groups depicted, how we understand and relate to what we read or see is inherently limited by our own minds and bodies. Even without consciously doing it we compare and contrast the scenes and "people" in a science fiction story, television show, or movie with our own experience and knowledge as individuals. Of necessity we use our own concepts of society and what it means to be human as the starting point for, if not necessarily judging, at least comprehending what is depicted in a science fiction work. In that sense, and updating the ancient Greek philosopher Protagoras, human beings truly are the measure of all things—at least to us human beings.

This intrinsic anthropic point of view also includes human biology. We use our own bodies and thought processes as the benchmark for understanding the nonhuman. Here too this does not necessarily mean making value judgments about the superiority or inferiority of modern-day humans vis-à-vis other life forms, including versions of humanity altered from the current "norm" by scientific means or evolution. Instead it can be a simple matter-of-fact observation that most of the patrons at the Mos Eisley cantina in *Star Wars: A New Hope* (1977) have bodies noticeably different from ours. Likewise the fact that the extraterrestrial Kloros in Isaac Asimov's short story "C-Chute" (1951) and some patients in James White's "Sector General" series of works breathe chlorine rather than oxygen leads us to the inevitable conclusion that their physiology must be radically different from the human variety.

But to understand how our bodies would respond in new environments, the ways they can be injured or altered, and how they compare to possible nonhuman sentient beings such as extraterrestrials or machines with artificial intelligence, we must first understand what our bodies are and how they work. Each of us normally thinks of his or her body holistically, as a single entity encased by a boundary of flesh and operated by what a person considers "me." While the literal truth of this basic "ghost in the machine" concept is an appropriate topic for philosophical and scientific critique, this is the way it typically *feels* to us when we stop to think about it and do not merely take what we "are" for granted.

However, just as a particular arrangement of metal, plastic, wheels, an engine, etc. can be conceptually lumped together into a single word, "automobile, " the term "human body" is our overall designation for a specific collection of many different materials and parts. The latter define what we are both physically and functionally, and they must continuously interact together in complex ways to sustain the process we call "life." This chapter will examine what it is that, in the most fundamental sense, makes us what we are.

1.1 The Stuff of Life

Learning how the human body works involves studying it at different levels of size and complexity. We can figuratively peel away each of its layers like an onion or take it apart like a nested matryoshka doll. By going deep enough we leave the world of biology behind and enter that of physics, chemistry, and astronomy.

Like everything else in the universe our bodies are composed of matter and energy. Both of these terms are more complicated than one might think. We intuitively think of matter as the solid, liquid, or gaseous "stuff" that makes up the world. However, at a scale far smaller than our unaided senses can detect, matter consists of "particles" called "atoms." These occur singly or in combinations called "molecules " [1, 2].

Atoms in turn consist of still smaller "subatomic" particles that include protons, neutrons, and electrons. Protons have a single positive electrical charge, electrons have an equal but negative charge, and neutrons have no charge at all. Protons and neutrons in turn consist of combinations of even smaller particles called "quarks." While protons and neutrons can exist independently of atoms, quarks are not seen in isolation.

An atom consists of a central "nucleus" surrounded by one or more "orbiting" electrons. This nucleus contains at least one proton and (usually) at least the same number (and often more) of neutrons. Nearly all the volume of an atom is empty space, with the nucleus and surrounding individual electrons being tiny compared to the atom's overall size. Atoms with different number of protons are called "elements." So far 118 elements have been identified, each with its own unique combination of chemical and other properties. Some of the most important elements involved in life processes include hydrogen (one proton), carbon (six protons), nitrogen (seven protons), and oxygen (eight protons).

"Isotopes" are atoms with the same number of protons but different number of neutrons. For example, the most common isotope of hydrogen has no neutrons, while another called "deuterium" has one neutron and a third, "tritium," has two. Atoms normally have an equal number of protons and electrons. Since these latter particles have equal but opposite electric charges such an atom is electrically neutral, with no net charge. However, atoms can either gain electrons, thus producing a net negative charge, or lose electrons and develop a net positive charge. This process is called "ionization" and the charged atom is an "ion."

The size and mass of an atom as well as its constituent parts are measured at scales far smaller than we normally use [3]. The size of a quark or electron is no more than about 1×10^{-18} m, and an electron has a mass of only about 9.11×10^{-31} kg. Protons and neutrons each have a size of about 1×10^{-15} m and similar masses—approximately 1.67×10^{-27} kg for a proton and minimally greater for a neutron. The average radius of the smallest type of atom, hydrogen, is on the order of 5×10^{-11} m.

A greatly oversimplified model of the atom compares it to our Solar System. The nucleus becomes analogous to our Sun and the orbiting electrons to planets. Some early twentieth century science fiction writers seized on this image and went one step further to postulate the existence of electrons really being subatomic worlds with plant, animal, and even human life similar to our Earth. This concept led to works like Ray Cummings's novel *The Girl in the Golden Atom* (1922) and Henry Hasse's novelette "He Who Shrank"

(1936). It also became a staple of comic book stories in the mid-twentieth century such as the Silver Age DC Comics superhero The Atom, whose ability to shrink as small as the plot required allowed him to visit these tiny worlds in stories such as "The World of the Magic Atom" (*The Atom* #19, June-July 1965) and "The Up and Down Dooms of the Atom" (*The Atom* #32, August-September 1967). Even his fellow Justice League of America colleagues Green Lantern and The Flash occasionally shrank down to meet friends and foes in thrill-packed adventures on subatomic planets, e.g. "Prisoner of the Power Ring" (*Green Lantern* #10, January 1962) and "Parasite Planet Peril" (*Green Lantern* #20, April 1963).

Unfortunately this idea has no basis in reality. An atomic nucleus with its protons and neutrons bound together is not analogous in its details to our Sun, which consists primarily of isolated protons and electrons, a much smaller proportion of helium nuclei (2 protons), and still tinier amounts of other ions in a high-temperature "plasma." Electrons, protons, and neutrons may behave under some circumstances like discrete particles but as mathematically described waves under others. Far from having single orbits like planets the distances of electrons from the nucleus vary over a range of possible locations. The inhabitants of an electron-world would also presumably have to consist of "atoms" much smaller than the electron itself. For these and other reasons any story involving a subatomic world or even characters shrinking to or toward atomic size such as the 1957 movie *The Incredible Shrinking Man*, however thought-provoking in other ways, is not based on real science.

While we perceive matter as something tangible, energy is better described as a process, characterized by actual or potential action and change. Energy is defined in the most general terms as "the ability of one system to do work on another system" [4], or "the capacity to do work" [2]. Work is done when "energy is transferred to or from an object by means of a force acting on the object" [5].

These rather abstract concepts can be better understood by giving examples of the many forms of energy involved in life processes as well as others encountered in everyday life. "Kinetic" energy involves the motion of an object, such as a car moving down the street or the pedestrian walking on a nearby sidewalk. A compressed spring has "potential" energy—release the spring and the energy is converted to kinetic energy. The food we eat and digest releases "chemical" energy from interactions among the atoms and molecules in our meal. "Electrical" energy and "magnetic" energy are familiar to us regarding what they do based on our everyday experience with light bulbs and electronic devices or when we use a compass.

"Nuclear" energy is the kind stored within an atom's nucleus itself and released under certain circumstances such as the explosion of an atomic bomb.

"Gravitational " energy includes the attraction that anything having mass exerts on other objects with mass and even "electromagnetic radiation" (e.g. light, see Chap. 6), which itself is a form of energy. "Thermal" energy involves motion at the atomic and molecular level and determines how "hot" or "cold" an object is, i.e. its temperature.

At the beginning of the twentieth century "classical" physics considered matter and energy to be distinct. Each was thought to be "conserved" in all interactions, that is, to be neither created nor destroyed. Several years later Albert Einstein postulated that matter could be converted into energy and vice versa, with the total amount of matter and energy in an interaction being conserved. This relationship is expressed quantitatively in his famous formula $E = mc^2$, where E is energy expressed in standard units of joules, m is the rest mass in kilograms, and c is the speed of light in a vacuum (slightly less than 3×10^8 m/s).

Going back to the very beginning of time, all existing matter and energy originates from the "Big Bang" some 13.8 billion years ago when the universe itself started to expand from a tiny initial state with extremely high temperature and density [6–8]. What caused that expansion is still uncertain but we can be reasonably sure it was not due to a spaceship's exploding fuel tank as suggested in the *Doctor Who* serial "Terminus" (1983). It ultimately resulted in the creation of two basic types of matter. Stars, planets, and us are all made of "baryonic" matter, the variety containing subatomic particles such as neutrons and protons that we consider "normal." Only about 4.9 % of the universe's total matter and energy consists of baryonic matter [8].

Roughly 26.8 % of the universe consists of "dark matter." In this context the adjective "dark" does not refer to matter that has been seduced by the wrong side of the Force and joined the Sith. Instead this is matter that we can detect only by its gravitational effects, such as how huge amounts of it surrounding our Galaxy affect the latter's rotation, and not by "seeing" it via light, radio waves, etc. Some dark matter may be baryonic and "dark" only due to our current inability to detect it due to its great distance or other factors. However, most is thought to be "non-baryonic," consisting of theoretical particles whose exact nature is still uncertain. The remaining approximately 68.3 % of the universe consists of "dark energy." It is responsible for making the universe expand more rapidly than it otherwise would. What dark energy actually "is" remains uncertain, but it may be an intrinsic property of "empty" space itself.

The baryonic matter present in the early universe could not produce life as we know it. Shortly after its birth our cosmos consisted overwhelmingly (about 92% of nuclei) of a single element, hydrogen, with the remainder being helium except for an extremely tiny percentage of lithium (three protons in its nucleus). During a period of several hundreds of millions of years after the Big Bang the first stars formed from that primordial hydrogen and later congregated into galaxies [7]. Stars significantly larger and more massive than our Sun ultimately explode as one type of "supernova." For some time before and especially at the time of that explosion those stars produce elements more massive than hydrogen and helium, including the carbon, oxygen, etc. that we need for life. Those elements then become part of the huge hydrogen clouds that form "younger" stars like our Sun, born some 4.57 billion years ago, as well as the planets around those stars.

Thus, all the elements that make life on Earth (including human beings) possible are the "leftovers" from the creation of our Sun and the rest of the Solar System. While many elements have important roles to play in keeping us and other living things alive, the critical one is carbon. It has the most complex chemical interactions of any element, being able to form bonds with so many other types of atoms in so many different ways that it has an entire branch of chemistry devoted to it, "organic chemistry." The common science fiction phrase "carbon-based life-form" does indeed describe us and every other living thing on Earth.

Besides carbon, the most common elements used by living things are hydrogen, oxygen, and nitrogen [9]. Other elements also have important roles to play. As we will see, sodium and potassium are critical for the functions of cells. Calcium and phosphorus are essential for bone development, and the latter also contributes to the types of molecules that generate the energy we need to live. Sulfur is a key component in other important molecules and iron is needed for production of our blood.

The single most important compound (defined as two or more elements chemically bonded in fixed proportions) [2] for terrestrial life is water. It consists of two hydrogen atoms bound with one of oxygen (H_2O). Similar to carbon, in its liquid state water has both physical and chemical properties that allow a wide range of reactions among many different elements and compounds. Life on Earth originated due to the presence of large amounts of liquid water, and it makes up about 70 % of an adult's body weight.

1.2 What Is Life?

Water, carbon, and many other elements and compounds are the basic building blocks of terrestrial life at the atomic and molecular level. But even if we pour packets containing all the individual chemicals that make up our bodies into a large vat of water, no amount of stirring would make a person emerge from it. Not even someone as technologically advanced as Marvin the

Martian in the cartoon "Hare-Way to the Stars" (1958) could really create "just add water" Instant Martians. The ways that the raw, individually lifeless chemicals that make up a living thing are organized and interact are also crucial in deciding whether their combination constitutes life.

While it is easy to see that a rock is not alive[1] and a raccoon is, the exact boundary that separates the inorganic from the organic is fuzzy. Both past and present terrestrial life has taken an astonishing range of forms. Although both people and plants are living things, what "alive" means for each of them is, in terms of both chemistry and capability, very different.

Describing what extraterrestrial life might be like is a common practice in science fiction [10]. Those alien beings may to some degree be modeled on or at least inspired by various terrestrial forms of life in form and/or how they act. Thus, aliens may be humanoid and typically bipedal in appearance, as is generally the case in the *Star Trek* universe. In Larry Niven's "Known Space" works the Kzinti are similar to felines, while Puppeteers are tripedal herbivores with a herd mentality, with the latter two traits shared by animals such as cattle and deer here on Earth. The Mesklinites of Hal Clement's novel *Mission of Gravity* (1954) have centipede-like bodies as an adaptation to their high-gravity world, while the Selenites of H. G. Wells's *The First Men in the Moon* (1901) are insectoid in appearance and have a society similar to an ant colony.

Aliens that differ markedly in appearance from anything terrestrial still may act in ways that humans can (although perhaps only with great effort) understand, often enough that some form of communication becomes possible with sentient extraterrestrials. Thus the titular character of Sir Fred Hoyle's novel *The Black Cloud* (1957), though sharing little more than the fundamental laws of physics with us as living organisms, turns out to be relatively talkative. Even the otherwise incomprehensible radio wave-based Waveries in Fredric Brown's story of that name (1945) show characteristics of being living such as moving volitionally, as do the unseen "aliens" of Robert A. Heinlein's story "Goldfish Bowl" (1942).

The intrinsic properties and relative abundance in the universe of carbon and water compared to every element except hydrogen and helium make them successful materials for creating life—a statement whose proof can be found in your mirror. Science fiction occasionally deals with aliens who use elements other than carbon and water as the basis of their structure and life processes. For example, the chemical properties of silicon are in some ways similar to carbon, and under environmental conditions radically different than Earth's (such as very high temperatures and/or pressures) might conceivably serve as

---

[1] Barring, of course, science fiction exceptions such as the *Outer Limits* episode "Corpus Earthling" (1963) or perhaps the Horta in the *Star Trek* episode "The Devil in the Dark" (1967).

a substitute. Even on our planet diatoms, a type of algae, protect themselves with walls made of silica (silicon dioxide).

Likewise liquid ammonia might be a reasonable replacement for water in organisms living at temperatures well below the latter's freezing point, although at the expense of slower chemical reactions. But overall the possibility of intelligent or even merely macroscopic life based on anything besides carbon and water remains speculative, with perhaps a slightly more optimistic possibility for microorganisms.

Given the myriad physical characteristics living things have or could have, defining what life is might be better approached indirectly by concentrating more on function than form—on what living beings do rather than merely how they are made. Traits associated with though not entirely defining life include having "…the capacity for growth, functional change, and continual change preceding death," [11] being "a self-sustaining chemical system capable of undergoing Darwinian evolution," [9], and the ability of living things to "take in energy from their surroundings and use it to maintain their structure and organization" [12].

The "Darwinian evolution" phrase includes the ideas that for a particular type of life to survive at least some of its kind must not only reproduce but create copies of themselves that are not quite identical in key ways from the "originals." Offspring having traits better adapted to living in a particular environment would thus be more likely to survive and reproduce than those that do not, in turn passing on those "adaptive" traits to their own offspring. This process is called "natural selection."

Still, even collectively such ideas cannot give a comprehensive or universal definition of life. In some ways the question goes beyond science and into the realm of philosophy [13]. Fortunately, after laying all this background, we can narrow our focus to the realm of the merely human type of life.

1.3 The Chemistry of Life

Although carbon and water form the core foundation for terrestrial life, including the human body, in themselves they are "inorganic." Neither a diamond nor the water in a glass would ever be considered living. However, certain chemicals consisting of carbon along with the two elements that form water, hydrogen and oxygen, and a few other elements such as sulfur and phosphorus form the biochemical infrastructure for our bodies. The four most important groups of chemicals used by terrestrial living organisms, including us, are carbohydrates, proteins, lipids, and nucleotides [14].

Carbohydrates consist of various combinations and arrangements of carbon, hydrogen, and oxygen atoms that in their simplest form have the formula $(CH_2O)_n$—that is, a ratio of one carbon atom to two hydrogen atoms to one oxygen atom [15]. Simple carbohydrates such as glucose and fructose are "sugars" or, more technically, "monosaccharides." Glucose is present in our blood and acts as an important energy source for our bodies. Fructose is the sugar that gives honey and many fruits their sweetness. Both glucose and fructose have identical formulas, $C_6H_{12}O_6$, but have different chemical properties because they are "isomers"—compounds with the same types and ratios of atoms but different structural or spatial arrangements of them.

Sucrose, or "table sugar, " is a "disaccharide"—a combination of one glucose molecule and one fructose molecule, with a formula of $C_{12}H_{22}O_{11}$. Lactose is found in the milk of nursing female mammals, including women, and is therefore called "milk sugar." It has the same chemical formula as sucrose but is a combination of one glucose molecule with another sugar, galactose, with these latter two simple sugars also being isomers of each other. Disaccharides such as sucrose and lactose are broken down in our bodies into their constituent monosaccharides (e.g. glucose), which are then used as fuel for metabolic processes in our bodies—in other words, to produce energy for life.

A large number of monosaccharides also can link to each other to create a "polysaccharide," a very long and chain-like molecule composed of many monosaccharide units. These "complex carbohydrates" include cellulose, glycogen, and starch. Cellulose is a major structural component in many plants, such as the wood of trees. Humans cannot digest the cellulose in the grains (e.g. bran) and vegetables we eat and thus use it for energy. However, the fibers and other forms of cellulose in those foods assist health by adding bulk to our stools (feces) and thus helping prevent constipation and other ills. Glycogen is a "storage" form of glucose present particularly in the liver and muscles. Plants use starches as an important energy source, and when we eat them in the form of potatoes and grains our bodies break them down into glucose molecules for our own fuel.

Proteins are critical to keeping us alive and, literally, in one piece. Without them our bodies would collapse into a watery and very dead mess. Some, such as collagen, give our bodies their structure, particularly within skin and bones. Enzymes are a large class of proteins that regulate our biochemical reactions. Proteins store some of the nutrients we need to live as well as transport substances (such as the oxygen in blood) to different areas of the body. Other varieties of proteins help form our muscles and make them work, while ones such as antibodies act to protect us from infections.

All proteins are assemblages of smaller units called "amino acids." Our bodies use 20 different types of amino acids [15]. They all contain various combinations of carbon, hydrogen, and oxygen atoms, and some also include nitrogen and even sulfur. The human body can produce 10 of these amino acids on its own, but the other 10 must be obtained from food. Besides producing proteins, individual amino acids can be involved in creating other important body substances such as some neurotransmitters needed for our brains and nerves to function.

All but one of these 20 amino acids can exist in either of two forms that are mirror images of each other—a "right-handed" or "left-handed" form. All life on Earth uses the left-handed form exclusively. Thus, an astronaut marooned on a distant planet whose plants and animals used the right-handed variety could not use them to provide nutritional value as food. It is thought that life on our world does not use a combination of left- and right-handed amino acids because that would produce proteins and other substances that could not function properly.

Why terrestrial beings use only left-handed amino acids is speculative. It could be a random event, with life just happening to start with that variety instead of the right-handed one. Amino acids can actually form in the environment of space and may do so in originally equal amounts of the left- and right-handed types. Meteorites containing them may have fallen on the very young Earth and formed the building blocks for the ultimate development of life.

However, ultraviolet light produced by starlight in the region of space where the Sun formed might have selectively destroyed more right-handed amino acids than the left-handed variety due to how the light was "circularly polarized" [9]. The light waves in polarized light vibrate in only one direction, while those in unpolarized light vibrate in all directions perpendicular to the direction the light travels. Polarized light can also be oriented at different degrees relative to a circle surrounding the direction light is traveling. If the light is polarized in a particular way its energy will be more readily absorbed by one variety of "handed" amino acid than the other and destroy it, leaving the other type to predominate. Thus, it is at least plausible that life on planets orbiting stars similar to our Sun might also have been more likely to select left-handed amino acids too, and thus be potentially (assuming they do not contain other, poisonous substances) edible to us.

Like carbohydrates, lipids are composed of carbon, hydrogen, and oxygen atoms arranged in a variety of configurations. However, while many chemicals in the body need to be dissolved in water to react and keep us alive, lipids are at most poorly soluble in water. Among other things this property enables them to form "barriers" to areas where the body needs to prevent or regulate the passage of water. This includes the membranes of our cells, and the "blood-brain barrier" that helps keep potentially harmful substances away from our brains.

Lipids include such classes as fatty acids, fats, oils, and steroids. Some types, such as phospholipids and glycolipids, have molecules with one end that is "hydrophilic" with a strong affinity for water along with a "hydrophobic" end that repels water. This characteristic allows them to interact in different ways with water simultaneously, both keeping water out of areas it should not go for a person to stay healthy and at the same time being able to have chemical and other interactions with it. For example, the phospholipids in a cell membrane help regulate how much water enters and leaves a cell. This "gate" function prevents too much water from either entering or leaking from a cell, either of which could make it not function properly or even destroy it.

Fats also serve as energy stores, such as in the form of adipose tissue—a type that some of us may have a bit more than we need. The human body uses glucose and other forms of carbohydrates as its main source of energy for routine metabolism, roughly comparably to how an automobile engine uses gasoline. During periods of increased exercise or other need such as environmental stress (e.g. being exposed to too much cold or heat) our bodies can, for a time, "step on the gas" by using greater amounts of carbohydrates to provide higher than baseline amounts of energy. Lipids act for the most part more as a long-term reserve fuel, used primarily to replenish other energy supplies as they are depleted.

Lipids also form the basis of triglycerides and cholesterol, two important chemicals that help transport fats through the body via the bloodstream. As we will see in the next chapter, too much of them in the blood can lead to serious problems, including blockage of blood vessels. However, having too little in the body would also be harmful. For example, cholesterol is part of cell membranes and needed to synthesize other important substances such as testosterone and estradiol, sex hormones that predominate in males and females respectively.

Water, carbohydrates, proteins, and lipids in various locations and combinations provide most of the raw material that comprise the human body. They are crucial to keeping it working as an efficient biological factory and doing the "grunt work" needed to keep us alive. But behind the scenes another class of chemicals common to all terrestrial life, nucleic acids, are working to make sure that our proteins are created as needed and made correctly so that they can regulate all the other substances within our bodies. The two primary types of nucleic acids, ribonucleic acid (RNA) and deoxyribonucleic acid (DNA), are required to produce all the cells in a human body [16]. DNA is also used as the "storage medium" for the instructions on how, under the proper circumstances, to make another human being. As we will see in later chapters, changes in DNA and the characteristics of the particular set an individual inherits are critical for evolution and natural selection.

The molecules comprising more complex carbohydrates, proteins, and lipids are made of chains of smaller subunits, such as the various amino acids that make up proteins [17]. Nucleic acids are composed of three basic parts—a sugar molecule, a "base," and a phosphate group (a phosphorous atom with four associated oxygen atoms)—linked in repeating units called "nucleotides." RNA uses the sugar ribose in its nucleotide units while DNA uses deoxyribose. Phosphate groups attach to the sugar component to link the entire nucleotide sections together.

Both RNA and DNA use one of four different kinds of bases for each nucleotide unit. The three bases they have in common are adenine (A), cytosine (C), and guanine (G). They differ in which fourth base they use, with RNA employing uracil (U) and DNA thymine (T). Adenine and guanine are members of one class of chemicals, purines, while cytosine, thymine, and uracil belong to another, pyrimidines. *Gattaca* (1997), a science fiction movie dealing in essence with how "well" nucleic acids function within individuals, used the abbreviations (A, C, G, and T) that denote the sequence in which bases appear within a strand of DNA as part of its title—i.e. G (guanine) A (adenine) T (thymine) etc.

RNA molecules usually consist of a single strand of different nucleotide subunits. As Francis Crick and James Watson reported in 1953, DNA normally consists of two strands entwined together as part of a spiraling structure—the famous "double helix." These paired strands are connected in a very specific, "complementary" way via their bases. Adenine in one strand couples only with thymine in DNA or uracil in RNA, while cytosine connects only to guanine. This property allows nucleic acids to code for the production of proteins.

In humans and other terrestrial life RNA and DNA are both involved in producing proteins. The information on how to construct a protein is included in a particular length of DNA based on the arrangements and types of bases in it. Making proteins involves production of "messenger RNA." One strand of DNA, the template strand, directs creation of a single-stranded length of messenger RNA (mRNA) via a process called "transcription" [18, 19] The template DNA strand has a pattern of bases complementary to what will be incorporated into the mRNA—e.g. adenine and cytosine in the DNA will correspond to uracil and guanine respectively in the mRNA. The second or coding DNA strand and the mRNA have identical sequences of bases except that the DNA strand has thymine where uracil is located in the mRNA.

To make proteins, a section of the two strands comprising the DNA helix temporarily separate from each other. The mRNA is then formed using the template DNA as a pattern for incorporating complementary bases into it, along with adding sugars and other needed components to complete the mRNA. A particular set and sequence of three nucleotides (a "triplet," or "codon") in the mRNA indicates which amino acid should be placed at a particular location to create a specific kind of protein.

A molecule called transfer RNA (tRNA) transports the amino acids needed to construct a protein to the mRNA. Each tRNA molecule is bound to a single amino acid. The tRNA in turn binds to the corresponding area on the mRNA that tells which amino acid should be put at a particular place in the future protein. The actual synthesis of the protein also involves ribosomal RNA (rRNA). In a process called "translation" the tRNA and rRNA combine their efforts to read the information in the mRNA and create a protein. The amino acids bound to tRNA are joined one after another in a sequential, linear fashion, like links in a chain, with the help of rRNA to eventually produce a complete protein.

1.4 Cells and Their Contents

The chemicals within our bodies are organized into two basic areas—cells and the "extracellular space" outside of cells [20]. All living organisms more complex than a virus consist of one cell or a collection of cells. Cells are self-contained in that they include all the components required to maintain their own life processes and replicate. A virus, while considered (at least nominally) living, cannot reproduce on its own but requires some of the machinery located within cells.

Single-celled organisms such as bacteria, protozoa, and some algae and fungi are "complete" in themselves. Each can independently interact with its environment by taking in nutrients, eliminating wastes, and reproducing. Some, such as cyanobacteria, form large colonies together consisting of many identical cells. Although life is thought to have originated on Earth perhaps as far back as 3.8 billion years ago, during most of that time it consisted of single-celled or very simple multicellular organisms. The emergence of a significant number of plants and creatures consisting of a more complicated collection of cells occurred during a period dating somewhere between only about 500 million to a billion years ago [21].

Complex, multicellular organisms such as humans are composed of different kinds of specialized cells. Each type contributes in its own way to the life processes of the whole body. One recent report estimated the total of cells in a 30-year old young adult weighing 70 kg (154 pounds) and 1.72 m (5 ft. 8 in.) tall as being at a minimum about 3.72×10^{13} (37.2 trillion) cells, with other reported estimates reaching up to 10 quadrillion (1×10^{16}) [22]. Some of our cells move and work individually, such as red and white blood cells. Others are part of larger structures and function collectively, e.g. in whole organs like the liver and heart. The various types of cells differ from each other in size, shape, degree of mobility, and the specific biochemical processes that go on within them. However, all share some common traits and components.

1.4.1 Cytoplasm

The outer boundary of a cell is a thin layer called the cell membrane [23]. This membrane separates and protects the cell from its external environment and encloses its contents. In a figurative sense, the cell membrane acts as a cell's "skin." It regulates what goes in and out of the cell, allowing nutrients and life-sustaining substances to enter and waste products as well as ions, atoms, and molecules related to cell function to leave. Plants, fungi, and bacteria have a cell wall outside this membrane that provides additional protection and a more rigid structure. However, animal cells lack this wall, giving them more flexibility but at the cost of greater vulnerability.

The cell membrane also acts as a "bag" to hold the liquid cytoplasm, composed predominantly of water, inside it. The cytoplasm contains the cytoskeleton, a network of tiny filaments and tubules that helps provide and maintain a cell's shape. The cytoskeleton holds the cytoplasm's internal structures in place, is involved in managing movement of substances into, out of, and within the cell, and helps the cell move and grow. The cytoplasm also contains various concentrations of elements dissolved in it. For example, most of our body's potassium is contained within cells, while a greater concentration of sodium is normally outside them in the extracellular space.

The cytoplasm also contains "organelles"—specialized structures that carry out particular functions within the cell. The "endoplasmic reticulum" (ER) forms a network containing tiny tubes and sacs within the cytoplasm [24]. One type, the "rough" ER, helps process and transport proteins, while another, the "smooth" ER, helps create lipids and glucose. The Golgi apparatus is a relatively large organelle that prepares protein for use and transfers lipids to different parts of the cell [25].

Vacuoles and vesicles are tiny "bubbles" within the cytoplasm [26]. While vacuoles typically hold mainly water, vesicles can act as "storage containers" for lipids and other substances. Vesicles can also store materials within the cell for later excretion into the extracellular space and move proteins inside the cell. Lysosomes, a specialized kind of vesicle, contain enzymes that break down various waste products, help recycle damaged and worn out organelles, and destroy "foreign" (e.g. invading viruses and bacteria) materials within the cell. Somewhat similar to a person's stomach, their contents are more acidic than the surrounding cytoplasm. The acid isolated inside lysosomes can "digest" food particles and other materials without damaging healthy parts of the cell.

Ribosomes contain rRNA [20]. They are the sites where proteins are constructed via the methods described previously. Ribosomes may be "free agents" moving within the cytoplasm, or they may be bound to membranes such as those found in the rough ER.

Mitochondria[2] are organelles that are critical for producing the energy our cells need to live [27]. They are the primary source for adenosine triphosphate (ATP), which acts like a molecular "battery" to transport chemical energy within a cell. Mitochondria produce ATP most efficiently via "oxidative phosphorylation," a process that can include the breakdown of glucose into waste products of carbon dioxide and water. They are also the site of beta oxidation of fatty acids and other processes that produce ATP.

1.4.2 The Nucleus

The earliest forms of life on Earth were "prokaryotes," a class that still exists today as bacteria and another kingdom of single-celled microorganisms called "Archaea." Prokaryotic cells lack mitochondria and a true cytoskeleton. Their DNA is condensed into a "nucleoid" that is located within the cytoplasm but not separated from it by a membrane [20, 23, 28].

Plants, fungi, and animals (including us) are all "eukaryotes." Our cells contain mitochondria and thus can produce more energy than a prokaryote can muster. Except for a small amount present in mitochondria, the DNA of eukaryotes is contained in a specialized structure, the nucleus. This organelle is separated from the cell's cytoplasm by a membrane system with two major components, an inner and outer nuclear membrane. Tiny pores in the nuclear membrane regulate passage of material between the nucleus and cytoplasm.

The shape of the nucleus is maintained via a system of filaments called the nucleoskeleton, somewhat analogous to what the cytoskeleton does for the rest of the cell. The nucleus itself also contains a specialized structure without a separate membrane called the nucleolus. It manufactures partially completed ribosomes that subsequently leave the nucleus for final assembly in the cytoplasm, where they play their essential role in creating proteins.

Most importantly, the nucleus and the DNA within it act as the command center for the cell by regulating the production of proteins. Single, coiled

---

[2] Not to be confused with the microscopic midi-chlorians mentioned by Qui-Gon Jinn in *Stars Wars: The Phantom Menace* (1999) as the intracellular "link" between macroscopic sentient life and the Force.

lengths of double-stranded DNA are bound to RNA and certain proteins, particularly a type called "histones," to form "chromosomes." [29, 30] Prokaryotes such as bacteria typically have a single large, circular chromosome. In humans and, as a rule, other eukaryotes chromosomes are linear structures. The DNA in chromosomes is coiled, compacted, and combined with certain proteins to make up a complex known as chromatin [31]. The net effect is that chromosomes are usually "balled up" within the chromatin and generally not seen as individual structures within the nucleus.

The number of chromosomes normally present in different species varies widely. A fruit fly has 8 chromosomes, a chicken 78, and one variety of butterfly has 268. Humans normally have 46 chromosomes. These include 22 different types of "autosomes," each one designated by a number of 1 through 22. With a few important exceptions our cells are "diploid"—that is, they each contain a pair of each type of autosome, for a total of 44. The remaining two chromosomes needed to reach the usual total of 46 are sex chromosomes. These come in two varieties, an X chromosome and a Y chromosome. The presence of two X chromosomes (XX) in an individual's cells produces a female, while the combination of one X chromosome and one Y chromosome produce a male (XY).

During cell division—"mitosis"—the chromosomes that are normally "lumped" together in the chromatin separate into individual strands [32]. Before a cell divides it produces a copy of each of its original chromosomes. When mitosis begins chromatin condenses into discrete chromosome pairs consisting of the original chromosome and its duplicate, joined together at a tiny area called the centromere. In animal cells, during mitosis the nucleolus disappears and the nuclear membrane breaks down. As the cell continues to divide each pair of chromosomes separates, so that when mitosis is complete each of the two "daughter" cells created has the same number and types of chromosomes as the original "parent" cell. After cell division is complete the nuclear membrane and a nucleolus reform in each of the two cells.

New cells produced by a different process, "meiosis," [33] are "haploid." Instead of the paired sets of chromosomes adding up to 46 that human diploid cells have, haploid cells contain only one set of 23 chromosomes. Gamete cells—oocytes and ova (immature and mature egg cells, respectively) in females and sperm cells in males—are haploid. They originate from a diploid cell containing two copies of each chromosome, one derived from the organism's female parent and the other from its male one. This cell then divides into two new but different cells, each containing a single set of chromosomes whose patterns of DNA are a blend of parts derived from each of the organism's parents. These two new cells then duplicate their chromosomes, with each pair again being a further mix of chromosomal material originating from the organism's parents, and then each cell divides into two different cells.

The net effect is that the original diploid cell can ultimately create up to four haploid gametes (egg cells or sperm cells), each of which is a different mixture of chromosomal material than the diploid cell possessed. The union of one ovum and one sperm cell, each individually haploid, produces a new cell that is diploid and contains a unique combination of chromosomal material. That latter cell can then divide and differentiate into an entire new organism. This is the basic mechanism for "sexual reproduction"—a process that holds interest for us humans beyond its core biological function of perpetuating a species.

"Genes" are individual sections of DNA within chromosomes that code for production of specific proteins and, to a much smaller extent, RNA [20, 30, 34, 35]. A gene may have multiple variants called "alleles." Each of these variants produces a protein with a slightly different structure, which in turn may also cause differences in how the protein functions. As we will see in Chap. 12, a change in how even a single protein acts can have major repercussions on a person's health.

Only a little over 1 % of our DNA actually codes for proteins. Current research indicates that noncoding areas, far from being the "junk DNA" they have sometimes been called in the past, play an important and dynamic role in regulating the activities of genes [36, 37]. An earlier hypothesis that, at least as a first approximation, each gene codes for only one protein has been found to be an oversimplification. Instead genes can code for "polypeptides," chemical subsections of what will eventually be incorporated into various types of complete proteins. Differences in "gene expression" involving processes such as alternative splicing (altering what coding information is incorporated into mRNA) can allow a single gene to create many varieties of a particular type of protein, with significant variations in how each one functions.

The "genome"—all the genetic information a cell contains—has been mapped in humans and some other species. This involves "sequencing" DNA—that is, determining the exact arrangement of nucleotide units within it, particularly those representing genes. The total number of genes present in humans is now estimated to be in the range of about 23,000 (give or take a few thousand). Even different individuals may vary regarding their total number of genes, with one person potentially having a little less than a hundred fewer genes than another [38].

At the species level the total number of genes in an organism's chromosomes does not necessarily correlate with its complexity. For example, a grape has more genes (a little over 30,000) than humans possess [38]. Factors such as what proteins a gene codes for as well as how gene activity is affected by influences other than the DNA sequence alone, "epigenetics," also come into play. Latter chapters will delve deeper into how genetic factors make us what we are, what happens when our DNA is damaged or differs from "normal," and how our genetic makeup can potentially be altered or even "improved."

1.5 Human Anatomy

Human anatomy—the ways our bodies are constructed from different parts—can be described at two basic levels. One is at the microscopic or cellular level, beyond or nearly beyond the range of our unaided vision. The other is at the level of gross[3] anatomy—the internal and external parts we can see and examine directly.

The cells in the human body are generally similar at the very fundamental levels just described. However, different types vary dramatically in their shapes, sizes, compositions, and the details of their actions at the biochemical level. By one classification scheme our bodies contain roughly 210 different types of cells [39]. These various types in turn have further subcategories based on how those cells function in different parts of the body and other factors.

For example, a variety of "keratinizing epithelial" cells are involved in creating skin and hair. Other cells have "exocrine" functions, creating a substance that is then released by way of a duct to somewhere within or outside the body. Examples include certain cells in our salivary and sweat glands. Cells with "endocrine" function produce "hormones"—chemicals that are released into the bloodstream by a gland in one part of the body to perform a physiological function somewhere else in it. Still other cell types line our lungs, esophagus, stomach, intestines and other organs, with functions that include acting as protective barriers or performing essential activities such as breaking down and absorbing food. Specialized nervous system cells provide us with vision and hearing, transmit electrical impulses to muscles and internal organs, and comprise our brains.

The cells in our body may be in different stages of development, ranging from stem cells that can divide and "differentiate" into one or more different kinds of other cells, to those that are "terminally differentiated"—that is, they have reached their final stage of development and growth (e.g. the neurons in our brains). Some cells, like those that ultimately form our red and white blood cells, start as "precursor cells" that, through different pathways, change into various intermediate cell types before reaching their "mature" or "final" form. Some cells are produced continuously or at least for long periods during

---

[3] Sometimes literally.

our lives, such as the fibroblasts that produce collagen and other materials that provide structure to our body. Others, like the mature myocardiocytes (also called cardiomyocytes) that comprise heart muscle, change in size during a person's development into adulthood but do not divide [40].

1.5.1 Tissues

Cells can be organized into "tissues"—groupings of specialized cells that together perform one or more particular functions within our bodies. The 4 basic types are connective, muscle, epithelial, and nervous tissues.

Connective tissue contains fibers and specific cells within an "extracellular matrix" consisting of certain proteins, carbohydrates (e.g. proteoglycans), and other materials that contribute to the connective tissue's structure. Connective tissue can act as a barrier or cushion between other tissues or organs. It may also support different body structures or (as its name implies) connect them.

For example, adipose tissue is a type of connective tissue with cells containing large amounts of fat. Cartilage can "cushion" bones when they come together at joints like our knees, help join our ribs together, give our nose and ears their flexibility, and separate the bones (vertebrae) of our neck and back. Bone too is a tissue. It contains a form of the mineral calcium phosphate that makes it rigid but also collagen that gives it a degree of elasticity—that is, over a limited range a bone will resume its original shape if it is deformed by a stress. Even blood is a type of connective tissue—a mobile variety circulating throughout the body.

Muscles contain cells with special protein filaments that can slide past each other and shorten the length of the cell, making it "contract." There are three basic types of muscle tissues. Muscles like those in our arms and legs are "skeletal" muscle. They are generally located parallel to many of our bones, such as the humerus in the upper arm or femur in the thigh. Skeletal muscles are connected to bones via tendons, primarily at joints. We can consciously make our skeletal muscles contract and move our joints, such as by raising an arm or walking.

"Smooth" muscle is present in internal organs, such as the stomach, intestines, bladder, and a woman's uterus. This type of muscle contracts and relaxes as required without us needing or even being able to control it. Our hearts contain the third kind of muscle, appropriately called "cardiac" muscle. Once again we (fortunately) do not have to continuously "will" our hearts to beat, but the myocardiocytes in them contract in rhythmic cycles without any deliberate effort on our part.

"Epithelial" cells are another type of connective tissue. There are many different kind of epithelial cells, with some optimized to provide specific functions in different parts of the body. Their most basic role is to line and protect the outside of the body (e.g. the outer layer of our skin, the epidermis, consists of epithelial cells) as well as internal structures and organs. Various types of epithelial cells form the inner linings of blood vessels, the stomach, intestines, gallbladder, bladder, and many other organs as well as some glands. Some epithelial cells also secrete hormones and other substances (e.g. sweat and mucus).

Nervous tissue contains two basic types of cells—neurons and glial cells. For example, the neurons in our brains receive and transmit electrical impulses that regulate activities throughout our bodies, process sensory information, and ultimately allow us to "think." Glial cells primarily provide nutritional and other support to neurons.

1.5.2 Organs and Organ Systems

The human body's organs are combinations of cells and tissues that work together as a group to perform one or more essential functions. Examples include the heart, brain, lungs, and liver. Organ systems consist of two or more organs that are all needed to perform a more general or overall function for the body.

The "nervous system" consists of two major components, the central nervous system and the peripheral nervous system. The central nervous system includes the brain and spinal cord. The peripheral nervous system is comprised of the nerves that arise from the spinal cord and the brainstem (the part of the brain that connects the rest of it to the spinal cord). The nervous system as a whole receives information from both outside and inside the body, processes the data obtained, and uses it to regulate internal functions and interactions with the external environment.

Nerves of the peripheral nervous system have both "sensory" and "motor" functions. For example, sensory nerve fibers in our skin can detect what the surrounding temperature is, the pressure of something being pressed against it, and disruptions of the skin like a cut that produce pain. Similarly other peripheral nerves are involved in hearing, smell, taste, and monitoring conditions involving internal organs such as the lungs and stomach. Those nerves act as the "wires" of an electrical system extending as a network throughout our bodies.

Sensory information from each of the 12 "cranial nerves" goes directly to different parts of the brain. There are two of each kind of cranial nerve, one for the right side of the body and the other for the left. These include an optic nerve connected to each eye, a pair of olfactory nerves for smell,[4] nerves for the "extraoccular" muscles of our eyeballs that let us move our eyes, the auditory nerves for hearing, and the hypoglossal nerves to our tongue. Most cranial nerves also have motor functions—that is, once the sensory information provided to the brain is processed it can then send commands back through the nerve to the muscle or other body part it controls or via other nerves to different areas of the body. Thus, if we see an object out of the corner of our eyes, that sensory data is sent via the optic nerve to the brain. It in turn can then send impulses to the muscles controlling eye movements so that we can move our eyeballs and look at the object more directly.

Sensory impulses from peripheral nerves other than cranial nerves go to the spinal cord.[5] From there they may travel to the brain for further processing. The brain can then send instructions to a body part based on the information it received via cranial nerves or back to the spinal cord and then into nerves arising from it. Alternatively some sensory impulses may go no farther than the spinal cord. The latter may then generate outgoing "motor" impulses back through the same or different nerves to a body part.

For example, a doctor can test the "knee-jerk" (more technically "patellar") reflex by tapping a ligament (a type of tissue connecting one bone to another) at the knee with a small rubber hammer. The tap stimulates nerve fibers connected to that patellar ligament (so named because it is attached to the patella, a small bone in the front of the knee), which send sensory impulses back to the spinal cord with the figurative message "Something hit me!" The spinal cord routes this data and generates motor impulses into a nerve that makes the quadriceps muscle at the front of the thigh contract briefly. The net effect is to make the leg "jerk."

An important part of the peripheral nervous system is the "autonomic nervous system." This includes nerves involved with functions that do not require our conscious control. These include regulating heart rate, blood pressure, breathing, digestion, perspiration, making our pupils wider and smaller, swallowing, etc.

Anatomically the brain is divided into three major parts—the cerebrum, the cerebellum, and the brainstem. The cerebrum is composed of the cerebral cortex—the convoluted collections of neurons that form the bulk of the brain as a whole—as well as many smaller structures. Humans possess a right and left "cerebral hemisphere" as well as a right and left cerebellum. The two cerebral hemispheres communicate through a bundle of neural fibers called

---

[4] The optic and olfactory nerves connect directly to the cerebrum, the "bulkiest" part of the brain that includes its two hemispheres. The other cranial nerves arise from the brainstem.
[5] Unlike the other cranial nerves, the olfactory and optic nerves are actually extensions of the brain itself, and therefore part of the central nervous system.

the "corpus callosum." Those two hemispheres have special sections or "lobes" where specific functions such as speech, vision, and hearing are at least relatively localized. They includes the "frontal lobes," regions involved with our ability to perform abstract reasoning, make conscious decisions based on current information and assessment of possible future consequences, and development of our personalities. The cerebral hemispheres are similar but not identical, with certain functions such as language processing more likely to be "dominant" in one hemisphere (e.g. speech centers are by far most commonly located in the left cerebral hemisphere).

The cerebellum is much smaller than the cerebrum and involved primarily with motor control—that is, the processes by which we voluntarily control our movements. It is especially important for performing fine movements (e.g. with our fingertips) and coordinating our movements so that they are done smoothly and the way we "want" them. The brainstem connects the cerebrum and cerebellum to the spinal cord. Besides being the site of origin for most cranial nerves, parts of the brainstem are involved with maintaining important internal regulatory control systems such as those for breathing, heart rate, blood pressure, and alertness.

As a cardiologist I can vouch for the importance of the "cardiovascular system." This organ system is the major means through which the body transports nutrients, oxygen, hormones and other chemicals, as well as the various types of white blood cells that are part of our immune system (more on it later).

The cardiovascular system consists of the heart and the blood vessels connected to it. They form a complete circuit through which blood circulates continuously. The heart is this system's electrically powered pump (or, more properly, set of two pumps) that keeps blood moving. Blood vessels are the "pipes" through which blood flows. "Arteries" are blood vessels that carry blood away from one of the two pumping systems of the heart, while "veins" return blood to them.

On average the heart weighs about 10–12 ounces in a man and a few ounces less than that in a woman. It is roughly the size of a person's fist and located mostly behind the lower part of the sternum (breastbone), with part of it extending behind the third, fourth, and fifth ribs on the left side of the chest. The human heart has four chambers—a right atrium, right ventricle, left atrium, and left ventricle—and four one-way valves. The right atrium and left atrium are located just above their corresponding ventricles and (at least in healthy hearts) are smaller than them. The two atria are situated side-by-side with each other and separated by the "interatrial septum." Similarly the two ventricles are also adjacent to each other and separated by the "interventricular septum" [41, 42].

The right atrium and right ventricle form one of the heart's two parallel pumping systems. They are separated by the tricuspid valve, which regulates blood flow between them. The right ventricle is the stronger pumping chamber of the two. It sends blood through the pulmonic valve into a large blood vessel, the main pulmonary artery. The latter divides into right and left branches that pass through and further branch out in the corresponding lungs, emerging from them as four small pulmonary veins.

The left atrium and left ventricle act as an even more powerful pumping system than the right-sided one. The pulmonary veins bring blood into the left atrium, where it then passes through the mitral valve into the left ventricle. The latter is the most powerful pumping chamber of a healthy heart, with the thickest and literally most "muscular" walls. Each contraction of the left ventricle in an adult resting quietly "squeezes" about 70 ml of oxygen-rich blood into the circulation.

This blood passes through the aortic valve into the body's largest artery, the aorta. Its first branches, just above its attachment to the left ventricle, are the two main coronary arteries that supply blood to the heart itself. After first rising towards the head the aorta soon makes a "U-turn." Three major arteries—the brachiocephalic, left common carotid, and left subclavian—arise from the inverted "U" or "arch" the aorta makes at its highest point in the chest. This trio of large arteries supplies blood to the brain and upper extremities (arms, forearms, wrists, and hands). The aorta then extends downward as the "descending aorta," its path roughly paralleling a person's backbone, as it gives off more arteries before splitting into two major branches at about the level of the umbilicus ("belly button"). Those branches continue to divide into smaller ones supplying blood to organs in the pelvis and extending into the lower extremities (thighs, legs, ankles, and feet).

All of these arteries diminish in width until they become arterioles, which in turn branch further into microscopic networks of still tinier capillaries. Blood leaves the capillaries via venules, which gradually increase in size to become veins. The inferior vena cava is the large vein that brings blood back to the right atrium from the lower part of the body. The superior vena cava brings venous blood back to that chamber from the head and arms.

Our lungs also play a critical role in blood circulation as part of a more general "cardiopulmonary" system. The lungs provide oxygen to and remove an important waste product, carbon dioxide, from the blood going through them. After passing through the left side of the heart and its arteries, oxygen-rich blood eventually reaches the capillaries. The latter allow oxygen, water, ions, and other substances to pass through them to reach cells, where they are used for metabolism and energy production. The blood then flowing into veins is oxygen-poor. The "oxygen saturation" of blood (roughly speaking, how much is bound to a particular protein, hemoglobin, in red blood cells) is reflected visually in the red, oxygen-rich blood carried by arteries outside the lungs and dusky bluish color of oxygen-poor blood in veins like those visible on the palm side of the forearm close to a person's wrist.

To pump blood efficiently all four cardiac chambers and valves must work together in a coordinated sequence. The "cardiac cycle" includes a period in which each chamber is actively contracting and pumping blood ("systole") and one in which it is passively filling with blood ("diastole"). With the tricuspid valve closed, the right atrium fills with oxygen-poor venous blood returning to the heart from the superior and inferior venae cavae. While this is happening the right ventricle is contracting and pumping blood through the open pulmonic valve into the pulmonary artery.

Meanwhile the left atrium and left ventricle are doing something similar. The left atrium passively receives blood leaving the lungs via the pulmonary veins. At the same time the mitral valve is closed as the left ventricle contracts, sending blood out through the open aortic valve into the aorta and rest of the body. This period during which both the right and left ventricles are contracting nearly simultaneously is "ventricular systole." Meanwhile the "relaxed," non-contracting atria are in "atrial diastole."

The mitral and tricuspid valves are closed at those times because the pressure in each ventricle exceeds that in its corresponding atrium. Heart valves normally consist of three "leaflets" except for the mitral valve, which has only two. These leaflets are small "flaps" of tissue that, when the valve is open, extend outward in the direction of blood flow. For example, when the tricuspid valve opens its leaflets move away from the right atrium and into the right ventricle. When a valve is closed its leaflets form a "seal" that inhibits blood from passing in the direction opposite it originally came, e.g. from the right ventricle back into the right atrium. Even in healthy people this seal may not be 100% effective in preventing "regurgitation" of blood, but barring development of anatomic and/or functional problems with the valve it normally does a good job of doing this.

After the ventricles finish contracting and pumping blood out, the pressure in each falls below that in its attached atrium. This causes the tricuspid and mitral valves to open and let a fresh supply of blood enter the two ventricles. Meanwhile the pressure in the pulmonary artery and aorta are greater than in the right and left ventricles, respectively, and the pulmonic and aortic valves close. This period of ventricular refilling is called "ventricular diastole." Late during that period the two atria normally contract, with this "atrial systole" or "atrial kick" sending a final spurt of blood into the ventricles. The ventricles then contract, generating enough pressure to close the tricuspid and mitral valves, open the pulmonic and aortic valves, and send another batch of blood through the lungs and rest of the body.

For the ventricles to contract at a healthy rate, however, they need to be stimulated by an electrical impulse. This normally arises from specialized "pacemaker" cells in the upper outer right atrium, a region known as the sinus node. For an adult at rest these impulses typically occur at a rate between about 60 to 100 times per minute, although (particularly in physically fit individuals) it may be somewhat slower. The sinus node is capable of increasing the rate at which it generates impulses based on the body's needs, such as increasing heart rate during exercise when our muscles in particular need more blood flow along with the oxygen and nutrients it brings.

The timing of systole and diastole are critical for the heart to pump blood most efficiently. To do this the "wiring" of the heart that carries impulses from the sinus node to the ventricles includes specialized fibers. These impulses normally also pass through an area of cells near the junction of the right atrium and right ventricle called the "atrioventricular" node. The latter slows conduction of the impulse so that it is delayed in reaching the ventricles until after they have filled adequately with blood. There are a variety of ways the heart's electrical system can figuratively "get its wires crossed," producing abnormal heart rhythms, or "arrhythmias." The more serious ones are commonly depicted (albeit often not accurately) in science fiction and other scenarios and will be discussed further in Chap. 2.

The cyclic pumping of blood through the body is also responsible for a person's "blood pressure." This has two components—the "systolic" blood pressure when the left ventricle is contracting and sending a "pulse" of blood through the open aortic valve into the aorta, and a lower "diastolic" blood pressure when the left ventricle has essentially finished contracting and the aortic valve is closed. Thus, a "normal" blood pressure of about 120/80 mmHg (millimeters of mercury) is reflected in the "120" as the systolic pressure and the "80" as the diastolic. Like heart rate, systolic blood pressure increases with exercise or other stresses, while diastolic pressure normally stays about the same or falls mildly.

The "pulmonary system" includes our two lungs and the lengths of "tubing" that connect them to the outside world. As noted previously it is involved in "respiration"—the process by which one gas, oxygen, is delivered for our cells' use and another gas produced by their metabolism, carbon dioxide, is removed from the body. The heart and lungs work in tandem to do this.

Atmospheric air contains about 78 % nitrogen, a biologically inert gas, along with roughly 21 % oxygen, 1 % argon, and 0.04 % carbon dioxide. Water vapor is also present, with its percentage varying with temperature—lower at lower temperatures and higher at greater ones—and described by the term "humidity." Air enters and leaves our bodies when we breathe it in and out.

While Time Lords may have respiratory bypass systems (as noted in the *Doctor Who* serial "Pyramids of Mars" from 1975), we mere humans are limited to using only our noses or mouth as portals for gas exchange. We can voluntarily not breathe through them for a limited time—no more than about a minute or two at best under typical circumstances. However, our bodies have built-in sensor systems that detect changes in our body chemistry when we are deprived of oxygen and soon activate a reflex that forces us to take a breath. Our metabolism is so dependent on continuously receiving an adequate amount of oxygen that covering our mouth or nose so that we cannot take it in will result in loss of consciousness within minutes.[6]

Moreover, while we can breathe through our mouths, the nose is better suited for this task. Besides being an essential component of our sense of smell, it helps reduce stress on the lungs by warming and humidifying air before it reaches them. The nose also has hairs that help filter out airborne particulate matter and microorganisms, reducing the chances they will reach the lungs to either irritate or infect them. Both the nose and mouth connect to a tubular structure, the pharynx. It divides in our neck into a pair of tubes—the larynx and the esophagus. The latter connects to the stomach, while the larynx soon turns into another tubular structure, the trachea, which bifurcates within the chest into still another type of tube, the bronchi. Bronchial tubes connect to and branch out extensively into smaller bronchioles throughout both lungs. This bronchial system acts as a conduit by which air can enter and leave the lungs.

Each lung resembles a sponge, with its interior honeycombed with many small cavities called "alveoli." The right lung is typically larger than the left, with both resting on top of a sheet-like muscle called the diaphragm that separates the thoracic cavity (the interior of the chest) from the abdominal cavity. The lungs act as bellows. When we inhale, the diaphragm contracts and moves downward, and the lungs expand. Exhaling has the opposite effects. The alveoli and tiny branches of the pulmonary arterial and venous blood vessels associated with them act together to extract carbon dioxide from blood that has returned to the heart and lungs, as well as replenish the oxygen supply of blood leaving them to go back to the rest of the body. The net effect is that exhaled air contains a lower percentage of oxygen (about 16–17 %) than the 21 % of inhaled air due its extraction in the lungs. Likewise, carbon

---

[6] This "design flaw" in the human body was exploited to fatal effect by the Doctor's archenemy, the Master, in the serial "Terror of the Autons" (1971).

dioxide is added to air that will be exhaled, with a resulting percentage of about 4%—about 100 times greater than in inhaled air.

The "gastrointestinal system" is primarily involved in digesting food, but it also serves other functions. The human body has minimal reserve when it comes to being supplied with oxygen, but it has a somewhat greater capacity to store the water and nutrients it needs. Nonetheless maintaining those supplies means we must take in food and water. That is usually done by eating and drinking, processes that obviously start by putting those substances into our mouths. Solid foods are "pre-processed" by chewing them with our teeth and mixing them with saliva, a fluid secreted by our salivary glands containing not only water but also enzymes that begin the digestive process.

Whatever we swallow passes into the part of the pharynx located in the back of the throat and from there down into the esophagus. Remember that the larynx, leading ultimately to the lungs, also comes off the pharynx. Everyone at one time or another experiences food or water accidentally "going down the wrong tube" when it is diverted into the larynx instead of the esophagus. Fortunately the larynx has a small flap of tissue, the epiglottis, that acts as a one-way valve for it. Swallowing makes the epiglottis cover the opening to the larynx so that what we eat or drink has only one forward path, to the esophagus. During breathing, however, the epiglottis leaves the larynx open.

The tube-like esophagus runs vertically in our chests, near our spines. It contains both smooth and skeletal muscle. Far from being a passive "pipe" its muscles actively contract and relax in a pattern called "peristalsis" that helps food get to its next destination, the stomach. The boundary between the esophagus and stomach is at the level of the diaphragm. The "gastro-esophageal sphincter" separates those two organs. It opens to allow passage of food and water from the esophagus but otherwise normally remains closed to prevent stomach contents from "refluxing" back into the esophagus. This is important not only because the body needs to have food proceed onward for further digestion and not back up, but also because the acid produced by the stomach to digest food is harmful to the esophagus.

The stomach is a saclike structure where swallowed food is temporarily stored. Besides producing enzymes involved in digestion it has specialized cells that secrete hydrochloric acid, used to help break down food into components the body can absorb as nutrients and also destroy many potentially harmful bacteria in food. Like the esophagus it has smooth muscle, used both to "churn" the food to assist digestion and to eventually help send it farther within the digestive tract. The stomach has a limited ability to absorb water, ethanol, and some medications, but otherwise does not send what we eat directly into the bloodstream.

Another sphincter separates the stomach from the duodenum, the first section of the small intestine. While the latter as a whole is quite long in an adult, stretching out to an average of about 7 m with an approximately garden hose-like diameter of about 2.5 cm and folded up as loops within the abdomen, the duodenum itself averages around only 25 cm long. It is responsible for further processing of food, both secreting enzymes of its own to do this as well as being connected by ducts to the liver, gallbladder, and pancreas, which provide other enzymes and chemicals used for digestion.

The rest of the small intestine is divided roughly equally into a middle section, the jejunum, and a final one, the ileum. These sections absorb water and nutrients from digested food while moving it through them via contractions of smooth muscle in their walls.

The ileum ends at the "ileocecal valve" where it inserts into the cecum. The latter is the first section of the large intestine (colon) and is located at the lower right side of the abdomen. The appendix is a short, thin pouch extending from the cecum. Its wormlike appearance provides the basis for its more formal name, the "vermiform (Latin for "wormlike") appendix." The appendix may act as a reservoir of bacteria the human body needs for health and be part of its system for resisting infection. The large intestine is significantly wider but also shorter than the small intestine, with an overall length of about 1.5 m. Beyond the cecum it first rises in the direction of the head (the ascending colon), then extends from the right to the left side of the upper abdomen (the transverse colon) before moving downward in the left side of the abdomen (the descending colon). Finally it curls into a short, roughly S-shaped section (the sigmoid colon) before leading out from the body through the rectum and anus.

The large intestine completes the process of digestion. It absorbs any remaining nutrients and water that have not been removed already by the small intestine. Any residual solid material that cannot be digested is ultimately expelled from it through the anus in the form of feces.

The large intestine also contains many hundreds of types of bacteria, part of what is called the "normal gut flora" located there and in the small intestine. These bacteria are "symbiotic," coexisting with us in a mutually beneficial arrangement. They use the human body as their "home" and source of the nutrients they need for growth, but they "pay rent" by providing us with important health benefits. Those bacteria help digest and absorb certain types of food, make some vitamins (e.g. Vitamin K, needed for blood clotting) we need, and help us resist infection from harmful bacteria by competing with them for nutrients and providing our disease-resisting (immune) system with a low-level stimulus to keep it primed for action.

The other major members of the gastrointestinal system are the liver, gallbladder, and pancreas. The liver is a large organ located in the upper right abdomen. It serves many functions, including storing energy sources such as glycogen and fat as well as some vitamins and iron. The liver also produces hormones and proteins, metabolizes alcohol and many drugs, removes some harmful substances and recycles others (e.g. old red blood cells), and aids in digesting food. The latter includes producing bile, which helps digest fats. Bile is stored in a special organ, the small bulb-like gallbladder, located just underneath the liver. When bile is needed the gallbladder can, somewhat like using a tube of toothpaste, "squeeze" it out through a small tube (duct) into the duodenum.

The pancreas is located roughly behind the stomach in the upper central section of the abdomen. That organ produces enzymes that aid in digestion of carbohydrates, proteins, and especially lipids (fats) and excretes those enzymes through a duct into the duodenum. Some cells in the pancreas produce secretions that are alkaline, with a high concentration of bicarbonate ions, to help neutralize the acid added to partially digested food before it leaves the stomach.

The "urinary system" also has multiple functions. It shares some anatomy as part of the more general term "genitourinary system" with the "reproductive system" in males. The latter system in both females and males has enough biological and other interest in its own right to merit a chapter of its own, Chap. 10, and will be discussed there.

The kidneys are the major components of the urinary system. They are paired bean-shaped organs located in the upper abdomen, near the back and on either side of the spine. The lower rib cage partially protects them. The kidneys filter blood and regulate the body's balance of water and ions such as sodium, potassium, and calcium. They are a major route by which some ingested medications as well as various waste products are excreted. The kidneys also play an important role in regulating blood pressure and help maintain the body's blood within a "normal" range of pH (about 7.35–7.45).[7]

As part of their filtering function the kidneys produce urine, consisting of water, waste products, and other substances being removed from the body. Urine leaves each kidney via a long thin tube called the ureter. Both ureters empty into the bladder, which acts as a reservoir for urine. As the latter collects in it the bladder wall is stretched, eventually triggering sensory nerves that send impulses back to the brain to let a person know this is happen-

---

[7] The pH is a measure of how acid or alkaline a substance is. A pH of 7.0 is neutral, lower numbers are increasingly acid and higher ones increasingly alkaline. For example, the hydrochloric acid produced by the stomach has a pH as low as about 1.

ing. When enough urine accumulates, the bladder can be voluntarily (or, if it becomes too filled, involuntarily) emptied. Urine passes out of the body through a single tube, the urethra. In women the urethra is short and has a direct route to the outside. In men the urethra is longer, surrounded at its origin from the bladder by the "prostate gland," and reaches the outside through a genital structure, the penis.

The "musculoskeletal system" includes the body's many skeletal muscles and bones. An adult's body contains roughly 650 skeletal muscles. Most of them, like those in our arms and legs, are under voluntary control—that is, we can control them with our brains to contract or relax. Some skeletal muscle contractions, however, are involuntary—not under conscious control—such as those associated with shivering when someone is cold, or unintentionally blinking to keep the visible parts of our eyes moist and to clear any irritating substances from them.

Some skeletal muscles in our elbows, knees, etc. attach to bones at those joints such that when the muscles contract the joint flexes. Relaxing those muscles returns the joint to an extended, "neutral" position. Other muscles act the opposite way. Glucose is used as the main fuel for muscle cells in energy-using processes that also require adequate supplies of oxygen—"aerobic metabolism." If the supply of oxygen is not sufficient to meet its needs the muscle can generate energy through a considerably less efficient means, "anaerobic metabolism." The latter produces more of a byproduct, lactic acid, than the body can immediately remove. An excess of lactic acid contributes to the feeling of soreness and aching in muscles after heavy exercise.

Bones give the body its form and rigidity. They also serve as a reservoir for calcium and phosphate, which can be released into the bloodstream if needed. Bones range in size from the largest one, the femur (located in each thigh), to tiny bones in the inner ear needed for hearing. At birth an infant has about 300 separate bones. Over time, however, many of those bones (such as the separate bony plates in a baby's skull) fuse together to produce a total of 206 in an adult.

Our skeletal system has two main divisions. The "axial" system includes the skull, spine (more technically, the "vertebral column"), and the rib cage. The "appendicular" skeleton encompasses the pelvis, the bones in our arms and legs, and the ones near our shoulders, the clavicles (collarbones) and scapulae (roughly triangular bones that form our "shoulder blades"). Without bones the human body would resemble a squishy but far less mobile and dangerous version of the titular "hero" of the 1958 movie *The Blob*. Certain bones, such as the skull and rib cage, also provide protection to vulnerable organs such as the brain, lungs, and heart. The vertebral column not only allows us to stand upright but also protects the spinal cord running through most of it.

The separate bones in a skeleton are overwhelmingly connected to each other only by other tissues—muscles, ligaments, cartilage, etc. After death bones typically remain intact long after those other tissues have decomposed. Thus, despite depictions in art and elsewhere, a skull will not have its well-known grin, because the lower jaw (mandible) will no longer be connected to the skull by ligaments and associated tissues at the two temporomandibular joints (located near the base of each ear). Likewise any magic powerful enough to animate skeletons, such as in the movie *Army of Darkness* (1992), would also have to provide those ligaments and other tissues again too, to avoid them collapsing into a heap of a bit over 200 separate moving bones.

While the name of the "integumentary system" may be unfamiliar many of its major components are easy to see. They include skin, hair, and nails. Based on its sheer extent the skin is actually the largest organ of the human body. Its primary function is to keep our "insides" from becoming "outsides." However, it also provides a layer of relatively waterproof protection to our internal parts, acting in a way as a (literally) skin-tight raincoat. Skin also helps with temperature regulation, particularly through sweating to help provide cooling in a hot environment. Exposure of skin to sunlight also helps our bodies produce Vitamin D.

Skin is comprised of three separate layers. The epidermis forms the outermost layer, the dermis is the middle one, and the deepest is the hypodermis, made up primarily of "subcutaneous" fat. While our eyes naturally appreciate the epidermis most, the dermis is more complex, containing nerve fibers, blood vessels, sebaceous and sweat glands, hair follicles, etc. Those nerve fibers are involving in sensing pressure, pain, temperature, and other sensations.

The "endocrine system" includes glands, tissues, and individual types of cells scattered throughout the body that secrete many kinds of hormones into the bloodstream. For example, the pituitary gland is a very small structure attached by a short stalk to the base of the brain close to the latter's center. It secretes a number of important hormones as well as other substances that stimulate release of additional types of hormones from distant glands. For example, the growth hormone it releases stimulates cell growth within internal organs and increases both bone and muscle mass. That hormone also promotes production of proteins and breakdown of fats. In children growth hormone contributes to their increase in height as they grow.

The pituitary gland also secretes thyroid-stimulating hormone, which (as its name suggests) stimulates release of thyroid hormone from the thyroid gland. The latter is located in the lower front of the neck. The thyroid hormones it releases are essential to help regulate many aspects of the body's metabolism and for healthy development of cells. The thyroid gland also releases another hormone of its own, calcitonin, which reduces calcium levels in blood.

Two other important pituitary hormones are follicle-stimulating hormone and luteinizing hormone, both of which are responsible for stimulating production of sex hormones such as estrogen and testosterone. Other cells in the pituitary gland secrete antidiuretic hormone, which helps regulate water balance in the body. Yet another product of the pituitary gland, adrenocorticotropic hormone, stimulates the two adrenal glands, each perched atop a kidney, to release their own hormones, which include cortisol and aldosterone. Cortisol has anti-inflammatory properties and has important effects on metabolism involving fats, proteins, and carbohydrates. Aldosterone helps the body maintain appropriate balances of water and electrolytes (ions of sodium, potassium, etc. dissolved in water). The adrenal glands also secrete "catecholamines," substances that include epinephrine ("adrenaline") and norepinephrine, as well as chemicals that the body uses to make testosterone and estrogen.

The parathyroid glands are located behind the thyroid gland. There are usually four of them. They produce parathyroid hormone, which increases blood levels of calcium by releasing it from bone into the bloodstream. This hormone and calcitonin act together to regulate calcium metabolism in the body.

Besides its role in digestion, the pancreas also acts as a gland. It releases hormones such as insulin, which decreases blood glucose levels, and glucagon, which increases them.

The "hematopoietic system" includes red blood cells (erythrocytes, or RBCs) to carry oxygen throughout the body, white blood cells (WBCs) that protect against infection, and other cells (e.g. platelets) required for blood to clot. These cells originate primarily in bone marrow, located largely in bones that include parts of the pelvis, the sternum (breastbone), and long bones such as the femur in the thigh and humerus in the upper arm. However, the liver, kidneys, and spleen (an organ located in the upper left abdomen, near the stomach) also contain areas producing those cells.

All these cells originate from "hematopoietic stem cells," which can differentiate into many different types of blood cells. "Erythroid" cells are those that ultimately form RBCs. They include "reticulocytes," an immature form of RBC released into the bloodstream that soon develops into a mature one. RBCs in blood almost always lack a nucleus, a characteristic distinguishing them from most cells in the body. An RBC's average lifespan in the bloodstream is about 120 days. RBCs contain hemoglobin, an iron-containing protein that both binds and releases oxygen. By far the most common type of hemoglobin after birth is Hemoglobin A, but many genetic variants of hemoglobin exist. One type, Hemoglobin F, is normally found in the developing fetus and binds oxygen more strongly than Hemoglobin A. Other types, such as Hemoglobin

S, are commonly associated with increased fragility and rupture ("hemolysis") of red blood cells.[8]

Interestingly, the reason why Vulcans are said to have green blood—as mentioned, for example, in the *Star Trek* episode "Obsession" (1967)—is that their blood uses copper instead of iron in their equivalent of hemoglobin. On Earth some invertebrates, but no mammals, use "hemocyanin," a protein analogous to hemoglobin using copper rather than iron to bind oxygen. These animals include octopi, lobsters, crabs, scorpions, and spiders [43, 44]. Unlike the hemoglobin found in RBCs, hemocyanin is not bound to cells in their circulatory systems but circulates freely in their bodies. Hemocyanin is not as efficient for carrying oxygen as hemoglobin at a normal human-level body temperature. However, it can be more efficient than hemoglobin in a cold environment and at low levels of oxygen [45]. As mentioned in another *Star Trek* episode, "Amok Time" (1967), that first characteristic is quite different from the climate of the planet Vulcan while the second is more consistent with it. On the other hand, unlike Mr. Spock's blood the color of hemocyanin is not green but blue when combined with oxygen and nearly clear when oxygen-depleted. Perhaps Vulcans use a different kind of carrier protein coupled with copper to give their blood its verdant hue.

Hematopoietic stem cells also produce "megakaryocytes." These latter cells become much larger than RBCs and "bud" off smaller cells called "platelets." When a blood vessel is damaged platelets aggregate there and stick together to form a plug. This is an important step in a complex process leading to blood coagulating and forming clots to seal off an injured area.

Some hematopoietic system cells are also part of the "immune system." It is responsible for recognizing foreign materials within the body and fighting infections. WBCs are a major part of the immune system. One class of WBCs, granulocytes, includes neutrophils, eosinophils, and basophils. Neutrophils are by far the most numerous cellular "warriors" against bacterial and other types of infections. They are able to engulf harmful bacteria—a process called "phagocytosis"—and destroy them. Eosinophils are particularly effective at dealing with infections by parasites. Basophils, the least numerous of this group, help initiate an "inflammatory" response and increase blood flow to an area with an infection. These effects help bring other infection-fighting cells and chemicals to the area to battle harmful bacteria and other pathogens.

Monocytes are another type of WBC. Both they and another variety of cell they can develop into, macrophages, also attack and destroy invading microorganisms by phagocytosis and other means. Macrophages typically are located in organs such as the liver, where they can engulf and destroy foreign and

---

[8] Chapter 12 will discuss important genetic issues involving Hemoglobin A and Hemoglobin S.

other particles, including debris from dead cells. After "swallowing" a harmful microbe a macrophage can signal other types of WBCs of the microbe's presence, thereby "calling in" those additional troops to see if that pathogen had company and battle the infection.

Lymphocytes are yet another kind of WBC. They are present in blood but also reside in "lymph nodes," small "lumps" of tissue that are part of the body's "lymphatic circulation." The latter is a system of vessels carrying a clear fluid called lymph that roughly parallels the "conventional" circulatory system of arteries and veins. The lymphatic vessels help return fluids to blood.

There are three major classes of lymphocytes—T-cells, B-cells, and one with the aggressive name "natural killer cells."[9] T-cells are primarily involved in "cell-mediated immunity." This includes killing cells infected with a virus, signaling other types of cells to battle infections, and destroying cancer cells. B-cells are associated with "humoral immunity." This involves production of "antibodies," a type of protein designed to identify and help destroy invading bacteria and viruses. B-cells can differentiate further into plasma cells, which secrete large amounts of pathogen-fighting antibodies.

Our bodies also contain tissues and organs that form part of the immune system. Besides the lymph nodes, "lymphatic tissue" containing large numbers of lymphocytes is present in organs such as the spleen, thymus, tonsils, and in scattered areas within the intestines. Although the spleen is not essential to life it performs helpful functions such as filtering blood, producing antibodies, and recycling red blood cells. The thymus is located behind the sternum and diminishes in size as we age. The tonsils are located in several areas in and around the back of the throat.

1.6 The Bottom Line

As was evident long ago, the material in this chapter falls well short of the page-turning, "Wow-I-can't-wait-to-read-what-happens-next!" nature of an exciting science fiction story. This information is, however, important as a backdrop to help understand the more dramatic changes that exotic environments and techniques—microgravity, radiation, suspended animation, genetic engineering, nanotechnology, etc.—can or cannot (except, perhaps, with great difficulty) make on the human body. It also emphasizes the point in future chapters that our bodies are far more complex and fragile and much less malleable than science fiction sometimes depicts.

---

[9] With a moniker like that any shady microorganism should clearly think twice before messing with one of them.

An extreme example illustrating the latter idea is the concept of a "shapeshifter"—a human or alien that can dramatically change its form. A few of many science fiction examples of shapeshifters include the alien in John W. Campbell's (writing as "Don A. Stuart") classic story "Who Goes There?" (1938), the 1953 movie *It Came from Outer Space*, Ray Bradbury's "The Martian" (1949), the Changelings introduced in the television series *Star Trek: Deep Space Nine*, and a wide variety of comic book characters that include Chameleon Boy of DC Comics' Legion of Superheroes and the Martian J'onn J'onzz. The changes in form such characters can perform typically seem to occur over a matter of a few seconds or minutes.

However, basic principles of real-life physics and biology clash dramatically with such rapid transformations. For example, a change into a different form with a considerably different mass than the original raises the question of where the excess mass either went or came from. Moreover, such changes in shape and appearance would require massive modifications in both large numbers and many different types of cells in the shapeshifter's body. These might include skin cells, the distribution of subcutaneous fat, skeletal structure, the location and size of some internal organs, changes in the courses of large and small blood vessels and nerves throughout the body, etc.

Such wholesale changes are also typically shown occurring under the voluntary control of the individual—a very dubious prospect in itself based on the complicated anatomy of muscles and other tissues at both the microscopic and macroscopic levels discussed earlier—and occurring in ways and at rates that cells are simply incapable of doing. For example, even at its most rapid cell division and transformation takes at least many minutes rather than the seconds required for common depictions of shapeshifting. Thus, such relatively mundane changes as the "new" form showing longer hair than the first one would also require the cells of hair follicles to work far quicker than any cell can.

For these and other reasons, even when used in a work with science fiction elements, the core idea of shapeshifting by a complex multicellular being (human or otherwise) falls firmly into the field of fantasy rather than science. Non-physical explanations, such as the shapeshifter merely "hypnotizing" others into seeing it in a different form (e.g. the *Star Trek* episode "The Man Trap" from 1966, or "The Man Who Was Never Born," a 1963 tale from *The Outer Limits*), have a bit more plausibility. However, they raise their own questions about the feasibility of "mind control" and similar mental "powers" that will be addressed in Chap. 8. Overall, shapeshifting is an example of a "holistic" concept—one that focuses only on the final and overall results while ignoring the details, intermediate steps, and basic scientific principles that would be required to actually make it "work." The reader or viewer should know that it is "impossible"—but still perhaps "accept" it as an enjoyable, entertaining, and/or thought-provoking element of a work.

This idea of a science fiction concept stretching or even breaking the limits of "realistic" science is one that will reappear in later chapters. More specifically, they will highlight how much of a gap exists between science fiction depictions of radiation exposure, suspended animation, genetic engineering, etc. and current scientific ideas and capabilities. Nonetheless, even if their "reality" is confined to our imaginations, concepts such as shapeshifting can work as "tropes" rather than "truths" in the context of good storytelling.

Finally, although this chapter's descriptions of how our bodies work may seem complex, it barely scratches the surface. The scope and detail of many, many research papers, medical textbooks, etc. far exceed anything described here. For example, a standard printed book on the cardiovascular system can contain about two thousand pages of tiny print and still not describe nearly all of what is known about even that single (albeit important) part of the human body. Such textbooks are actually physically dangerous—massive enough to break bones and crack skulls when dropped from only a few meters above a vulnerable part of the victim's anatomy.[10] The medical and other scientific information provided in this chapter and those that follow is extremely condensed. However, even in such a highly distilled form it can still highlight how much the science of medicine is actually reflected (or not) in science fiction. Please keep that in mind as we focus on particular topics.

References

1. Tro N. Chapter 2. Atoms and elements. Chemistry. A molecular approach. 2nd ed. Upper Saddle River: Prentice Hall; 2011. pp. 42–77.
2. Tro N. Chapter 1. Matter, measurement, and problem solving. Chemistry. A molecular approach. 2nd ed. Upper Saddle River: Prentice Hall; 2011. pp. 1–41.
3. Smith G. The Scale of the Universe. 1999. http://cass.ucsd.edu/archive/public/tutorial/scale.html. Accessed 5 April 2015.
4. Energy. In: Parker SB, editor. McGraw-Hill encyclopedia of physics. 2nd ed. New York: McGraw-Hill; 1992.
5. Walker J. Kinetic energy and work. Halliday & Resnick. Fundamentals of physics. 9th ed. Hoboken: Wiley; 2011. pp. 140–65.
6. Chapter 6. Big Bang cosmology–the evolving universe. In: Jones MH, Lambourne RJ, editors. An Introduction to galaxies and cosmology. Cambridge: Cambridge University Press; 2007.

---

[10] My copy of the 8th edition of *Braunwald's Heart Disease* weighs 5 kg (11 pounds). While it would be ironic and certainly not recommended to use it in this way, the book could potentially be employed to inflict serious blunt trauma (see Chap. 2) on some unfortunate individual.

7. Stratmann HG. Galactic cannibalism: who's on the menu. Analog Sci Fict Fact. 2013;133(7/8):22–32.
8. Planck Mission Brings Universe into Sharp Focus. 2013. http://www.nasa.gov/mission_pages/planck/news/planck20130321.html. Accessed 5 April 2015.
9. Sephton MA. Origin of life. In: Rothery DA, Gilmour I, Sephton MA, editors. An introduction to astrobiology. Cambridge: Cambridge University Press; 2011. pp. 1–41.
10. Schmidt S. Engineering organisms: alien bodies and minds. Aliens and alien societies. Cincinnati: Writer's Digest Books; 1995. pp. 68–101.
11. Life. The Oxford dictionary and thesaurus (American edition). New York: Oxford University Press; 1996.
12. Schmidt S. Biochemical basics. Aliens and alien societies. Cincinnati: Writer's Digest Books; 1995. pp. 50–67.
13. van Inwagen P. Chapter 9. The proposed answer. In: van Inwagen P, editor. Material beings. Ithaca: Cornell University Press; 1990.
14. Pollard T, Earnshaw W, Lippincott-Schwartz J. Molecules: structures and dynamics. In: Pollard T, Earnshaw W, Lippincott-Schwartz J, editors. Cell biology. 2nd ed. Philadelphia: Saunders Elsevier; 2008. pp. 33–56.
15. Tro N. Chapter 21. Biochemistry. In: Tro N, editor. Chemistry. A molecular approach. 2nd ed. Upper Saddle River: Prentice Hall; 2011. pp. 952–85.
16. Sen S, Kar D. Nucleic acid. In: Johri B, editor. Cytology and genetics. UK: Alpha Science International Ltd.; 2005. pp. 59–92.
17. Lewin B. Introduction: cells as macromolecular assemblies. In: Lewin B, editor. Genes VI. New York: Oxford University Press; 1997. pp. 3–48.
18. Corden J. Gene expression. In: Pollard T, Earnshaw W, Lippincott-Schwartz J, editors. Cell Biology. 2nd ed. Philadelphia: Saunders Elsevier; 2008. pp. 253–78.
19. Tollervey D. Eukaryotic RNA processing. In: Pollard T, Earnshaw W, Lippincott-Schwartz J, editors. Cell biology. 2nd ed. Philadelphia: Saunders Elsevier; 2008. pp. 279–96.
20. Pollard T, Earnshaw W, Lippincott-Schwartz J. Introduction to cells. In: Pollard T, Earnshaw W, Lippincott-Schwartz J, editors. Cell biology. 2nd ed. Philadelphia: Saunders Elsevier; 2008. pp. 3–16.
21. Gilmour I. A habitable world. In: Rothery DA, Gilmour I, Sephton MA, editors. An introduction to astrobiology. Cambridge: Cambridge University Press; 2011. pp. 43–84.
22. Bianconi E, Piovesan A, Facchin F, Beraudi A, Casadei R, Frabetti F, et al. An estimation of the number of cells in the human body. Ann Hum Biol. 2013;40(6):463–71.
23. Sen S, Kar D. Cell. In: Johri B, editor. Cytology and genetics. UK: Alpha Science International Ltd.; 2005. pp. 5–36.
24. Pollard T, Earnshaw W, Lippincott-Schwartz J. Endoplasmic reticulum. In: Pollard T, Earnshaw W, Lippincott-Schwartz J, editors. Cell biology. 2nd ed. Philadelphia: Saunders Elsevier; 2008. pp. 345–63.

25. Pollard T, Earnshaw W, Lippincott-Schwartz J. Secretory membrane system and the golgi apparatus. In: Pollard T, Earnshaw W, Lippincott-Schwartz J, editors. Cell biology. 2nd ed. Philadelphia: Saunders Elsevier; 2008. pp. 365–89.

26. Pollard T, Earnshaw W, Lippincott-Schwartz J. Degradation of cellular components. In: Pollard T, Earnshaw W, Lippincott-Schwartz J, editors. Cell biology. 2nd ed. Philadelphia: Saunders Elsevier; 2008. pp. 409–20.

27. Pollard T, Earnshaw W, Lippincott-Schwartz J. Mitochondria, chloroplasts, peroxisomes. In: Pollard T, Earnshaw W, Lippincott-Schwartz J, editors. Cell biology. 2nd ed. Philadelphia: Saunders Elsevier; 2008. pp. 331–44.

28. Pollard T, Earnshaw W, Lippincott-Schwartz J. Evolution of life on earth. In: Pollard T, Earnshaw W, Lippincott-Schwartz J, editors. Cell biology. 2nd ed. Philadelphia: Saunders Elsevier; 2008. pp. 17–28.

29. Sen S, Kar D. Chromosome. In: Johri B, editor. Cytology and genetics. UK: Alpha Science International Ltd.; 2005. pp. 37–58.

30. Pollard T, Earnshaw W, Lippincott-Schwartz J. Chromosome organization. In: Pollard T, Earnshaw W, Lippincott-Schwartz J, editors. Cell biology. 2nd ed. Philadelphia: Saunders Elsevier; 2008. pp. 193–208.

31. Pollard T, Earnshaw W, Lippincott-Schwartz J. DNA packaging in chromatin and chromosomes. In: Pollard T, Earnshaw W, Lippincott-Schwartz J, editors. Cell biology. 2nd ed. Philadelphia: Saunders Elsevier; 2008. pp. 209–30.

32. Pollard T, Earnshaw W, Lippincott-Schwartz J. Introduction to the cell cycle. In: Pollard T, Earnshaw W, Lippincott-Schwartz J, editors. Cell biology. 2nd ed. Philadelphia: Saunders Elsevier; 2008. pp. 731–46.

33. Pollard T, Earnshaw W, Lippincott-Schwartz J. Meiosis. In: Pollard T, Earnshaw W, Lippincott-Schwartz J, editors. Cell biology. 2nd ed. Philadelphia: Saunders Elsevier; 2008. pp. 815–32.

34. Lewin B. Part 2: From gene to protein. In: Lewin B, editor. Genes VI. New York: Oxford University Press; 1997. pp. 151–278.

35. Lewin B. Part 1: DNA as information. In: Lewin B, editor. Genes VI. New York: Oxford University Press; 1997. pp. 49–150.

36. Maher B. The human encyclopedia. Nature. 2012;489:46–8.

37. Kellis M, Wold B, Snyder MP, Bernstein BE, Kundaje A, Marinov GK, et al. Defining functional DNA elements in the human genome. Proc Natl Acad Sci U S A. 2014;111(17):6131–8.

38. Pertea M, Salzbert S. Between a chicken and a grape: estimating the number of human genes. Genome Biol. 2010;11:206–12.

39. Cells of the Adult Human Body: A Catalogue. 2015. http://www.bioon.com/book/biology/mboc/mboc.cgi@code=220801800040279.htm.

40. Senyo SE, Steinhauser ML, Pizzimenti CL, Yang VK, Cai L, Wang M, et al. Mammalian heart renewal by pre-existing cardiomyocytes. Nature. 2013;493(7432):433–6.

41. Stratmann HG, Stratmann M. Sex and your heart health. Starship Press, LLC; 2007.

42. Burkhoff D, Weisfeldt M. Cardiac function and circulatory control. In: Goldman L, Bennett J, editors. Cecil textbook of medicine. 21st ed. Philadelphia: W.B. Saunders Company; 2000. pp. 170–7.
43. Markl J. Evolution of molluscan hemocyanin structures. Biochim Biophys Acta. 2013;1834(9):1840–52.
44. Miller K, Cuff M, Lang W, Varga-Weisz P, Field K, van Holde K. Sequence of the octopus dofleini hemocyanin subunit: structural and evolutionary implications. J Mol Biol. 1998;278:827–42.
45. Strobel A, Hu M, Gutowska M, Lieb B, Lucassen M, Melzner F, et al. Influence of temperature, hypercapnia, and development on the relative expression of different hemocyanin isoforms in the common cuttlefish sepia officinalis. J Exp Zool. 2012;317(8):511–23.

2
Hurting and Healing Characters

As to diseases make a habit of two things—to help, or at least, to do no harm.

Hippocrates
Epidemics

HUMANOID. Serial No. 81-H-B-27. The Perfect Mechanical. "To Serve and Obey, and Guard Men from Harm."

Jack Williamson
"With Folded Hands" (1947)

Humans generally prefer to be healthy and without pain. We can increase the odds of staying that way by eating proper amounts of nutritious foods, getting enough exercise, and not engaging in optional behavior that is likely to harm us. Nonetheless we cannot entirely avoid illness or injury, and barring major medical advances we will all ultimately age and die.

Risk is an intrinsic part of human existence, present in even the simplest decisions of everyday life. Despite the possibility of food poisoning, we must still eat food because the ultimate results of starving are worse and far more certain to cause harm. No matter how carefully we drive there is always the chance a runaway truck could smash into us. As Hippocrates knew all too well, every time a physician prescribes a new medication or performs an operation there is a nonzero chance of making the patient worse. An ongoing goal of medical practice is to find new ways to tilt this "risk-benefit ratio" as far away as possible from the risk side and toward the benefit one.

Even just enjoying life and trying to accomplish any goal involves some degree of physical or psychological risk, including that of failure. Whether choosing how to spend our money or who to marry there is always the possibility of making a wrong decision, yet we do it in the hope that things will work out well and our lives will be improved. After getting out of bed in the morning a person might slip in the shower and break a leg, catch a cold from a fellow employee at work, or be hit in the head by a foul ball at a nighttime baseball game. But if that individual elects to stay in bed he or she will also accomplish and enjoy much less than by taking such risks.

Likewise, astronauts and cosmonauts know how risky their ventures into space are. Nonetheless they still put their lives on the line and, as Chap. 3 will describe, in a few cases have lost them in pursuit of what the typical science fiction reader will agree is a noble goal. A world such as that depicted in Jack Williamson's classic novelette "With Folded Hands" (1947) where *all* avoidable pain and risk is eliminated by overprotective robots might increase how long we live, but at the cost of turning life into mere existence.

Science fiction excels at presenting characters who either willingly or are forced to take risks in imaginatively original settings. Protagonists may encounter dangers no human has ever encountered before as well as deal with new ideas that challenge their preexisting worldviews and conceptions. When reading or watching those characters' adventures we have the opportunity to identify with their struggles, share their pain and joys, and live for a time in a world with more thrills, tragedies, and triumphs than our own.

But while many might enjoy seeing characters persevere and win against overwhelming odds in the movie *Independence Day* (1996) or Larry Niven and Jerry Pournelle's *Footfall* (1985), fewer would wish to actually live through the devastating alien invasions those works depict. In a story the author is in complete control of what happens, and however many imaginary millions are injured or killed in it we readers are left physically unharmed. But in the real world success depends on what *we* do, with no guarantee of survival or a "respawn" button to use after failure. Thus, there can be a huge gap between how dangerous we want our own lives to be and those that characters in science fiction works live. We hope that flesh-and-blood astronauts have space missions that are as safe as possible, while a story in which fictional ones were not subjected to terrible dangers would be boring.

Besides those still restricted to science fiction such as falling into a black hole or teleportation accidents, there are many realistic ways to inflict or threaten to inflict sickness, injury, and other medical mayhem on characters before (perhaps) healing them. However, science fiction may be less than completely accurate in how it shows them being hurt or depicting how medical care is delivered and healthcare providers act. A little "fudging" with medical details might be justified in the interests of dramatic license and moving the story's plot along. On the other hand, a science fiction work would be weakened if it contains an egregious medical error that is *not* needed for story purposes. This chapter will suggest ways to inject at least some degree of realistic medicine into science fiction.

2.1 Common Illnesses

Murphy's Law applies to every part of our bodies. So many things can go wrong with it that not even the most educated and experienced physician could know all their details by rote. Those many diseases and dysfunctions can be grouped into a smaller number of general categories. These include infections, autoimmune diseases (one in which the body attacks its own tissues), disorders of organ function, metabolic abnormalities (e.g. reduced production of insulin in Type 1 diabetes mellitus), nutritional deficiencies (such as those associated with lack of specific vitamins), "wear-and-tear" damage to body parts (e.g. the cartilage in our knees), diseases that directly damage organs and tissues, processes that target nerves and the inner lining of blood vessels, genetic disorders, cancer, and many more.

2.1.1 Common Causes of Death

Some diseases and injuries may run only a short course leading to complete recovery. Other types cannot be completely cured and cause chronic disability. Some are so serious they ultimately prove fatal. The most common causes of death differ based on age. Statistics for 2010 from the Centers for Disease Control and Prevention in the United States show that babies less than 1 year old are most likely to die of "congenital anomalies"—abnormalities in development—with premature birth being the second most common cause of death [1]. Between the ages of 1 and 44 years old the most common cause of death is not any disease, but accidents. The second most common cause of death between the ages 1–14 and 35–44 is cancer, while the second and third most common causes between ages 15 and 34 are homicide and suicide. It is only after age 45 that actual, acquired diseases take the top spots as causes of death. Between the ages of 45 and 64 cancer is the most common cause of death, followed by heart disease. After age 65 those two diseases reverse their order, with heart disease being the leading cause of death followed by cancer.

Babies and children are particularly vulnerable to infections. At the beginning of the twentieth century pneumonia, tuberculosis, enteritis (infections involving the intestines), and other infectious diseases were the most common cause of death in that age range. Improvements in sanitation and hygiene, better nutrition, and development of antibiotics and vaccines have dramatically lowered mortality during infancy and childhood. In 1900 30.4 % of all deaths in the United States occurred in children less than 5 years old, while near the end of the twentieth century in 1997 that rate had fallen to 1.4 % [2].

2.1.2 Cardiovascular Disease

The high rates of homicide and suicide in the 15–34 year age range are associated with injuries to the body that will be discussed in the next section. In older adults both heart disease alone and the more general category of cardiovascular disease, which also includes problems with blood vessels, are major causes of morbidity and mortality. The heart and its valves can be damaged due to gradual or sudden loss of their blood and oxygen supplies, infections, inflammation, infiltration by materials that reduce their ability to function properly, "wear-and-tear" from aging, exposure to toxic substances, tumors, etc. The heart's complex electrical system can "short circuit" in a variety of ways that can significantly reduce cardiac function and potentially lead to death.

Blood vessels become "stiffer" with age, thus contributing to development of high blood pressure, or "hypertension." The latter in turn can put increased strain on the heart and damage it as well as other organs (e.g. the brain and kidneys). Sections of major arteries such as the aorta may enlarge and even rupture, and smaller arteries can become inflamed. Blood clots ("thrombi") can form in both arteries and veins. A thrombus can potentially travel through those blood vessels ("embolize," with that mobile blood clot being called an "embolus") to distant parts of the body where it can cause damage by blocking the blood supply to important organs.

For example, an arterial embolus going to the brain can cause either a significant temporary reduction of blood to part of that organ (a "transient ischemic attack," or TIA), or permanent damage (a "cerebrovascular accident," or CVA—more commonly called a "stroke"). One or more emboli passing through the venous system to the right side of the heart can lodge in the pulmonary arterial system in the lungs, a "pulmonary embolism." This reduces the affected sections of the lungs' ability to exchange oxygen and carbon dioxide. Depending on how much of one or both lungs are involved, the effects could include a person experiencing only mild shortness of breath, or dying due to inability to breathe effectively.

Arteries in many different parts of the body can develop "atherosclerosis" [3]. This process involves injury to the inner lining (intima) of the aorta and smaller arteries. Damage inside those arteries can progress to development of raised areas called "plaques." High blood pressure, elevated cholesterol (particularly of a type called "low-density lipoprotein," or LDL), diabetes, and smoking are all risk factors for developing atherosclerosis. Atherosclerotic plaques typically start out as yellowish fatty streaks along the intima and can be present as early as age 10. Over a period of years damage to the artery could increase due to inflammation and accumulation of both muscle cells and white blood cells containing fat ("foam cells") within the intima.

The most advanced stage of atherosclerosis is the "fibrous plaque." This is a white, raised area that at least partly blocks off the artery itself. Fibrous plaques are usually fairly soft. However, especially in older adults and those with diabetes those plaques can also accumulate calcium and become hard and rock-like.

Atherosclerosis in the arteries supplying the heart causes "coronary artery disease," or CAD. If an atherosclerotic plaque blocks off most of the interior ("lumen") of a coronary artery, the area of the heart supplied by that artery may not receive as much blood and oxygen as it needs either when demand for them is higher with exercise or even when a person is at rest. The resulting "myocardial ischemia" may cause various symptoms. A common one is "angina pectoris," typically felt as a pressure or "heavy" sensation beneath the sternum (breastbone). This "chest pain" may spread into the neck, jaw, or either arm (most typically the inner side of the left arm). It may also be felt as "indigestion" and accompanied by other symptoms such as shortness of breath, sweating, nausea, or lightheadedness.

Angina may be "stable," occurring in a reasonably predictable pattern in terms of how often it occurs, how long it lasts (typically only a few minutes), and when it happens (e.g. occurring with exertion and relieved by rest.) As one or more fixed blockages in the coronary arteries become severe, leaving only a relatively small opening for blood to go through, angina may occur at rest or wake a person up at night. Also, if the pattern of angina changes in some way—for example, occurring at rest when before it occurred only with exertion—it comes "unstable angina." The latter can be a warning that a person is at increased risk of an even more serious event—a "myocardial infarction," "MI," or "heart attack." This occurs when an area of the heart (most commonly part of the left ventricle) is deprived of blood long enough (about 20–25 min) that permanent damage to it starts to occur. This is most commonly due to sudden rupture of a fibrous plaque, resulting in rapid formation of a blood clot and partial or complete occlusion of the artery at that location. Such plaques may originally have only mildly blocked the artery and produced no prior symptoms such as angina. In fact, an MI may be the first sign that a person even has CAD.

An individual having an MI may have no symptoms, vague nonspecific symptoms such as a feeling of indigestion that are not recognized as coming from the heart, more "classic" ones such as prolonged chest pain (although about one-third of people do not have it), or even sudden collapse and death. An MI may be so severe that it could at least temporarily impair a person's heart function and/or permanently damage it enough that the heart cannot supply an adequate amount of blood to the rest of the body. This can be associated with problems such as severe shortness of breath, a dangerous drop in blood pressure, lightheadedness and loss of consciousness, or even prove fatal.

However, the most immediate risk of an MI is that the heart can become electrically unstable even if only a relatively small part of it is deprived of blood or damaged. Individuals with an acute MI may experience "sudden death" due to development of a life-threatening abnormal heart rhythm (arrhythmia), often within the first hour after the MI occurs. One of these arrhythmias, ventricular tachycardia (VT), is caused by electrical impulses arising from a ventricle at a faster rate than the ones at their normal point of origin, the sinus node. VT can produce a heart rate as low as 110 beats per minute, with a person usually having only mild or no symptoms with that rate. However, VT can also occur at a rate as high as 300 beats per minute or more—too fast for blood to move in and out of the heart to supply the rest of the body or generate a detectable pulse.

An MI can also cause an even deadlier arrhythmia, ventricular fibrillation (VF) [4]. With VF the ventricle is producing impulses that are so frequent and chaotic (e.g. 400–500 times per minutes) that the heart just "quivers," with no effective pumping action at all. If either VF or the faster types of VT are not treated within minutes the person can suffer permanent brain damage or even die due to lack of blood to that organ and the rest of their body.

Both VT and VF are diagnosed by their characteristic patterns of the heart's electrical activity on an electrocardiographic monitor or printed electrocardiogram ("ECG" or "EKG"). They can be treated by delivering one or more electrical shocks to the chest from a "defibrillator." In hospitals and many other settings a monitor and defibrillator are combined into a single device. An "automated external defibrillator" does not include the monitor but is programmed to detect ventricular tachycardia or ventricular fibrillation and deliver an electric shock if indicated. Both types of defibrillator require that two adhesive pads or, as an alternative with a monitor-defibrillator unit, a pair of paddles be placed firmly on the patient's bare chest to deliver a shock. Most commonly one patch or paddle is placed just to the right of the upper sternum (breastbone) while the second is placed near the lower left edge of the rib cage. The person delivering the shocks and any helpers need to stay out of direct contact with the patient to avoid receiving part of that "jolt" too—hence the "Clear!" and other phrases used to warn others away before giving the shock.

The full details of "cardiopulmonary resuscitation," including chest compressions, administering medications, supporting a person's respirations with tubes placed into the throat, etc. are too involved to go into here. I will, however, return to this subject later in this chapter to critique how these procedures are sometimes portrayed in science fiction and elsewhere.

2.1.3 Infections

The drama of a character having an MI or the idea of a weakened or aged person having to stay in the microgravity of space because the "heart wouldn't take" coming back to Earth's gravity have a valid place in science fiction (although how those risks are portrayed may not be entirely accurate). Similarly, inflicting a serious infectious disease on someone or an entire population can be used to create dramatic situations in science fiction.

However, such characters are unlikely to be depicted as suffering from common infections such as colds, or a very minor case of cellulitis (a bacterial infection of the skin, associated with redness and perhaps swelling) associated with a hangnail. Instead they seem more likely to encounter new, exotic infectious diseases, whose mortality rate might make that of the medieval Black Death and the modern-day Ebola virus appear closer to that of a bad case of the sniffles by comparison. Those horrific infections in science fiction may be terrestrial in origin, such as the Blue Death in C. L. Moore's and Henry Kuttner's "Vintage Season" (1946), or in George R. Stewart's 1949 novel *Earth Abides*. Alternatively the infection might be of extraterrestrial origin, such as in Michael Crichton's 1969 novel (and later movie) *The Andromeda Strain* or the 1953 television serial *The Quatermass Experiment*. In some classic works, like H. G. Wells's *The War of the Worlds* (1898) or several of Ray Bradbury's *The Martian Chronicles* stories, the direction of infection is reversed, with human pathogens killing aliens.

The possibility of a widespread epidemic killing large numbers of people is, unfortunately, firmly rooted in reality. Whether the infectious organism in a science fiction work develops naturally, or is either accidentally or deliberately unleashed after misguided scientists develop it, there are historical analogs. The aforementioned Black Death, mass casualties in Native American groups due to their lack of immunity to diseases such as smallpox carried by European explorers after the first voyages of Christopher Columbus, the 1918 influenza pandemic that infected an estimated 500 million people (about 30 % of the entire world's population) and killed up to 50 million of them, and more recent widespread infections involving the human immunodeficiency virus (HIV), the causative agent for AIDS (acquired immunodeficiency syndrome), and Ebola virus, are all too real [5, 6].

However, where science fiction tends to deviate from realism is by exaggerating the numbers of victims and the specific effects of the diseases. Based on the total number of people who died, the deadliest pandemic in history was the 1918 influenza pandemic. It killed about 3 % of everyone living at the time—a horrendous loss of life, but far from the near-extinction of humanity that occurs in *Earth Abides*. Likewise the vampire-like symptoms that afflict the infected in Richard Matheson's 1954 novel *I Am Legend* and its multiple film adaptations bear at most very tenuous relationships to any "real life" infections.

Moreover, the extreme mortality rates and incredibly rapid demises of individuals exposed to the titular microorganism of *The Andromeda Strain* are not comparable to even the deadliest terrestrial diseases. The suggested cause for that fictional microorganism's deadly effects on humans was that it produced "disseminated intravascular coagulation" (DIC). This term (roughly) means "blood clots form in vessels throughout the body." DIC is an actual condition associated with infections, severe injuries, and other causes. However, while certainly serious, in real-life medicine DIC is by no means uniformly fatal and certainly not nearly instantaneous in its effects, including death.

A potential "advantage" that science fiction has over using only known types of infectious (and other) diseases is that the writer has some flexibility in depicting how an imagined one is transmitted, the effects it has on humans, and other characteristics based on the story's plot. Thus, in *The Andromeda Strain* the pathogen was highly contagious but did not reproduce under very specific physiological conditions in humans—points critical both to how dangerous it was and how it was ultimately contained.

Another issue is whether an alien microorganism could actually cause disease in a human, or if a terrestrial variety could do that to an alien. For obvious reasons this question cannot be definitively answered at present. However, based on general biological principles the likelihood of this happening would overall seem extremely low. Despite the obvious differences among all the myriad life forms that have evolved on Earth they share many characteristics at the atomic and molecular level. As described in Chap. 1, these include use of similar DNA and RNA, only the left-handed type of amino acids, and a range of specific varieties of proteins, carbohydrates, lipids, and many other substances.

Life originating in another star system might be based on variations of at least some of these general categories of chemicals. However, it is most likely that at least some of the *particular* ones used by the alien equivalent of bacteria and other pathogens would be dissimilar enough from terrestrial types to be incapable of causing infection in humans.

On Earth pathogens are generally specific for the cells and metabolic environments of particular species. Harmful microorganisms can sometimes cross species, such as the way HIV is thought to have spread from other primates to humans. However, the biological dissimilarity between terrestrial animals and alien ones would be expected to be far, far greater. Moreover, many bacteria such as the normal "gut flora" we possess or the adeno-associated virus that will be discussed in Chap. 12 can infect us but do not actually cause disease. Thus, even if an alien microorganism entered our bodies it might not only do no harm but die of starvation due to our biochemical pathways being too different from it to allow it to "feed" off or do anything else with us. On the other hand, if these organisms were somehow able to survive within us, potentially they could cause harm by producing substances toxic to us as part of their life processes without actually causing an infection.

Compared to the potential risks posed by unearthly bacteria, those from an extraterrestrial virus seem even lower. Unlike bacteria, all viruses are pathogens. They cannot reproduce on their own and need to take over the resources of a cell to do that. Viruses are basically a protected strand of DNA or RNA. Here too, however, a virus would need to be genetically and biochemically compatible enough with the cell it infects to replicate and cause an infection. Also, a virus could not develop and reproduce in the absence of cells. If the latter have significant biochemical and other differences from our cells, the viruses that infect those alien ones will not be able to do the same to ours.

Viruses are considered to be at the boundary between living and nonliving things. Still simpler infectious agents, prions, are considered nonliving but can cause diseases in humans, such as bovine spongiform encephalopathy (aka "mad cow disease") and kuru. Prions consist only of protein, without DNA or RNA. They can damage the brain as part of their replication process. However, whether prions could develop in an extraterrestrial environment, much less whether they could do so in a way to make them infectious, is highly speculative.

A potential exception to alien microorganisms being most likely "benign" is that any microbial life elsewhere in our own Solar System might at least somewhat resemble the terrestrial variety. As noted in Chap. 1, Earth's organisms' exclusive use of left-handed amino acids might be due to physical conditions associated with our Sun and its development, and thus could also apply to life developing on its other worlds.

For example, one theory for the origin of life on our world is that it first evolved on Mars roughly 4 billion years ago when the Red Planet briefly (in geological terms) might have had enough liquid water on its surface and other "ingredients" for life to originate there. Thus, if those microorganisms (or at least the chemical precursors needed for later development of life) arrived on Earth from Martian rocks blasted into space by large asteroid impacts, any present-day microbes on Mars might be somewhat similar to their distant "descendants" on our world. However, despite some possible chemical similarities those Martian "bacteria" or "viruses" would never have had the opportunity or the necessity to adapt themselves in ways that could either infect humans or cause disease. Still, the remote possibility that future astronauts could bring back hostile (though microscopic) invaders back from Mars to Earth is at least being seriously considered [7].

Another type of "infection" sometimes depicted in science fiction is the idea that intelligent alien organisms could either act as parasites to control our minds or symbiotes to enhance their hosts in some way. Examples of the former include the slug-like aliens in Robert A. Heinlein's 1951 novel *The Puppet Masters*, the "ear worms" used by Khan for mind control in *Star Trek II: The Wrath of Khan* (1982),[1] and most of the Goa'uld from the *Stargate SG-1* television series, while the humanoid Trill in the later *Star Trek* series meld with their hosts in a more benign relationship.

However, any alien organisms planning to "tap into" the human brain by attaching themselves to our central nervous systems would face formidable biological challenges. Besides issues regarding compatible amino acids, proteins, etc. described previously, the human body has a strong system for rejecting foreign material that enters it. Transplanted organs like kidneys and livers must be "matched" closely with the recipient to make sure they share as many different important "antigens" as possible that help the body decide what belongs in it and what does not. Having more such similarities reduces the chance the recipient's immune system will identify the new organ as "foreign" and try to destroy it. An alien organism living near or under the skin on one's back would be expected to be a red flag to a person's immune system that something is very much amiss.

The feasibility of creating an interface between the alien's nervous system and a human's would also be questionable. It would require many extensive connections between two entirely different sets of cells working in a highly coordinated manner throughout both the alien's and human's central nervous systems. Worse, the biological "circuitry" involved would have to carry two-way information well enough to allow one mind to control the other or for a "blending" of minds. Considering the formidable challenges of trying to do something this extensive between two human brains (a subject for later chap-

---

[1] The drama of the scene in which Commander Pavel Chekov and Captain Clark Terrell are "infected" by the alien organisms can overpower the biological questions it raises about how exactly those "worms" actually exert their will-numbing effects on their hosts. Once placed in the latter's external auditory canals, do those creatures whisper Khan's commands to the Starfleet officers? Less facetiously, do they secrete a substance that could reduce a human's ability to think for himself? Hopefully they do not actually burrow directly into the victim's brain. This would not only require rupturing the tympanic membrane ("eardrum") and destroying the delicate anatomical structures in the middle and inner ears, leaving the individual deaf on that side, but would also cause significant, irreversible damage to the cerebral cortex. The worm would also have to traverse a comparatively long distance through that cortex to presumably reach one or both frontal lobes, the parts of the brain most associated with volitional activity. Maybe it's better not to think too deeply about such things and just enjoy the movie…

ters), the chance of an alien one being able to accomplish this via a "natural" process seems incredibly small.

The example given previously of the Trill ameliorates some but not all of these issues. That humanoid species and their symbiotes presumably evolved on the same planet, so at the most basic level their tissues might be compatible. However, even if some way were found to address the rejection issue, the interface issues remain. Neither the human nor the Trill central nervous system would have the functional equivalent of a computer's Ethernet port that another organism could plug into and at least share control and "thoughts." Thus, while the concept of a shared consciousness and either unilateral or shared control is simple and useful for story purposes, the details of how that relationship would work biologically speaking are unclear.

2.1.4 Cancer

Another major health issue, cancer, might be used as a way of explaining why a character in science fiction has only a limited time to live. However, the general term "cancer" encompasses a wide range of diseases with a broad spectrum of different symptoms, effects on the body, prognoses, and treatments. All types of cancer involve particular cells becoming abnormal and undergoing uncontrolled division and growth. This can result in tumors that can invade and grow large enough to destroy and replace normal tissue. Tumors can grow locally, where they can interfere with the function of the organ or other tissue where they originated, or may also spread (typically through the bloodstream) to distant areas of the body ("metastasize"). Some can produce a "paraneoplastic syndrome" by secreting excess amounts of hormones and other substances that can have effects on other parts of the body, such as the parathyroid hormone-related protein secreted by some types of lung and breast cancers that could cause increased levels of calcium in the blood.

The term "tumor" is also not synonymous with "cancer." A tumor may be "benign" and "noncancerous" in the technical sense that its abnormal cells neither invade normal tissue nor spread to other parts of the body. However, such a tumor can still cause problems by compressing normal tissue. For example, most "meningiomas"—a type of tumor arising from the meninges, thin tissues covering the brain and spinal cord—are benign in the above sense. However, if one located in the skull grows large enough it can compress the brain with results that definitely are not benign in the more general meaning of the word. Also, one group of cancers, leukemia, is associated with abnormalities and overproduction of certain types of (abnormal) white blood cells rather than "solid" tumors.

Many different types of cells within the human body can become cancerous, and the individual characteristics of each kind of cancer vary tremendously. Some varieties of cancers are extremely slow growing, rarely metastasize, and occur in areas where they have minimal impact on normal tissue if treated early in their course. The most common type of skin cancer, basal cell carcinoma, has these characteristics. On the other hand, a less common but far more serious type of skin cancer, malignant melanoma, is much more aggressive and deadly, often spreading to the lungs, brain, and other distant parts of the body.

Organs also contain multiple types of cells, and the causes, characteristics, and treatments for the specific kinds of cancer each can cause may be very different. For example, four major types of lung cancers are adenocarcinomas, squamous cell carcinomas, large cell carcinomas, and small cell carcinomas. Smoking is a risk factor for all four, but small cell carcinomas are generally faster growing and more likely to spread to other parts of the body than the other three. Small cell carcinomas are also more likely to secrete different substances than the other types of lung cancer as part of paraneoplastic syndromes [8, 9]. Potential treatments for each type of lung cancer, which can include surgical resection, chemotherapy, and radiation therapy, also vary greatly.

Likewise, various organs and tissues differ in their chances of developing cancer. The heart has a low likelihood of developing either benign (e.g. myxomas) or malignant tumors, while breast cancer in women and prostate cancer in men are all too common. Certain tumors such as neuroblastomas occur far more often in children than adults, while seminoma (a type of testicular cancer) is the most common type of malignancy in men between the ages of about 15 and 35.

In short, the word "cancer" encompasses a very heterogeneous group of illnesses with a great range of morbidity and mortality. Its spectrum stretches from very slow growing and non-aggressive malignancies that can be very successfully treated via current methods, to ones that grow and spread quickly, have very limited treatment options, and can kill within only months or less after diagnosis despite the best available efforts. The use of the term "cancer" in science fiction typically implies that the particular kind a character who says he or she is dying from it has is indeed one of the more nasty types, although the particular one is typically not named. The key point here is that, although having a character with a limited time to live due to cancer (or perhaps some other incurable infection or disease) can certainly add drama to a story, not all types of cancer are "deadly" enough to use for that plot element.

Moreover, in real life a person in the terminal stage of cancer (or, for that matter, many other fatal diseases) is typically very ill and likely to be "cachectic," with a body that is wasted and weak. Such characters would be unlikely to either appear healthy or be physically able to engage in strenuous activities as part of a science fiction story's plot before, perhaps, dying heroically to save the world (after all, they were going to die soon anyway…). Also, barring significant advances in medical technology that would make such an estimate accurate, the cliché that the character has only "(fill in the blank) days/weeks/months to live" may serve a dramatic function but is medically dubious. A modern-day physician can give a general estimate based on statistics of *average* time of survival for a person with a cancer (or other life-threatening disease) at a certain point in its course and taking into account the individual patient's overall condition. However, except at the very end of life there are usually far too many variables involved to predict with absolute certainty when a *particular* person will die, either significantly sooner or later than that average.

2.1.5 Miscellaneous Health Issues

Going beyond "high drama" medical events like heart attacks, cancer, and diseases such as infections that can produce unusual effects in characters (e.g. an infection actually taking over the human victim's body as in *The Quatermass Experiment*), science fiction typically mentions other types of actual diseases only in passing. A "mundane" ailment such as heartburn may be talked about in an aside as in the *Star Trek: Voyager* episode "Message in a Bottle" (1998), but it does not become the primary focus of the plot. Similarly, unlike the rest of us, science fiction characters rarely seem to get headaches, constipation, or tinea pedis (aka "athelete's foot").[2] Moreover, many common illnesses that are major causes of disability and death such as diabetes mellitus, various types of lung disease such as emphysema, thyroid disease, liver and kidney failure, rheumatological diseases such as systemic lupus erythematosus, inflammatory bowel disease, etc. rarely seem to "make the cut" for inclusion in science fiction. Likewise a character may be said, for example, to need a new liver or kidney but not be depicted as showing any of the symptoms associated with end-stage dysfunction of those organs.

Some neurological and musculoskeletal problems may be an exception to this. In the *Star Trek: The Next Generation* episode "Sarek" (1990) the latter, Mr. Spock's father, is discovered to be suffering from "Bendii Syndrome," a reasonably portrayed form of senile dementia in Vulcans. My own story "The Best is Yet to Be"[3] deals in large measure with the devastating effects

---

[2] My story "The Human Touch" (*Analog Science Fiction and Fact*, May 1998) does, for satiric purposes, include characters with medical problems such as hemorrhoids and pediculosis pubis (infestation of pubic hair by lice). However, I doubt these particular diseases will catch on as popular ones in other science fiction.

[3] *Analog Science Fiction and Fact*, December 1996

of Alzheimer's disease. The title character in Robert A. Heinlein's (writing as "Anson MacDonald") 1942 story "Waldo" has a disease consistent with myasthenia gravis (or more specifically, since he is depicted as being born with his condition, "congenital myasthenic syndrome") that produces profound muscle weakness. Miles Vorkosigan, the protagonist of many works by Lois McMaster Bujold, is born with brittle and easily fractured bones whose presence play an important role in the subsequent growth and development of both his body and character.

Perhaps one reason why many illnesses fail to be mentioned in science fiction, at least in stories involving "futuristic" technology of human or alien origin, is that they have already been minimized or eliminated by yet-to-be-discovered medical advances. Thus, a story set in 1700 could reasonably include one or more characters infected with or who survived smallpox. In the twenty-first century, however, that infection is (fortunately) essentially restricted to fiction involving potential bioterrorism and not an everyday experience. Moreover, unless they are pivotal to the story (e.g. Anton Chekov's "The Sneeze" or Isaac Asimov's 1957 story "Strikebreaker) many common biological functions—e.g. eructation (aka "belching"), borborygmus ("rumbling of the stomach," associated with movement of gas inside the intestines), and the periodic need for excretory functions—do not need to be mentioned to explore a science fiction or other type of work's primary themes. And, of course, the dramatic potentials inherent in a science fiction character suffering a heart attack are somewhat greater than if that individual must deal instead with the discomfort of athlete's foot.[4]

2.2 Common Injuries

The human body can be damaged externally and internally in myriad ways. Skin can be scratched, scraped, punctured, bitten, clawed, bruised, cut, burned, and exposed to extremes of pressure and cold. Internal organs can be lacerated by penetrating injuries as well as bruised and ruptured by blunt trauma from being struck. Bones can be broken, muscles and their ligaments and tendons stretched and torn, blood vessels and nerves sliced and crushed.

---

[4] However, there are exceptions to this. The novel *Mars Crossing* (2000) by Geoffrey A. Landis describes an expedition to Mars that ultimately leads to disaster when a single crewmember develops what is said to be a case of this typically nonfatal fungal infection. The fungus not only spreads to and sickens the spacecraft's other members but only starts to grow on vital internal parts of the craft itself. Ultimately the fungus infiltrates the latter's fuel-controller electronics, causing a short circuit that makes the spacecraft explode, killing the entire crew. In this particular case, on balance it actually would have been "better" and less dramatic for an astronaut to "only" have a heart attack…

Individual body parts such as fingers, whole limbs, the nose, and external ears can be ripped off (or, to use the medical term for this, "avulsed").

Similarly a man's external genitalia and every person's eyes, brain, and spinal cords are particularly vulnerable to injuries that can quickly incapacitate or kill. Poisons and toxic substances can rapidly cause harm or death by being swallowed, placed on the skin, or inhaled. Traumatic injuries may result in many short- and long-term complications, including internal and external bleeding, difficulty breathing, infections, organ failure, and inadequate healing of wounds and fractured bones.

As with diseases, the many ways trauma affects the human body can be organized into general categories. This chapter will concentrate on penetrating injuries, blunt trauma, explosions, burns, electrical injuries, and exposure to poisons—all potential risks for science fiction characters. Chapter 3 will cover exposure to vacuum, acceleration and deceleration injuries, and other forms of trauma more specific to the space environment. Chapters 6 and 7 will, respectively, describe how radiation and excessively cold conditions damage the human body.

2.2.1 Penetrating Injuries

Penetrating injuries include those caused by conventional weapons/projectiles like bullets, knives, swords, spears, and arrows as well as ones more specific to science fiction such as laser beams and lightsabers (ask Qui-Gon Jinn about that last one). The effects of penetrating injuries vary depending on factors such as how wide and deep the penetration is, what does the penetrating, and what body part is penetrated. The net results can include quickly killing, temporarily or permanently incapacitating, or merely annoying a character.

The skin has sensitive pain receptors stimulated by even trivial punctures, such as when someone has blood drawn with a needle and syringe. If the area of skin penetrated is relatively small (e.g. the diameter of a bullet or stiletto) the pain at skin level is significant but not necessarily sufficient by itself to put the victim out of action. Such damaged skin may start showing inflammation within several hours, and over the next few days can become significantly infected. However, those effects are delayed far too long to prevent a science fiction character attacked in this way from retaliating immediately, assuming the weapon has not reached one of the vital internal body parts that will be discussed later.

As with the value of real estate, the major immediate effects of a penetrating injury are based on location. The arms and legs have a limited repertoire of major structures that can be damaged—primarily bones (including joints), muscles, blood vessels, and nerves. Penetrating "just" a muscle with a projectile, sharp pointy thing, or energy beam will, even if the area penetrated is small, cause pain due to sensory nerves in it. However, loss or damage to a significant bulk of large muscles such as those in the thighs or upper arms is required to significantly affect the injured region's function, e.g. make it difficult for a person to stand or move the arm.

There are several caveats with this. If a bullet or similar projectile lodges in a limb's muscle it may cause further pain with movement of that extremity and thereby make the person at least "favor" it. Also, the amount of injury to both the muscle and surrounding structures will depend on the speed of the bullet, its mass, and whether it fragments or deforms when it enters the muscle. A high-speed bullet may pass relatively cleanly through a muscle without dissipating too much energy along the way. A lower-speed one that fragments and stays within the muscle can provide more net destructive energy to the muscle as well as surrounding tissues. Also, if the bullet still has enough energy left after its passage through the muscle to "punch through" the skin again, the exit wound produced may be larger and more ragged than the entrance one.

By comparison a rigid penetrating object with a uniform diameter like a rapier would be expected to produce similarly sized entrance and exit wounds. Penetration by an object that can also transfer energy in the form of heat to the muscle, such as a high-energy laser or a lightsaber, could also damage muscle via thermal injury over an area wider than the width of the weapon itself.

The major skeletal muscles in our limbs have a rich supply of blood vessels to provide nutrients, oxygen, and other substances they need to work. Even if a penetrating injury misses a major artery or vein in a muscle there are enough smaller blood vessels in it that, depending on the size of the "hole" produced or additional injury around the path of penetration (e.g. from a fragmenting bullet), the damaged muscle is likely to bleed significantly.

Bleeding can occur within the muscle itself and cause swelling, or blood can go to the outside. If the rate of external bleeding is slow enough it can be stopped or at least reduced by covering and compressing the "leaking" area. Bleeding within the muscle may be more difficult to control, leading within perhaps minutes to additional soreness, pain, and difficulty using the muscle. If that bleeding is great enough it can compress larger arteries and veins within the muscle that were missed by the penetrating object. This in turn could reduce their ability to supply enough blood and thus oxygen, nutrients, etc. to the damaged muscle, causing an "ischemic" injury.

Penetrating injuries to bones in limbs may take many forms. Bones are considerably denser and less "yielding" to a bullet or knife than skin or muscle. Depending on factors such as how much energy a projectile has, the force behind the thrust of a knife blade or similar weapon, and the mass, "hardness," or shape of the penetrating object itself, a bone may stop, deflect, or be fractured by it. Bones have nerve fibers carrying pain sensations in both their outer lining, the "periosteum," and internal cavities. The degree of pain produced by something striking it will be roughly proportional to the extent of injury, with a "nick" expected to be less painful than a full fracture.

A fracture can be incomplete, such as a "crack" in the bone; complete, with the bone broken into two or a few major parts; or so severe the bone is shattered into multiple fragments. While an incomplete fracture may be painful the bone might still be mechanically stable enough that a person could continue to use the limb to a limited degree. However, applying enough stress to an incomplete fracture could make it buckle more and turn it into a complete fracture.

A penetrating injury to a large joint (e.g. the knee, shoulder, or elbow) could be particularly devastating and immediately incapacitating.[5] These joints represent "soft spots" for injury—areas where two bones meet, connected by specialized areas at the ends of muscles called tendons and attached directly to each other by strap-like ligaments. Joints can also contain various soft tissues and structures, such as the two cartilage-containing disks ("menisci") that act as cushions between the ends of the femur and tibia, the longest bones in the thigh and lower leg, respectively. The human body has only a limited ability to heal on its own from severe joint injuries. In the absence of appropriate current (or futuristic) medical care they can cause marked long-term disability.

Penetrating injuries occurring at or near smaller joints (e.g. those involving the bones of the wrists, toes, and fingers) can also be incapacitating to varying degrees. A bullet or other high-energy projectile could penetrate a toe or fingertip forcefully enough to avulse it. Our wrists contain eight small "carpal" bones and our ankles seven overall larger "tarsal" bones. Penetrating injuries to the wrist can damage one or more carpal bones or the ends of the two long bones located there, the ulna and radius, thereby limiting use of that joint. Moreover, the wrist also contains several relatively large blood vessels (such as the radial artery, used to check for a pulse, and the ulnar artery) and nerves, all confined in a relatively small volume. Injuring those blood vessels and nerves could cause more net damage via bleeding and other means than injuries to the carpal bones themselves.

A penetrating injury to or near the ankle joint is more likely to strike a bone directly, such as the end of the two long bones in the lower leg, the tibia and fibula, or the largest of the tarsal bones, the talus. These bones are generally "sturdier" than those in the wrist and more resistant to injury. However, too

---

[5] This is something characters encountered in the fantasy role-playing game *The Elder Scrolls V: Skyrim* who have "taken an arrow to the knee" know all too well.

much damage will make it difficult if not impossible for a person to put a normal amount of weight on the affected lower extremity.

Disrupting any of these joints by a penetrating (or other traumatic) injury will also cause pain and bleeding within it, further incapacitating a character. An injury to one of the large joints in a lower extremity—the hip, knee, or ankle—will literally make a character "not have a leg to stand on." In some cases the joint itself does not need to be injured to be put out of action. A clavicle (collarbone) runs from near the top of the sternum (breastbone) horizontally to each shoulder. If one is fractured by a penetrating or other injury, the corresponding shoulder will be immediately limited in its motion. Fractures involving the bony part of the pelvis can likewise affect how well the nearby hip joints can support a person.

A penetrating injury to a major artery or vein in a limb will cause significant internal and probably external bleeding. A puncture, tear, or cut to a major systemic artery[6] in a limb (or, for that matter, elsewhere in the body) will result in blood spurting from it due to the force generated by each heartbeat, reflected in the systolic blood pressure. This blood's hemoglobin is highly saturated with oxygen and therefore red.

The artery will continue to spurt blood until it is compressed either externally, such as by a person's hand, or internally such as by surrounding tissue and blood, leading to clotting. However, a severe injury to the artery (such as being completely severed) may make it impossible to stop the bleeding by such simple means. A medical maxim is, "All bleeding stops"—and unfortunately the artery may cease spurting blood because the body has lost too much of it from the circulation and/or the heart has stopped as an end result of "hemorrhagic shock," a life-threatening drop in blood pressure due to severe blood loss.

Conversely, blood in even large veins is at considerably lower pressure than in systemic arteries. Instead of spurting, venous blood "oozes," and so this type of bleeding can be more easily controlled by compressing the site and allowing the injured vein(s) to seal off by clotting. Venous blood (except in the pulmonary veins leading from the lungs) is bluish-purple due to its hemoglobin having reduced oxygen saturation, but it can turn more reddish with exposure to oxygen in air after external bleeding.

Under some circumstances, such as when bleeding cannot be adequately controlled by direct pressure alone, careful and temporary use of a tourniquet applied before the area of severe arterial or venous bleeding in a limb can also serve as a stopgap measure until definitive care can be given. To be effective a

---

[6] A systemic artery is one that ultimately originates from the left ventricle, as opposed to the "pulmonary" arteries that arise from the right ventricle and pass through the lungs.

tourniquet has to be tight enough to restrict both venous and arterial blood flow in the limb. However, its use can be painful, and if the tourniquet is left in place too long (e.g. perhaps 1–2 h) the limb may suffer potentially permanent damage to muscles, nerves, etc. from ischemia [10].

The large nerves that carry sensory (e.g. pain) and motor (stimulating a muscle to move) information in the extremities are relatively thin. It would take a somewhat "lucky" single shot, thrust, or laser blast to hit one of them directly. However, a fragmenting bullet, or a penetrating injury covering a wide enough area could injure them. Completely cutting a major nerve would result in loss of pain and other sensation to the muscles it supplies as well as the ability to move them. Even if it escapes direct damage the nerve could be compressed by surrounding bleeding and swelling caused by the injury, thereby impairing its ability to work. The character may experience this as pain, tingling, numbness, and weakness in the affected limb.

Like the limbs, other body parts vary in their vulnerability to penetrating injuries. Starting from the top, our skulls overall act as a reasonably effective bony shield to lower-energy weapons such as knives, but less so to high-speed projectiles or high-energy weapons such as a lightsaber. If something does get through the skull to reach the brain it will cause direct damage along its path. Some areas of the cerebrum, which forms the bulk of the brain, are (relatively speaking) less "critical" for major sensory, motor, and "thinking" functions.

Depending on what areas and how much of the cerebrum is involved by the injury, the effects on a person could range from a small possibility of only mild impairment to (far more likely) severe dysfunction, including rapid loss of consciousness. Damage to parts of the brain controlling critical body functions and its connections to the spinal cord (e.g. in the brainstem) will cause death quickly. However, while a "bullet to the head" or other penetrating injury severe enough to reach the brain is by far most likely to cause serious if not soon fatal injury, it may not necessarily do so.

Our skulls also have obvious weak points where they offer minimal protection, e.g. our eyes, ears, and mouths. Our facial bones—roughly speaking, those located lower than our foreheads—as well as some bones near our temples are thinner than the rest of the skull, and therefore more vulnerable to injury.

If a bullet or other penetrating injury reaches the brain by some route but manages to miss a critical part of it, the cerebral cortex itself has little pain sensation. However, bleeding within the brain (cerebral hemorrhage) or from blood vessels in tissues near it can (among other bad effects) cause increased pressure inside the skull. This and any swelling associated with the penetrating injury can produce a mild to severe headache. If this pressure is high enough it can actually force part of the brain through openings in the skull, such as the foramen magnum at its base where the spinal cord connects to it. If a critical part of the brain "herniates" in this way it can be quickly fatal.

As mentioned previously, the meninges are membranes that cover the brain and spinal cord. The outermost of their three layers is called the "dura mater." If a penetrating injury or, more likely, a blow to the head causes bleeding between the dura and the skull, both clotted and non-clotted blood accumulating between them can compress the brain, causing an "epidural hematoma." This bleeding is usually due to injury to arteries and can rapidly cause symptoms based on what part of the brain is most compressed and whether herniation occurs. Symptoms can include headache, nausea, vomiting, seizures, numbness, weakness in limbs, confusion, or visual problems. The victim may be conscious and lucid immediately after the injury. However, as arterial blood rapidly accumulates within the skull, symptoms may develop within minutes, potentially terminating in unconsciousness and death.

A "subdural hematoma" occurs with bleeding between the dura and the middle layer of the meninges, the "arachnoid mater." It can produce symptoms similar to an epidural hematoma. An "acute" subdural hematoma is more likely to be due to blunt trauma to the skull (e.g. someone being hit by a pipe or bat) than by a penetrating injury. It may cause death rapidly. However, subdural bleeding is usually from low-pressure veins rather than high-pressure arteries. If bleeding is slow enough it can develop over a period as long as days into a "chronic" subdural hematoma. Symptoms of a subdural hematoma may be similar to that of an epidural hematoma.

A "subarachnoid hemorrhage"—bleeding between the arachnoid matter, and the meninges' innermost layer, the "pia mater"—can also result from a penetrating injury or, more commonly, blunt trauma to the skull. It too can be associated with rapid onset of headache, vomiting, weakness on one side of the body, and loss of consciousness, potentially ending in death.

The vertebrae of our neck and back are bones of the "vertebral (spinal) column" that protects our spinal cord from injury. Seven of these bones are present in the neck (cervical vertebrae), twelve in the upper back (thoracic vertebrae), and five in the lower back (lumbar vertebrae). These vertebrae are "stacked" on top of each other in a mildly curving (primarily at the lower back) column. They are both separated and connected by "intervertebral disks" formed of cartilage and fibrous tissue that help give the spine some mobility.

The vertebral column continues with five fused bones that compose the sacrum, which forms the back portion of the bones that comprise the pelvis, and ends with the four tiny fused bones of the coccyx, or "tailbone." The cervical, thoracic, and lumbar vertebrae are roughly ring-shaped with a central opening through which the spinal cord passes. The latter stops at about the second lumbar vertebra, although a bundle of nerve fibers called the "cauda equina" (Latin for "horse tail") extends from its end, as does a filament that helps anchor the spinal cord to the coccyx bones.

Individual nerves originating from the spinal cord leave it through openings between the vertebrae except for the first spinal nerve, which passes between the base of the skull and the first cervical vertebra. Spinal nerves are grouped into pairs and generally named after the vertebra above which they emerge from the vertebral column. Thus, there are eight cervical spinal nerves (C1-C8), with C8 emerging between the last of the seven cervical vertebrae and the first thoracic vertebra. Similarly there are twelve thoracic spinal nerves (T1-T12), five lumbar spinal nerves (L1-L5), five sacral spinal nerves (S1-S5), and a single pair of coccygeal spinal nerves. Although, as mentioned previously, the spinal cord itself ends at about second lumbar vertebra, the spinal nerves extend from near its end to emerge lower in the vertebral column.

The bones of the vertebral column help protect the spinal cord against penetrating and other injuries. However, a bullet, projectile, or a powerful enough energy beam could damage those bones and injure the spinal cord or peripheral nerves extending from it. Damage to the spinal cord itself could result in loss of both sensory and motor function to parts of the body supplied by nerves originating anywhere below the level of injury.

Thus, a severe injury to the spinal cord at the level of the lumbar nerves can significantly impair the ability to walk or feel sensations in the lower extremities, and markedly reduce a person's ability to control bladder and bowel function. Injuries higher in the back, at the level of the thoracic nerves (T1-T12), will result in an individual having little or (more likely) no ability to walk ("paraplegia"), although arm strength will (except for injuries very high in the thoracic level of the spinal cord) be preserved. In addition to this, damage to the lower cervical spinal nerves (C5 to C8) will cause increasing weakness of the arms as the level of injury to the spinal cord goes higher. Individuals with damage above C4 will not be able to move the arms or legs (quadriplegia), and very high cervical injuries will result in a person not even being able to breathe adequately due to (among other things) loss of function involving the paired "phrenic" nerves supplying the diaphragm.

The rib cage and pelvis protect organs within the "thoracic cavity" and "pelvic cavity" respectively. The former contains the heart, most of the aorta, and the lungs. The sternum and (usually) 12 pairs of ribs comprise most of this cage. The bony part of the ribs attach to the thoracic vertebrae at one end, then curve around to the front of the chest. Starting from the upper back and going downwards, only the first seven pairs attach to the sternum. The next three on each side typically attach to the seventh rib and to each other via cartilage instead of bone-to-bone attachments like the first seven. The lowest two pairs of ribs are "free floating," attached to vertebrae at one end and not attached to anything at the other.

As expected, the ribs and sternum are the structures that offer the greatest protection to a penetrating injury. Those bones can be pierced and fractured, but it takes a relatively high-energy projectile like a bullet or massive object to do it. A knife or, unless wielded with great force, even a sword or battle axe might just bounce off the sternum and ribs or "merely" fracture them without directly reaching the vulnerable heart and other structures they protect. However, the spaces between the ribs—the "intercostal spaces"—contain cartilage instead of bone. They offer much less resistance to punctures. A knife whose blade slides between the ribs would meet much less resistance than from the ribs themselves.

The heart is located partly behind the lower sternum, with its "bottom" located at about the level of the "xyphoid process," the small "tip" at the end of the sternum. However, most of the heart normally extends along the left side of the chest and is protected by ribs.[7] The right atrium and right ventricle are positioned so that they form most of the "front" of the heart, directly behind the sternum, while the left atrium and left ventricle form the heart's "back." There is also a "soft spot" to reach the heart through the "epigastrium," the inverted-V shaped region just beneath the middle of the chest where the lowermost ribs on each side that attach to the sternum meet and the xyphoid process is located. Something sharp thrust there will encounter only easily yielding soft tissue.

A puncture wound involving the heart could—but not necessarily—be quickly fatal. Its effects would depend on how much of the heart was pierced and damaged along with the degree and rate of bleeding from it. A thin bag-like tissue called the "pericardium" surrounds the heart. If the latter bleeds into an at least relatively intact pericardium, enough blood could quickly accumulate in that closed "bag" to compress the heart and interfere with its ability to fill during diastole. This leads to "pericardial tamponade," and if it is severe enough a person's blood pressure can rapidly fall enough to cause unconsciousness and perhaps death. But if the puncture wound to the heart and associated bleeding from it is small, the pericardium might instead compress and help stop the bleeding.

---

[7] Hopefully to no one's surprise, these anatomic details render the "manual cardiotomy" performed on someone who probably did not sign a consent form for the procedure depicted in *Indiana Jones and the Temple of Doom* (1984) "magical" rather than "medical" in its level of realism. I also suspect that the "surgeon" doing the procedure may have been practicing medicine without a license, and he certainly did not use standard sterile technique for the operation. However, a conventionally trained cardiothoracic surgeon might well admire that character's efficiency in performing the surgery.

The ascending aorta lies mainly behind the sternum and is thus protected by it. In the chest the descending aorta first lies near the back, close to the vertebral column and just to the left of the sternum, and is also protected by ribs. However, in the abdomen the aorta is shielded only by soft tissue (e.g. skin, fat, and internal organs). A penetrating injury to any part of the aorta can make blood pump out of it with every heartbeat rapidly enough to cause unconsciousness and death within minutes. The pulmonary artery represents a smaller and more localized target than the aorta and is better protected by the sternum. If something penetrates the pulmonary artery its rate of blood loss will, all else being equal, be slower than from the aorta due to the lower pressure within it compared to the latter. However, that bleeding could still be enough to ultimately cause loss of consciousness and even prove fatal.

The largest targets inside the chest are the lungs. A penetrating injury to one of them may damage only a small amount of lung tissue. However, a lung injury can have serious secondary effects that can soon incapacitate a person. Air normally confined to the lungs could go instead out into the thoracic cavity and possibly increase in amount with each breath. This type of injury can cause an "open pneumothorax" if the free air has a way of getting out (e.g. through the entry wound). An individual with an open pneumothorax will typically be short of breath and have a fast heart rate, with the affected lung likely to be partially collapsed. If the "opening" to the chest created by a penetrating injury is large enough it can cause a "sucking" chest wound, with more air moving in and out of the "hole" with each breath than through the mouth and nose. This can add further to the victim's respiratory distress.

However, if air can enter the damaged lungs but enough cannot get out through the penetrating wound or the mouth and nose, increasing amounts of air will be trapped within the thoracic cavity and lead to a "tension pneumothorax." This can compress not only the damaged lung but might eventually shift the positions of, compress, and thereby affect the functions of the nearby heart and opposite lung. A tension pneumothorax could potentially incapacitate or kill a person within minutes.

A pierced lung can also bleed into the thoracic cavity, producing a "hemothorax." If bleeding is severe enough blood can compress the lung and lead to lightheadedness, loss of consciousness, and death. Worse, a punctured person could have a combination of hemothorax and pneumothorax, a "hemopneumothorax."

The uppermost part of the esophagus runs though the neck, where it (like other structures there) is highly vulnerable to penetrating injuries. However, once it gets past the neck it is protected on all sides by at least some bones (e.g. the sternum and ribs). The esophagus runs through the chest roughly behind the sternum and in front of the spine. A penetrating injury to the esophagus would most likely be associated with relatively mild bleeding and, at least initially, be unlikely to incapacitate a person. However, within a few days or less afterwards, such an esophageal injury could result in life-threatening infections and other complications.

The abdomen contains many organs and blood vessels vulnerable to penetrating injury. While those in the upper abdomen are protected to varying degrees by the lower rib cage, ones in the lower abdomen are shielded only by skin and soft tissue (e.g. fat, muscles, etc.), thus making them especially vulnerable to projectiles and sharp objects.

The liver is the largest single internal organ. It is located mainly in the right upper abdomen, although part of it extends into the left upper abdomen and by the stomach. The liver is positioned high enough in the abdomen that most of it is protected by the lower rib cage on the right. It contains a rich system of blood vessels, including some of the largest arteries and veins in the body. A penetrating injury to the liver could cause rapid bleeding into it and the abdomen. However, the liver itself has a large "functional reserve"—that is, it can still function at a normal level even if most of it were damaged. It also possesses an excellent capacity to repair itself and, unlike our other internal organs, can regenerate. Even if a relatively large section of the liver is removed or destroyed its remaining cells can divide, proliferate, and eventually perhaps grow the liver back to its normal size.

Both kidneys are normally located on either side of the spine, in the upper back section of the abdomen (the "retroperitoneal space"). They are protected by the rib cage except for their lowermost sections, where here too only skin and other soft tissue lie between them and the outside of the body. The right kidney is usually lower than the left, and thus less protected by the rib cage. The kidneys too have a good degree of functional reserve. Overall "renal" function remains normal even if one kidney is removed and the other is healthy.

Either kidney can bleed significantly after a penetrating injury, although generally not as much as the liver does. Blood from a penetrated kidney can reach the skin at the lower back and cause what looks like a large area of bruising there.

Each kidney connects to the lower back of the bladder via a thin tube, the ureter. The latter represents a relatively long but narrow target for penetrating injuries. However, should the ureter be punctured or severed urine will leak into the abdomen. Here too the most serious effects on a person will be delayed for hours to days, with bacteria in the urine potentially causing life-threatening peritonitis, septic shock, and other infections over a time course similar to penetrating injuries to the intestines.

The bladder lies protected by the bones of the pelvis when empty. However, as it fills with urine and distends its upper portion rises into the abdomen, making it more vulnerable to a penetrating injury. The latter might cause significant but not necessarily immediately incapacitating pain. However, as with puncturing or severing a ureter, the major morbidity and mortality with a ruptured bladder is that leaking urine might, within days, cause an overwhelming infection.

The stomach and large intestine are mainly and the small intestine almost entirely unprotected by bones. The rib cage protects the uppermost parts of the stomach and large intestine, and the very end of the latter is shielded by the pelvis. Otherwise they and the small intestines are protected only by easily penetrated soft tissue.

Outside of pain and psychological stress-related effects, sending a bullet or sword through the stomach and/or intestines may not incapacitate a person immediately. Those structures are less likely to bleed as severely as the liver or other abdominal organs. However, within hours to few days afterwards pain can increase markedly due to complications from the injury. These include free air entering the abdomen and, especially with the intestines, problems with dysfunction, inflammation, and death of a portion of the injured organ. Bacteria that normally reside peacefully in the intact intestines can, when the latter are injured, produce "peritonitis" and other infections. These can progress to "septic shock," with fever and a drop in blood pressure. Without treatment (e.g. antibiotics, surgery, etc.) the character will likely die within several days—and the mortality rate even with the best present-day care is still significant.

The pancreas is a localized target in its roughly central location in the upper back of the abdomen. It is mostly but, especially in its middle section, not entirely protected by the rib cage. Most of the pancreas is located in the retroperitoneal space, with only its leftward pointing "tail" in the left upper abdomen. A penetrating injury will most likely cause retroperitoneal bleeding, but not necessarily enough to cause major blood loss. However, with such an injury the digestive enzymes the pancreas produces can be released into and damage it as well as nearby tissues. A penetrating injury to the pancreas is not likely to cause a character's immediate demise, but it can prove fatal within days due to subsequent complications such as severe inflammation of the pancreas (pancreatitis), infection, etc.

The spleen is roughly the size of a kidney and located on the far left side of the upper abdomen, nearly completely shielded by the rib cage. A penetrating injury to it can cause bleeding into the abdomen, ranging in severity from mild enough to not appreciably affect a character to sufficient (typically over the course of hours) to cause hemorrhagic shock and death.

While penetrating injuries to any internal organs in the chest and abdomen can be serious, those most likely to cause unconsciousness and death within minutes are ones to the cardiovascular system. A severe injury to the heart can cut off blood supply to the brain quickly enough that a character could collapse within seconds and die within minutes. Likewise, as noted previously a penetrating injury to the aorta can cause a large volume of blood to be pumped out within minutes into the chest or abdomen and prove equally fatal.

Venous blood is carried back to the heart from the lower body by the inferior vena cava. It runs roughly parallel to the aorta in the abdomen. While it would take somewhat longer due to the lower pressure of blood in it compared to the aorta, a penetrating injury involving the inferior vena cava could also result in serious and potentially deadly bleeding into the abdomen. Worse, the proximity of the aorta and inferior vena cava to each other in the abdomen means that both might fall victim to a single penetrating injury such as a bullet or sword thrust.

Overall, the abdomen is one of the most vulnerable parts of the body to penetrating injuries. For example, in *Star Wars: The Phantom Menace* Qui-Gon Jinn's death is due to such an injury caused by Darth Maul's lightsaber passing through what appears to be the upper middle of the Jedi Master's abdomen from the front. Qui-Gon falls almost immediately, and dies within minutes after giving what in retrospect might be considered unwise advice regarding one Anakin Skywalker.

During its entry into his abdomen the Sith's lightsaber probably burned through part of the Jedi's small intestines, but that by itself would not have killed him so quickly. Instead the rapidity of his demise suggests that it also burned through his descending abdominal aorta and perhaps also his inferior vena cava. Internal bleeding, especially from the aorta as it continues (before too much blood loss occurs) to pump blood into his abdomen, could indeed have led to a quick loss of consciousness and death. As mentioned previously, those large blood vessels also course near the spine and spinal cord. The unstoppable lightsaber blade could also have damaged those structures, quickly rendering Qui-Gon unable to even stand due to loss of nerve control to his legs.

Finally, the neck is another highly vulnerable area. It has many structures susceptible to penetrating injuries as well as damage from blunt trauma, lacerations, dislocations, and torsion ("twisting") injuries. The cervical vertebrae and their ligaments, muscles, and other supporting tissue help give the neck more mobility that the rest of the spine. Unfortunately they also make it more vulnerable to injury from causes such as sudden acceleration or deceleration as well as torsion. At one end of the injury spectrum, a superficial cut or muscle strain will not incapacitate a character. At the other, sudden decapitation will result in loss of consciousness within seconds and is not a "treatable" injury by current or conceivable near-future medical practice (the 1962 movie *The Brain That Wouldn't Die* notwithstanding—see Chap. 7).

A scene depicted in movies of one character killing another instantly by "snapping" the latter's neck is, however, unlikely in the absence of superhuman strength (which is a possibility in science fiction, such as the 2013 movie *Man of Steel*). The force applied would not necessarily require breaking the cervical vertebrae but dislocating them in such a way that they severely injure or sever the spinal cord. However, that force would have to be both great enough and applied in the appropriate direction to do this—a not impossible but difficult task using "standard" human strength. Also, to cause death the spinal cord would have to be markedly damaged or cut very high in the neck, above the level where the spinal nerves required for breathing, C3-C5, emerge. Doing it lower than that level in the spinal cord would make the person quadriplegic but still alive and able to breathe. Even if the injury were above C3-C5 and the ability to breathe suddenly lost, the victim would not necessarily lose consciousness immediately.

The neck also contains large arteries and veins that run in pairs up each side of the neck toward its front. The right common carotid artery originates in the neck as a branch of the brachiocephalic artery, the first large artery coming off the aortic arch. Conversely the left common carotid artery arises directly from the aortic arch, just after the origin of the brachiocephalic artery. Both common carotid arteries divide at about the level of C4 into two branches— the internal carotid artery, which supplies blood to most of the brain, and the external carotid artery, which perfuses the face and scalp. Both common carotid arteries and their branches are located too deep within the neck to be seen directly. However, pulsations of each common carotid artery in the lower neck can be seen and felt.

Similarly the neck also contains internal and external jugular veins running along both sides of it. External jugular veins are close to the skin and vary greatly in size. They may or may not be visible in a particular person. The internal jugular vein is, like the common carotid artery and its branches, located deeper in the neck and not normally visible from the outside.

The carotid arteries and jugular veins are protected only by muscles and other soft tissues in the neck. They are all relatively close to the skin, particularly the external jugular vein. It is possible to lacerate or sever one or more of these vessels with a deep enough cut, leading to rapid and very messy external bleeding. Here too, however, short of near or complete decapitation it might take the victim seconds to minutes to lose consciousness and ultimately die depending on the severity of injury to these vessels. If a jugular vein alone is punctured or partially cut, bleeding could potentially be controlled by enough compression, whereas similar damage to a carotid artery would be more difficult to control due to the higher pressures within it.

Cutting off blood flow from an internal carotid artery could cause symptoms similar to those associated with a stroke. These might include weakness or paralysis along with loss of sensation on the side of the body opposite the side of the damaged artery, visual problems, difficulty with either understanding speech or speaking, and reduced consciousness. In general, the left cerebral hemisphere is responsible for sensory and motor functions on the right side of the body, and vice versa. As mentioned in Chap. 1, certain functions, such as those associated with speech, are localized to one hemisphere. Therefore, a one-sided interruption of blood supply to a particular cerebral hemisphere will have effects on the functions and side of the body that hemisphere controls.

Dracula-style vampires could presumably use their prominent canine teeth to puncture a jugular vein and make it ooze blood, or perhaps a spurting carotid artery for "fast food." However, in a particular person the external jugular may be collapsed so much that it is hard to puncture. Also, the internal jugular vein is deeper in the neck, varies from person to person in its exact course, and is shielded well enough by overlying muscles and tissue that it might make a difficult target to reach. The carotid arteries present similar access problems to being bitten and drained.

Even successful puncture of a jugular vein presents potential "problems" for the vampire. A small puncture might bleed into the neck where the latter cannot imbibe blood, or ooze so slowly that it could provide "nourishment" no more than a sip at a time. The puncture wound might also form a clot, shutting off the supply of blood and requiring another bite. Conversely, if the vampire were able to inject a chemical that impairs clotting (an "anticoagulant") into the blood, such as in the venom some poisonous snakes have, it would increase the chances of maintaining a "free-flowing" meal. As a practical point, the right internal jugular is usually larger than the left, thus making the right side of the neck a potentially "better" target.

Other arteries in the neck supply blood to the "back" (posterior part) of the cerebral cortex, the midbrain, and the cerebella. The right and left vertebral arteries are branches of the right and left subclavian arteries. The right subclavian artery, like the right common carotid artery, branches off from the brachiocephalic artery shortly after it originates from the aortic arch. Conversely, the left subclavian artery arises directly from the aortic arch, after the brachiocephalic and left common carotid arteries.

Both vertebral arteries course up the back of the neck on either side, joining together shortly after entering the skull to form the single basilar artery. The latter provides blood to the aforementioned parts of the brain, joining the internal carotid system in a roughly circular vascular structure within the brain called the "circle of Willis." A penetrating injury that directly or indirectly interrupts blood supply in particular to the basilar artery can be associated with symptoms that include headache and vertigo, inability to move nearly all of the body, and/or impaired consciousness and death.

A penetrating injury to the eye is also a very bad thing. Rupture of its globe can cause irreparable damage to or loss of the eye itself. Chapter 14 includes a more detailed look at the anatomy of the eye, but in simplest terms this important structure can, overall, be compared to a jelly-filled balloon. The eye is in line with only a very small part of the cerebral cortex that forms the bulk of the brain. A relatively thin plate of bone where the back of the eye sits in a hollowed-out part of the skull (the orbit) also offers only mild protection against a projectile, stiletto, arrow, or any other high-speed and/or pointy object from going deeper into the skull. However, if an object breaks through the back of the orbit, it would mostly likely reach the non-cortical structures at the base of the brain or the midbrain. A serious penetrating wound to those structures with associated bleeding could be quickly fatal.

2.2.2 Blunt Trauma

Blunt trauma refers to injuries caused by an object striking a part of the body (or vice versa). The impact of a club, debris from a falling building, or an attacker's fist can all cause damage. The severity of injury correlates with factors such as the vulnerability of the region struck and the force of the impact. Alternatively, a person's body can strike something else with tremendous force, e.g. hitting the ground after falling from a great height or hitting the dashboard of a car crashing into a wall at high speed.

The latter types of blunt trauma also involve deceleration injuries. These involve a person's whole body or particularly vulnerable parts of it (e.g. the head and neck) moving at high velocity and then stopping suddenly. Internal organs, mobile parts such as the neck, and large blood vessels (e.g. the aorta) continue to move forward after the external part(s) of the body have stopped. These structures can be torn from supporting tissues inside the body, the vertebrae of the neck can be dislocated and injure the spinal cord, and the aorta itself can receive a fatal tear. Blunt trauma to the sternum or the left side of the chest can produce "commotio cordis," with the heart being "stunned" and going into ventricular fibrillation. This injury is actually responsible for 20 % of sudden deaths in young athletes within the United States, such as after one is struck by a baseball, hockey puck, bodily contact, etc [11].

Blunt trauma to the skin can cause bruising—more technically, a "contusion"—due to rupture of tiny blood vessels beneath it. Sufficient force can fracture bones in the extremities, rib cage, etc. or the skull itself. Fractures can cause bleeding and swelling in the affected area. The sharp, irregular end of a completely fractured rib might puncture the adjacent heart or lung. Although they have some resiliency, bones in the arms, legs, spine, etc. can be fractured by sufficient blunt trauma. As with ribs, a complete break in those bones can produce a sharp end that could cut into a large blood vessel or nerve, causing severe and (especially in the case of any artery) potentially fatal secondary damage. A severe blow can damage or dislocate joints such as the shoulder or knee, rendering them unable to function normally.

The brain is cushioned against external forces by "floating" within a thin layer of "cerebrospinal fluid." However, it has only a very mild degree of mobility within the skull. A strong blow to the head can cause a contusion to both the part of the brain beneath the blow and a "contrecoup" injury to the opposite side of the brain. The contrecoup injury is caused by a blow forcing the brain to move forward and strike the part of the skull opposite the area hit directly.

The immediate and long-term effects on a person caused by blunt trauma to the skull vary widely. A "mild traumatic brain injury," which includes the more commonly used term "concussion," may not necessarily be associated with loss of consciousness. If the latter occurs it may last for no more than a few seconds to minutes. The person involved can afterwards show confusion and disorientation, not remember what happened, and have symptoms such as headache, ringing in the ears ("tinnitus"), nausea, and difficulty standing and balancing.

The duration of loss of consciousness and symptoms correlates very roughly with the severity of injury. If a person loses consciousness it may take minutes, hours, or longer to fully recover from the blow. The convention used in many (particularly older) movies and television shows of depicting a character immediately losing consciousness after a punch or other blow to the head, then immediately recovering fully like someone waking up from a nap just in time for a plot twist, is not medically realistic.

For example, the cliffhanger in many chapters of classic Republic Pictures serials produced in the 1930s through 1950s involves the hero being immediately knocked out after a cowardly blow to the back of the head by a villain. The protagonist wakes up fully recovered just in time to notice he or she is in a runaway car about to go over a cliff, and then jumps out of the moving vehicle to safety (or at least as safe as doing that at a dangerously high speed would allow). Unfortunately, in real life the awakening individual would probably take much too long to recover enough to realize that life-threatening predicament and escape.

More severe blunt trauma to the head is typically associated not only with longer duration of symptoms but with other, potentially life-threatening complications such as bleeding into and swelling of the brain (cerebral edema). If the pressure inside the skull ("intracranial pressure") rises too high it could result in prolonged coma and ultimately death. Depending on the severity of injury the person may initially recover consciousness, but then lapse into a coma within hours or less as these other, longer-term effects occur. Even repeated mild to moderate trauma to the brain, as may be associated with some competitive athletic activities, might ultimately lead to permanent damage and dysfunction to it.

Blunt trauma to the front of the face can involve injuries such as fractures, abrasions ("scrapes"), lacerations, bruising, and swelling. Damage caused by person-to-person violence (e.g. punching), relatively low speed vehicular accidents, single blows by clubs or similar objects, etc. may not be directly life-threatening but can significantly incapacitate a character. A blow to the eye with possible fracture to part of the skull's orbit can cause temporary or permanent loss of vision. A dislocated and/or broken jawbone (mandible) could make it difficult to speak or eat.

Likewise, bleeding from the nose and mouth caused directly or indirectly (e.g. from teeth being knocked out or cutting the inside of the mouth) is usually not severe enough to be life-threatening by itself. However, bleeding in possible combination with swelling of the back of the throat could block passage of enough air to the pharynx to make it impossible to breathe adequately.

As with penetrating injuries to them, blunt trauma to the abdomen and pelvis can also damage the internal organs they contain and cause potentially serious, even fatal bleeding. The liver, spleen, kidneys, pancreas, gastrointestinal tract, and bladder can be injured directly, or indirectly such as by changes in pressure within the abdomen caused by a blow to it or rapid deceleration. Depending on how the blunt trauma is delivered and its severity these organs can be cut, ruptured, or torn. Signs of significant abdominal trauma seen within hours after the event include abdominal distention and pain as well as bruising of the skin over the injured area.

2.2.3 Burns

Burn injuries involve more than just exposing the skin to fire, a hot object, etc. They also include damage to other tissues, including internal organs, caused by "wet heat" (e.g. steam or a scalding liquid), chemicals (e.g. strong acids), electrical currents, and radiation.[8] Depending on their location and severity burns can do anything from irritating to incapacitating to killing a character.

The skin may be the main or sole body tissue involved with a burn. A burn's severity is usually categorized by how much of the body's surface area is involved and the depth of injury to the skin. A mild case of sunburn is an example of a "first-degree" burn. This type involves only the epidermis and is associated with redness and pain. The skin remains intact and dry, with healing and complete recovery expected over a matter of days.

"Second-degree" burns involve the outermost part or the entire layer of the dermis, the layer of skin tissue beneath the epidermis. More superficial burns involving the dermis are associated with pain and blistering, as well as possible "wetness" of the area due to fluid leaking into the damaged tissue. Second-degree burns may take several weeks to heal and are at risk of developing infections associated with loss of skin integrity. Scarring is uncommon but can occur

A second-degree burn involving the deeper parts of the dermis may be less painful due to destruction of nerve fibers in the skin that carry pain sensations to the brain. However, these more serious burns are likely to take many weeks to heal and result in significant scarring.

"Third-degree" burns extend all the way through the dermis, and "fourth-degree" burns involve tissues (bone, muscle, etc.) beneath the skin. Pain sensation is lost in the skin in both types of burns from destruction of nerve fibers and healing is incomplete, with significant scarring. These burns may require surgical removal of dead tissue, skin grafting, and amputations of limbs to treat.

Electricity with a sufficient combination of current and voltage can cause burns to both skin and internal organs depending on the path it takes through the body and how long it is applied. Current reflects the total flow of electrons in a circuit while voltage is related to the "push" behind them. A very high voltage associated with an extremely low current, such as produced by a science lab's Van der Graaf generator, might cause a person's hair to stand on end but not be enough to cause pain or injury.

An alternating current of about 1 mA (a thousandth of an ampere) might be detected as a mild "shock," while one 5–10 times greater than that can cause muscles to contract—perhaps enough that a person grasping an object with that much current will be unable to let go of it. Still higher currents could cause internal burns depending on what body tissues form part of the circuit electricity passes through. Too much current, beginning at a range of about 30 mA, can even shock the heart into ventricular fibrillation.

---

[8] See Chap. 6 for more about the effects of radiation.

Lightning strikes have a brief duration but can have voltages in the millions and currents in the tens of thousands of amperes. Considerably less voltage and current levels than those maximums can burn muscles and visceral organs (e.g. the stomach or heart) as electricity courses through the body.

Potentially life-threatening internal burns could also result from breathing in hot air (e.g. that produced by a fire), with damage to the upper airway and lungs, or ingesting a caustic substance such as lye. These kinds of burns can be fatal after hours or days. Chemical burns to the skin and other tissues involves "denaturing" the proteins in them—that is, altering their physical and chemical properties so they no longer function properly. For example, boiling an egg denatures the proteins in its white and yoke and turns it "hard-boiled." Strong bases like lye can also "saponify" fats in body tissues, turning them into a more liquid "soap" (*not* the kind used for washing) and further damaging them.

2.2.4 Explosions

Modern action and science fiction movies are technically sophisticated at depicting explosions ranging from conventional or nuclear devices in a terrestrial environment, to asteroid-sized spacecraft and even whole worlds in an extraterrestrial setting (e.g. the first Death Star and Alderaan in 1977's *Star Wars: A New Hope*). Explosions typically involve the sudden release of large amounts of energy via chemical (e.g. combustion) or nuclear reactions. This energy can be released in the form of light, heat, high-velocity gases produced by the reaction, and shock waves moving outward from the explosion. Initiating a chemical explosion in particular within a confined container or space can increase its intensity.

A person close to an explosion can be injured in many ways. Shock waves produced by an explosive blast cause a temporary but dramatic increase in local atmospheric pressure. The difference in pressures between that caused by the explosion and the much lower ones normally present outside and inside the human body can rupture eardrums and air-filled internal organs such as the lungs, stomach, and intestines [12, 13]. Rupture of the lungs is an especially common and serious problem. It can cause difficulty breathing due to direct damage to the lungs and development of a pneumothorax and/or hemothorax in one or both of them. Free air released from damaged lungs can enter the bloodstream as an "air embolism" and block off critical blood vessels, e.g. the arteries supplying blood to the brain.

The shock wave and expanding gases released from an explosion initially displace surrounding air. As the high positive pressure they produce passes, a negative pressure zone is created into which that displaced air then rushes back. The speed of the air in this "blast wind" can reach over 2400 km/h—far greater than the typical speed of about 200 km/h in hurricane-force winds [13]. This blast wind is powerful enough to potentially cause traumatic amputation of limbs and evisceration. Alternatively the initial shock wave and subsequent blast wind may produce little evidence of external injury but possible fatal perforation, rupture, and tearing of internal organs and blood vessels. If the victim is near a structure such as a building the severity of injury is typically much greater than in an open field due to reflection of shock waves from those structures.

The brain can be injured by pressure changes associated with the initial shock wave and the subsequent blast wind, as well as by the rest of the body transmitting kinetic energy from the explosion to it. The range of injuries is similar to that described previously for blunt trauma. The brain may suffer contusions from both direct effects on the skull and contrecoup impacts with it. The degree of injury and symptoms can range from that of a mild concussion to potentially fatal bleeding and swelling within and/or around the brain.

All the effects described so far are "primary" blast injuries, caused by the explosion itself. "Secondary" blast injuries are caused by debris and projectiles propelled at high velocities by the blast (e.g. shrapnel) that can in turn cause penetrating injuries. The shock wave and blast wind produced by an explosion can be powerful enough to hurl a person away from the explosion and strike a nearby object like a wall, causing "tertiary" injuries. Other potential injuries include "flash burns" caused by release of thermal energy (heat), and crush injuries such as might be produced by falling debris from a building damaged by the explosion.

Besides physical injuries, an individual who survives an explosion can experience immediate and long-term psychological effects. These include development of post-concussion syndrome, with symptoms such as headache, dizziness, fatigue, memory problems, and anxiety that can persist for weeks after the event. Acute stress disorder and post-traumatic stress disorder can also occur, characterized by some of those same symptoms along with depression, agitation, and rapidly changing moods. Duration of these problems is up to about a month with acute stress disorder and longer than a month with post-traumatic stress disorder.

Thus, any scene in a TV show, movie, or written work in which a character near a massive explosion receives only minor, perhaps cosmetic injuries such as superficial abrasions and no obvious psychological effects represents the least likely result of that event. It is similar to the blow-to-the-head-with-immediate-recovery story convention discussed previously. Neither is completely impossible, but both are improbable. Although these depictions are useful as writing conventions, they are less than realistic.

Explosions in space are different in that the only expanding gases involved are in any spacecraft there. The effects of an explosion within a spaceship would, as on Earth, depend on factors such as its power, how close individuals are to it, the size of the compartment it occurs in, and whether it leads to rupture of the outer wall of the craft itself. The explosion of an oxygen tank in the Service Module during the Apollo 13 lunar mission while its three astronauts were in the Command Module did not injure them directly. However, such an explosion with an astronaut nearby might have done this.

The relatively small size of present-day and near-future spacecraft increases the risk a person being injured by an explosion inside it due to all the factors stated above. Moreover, even if an individual is not hurt directly the damage to the spacecraft and its supplies could imperil the astronauts indirectly by reducing food and water supplies to critical levels, impacting the ability to maneuver or land the craft, etc.

2.2.5 Poisons and Toxins

Although perhaps more common in crime fiction than science fiction, one way of threatening or taking the life of a character is with a poison or other dangerous substance. Poisons harm a person's health by interfering in some way with one or more important biological functions. How "poisonous" a particular substance is depends on factors such as the dose required to produce a harmful effect either acutely or cumulatively, how it is delivered into the body, and how much is given. For example, some harmful substances such as lead and mercury accumulate within the body, while repeated exposure to less than lethal doses of arsenic can lead to chronic poisoning. Potassium is a necessary element for cell function. However, injecting a solution with a high concentration of it as a bolus into a vein can make the heart stop by interfering with its electrical functions and the ventricles' normal ability to contract and relax. Even breathing 100 % oxygen at normal atmospheric pressure can damage the lungs, and water can be "poisonous" if very large quantities are ingested over a short time frame.

Poisons can be ingested directly or indirectly by eating animals or plants containing them, inhaled as gases, absorbed through the skin or mucus membranes (e.g. those inside the mouth), or injected into the body. For example, Isaac Asimov's 1954 novella "Sucker Bait" deals with colonists on an exoplanet developing beryllium poisoning by eating plants containing it.

How rapidly a poison works varies widely. Swallowing a lethal dose of potassium cyanide or breathing in enough hydrogen cyanide gas can cause death within minutes by blocking the body's ability to use oxygen and generate sufficient energy within its cells. So-called "nerve agents" can act and kill within minutes by interfering with nerves that are part of the system that controls the functions of our internal organs. They can be ingested, breathed in as an aerosol, or absorbed through skin. These agents can kill by suppressing respiration, and survivors may have long-term neurological damage.

2.3 Delivering Medical Care

Physicians, nurses, and other healers can be supporting or, less frequently, main characters in science fiction. Each of the television entries and movies in the *Star Trek* universe has its own prominent medical practitioners, starting with Dr. Leonard McCoy and Nurse (later Doctor in 1979's *Star Trek: The Motion Picture*) Christine Chapel in the original series. In the world of *Star Trek: The Next Generation* both Dr. Beverly Crusher and, for its second season, Dr. Katherine Pulaski tended their crewmates and alien patients. *Deep Space Nine*'s Dr. Julian Bashir, the holographic "Doctor" of *Voyager*, and the extra-terrestrial Phlox on *Enterprise* all provided healthcare in their own individual styles.

Likewise, if she had been born several centuries in the future, Dr. Janet Frasier from the television series *Stargate SG-1* could have served in the USS *Enterprise*'s sickbay. In written science fiction the myriad medical personnel in James White's "Sector General" tales are as dedicated as their screen counterparts to the cause espoused by The Three Stooges in their Academy Award-nominated film *Men in Black*,[9] "For duty and humanity!" (and, in the case of science fiction works, the health of aliens too). In the movie *Forbidden Planet* (1956) the C-57D's medical officer, Lt. "Doc" Ostrow, heroically gives his life to gain knowledge that helps saves the lives of others.

Individual "space doctors" star as the protagonists in the "Med Service" stories by Murray Leinster (the pen name of William F. Jenkins) and the "Ole Doc Methuselah" ones of L. Ron Hubbard (writing as "Rene Lafayette"). Each of those two series involves a solo practitioner traveling in his own personal faster-than-light starship accompanied by a pet-like alien companion. Both physicians cure disease, treat injuries, courageously battle the machinations of nefarious villains, and in general save the day from both a medical and overall story standpoint on distant worlds. Thus, in Leinster's first entry in his series, "Med Service" (1957, later published under the title "The Mutant Weapon"), the good doctor Calhoun not only eradicates a deadly plague on a human-colonized alien world but defeats the criminals who deliberately introduced it.

---

[9] Not to be confused with the 1997 science fiction film with that same title.

Stephen Thomas Methridge, the "Ole Doc" of L. Ron Hubbard's stories, has even more impressive credentials. Besides being some 750 years old but with the healthy, youthful appearance of a man about 30 due to life-extension treatments, he is a member of the "Soldiers of Light." These are the 700 elite members of the Universal Medical Society, a powerful future organization of physicians dedicated to developing and practicing the most advanced medicine throughout the Galaxy. "Ole Doc" is, of course, the best of the best, venturing from world to world performing medical miracles with ease, rescuing beautiful damsels in distress, and triumphantly engaging in blaster battles and two-fisted physical action against evildoers.[10]

Physicians serving as main characters in science fiction need not be human. Dal Timgar is a humanoid hailing from Garv II who yearns to be a medical doctor in the novel *Star Surgeon* (1959) by Alan E. Nourse (an author who was the relatively rare combination of practicing physician and science fiction writer—something I can relate to). And like the ensemble of many medical personnel depicted in the "Sector General" series, individual physicians such as the psychiatrist-protagonist of recent stories by Rajnar Vajra beginning with "Doctor Alien" (2009) can also specialize in treating extraterrestrials.

Although such characters may have minor quirks and flaws in their personalities, in their roles as physicians they are often portrayed essentially as wizards, able to deal with any plot-required medical challenge incredibly quickly if not always easily. Whether Dr. McCoy is treating a never-before-seen alien (e.g. "The Devil in the Dark" from 1967), finding a cure for a newly encountered disease ("Miri," 1966), successfully performing his first brain transplant ("Spock's Brain," 1968), or merely confirming yet another red-shirted crewmember's gruesome demise (too many episodes to cite) he consistently practices his profession at its highest and most inspiring level. Physicians in this category, at least in the context of their art, are scientific wonder workers and, outside perhaps of being prone to bouts of irascibility, saintly.

Conversely, other science fiction replaces physicians with technology. In *Star Wars: The Empire Strikes Back* (1980) medical droids heal Luke Skywalker's wampa-inflicted wounds on Hoth and perform off-screen surgery to give him a bionic hand. The autodocs in Larry Niven's "Known Space" works reduce or eliminate the personal touch of a physician, being literally a "doc-in-a-box."[11] C.M. Kornbluth's short story "The Little Black Bag" (1950) depicts

---

[10] My reaction to reading this physician's space opera adventures and the perks of being a "Soldier of Light" in the Universal Medical Society is to ask, "Where do I send my dues to join the UMS?" While the American College of Cardiology and other medical organizations I belong to are certainly fine too, they do not yet offer interstellar career opportunities comparable to the ones that futuristic professional group does.

[11] The challenges involved with actually creating an autodoc are described in Chap. 5.

both a physician who, before his redemption, fell from grace via excessive fondness for alcohol as well as a future in which individuals merely "play doctor" using idiot-proof medical instruments.

Even worse, in a subset of the "mad scientist" genre of science fiction, doctors may be presented as grossly misguided in their goals or simply evil. A quintessential example is Victor Frankenstein, whose medically related studies in Mary Shelley's 1818 novel *Frankenstein* inspire him to perform experiments leading to ultimate tragedy. The titular physician in H.G. Wells's *The Island of Dr. Moreau* is also not one it would wise to choose as one's primary practitioner even if he is in your health insurance plan's network.

Given my medical background it should be no surprise that doctors frequently appear as either supporting or main characters in my own science fiction stories. They run a gamut similar to the one I have just described. "Tin Angel,"[12] co-authored with G. David Nordley, includes a dedicated doctor as the protagonist and another who, though technically competent, is not a pleasant person. Dr. Renard in "The Best is Yet to Be"[13] is another compassionate primary care physician, and Dr. Young in "The Eumenide"[14] is an equally caring psychiatrist. Dr. Stone in "Hearts in Darkness"[15] is an astronaut-cardiologist who, though dedicated to his craft, has to wrestle with his own psychological shortcomings and finds himself strained nearly to the breaking point.

On the other hand, Dr. Steaman's research in "The Day the Music Died"[16] unwittingly unleashes a deadly menace on the world. In "Primum Non Nocere"[17] I envisioned Dr. Hans Schuller as my "evil twin" (right down to his initials), depicting him doing things opposite to what I would do, though I perversely put some words in his mouth that are valid from a medical standpoint. Perhaps worst of all is the hospital of the future shown in "The Human Touch,"[18] where the quest for money consistently trumps patient care and human doctors and nurses have been replaced by machines devoid of bedside manner or compassion.

Besides portraying medical personnel, another factor in some science fiction depictions of healthcare is what medical supplies and equipment are available in a particular situation. As Chap. 5 will describe, in current and near-future space missions only a limited amount and variety of medications,

---

[12] *Analog Science Fiction and Fact*, July 1994.
[13] *Analog Science Fiction and Fact*, December 1996.
[14] *Analog Science Fiction and Fact*, January 1998.
[15] *Analog Science Fiction and Fact*, March 2002.
[16] *Analog Science Fiction and Fact*, May 2010.
[17] *Analog Science Fiction and Fact*, December 2010.
[18] *Analog Science Fiction and Fact*, May 1998.

bandages, needles, syringes, etc. can be brought. Once those are used up it may not be possible to replenish them from Earth in time to help an injured or ill astronaut, nor feasible to evacuate that individual back to our world. Likewise the full range of medical expertise, diagnostic devices such as computed tomography (CT) scanners, operating suites, etc. available in a large modern hospital will not be available.

These issues can be more reasonably ignored in stories set far enough in the future. Genetic engineering, nanotechnology, and other means discussed in later chapters might render much of current medical care easier as well as more effective and "portable." Medical resources available on a large space habitat, in a populous Martian colony, or well-developed extrasolar world could be far greater than present capabilities on the International Space Station and deep space missions.

One common problem in the way medical care is depicted, particularly in science fiction movies and television, is how a medication is given and the time it takes to act. Medicines can be delivered into the human body by routes that include oral, rectal, transcutaneous (through the skin, e.g. via a patch), sublingual (beneath the tongue), topical (directly on the skin), intradermal (into the dermis), subcutaneously (into the layer of skin beneath the dermis), intra-articular (into a joint), intramuscular, intravenous, inhalation, intra-arterial, epidural (into the space just outside the spinal cord's dura mater), and spinal (through the spinal cord's dura mater and into the cerebrospinal fluid of the subarachnoid space). Some medications are given by only one of these methods. Others can be delivered in multiple ways.

How rapidly a medication takes effect depends on factors such as the route of administration, and whether it acts directly or must be metabolized first in the body to a more active form. Generally speaking medications are absorbed and act more slowly when given orally, more quickly when injected into tissues such as muscle, and faster still if delivered directly into a blood vessel or, as in the case of gases used to produce general anesthesia, inhaled. Intravenous injections are far more common than intra-arterial ones for several reasons. These include peripheral veins (e.g. those at the elbow or forearm) generally being more accessible than arteries, easier control of bleeding due to those veins having lower pressures and thus being more compressible than arteries, and the greater risk associated with a thrombus forming in an artery compared to a vein following puncture of a blood vessel.[19]

The *Star Trek* hypospray is far more versatile for delivering medicines than present-day methods. It apparently can give whatever medication is needed directly through skin and even clothes over any part of the body. That instrument is much easier to use than the sometimes difficult task of finding a suitable vein for direct injection or placing an intravenous line, cleansing the surrounding skin with an alcohol wipe, then carefully inserting a needle or short plastic tube into the vein before injecting a medication with a syringe.

Modern day "jet injectors" use high pressure (e.g. from a gas cartridge) to inject a very small amount of vaccine or other medicine in liquid form through the skin into subcutaneous tissue or a muscle. A hypospray presumably works on a similar principle. That device is often shown being used over the neck and, especially when employed for sedation, delivering medications that work almost instantly.

However, there are real-world issues with the hypospray and how it is used. To deliver medications into the body a hypospray must break the barrier of the skin, potentially sending harmful bacteria and other contaminants on the skin as well as perhaps even bits of overlying clothing into a person. Dr. McCoy does not prepare the skin using an alcohol swab or other obvious method before giving an injection to reduce risk of infection. Worse, at times the same hypospray is used serially on multiple individuals, despite the risk this non-sterile technique has of causing cross-contamination (e.g. spreading disease from one person to the other). This latter risk has actually reduced the use of jet injectors to vaccinate large groups of people.

Also, if the hypospray delivers an intramuscular injection as current jet injectors can do, the neck would not be the best place to do it. The mass and thickness of muscles there are less than in the upper arm, thigh, or upper buttocks, which are typical sites for an intramuscular injection. There is a greater chance the high-pressure jet of medicine it delivers could adversely affect structures in the neck such as nerves, or go through a muscle to deeper structures like the trachea. Also, the rate of absorption of a medication following an intramuscular injection is relatively slow, typically measured in at least minutes and significantly slower than an intravenous injection.

---

[19] Overall, a thrombus (blood clot) blocking an artery is more likely to damage or destroy tissue it provides blood than one in a vein. If even a very small thrombus in an artery breaks off into the bloodstream and becomes an "embolus" it can injure an organ by occluding a smaller artery "downstream" (e.g. causing a stroke if that artery supplies blood to the brain). A similarly sized embolus in a vein will likely pass through the right side of the heart and block off a small branch of the pulmonary artery—not a good thing, but generally less significant than what an arterial embolus could do.

Moreover, no matter how potent future medicines are it will still take time for the human body to absorb them into the bloodstream, in some cases convert them within the body from a less to more active form, etc. regardless of how they are delivered. It would also take more than a few seconds for the medication to circulate after an intramuscular injection (or even an intravenous one) and reach the brain to cause sedation or other effects. Thus, instantly sedating someone via a hypospray injection is simply not possible due to how our bodies work.

Individuals also vary greatly in their sensitivity to medications and what constitutes an appropriate dose. For example, midazolam, a medication used intravenously as part of "conscious sedation" during some medical procedures, can take as little as a minute or so to start making an adult patient drowsy. However, while some patients can achieve adequate sedation with a dose of 2 mg, others require a total of 10 mg or more. Because there is a risk of giving too much of that medication and causing serious side effects such as reduced breathing rate, doses are "titrated." This means that a small dose is given, the effects on a person observed (Is he or she becoming drowsy or having problems caused by the medication?), and another dose then given if indicated. Assuming the human body itself has not changed by the twenty-third and twenty-fourth centuries, giving single doses of some medications via a hypospray will still not be good medical practice.

While injecting a medication into a blood vessel will make it work more rapidly than an intramuscular injection, there are potential problems with doing this via the high-pressure jet of a hypospray or a contemporary jet injector. Arteries and veins deep within the neck and arms are not easy targets for a conventional needle—much less a hypospray—due to their relatively small diameters and somewhat variable courses through body tissues.

To reach the bloodstream directly a jet of medication would also have to penetrate the blood vessel's wall, thereby increasing the risk of bleeding from it. Too little pressure and the jet cannot do this and becomes perhaps an intramuscular injection instead, with significantly slower onset of action by the medicine. Too much pressure and the jet passes through the other side of the blood vessel and does not directly or at least entirely enter the bloodstream, while also increasing the risk of bleeding from what now becomes two puncture sites rather than one. For the jet to have just enough pressure to inject medication into the typically small lumen (usually measured in at most a few millimeters) of a blood vessel would be very difficult indeed to fine-tune.

2.4 "He's Dead, Jim."

Death in science fiction can be quick and certain. An unfortunate individual or being taking the brunt of a phaser set to its highest setting simply disappears, although the question as to where all the matter and energy composing that individual actually goes is unclear.[20] Likewise individuals on the receiving end of a blast from the handheld weapons used by the titular aliens of the late 1960s television show *The Invaders* take only seconds to incinerate, leaving no question of their demise. The latter show also depicted novel, medically themed ways to terminate troublesome humans. A disk-shaped weapon placed on the back of the neck could quickly induce a fatal cerebral hemorrhage (bleeding into the brain). Subjecting an earthling to the painful effects of one of the aliens' regeneration chambers mimicked death from a heart attack.

In other science fiction franchises, the death rays used by Daleks are notoriously effective, although it is unclear exactly what the specific harmful molecular, cellular, and other biological effects they produce actually are or how those rays produce them. And though they are not directly shown, there is no doubt that the destruction of both Death Stars resulted in mass fatalities due to direct or indirect injuries caused by the explosion itself, explosive decompression into the vacuum of space (see Chap. 3), and other traumatic effects.

At the opposite end of the spectrum, physicians in science fiction may seem to (at least figuratively) "pull the sheet" over a character much too quickly. On futuristic shows like those of the *Star Trek* series it sometimes appears that the art of performing "cardiopulmonary resuscitation" (CPR) to try to revive a victim has been lost over the centuries. Likewise, short of clearly fatal injuries such as decapitation, declaring someone is dead after spending only a few seconds checking for a pulse is not up to even current medical standards. Looked at more generously, perhaps future diagnostic equipment will be so good it can tell immediately whether trying to save someone would be futile despite CPR or other procedures and treatments, thus justifying a quick pronouncement of death.[21]

Television shows (science fiction or otherwise) may show a character lying on a bed in a medical setting such as an intensive care unit (ICU) or the equivalent. Suddenly a cardiac monitor showing his heart rhythm wails an alarm and someone cries, "His heart has stopped!", "We're losing him!", or

---

[20] See Chap. 8 for more about this issue.

[21] I will mention in passing that the untimely end of the 7th Doctor at the hands of a cardiologist understandably unfamiliar with his (by terrestrial though not Gallifreyan standards) distinctive cardiovascular system is a plausible plot device in the 1996 *Doctor Who* movie. However, I will refrain from filling many paragraphs about the myriad questionable (to put it mildly) details about how medicine is otherwise depicted and practiced in that presentation.

something equally dramatic. When watching such scenes of a "code"—medical slang used to describe a "cardiopulmonary arrest," in which a person's respiration (roughly speaking, "breathing") and blood circulation are inadequate to sustain life—I look at the monitor and watch what the medical personnel nearby do next.

Far too often they conduct what I call the "Fifteen Second Code," which may be even shorter than that. The first step in assessing someone who may be in cardiopulmonary arrest is to check for responsiveness, such as by asking, "Are you okay?" If that individual makes any kind of reply, then he or she may still be having a serious medical problem but is not in cardiopulmonary arrest. Exceptions to this might include hospital settings where a patient is already sedated or comatose, perhaps intubated (having a "breathing tube" or, more technically, an "endotracheal tube" in the throat), and who thus could not respond or speak. If an adult patient does not respond for whatever reason, a healthcare provider would check to see if the person was breathing and feel for a carotid pulse. If the latter is absent, then chest compressions would be started.

Among other things chest compressions involve placing the hands in an interlocked position over roughly the center of the patient's sternum, with the rescuer's arms perpendicular to it, and depressing the sternum to a depth of at least 5 cm (2 in.). Current recommendations include doing this at a minimum rate of 100 times per minute. Because doing chest compressions on a conscious person (e.g. an actor) is dangerous, performing the considerably less-than-ideal compressions depicted in this setting on television or in movies is understandable as a story convention.

The situation is different, however, for other components of cardiopulmonary resuscitation. In real life the monitor would be checked immediately to evaluate the patient's heart rhythm and a defibrillator obtained if needed as soon as possible. As described previously that monitor shows an EKG display in one or more "leads" of the heart's electrical activity. The information it provides includes the heart rate, whether the heart rhythm is regular or irregular, what part of the heart the rhythm is originating from (e.g. its normal location in the right atrium's sinus node or another such as a ventricle), and if there are "extra" heartbeats originating from abnormal locations.

Different heart rhythms have characteristic appearances on the EKG and require different treatments. For example, rapid arrhythmias such as ventricular fibrillation or pulseless ventricular tachycardia that would be fatal if untreated require quick use of a defibrillator. Conversely, a defibrillator should not be used to treat any form of sinus rhythm, dangerously slow heart rhythms, or if the heart essentially stops ("asystole").

Unfortunately the EKG analysis done by physicians in such scenes can leave much to be desired. Too often one of them will shock a patient whose monitor shows sinus rhythm or another rhythm for which an electrical shock is not only contraindicated but could actually cause a life-threatening one such as ventricular fibrillation. Likewise the monitor may suddenly go from sinus rhythm to "flatline," shown as a horizontal line going across the monitor. That pattern is commonly due to two major causes. The more benign (and a frequent) cause is that it is an artifact caused by an EKG lead attached to the patient by a sticky patch coming loose or falling off, the cable connecting those patches to the monitor becoming disconnected, issues with the monitor itself or its settings, etc. It is second nature for real-life nurses and physicians to immediately check for these technical causes, but apparently not for most fictional ones.

The more serious possibility is that the patient's heart has indeed stopped and the rhythm is asystole. However, in real life such a sudden change from another rhythm directly to asystole is very atypical. Instead asystole usually occurs more gradually. This includes an ominous pattern seen in patients with severe respiratory distress in which sinus tachycardia (a rapid rhythm originating in the sinus node, producing a heart rate of over 100 beats per minute) gives way over a minute or so to progressively slower sinus rates and then to only very rare "complexes" (corresponding to heartbeats) on the heart monitor and ultimately asystole.

Unlike safety issues with accurately showing chest compressions, it would not diminish the drama of a person "coding" nor are there practical reasons to preclude showing arrhythmias and their treatment accurately. Such depictions are simply and unnecessarily wrong, and could easily be made right by getting a little expert advice. Granted, this particular type of error is likely to be noticed by only a small fraction of the target audience and annoy even fewer (e.g. cardiologists like me). However, the more such avoidable medical mistakes creep into a presentation or story the more likely it is that a larger number of readers or viewers will recognize them as such, thus detracting from what otherwise may be a fine work. The overall suggestion here is for science fiction and other writers to avoid egregious medical errors and employ "real" medicine wherever possible, "fudging" the strict facts only when needed for story purposes or other considerations.

Indeed, the Fifteen Second Code with its startlingly brief duration contrasts sharply with real life ones that may last many minutes and involve far more activities by rescuers than those shown on television or in movies. Besides chest compressions and defibrillation, resuscitation can require complex sequences of administering medications through an intravenous line, down an endotracheal tube, or even via a needle inserted into the marrow of a bone, as well as techniques like electrically "pacing" a slowly beating heart. The fact that defibrillation requires that the patient's chest be bared and paddles or adhesive patches applied directly to skin means that the suggested audience rating would change if this were shown with a coding female character.

Also, in men with hairy chests defibrillation is more effective if that hair is removed. With time at a premium, one method to do this is to place one pair of those large, sticky defibrillator pads on the man's chest, then rip them away, hopefully taking most of that hair with them. A second set of defibrillator pads is then placed on those now depilated areas. A person in cardiopulmonary arrest will not experience pain with that (at least at the time), and when the alternative is between becoming a hairy-chested corpse or surviving with the need to grow back some hair, the latter seems the better choice. However, there is no lifesaving need for a male actor with a hirsute chest to undergo this procedure for purposes of a television show or movie.

2.5 The Bottom Line

Depictions of hurting or healing characters in science fiction can be mixtures of realistic medical principles and practice, useful but inaccurate writing conventions, and avoidable factual errors. A non-medically trained writer can perhaps be forgiven an occasional slip in that last category, although such tolerance might diminish as the number and seriousness of a work's mistakes increase. "Classic" conventions such as the dying character who slips suddenly and quietly away just after finishing a lengthy, eloquent, noble, heartbreaking, dramatic, plot-twist revealing, etc. speech can, even if recognized as clichés, still be effective.

Moreover, even the most realistic science fiction or "medical procedural" work needs to be edited and condensed for dramatic purposes. Few readers would be interested in reading the not-so-gripping but (speaking from personal experience) common scene in which a physician on call is awakened at home at 3 a.m. by a nurse's appropriate telephone request for a verbal order to give a patient a sleeping pill. Instead, a good science fiction work could include some combination of medical realism, verisimilitude, and imaginative speculation integrated into a dramatic framework with other good story elements such as an interesting setting, exciting plot, absorbing characters, etc. The bar is set higher for me and other science fiction writers with medical training to avoid frank medical errors and to avoid or at least be judicious about using not-entirely-accurate conventions such as the blow-to-the-head-with-complete-and-rapid-recovery even for valid story purposes. Writers without that background can be reasonably given more dramatic license, although keeping realistic elements in the story that do not unnecessarily bog down its telling is still desirable.

Like the descriptions of how the human body works in Chap. 1, this chapter too can only provide general ideas on how to traumatize and treat science fiction characters. Hopefully this information drives home the message that, however medical technology changes in the future or the human body is altered via genetic engineering or other means, it is desirable for science fiction to include or at least acknowledge as a baseline how medicine is currently practiced and our "non-modified" bodies actually work.

Straying too much from this idea could push a story from science fiction closer to science fantasy or just fantasy. Those latter two are fine genres in themselves, and it is reasonable for their standards for biological and medical "realism" to be looser due to the nature of their subject matter. However, a work of *science* fiction should be expected to have some grounding in actual medical science, even if that hold (as in the example of the hypospray given earlier) is tenuous.

References

1. 10 Leading Causes of Death by Age Group, United States–2010. 2014. http://www.cdc.gov/injury/wisqars/pdf/10lcid_all_deaths_by_age_group_2010-a.Pdf. Accessed 15 April 2015.
2. Achievements in Public Health, 1900–1999: Control of Infectious Diseases. 2014. http://www.cdc.gov/mmwr/preview/mmwrhtml/mm4829a1.htm. Accessed 15 April 2015.
3. Libby P. The vascular biology of atherosclerosis. In: Libby P, Bonow R, Mann D, Zipes D, editors. Braunwald's heart disease. 8th ed. Philadelphia: Saunder Elsevier; 2008. pp. 985–1002.
4. Bougouin W, Marijon E, Puymirat E, Defaye P, Celermajer DS, Le Heuzey JY, et al. Incidence of sudden cardiac death after ventricular fibrillation complicating acute myocardial infarction: a 5-year cause-of-death analysis of the FAST-MI 2005 registry. Eur Heart J. 2014;35(2):116–22.
5. Worobey M, Han GZ, Rambaut A. Genesis and pathogenesis of the 1918 pandemic H1N1 influenza A virus. Proc Natl Acad Sci U S A. 2014;111(22): 8107–12.
6. Watanabe T, Kawaoka Y. Pathogenesis of the 1918 pandemic influenza virus. PLoS Pathog. 2011;7(1):e1001218.
7. Warmflash D, Larios-Sanz M, Jones J, Fox G, McKay D. Biohazard potential of putative martian organisms during missions to Mars. Aviat Space Environ Med. 2007;78(4 Suppl.):A79–88.
8. Heinemann S, Zabel P, Hauber H. Paraneoplastic syndromes in lung cancer. Cancer Ther. 2008;6:687–98.

9. Pelosof LC, Gerber DE. Paraneoplastic syndromes: an approach to diagnosis and treatment. Mayo Clin Proc. 2010;85(9):838–54.
10. Lee C, Porter KM, Hodgetts TJ. Tourniquet use in the civilian prehospital setting. Emerg Med J. 2007;24(8):584–7.
11. Balady G, Ades P. Exercise and sports cardiology. In: Libby P, Bonow R, Mann D, Zipes D, Braunwald E, editors. Braunwald's heart disease. 8th ed. Philadelphia: Saunders Elsevier; 2008. pp. 1983–91.
12. Burgess P, Sullivent E, Sasser S, Wald M, Ossmann E, Kapil V. Managing traumatic brain injury secondary to explosions. J Emerg Trauma Shock. 2010;3(2):164–72.
13. Cernak I, Noble-Haeusslein LJ. Traumatic brain injury: an overview of pathobiology with emphasis on military populations. J Cereb Blood Flow Metab. 2010;30(2):255–66.

3

Space Is a Dangerous Place

The eternal silence of these infinite spaces terrifies me.

Blaise Pascal
Pensées

In space no one can hear you scream.

Tagline for the movie *Alien* (1979)

Sending human beings into outer space is a staple of science fiction. Early works such as Jules Verne's *From the Earth to the Moon* and H.G. Wells's *The First Men in the Moon* used slightly realistic or purely hypothetical techniques (a huge cannon in the former and "cavorite," a substance that can "cut off … the gravitational attraction of the Earth," in the latter) to send explorers beyond our world. Science fiction stories in the 1930s and 1940s often depicted voyages within and far beyond our Solar System as becoming routine, analogous to the development of international air travel during those decades. Konstantin Tsiolkovsky, Robert Goddard, and other early twentieth century visionaries pioneered techniques in rocketry that led to a real Space Age, highlighted by the 1957 launch of Sputnik, the first human ventures into space starting in 1961, and the Apollo missions. To date over 500 humans have flown into space [1]. The current record for single longest stay in space is 438 days, set by cosmonaut Valery Polyakov on Mir in the mid-1990s. Another cosmonaut, Sergei Krikalev, holds the record for the longest cumulative time in space, with multiple flights totaling 803 days.

But while science fiction continues to use extraterrestrial settings there are two major reasons why you still cannot book a flight on a rocket to Mars for your next vacation. First, spaceflight is much more expensive and difficult than writers of the mid-twentieth century depicted. In classic writings like Robert A. Heinlein's "The Man Who Sold the Moon" (1951) and his related Future History works or Ray Bradbury's *The Martian Chronicles* stories, once human spaceflight begins, travel within the Solar System grows rapidly in its capabilities and becomes commonplace. In more fanciful tales Dr. Zarkov can build a spaceship in his backyard to carry Flash Gordon, Dale Arden, and him to adventures on the planet Mongo.

But in real life the Mercury, Gemini, and Apollo space programs that ultimately led to human landings on the Moon involved flights over a period of 12 years, cost about $ 23 billion, and required hundreds of thousands of personnel to support. Since the Apollo missions, human spaceflight has been confined to low-Earth orbit (LEO, an orbit between about 160–2000 km above Earth's surface). Flights have included three missions to the Skylab space station in the early 1970s; the Apollo-Soyuz mission in 1975; the long-running Space Shuttle program; and many flights in the Soviet/Russian space program using Soyuz capsules, including trips to various Salyut space stations and Mir. The International Space Station (ISS) is the only current destination for human spaceflight, with the timeframe for possible return trips to the Moon and new ones to nearby asteroids or Mars still uncertain. The ISS's habitable volume of about 1200 m^3 provides (literally) far less room for extraterrestrial exploits than the vast majority of spacecraft and worlds depicted in science fiction, but it is the best we have for now.

Unlike Dr. Zarkov's spaceship (which amazingly never seems to run out of fuel and travels to its destination with astounding rapidity), or ones depicted in movies such as *Destination Moon* (1950) with yet-to-be-realized technologies such as nuclear engines powering single-stage-to-orbit vessels, actual human-rated spacecraft are expensive to build, consist partly or entirely of expendable parts, and can fail catastrophically. The multistage Saturn V Apollo Lunar vehicle system had an initial launch weight of 2,621,000 kg, consisting largely of the fuel needed to reach orbit and beyond. Only 45,350 kg was accelerated away from Earth orbit to the Moon, and the Command Module with the crew of three astronauts that returned to our world weighed only 5,670 kg—roughly 0.2 % of the original craft's weight [2]. This is a far cry from Dr. Zarkov's fly-to-Mongo-in-one-piece craft, and it makes one wish the good doctor had been more forthcoming about exactly he was able to make his ship's capabilities outshine those of the later Saturn V.

The infrastructure to support "routine" human spaceflight, such as a system of orbiting space stations, large bases on the Moon, or other refueling and supply stations in space, etc. is also still confined to science fiction due to cost, complexity, and other considerations. While robotic space probes with no need to provide a crew with complex life-support systems have ventured throughout the Solar System, humans have not.

The other major reason for the unfulfilled science fiction dream of colonizing the Solar System is that the space environment and conditions on its other worlds are far more dangerous and hostile than once thought. Unlike John Carter and other intrepid space heroes we will not have to battle hostile sentient aliens or unearthly monsters there.[1] Nor will we tread through the waterlogged jungles of Venus as depicted in Bradbury's "The Long Rain" (1950) or stroll in the cool, dry, but otherwise "shirtsleeve" environment of his Martian tales. This chapter will describe how our own bodies limit how well we can live away from Earth.

3.1 The Weakest Link

The human body evolved on Earth's surface and is optimized for physical conditions here. Those include an atmosphere with a sea level pressure of about 760 mmHg (millimeters of mercury) and containing 21 % oxygen, temperatures averaging about 22 °C, a surface gravitational force with an acceleration of 1 g (a change in velocity of about 9.81 m/s every second, or 9.81 m/s^2), and very low levels of background radiation [1]. The body systems described in Chap. 1 also need adequate supplies of nutrients and water to maintain life, all of which are (at least on a global scale, though not always locally) abundant on our world.

The range over which we can stay healthy or even alive when those optimal conditions are not present varies. Astronauts could theoretically live in microgravity indefinitely, although with the health risks outlined in Chap. 4. An adult can stay alive with little or no food for weeks and even months or go without water for about a week, although with serious effects on health. Individuals residing in the Andes Mountains have adapted to altitudes as great as about 5000 m, where atmospheric pressure is reduced to about 475 mmHg. The limits of how well our bodies tolerate cold will be detailed in Chap. 7. Environmental temperatures more than the world record for the highest on Earth (around 56.7 °C in the aptly named Death Valley) would be very difficult to survive for long.

Conditions beyond Earth's atmosphere are far harsher than any of these. There an unprotected individual would suffer fatal injuries within about 60–90 s. To live in space we must bring the essential requirements we need for life with us. We already have the technology to do that in some places, like LEO, the Moon, and Mars. Unfortunately all the other real estate in the Solar System that figured prominently in earlier science fiction stories such as Mercury, Venus, and the four largest moons of Jupiter are too dangerous for human visitors for the foreseeable future. While our probes can journey to

---

[1] It remains possible that nightmarish aliens like the classic BEMs (bug-eyed monsters) of pulp science fiction, with their puzzling predilection for lusting after scantily clad human females, may be lurking on worlds far away from our Solar System. However, the only way to find if they or more benign extraterrestrials like Vulcans exist is to wait for them to contact or visit us—or do that ourselves.

them we do not yet have practical means to adequately protect our far more fragile bodies from the lethal radiation levels near Jupiter or the extremely high temperatures (about 460 °C) and atmospheric pressure (about 90 times greater than Earth's) on the surface of Venus.

Even where we can live, like the ISS, astronauts can be injured or killed quickly if their protective technology fails. Let us see how Murphy's Law applies to people in space.

3.2 Atmosphere and Pressure

We need a continuous, adequate supply of oxygen and appropriate external atmospheric pressure to live. Because our brains are so dependent on oxygen to function, being suddenly deprived of it will typically cause a person to lose consciousness within no more than around 10 s and will start to cause permanent brain damage after about 4 min.

A slower, smaller reduction of oxygen is also unhealthy. The oxygen in our lungs normally has a pressure of about 104 mmHg (2.01 pounds per square inch, "psi") at sea level. Although individuals living continuously at high altitudes can function when that "partial pressure" of oxygen is as low as 54 mmHg (1.05 psi), those of us not adapted to such great elevations would begin to become mentally impaired and eventually confused as it continues to decrease below about 85 mmHg (1.65 psi). A person with that latter partial pressure in the lungs will feel short of breath and have rapid breathing and heart rates. If oxygen pressure were to fall further to about 35 mmHg (0.67 psi) or less that individual would soon become unconscious. Within minutes both breathing and heartbeat would slow and stop, leading to death.

There are also limits as to how much using supplemental oxygen can protect us from lower than normal external pressures and oxygen concentrations. Breathing 100 % oxygen in an unpressurized cabin is practical only to an altitude of about 10,400 m (34,000 ft.), where the external atmospheric pressure is about 187 mmHg. At a higher elevation of about 15,240 m (50,000 ft.) the ambient atmospheric pressure of 87 mmHg is so low that gas exchange cannot take place within our lungs, and survival in an unpressurized cabin requires a pressure suit or spacesuit.

At about 19,200 meters (63,000 ft., the "Armstrong" limit or line), with an external pressure of 47 mmHg, water and other fluids on a person's unprotected skin, tongue, or mucus membranes (e.g. inside the mouth) would actually "boil" at normal body temperatures. An even lower pressure would actually be required for blood itself to boil, since it is confined within blood vessels that provide some pressure on it. However, dissolved gases in blood (e.g. nitrogen) would still be released as bubbles and, as we will soon see, cause injuries [3, 4].

Earth is unique in the Solar System as the only place that currently has abundant free oxygen on its surface. But for our bodies to remain healthy we also need an appropriate amount of pressure from gases in the atmosphere on them. Before it was found that the atmospheric pressure on Mars is a little less than 1 % of Earth's at sea level (about 4.5 mmHg) some stories and movies (e.g. *Robinson Crusoe on Mars*, 1964) depicted astronauts walking on its surface without spacesuits, carrying only supplemental oxygen due to the Red Planet's lower than terrestrial levels of that gas.[2] However, oxygen from those breathing devices would only briefly delay those characters' demise due to the very low atmospheric pressure. That value of 4.5 mmHg is far below either that required for gas exchange in the lungs or the Armstrong limit, so even breathing 100 % oxygen would not keep a person from essentially suffocating.

Atmospheric conditions are even less hospitable elsewhere in the Solar System. There is no appreciable atmosphere on Mercury or on any asteroid or moon (including ours) with the sole exception of Saturn's largest moon, Titan. The latter's atmospheric pressure is a survivable 50 % greater than Earth's at sea level. Unfortunately Titan's atmosphere at its surface is composed of roughly 95 % nitrogen, up to about 5 % methane, trace constituents such as the decidedly unhealthy hydrogen cyanide, but no appreciable oxygen. However, in theory a space traveler there, if provided with oxygen and protected against that moon's average temperature of – 179 °C, could walk on Titan's frigid surface.

Besides exerting the tremendous pressure described previously, the atmosphere of Venus is about 96 % carbon dioxide and 4 % nitrogen. The largest planets in the Solar System—Jupiter, Saturn, Uranus, and Neptune—have thick atmospheres consisting overwhelmingly of copious amounts of gases other than oxygen, predominantly hydrogen with much smaller amounts of helium, methane, etc. Pressures also rise rapidly to crushing levels after entering their atmospheres, reaching about 3 million times that of Earth at about 20,000 km beneath the top of Jupiter's "clouds" [5]. Since the radius of Jupiter is approximately 71,490 km at its equator, atmospheric pressures even "deeper" are still higher—enough to turn hydrogen itself into a liquid metal. Unfortunately those conditions mean the humans and "Lopers" living on the possible rocky "core" of Jupiter in Clifford D. Simak's classic short story "Desertion" (1944) are forever confined to the realm of science fiction.

---

[2] The actual atmosphere on Mars is about 96 % carbon dioxide, 2.1 % argon, 1.9 % nitrogen, and only 0.145 % oxygen.

Despite what you may have read or seen in movies, exposing an unprotected person to the vacuum of space will not make that individual explode or freeze solid. The sudden decrease in pressure will make fluids leak into soft tissues, causing the skin to pucker and swell. This will make the victim look puffy but not blow up. If the person is wearing a spacesuit the latter will physically restrict this overall body swelling. Though I hate to contradict a seasoned space veteran like Buzz Lightyear from the 1995 movie *Toy Story*, eyeballs will also not pop out of their sockets. Eyes will, however, become bloodshot due to rupture of small blood vessels and feel gritty due to the normal thin film of tears on them forming tiny bubbles on their surfaces due to lack of external pressure. Water vapor from exhaled air will freeze, but overall body heat will be lost to space slowly enough that the victim will be dead well before cooling becomes an issue.

Unprotected exposure to vacuum will make a person become lightheaded and lose consciousness within an average of roughly 5–10 s and no more than around 15 s due to lack of oxygenated blood reaching the brain [1]. The unfortunate individual may have convulsions shortly afterwards. Heart rate rises at first and then both it and blood pressure fall until eventually the heart itself stops. Death occurs within about 90 s. The scene in *2001: A Space Odyssey* (1968) when the helmetless astronaut Dave Bowman manages to close the pod bay door after exiting his small craft represents about the longest time someone at least partially protected with a spacesuit could remain conscious and act effectively.

Exposure to vacuum for about 10–30 s is expected to be survivable with minor injuries. It could, however make air in the lungs expand enough to rupture and cause bleeding in their alveoli, so it would be important for an individual (e.g. Dave Bowman) to exhale as much air as possible before this exposure. Past around half a minute the extent of injuries would be expected to increase rapidly with greater durations of exposure, to the approximate uppermost survival limit of about 90 s. The latter may, however, be a conservative estimate. Studies involving dogs have shown high survival rates even after exposure to near-vacuum for 2 min. One study involving chimpanzees pretreated by breathing a high level of oxygen for 4 h demonstrated a high survival rate even after 210 s of exposure to near-vacuum conditions [6]. However, for obvious reasons comparable studies in humans are not recommended.

More typically, however, cinematic presentations of exposure to vacuum or near-vacuum trade scientific accuracy for visual drama. In the movie *Mission to Mars* (2000) a character's head apparently freezes almost instantly after his helmet is removed while floating in space above Mars in direct sunlight. Unfortunately this depiction not only contradicts how long a person could remain conscious in a vacuum but also reinforces the myth that space would be "cold" there—a concept that will be explored in the next section.

Likewise, in the 1990 version of *Total Recall* characters exposed to the minimal atmosphere on Mars develop swollen eyeballs that nearly pop out of their sockets, their faces become grotesquely swollen, and they retain consciousness for significantly more than 15 s while *audibly* gasping with pain (please note that sound would not be transmitted in this way at such a low atmospheric pressure). However, they recover completely and almost instantly when enough oxygenated air is released into the atmosphere near them. Not only does their facial swelling resolve immediately but their eyes don't even appear to be bloodshot. Moreover, the air around them is suddenly so apparently "normal" in oxygen concentration, pressure, temperature, etc. that they have no problems breathing or carrying on a conversation.

While not a comprehensive list, errors in this depiction include (a) Once again, eyeballs will not pop of their sockets[3] but eyes will be bloodshot; (b) consciousness would be lost more quickly; (c) facial swelling would not be nearly as pronounced but would also take much longer to resolve; (d) the (presumably) oxygenated air that erupts from the ground to save the characters would, in real life, probably not be within the narrow range of pressures etc. needed to save the characters nor be sustainable in their vicinity for more than a very short time as it spreads along the surface and is "diluted"; and (e) a sudden reduction in atmospheric pressure can cause "barotrauma" with a variety of potentially serious and even fatal internal injuries.

Barotrauma occurs due to gases within body cavities and blood suddenly expanding when outside pressure falls too much or too rapidly. While sudden unprotected exposure to vacuum would be the ultimate cause for this, much lower changes in external pressure (decompression) can also cause injuries, particularly and most acutely to the lungs. This latter risk would be greatest if the person were to keep air trapped inside the lungs by holding his or her breath and not exhaling it out quickly. Moreover, it could occur even during involuntary breath holding, such as with yawning or swallowing. Animal studies have indicated that a pressure difference between the inside and outside of the lungs of about 80–100 mmHg for 0.1–0.2 s is enough to rupture alveoli. Studies using human cadavers showed that an even lower gradient of about 70–81 mmHg was sufficient to do this [6, 7].

Air expanding in the lungs could also enter other nearby structures, such as "leaking" out into the mediastinum (the space beneath the breastbone) or under the skin. It could also cause a tension pneumothorax or a pneumopericardium (see Chap. 2), thus potentially compressing the lungs and heart.

Expanding air in the stomach and intestines (primarily the large intestine) could cause abdominal pain, belching, vomiting, and defecation [8]. Too

---

[3] Perhaps the reason Buzz Lightyear got the erroneous idea this would happen is that *Total Recall* was shown as a training film when he was still a space cadet at Star Command.

much distention of these viscera could even elicit a "vasovagal" reflex with a drop in heart rate and blood pressure, potentially causing lightheadedness and loss of consciousness. The pain you feel behind the eardrums when the jet you are flying on increases altitude is caused by only a mild reduction in outside pressure. A far greater and/or too rapid fall in pressure, such as being ejected into space without a spacesuit, could cause your eardrums to rupture. Air trapped in sinuses can also be extremely painful, although this is more likely to occur during repressurization after decompression rather than during the latter itself. Dental caries ("cavities"), loose fillings, and other tiny pockets of air involving your teeth might also cause toothaches.

Smaller decreases in pressure can produce the "decompression sickness" experienced by divers who rise too rapidly from the higher pressures present in deep water. The air that both the astronauts on the ISS and we on Earth breathe is nearly 79 % nitrogen. This gas is normally dissolved in our blood and causes no problems. However, a sudden loss in pressure makes nitrogen come out of solution and form bubbles in the bloodstream—a process called "ebullism" [3, 4, 6, 8].

Nitrogen bubbles can interfere with blood flow to various parts of the body, causing pain and injury. This includes "the bends" (pains in joints and muscles), difficulty breathing ("the chokes"), nausea, vomiting, and relatively mild neurological symptoms such as headache, weakness, numbness, or tingling in the arms and legs. They can also cause more severe problems such as temporary blindness, paralysis, confusion, convulsions, stroke, and unconsciousness [8]. Those latter effects are primarily due to the brain itself not getting enough blood, most likely due to bubbles traveling to it via the bloodstream and reducing its blood supply.

Nitrogen bubbles are more likely to form in veins than in arteries. This is related to the normal pressure in veins being significantly lower than in arteries. Bubbles forming in veins are also less likely to directly interfere with the brain's blood supply than if they form in arteries. However, a reported 25–35 % of people have a "patent foramen ovale," a (usually) at most very small opening between the heart's two upper chambers, the right atrium and left atrium [9]. If a patent foramen ovale is present it could allow bubbles formed in the veins to pass into the systemic arterial system, where they could then travel to the brain or other parts of the body [4].

There have been accidents during space programs involving brief or fatal exposure to vacuum. In 1965 an individual was exposed to extremely low-pressure conditions in a vacuum chamber during a spacesuit test. He lost consciousness after about 14 s, but regained it roughly 3 min later as the chamber was rapidly repressurized. Fortunately he suffered no long-term effects. In another incident a man working in a vacuum chamber was exposed for several minutes to reduced atmospheric pressure as low as that equivalent to an altitude of 22,555 m (74,000 ft.). His injuries included a ruptured lung, ebullism, and decompression sickness. He took about 5 h to regain consciousness after repressurization and treatment in a hyperbaric chamber, and had laboratory tests consistent with soft tissue injury, including muscle damage. Fortunately he too recovered [6, 10–12].

The three cosmonauts aboard Soyuz 11 in 1971 were much less "fortunate." They died during reentry when their craft accidentally depressurized. Due to the relatively small size of the capsule they were not wearing spacesuits when a ventilation valve opened prematurely, letting their air escape into space. Review of recordings from biomedical sensors attached to one of the cosmonauts showed his heart stopped about 40 s after pressure in their capsule was lost [6, 13].

On the ISS alarms sound if a pressure differential of more than 1 psi/h (57 mmHg/h) is detected and pressure drops by 0.4 psi (21 mmHg.) This condition is one of only three emergency (Class I) alarms, the other two being detection of a fire or a toxic atmosphere [14]. If an event (e.g. a major impact with a meteoroid or artificial space debris) were to occur with a craft, rapid depressurization could occur. Depending on the size of the hole(s) produced, the loss of air into vacuum could result in noise varying from a hissing sound to a loud bang ("explosive" decompression). Physical effects within the craft might also include flying debris, with items near the new opening being sucked outside it, as well as a rapid decrease in pressure and temperature, with the latter producing fog due to condensation of water vapor in the remaining air.

Survival of the crew in such a situation would depend on many factors. These include the size of the hole and associated rate of decompression; how long the astronauts could remain conscious (assuming they were not in spacesuits) as oxygen and pressure levels fell; how quickly they could react to the situation; if they were directly injured by the event (e.g. struck by an impacting meteoroid); and whether they could either repair the hole or escape the affected area (such as moving to a different module on the ISS and sealing the damaged one) [14].

When confronted with a hull breach or similar event involving sudden decompression the amount of "effective performance time" or "useful consciousness" an astronaut would have to deal with it effectively is related to the degree and rate of oxygen and pressure loss. The time of useful consciousness would be about 20–30 min if oxygen and atmospheric pressure were reduced to those at an altitude of about 5500 m (equivalent to a very high mountain, with atmospheric pressure of about 394 mmHg and oxygen levels 52 % of those at sea level). That time would fall to as little 9–12 s if those values were equivalent to an altitude of 13,100 m (atmospheric pressure 137 mmHg and 18% of sea level oxygen) or greater. If decompression leading to such levels were rapid rather than slow these times for effectiveness consciousness could be reduced further by as much as 50% [3, 15].

The human body's difficulty in adjusting to sudden reductions in pressure also applies to using spacesuits. Instead of the common depiction in science fiction (e.g. the movie *Destination Moon*) of astronauts immediately going outside their craft to deal with an emergency repair, such "extravehicular activity" (EVA) now typically requires a great deal of preparation beforehand. Astronauts in the first US spacecraft could do this because they breathed 100% oxygen at a pressure as low as 3.1 psi (160 mmHg) while in them. Current American spacesuits use a pressure of 4.3 psi (222 mmHg), while Russian ones use 5.7 psi (295 mmHg). Any higher pressures in current versions of these suits would make it too hard to move their joints and work in them [16].

However, while the astronauts in the Gemini program and cosmonauts in early Soviet spaceflights could perform EVA immediately because they were already at appropriate levels of oxygen and pressures, different conditions on the ISS require a lengthy preparation beforehand. The retired Space Shuttle used and the ISS still employs an atmosphere of 79% nitrogen and 21% oxygen at 14.7 psi (equivalent to 760 mmHg), similar to that here on Earth. Prior to EVA an astronaut must purge nitrogen from the bloodstream to reduce the risk of ebullism at the lower pressure of the suit [1].

Protocols to do this all require hours to implement before the astronaut can safely enter space. One option is to "pre-breathe" 100% oxygen in the spacesuit for about 4 h at a standard pressure of 14.7 psi. An alternative for a scheduled EVA is to pre-breathe oxygen with a mask for about an hour, reduce local pressure (e.g. in the airlock of the ISS) to 10.2 psi for 12–24 h, followed by a final pre-breathe of oxygen in the spacesuit of around an hour. A much quicker method requires only about 2 h and 20 min of preparation. It involves pre-breathing pure oxygen via a mask for 50 min and an airlock pressure of 14.7 psi, with the astronaut performing vigorous exercise on a stationary bicycle for the first 10 min of that period. Airlock pressure is then reduced to 10.2 psi, and after an additional 30 min the spacewalker dons the spacesuit and breathes 100% oxygen in it for an additional hour before starting the EVA.

While not using 100% oxygen as the usual "atmosphere" on the ISS and other craft prolongs the time needed to start an EVA, it avoids greater risks. Breathing pure oxygen at normal atmospheric pressure (14.7 psi) for more than about 6 h can damage the lungs, with clinical evidence of that injury (e.g. coughing, chest discomfort, and shortness of breath that can become severe) occurring after about 1–3 days [3, 4]. Using 100 % oxygen in space-suits at the much lower pressures described previously is safe. Moreover, in the confined space of the ISS or other craft an atmosphere containing only oxygen at a normal terrestrial pressure of about 760 mmHg presents a high risk of fire, as occurred in 1967 during a ground test of the Apollo 1 cabin that killed three astronauts [3].

Treatment options for astronauts who, despite all precautions, develop significant decompression sickness are limited even on the ISS. Original plans for it included placement of a component that could have functioned as a hyperbaric chamber, to provide more than normal atmospheric pressures for a time to reduce ebullism and other issues. However, that chamber was eventually cancelled. Hydration and keeping an astronaut in a pressurized spacesuit and delivering up to 100 % oxygen could be employed as treatment. The only medication likely to be effective is acetazolamide, which is also used by mountain climbers prior to high ascents [14]. Acetazolamide is usually given as a twice-a-day dose of 250 mg or a single 500 mg dose of a slow release form. Unfortunately this medication has potential side effects, such as development of urolithiasis ("kidney stones" that block a ureter, the tube leading from the kidney to the bladder),[4] gout, nausea, drowsiness, etc. Although the incidence of any single side effect is low, the chances of a person developing at least one significant type are worrisome.

3.3 Temperature, Toxins, and Trauma

Astronauts can be injured or killed by other hazards in space and their spacecraft. As Chap. 6 will describe, the human body tries to maintain a core (internal) temperature of about 37.6 °C (98.6 °F). However, it has only a limited capacity to do this when exposed to temperatures that are much below or above a "comfortable" external temperature of around 22 °C (72 °F). Near Earth the temperature a spacefarer is exposed to depends on what sources of heat are present. The Sun is the main overall source of heat, radiating it directly out into space. In LEO or on the Moon's surface the effective "temperature" in space is about 110 °C (230 °F) in direct sunlight, but falls to as low as about −178 °C (−288 °F) on the night side of the Moon.

Standard spacesuits are insulated and keep body heat inside them. Those suits employ a cooling system using circulating water to keep an astronaut from getting too warm during EVA and the heat-generating exertions

---

[4] As we will see in Chap. 4 astronauts are already at increased risk of developing kidney stones due to the effects of microgravity itself on the human body.

associated with it. Without such a cooling system the astronaut could "overheat" within minutes of entering space if exposed to the Sun at Earth's distance from it [17].

Spacecraft are designed to either conserve or radiate heat to keep occupants at comfortable temperatures. The Space Shuttle kept its cargo bay doors open during much of its stay in orbit to lose heat produced by the people and equipment on board it as well as heat gained when it was in sunlight. Craft can also rotate slowly while exposed to the Sun to even out heat distribution—the same principle as a rotisserie. On the other hand, during Apollo 13's return to Earth cabin temperatures fell to a little above freezing due to reduced electrical power output and heat following the explosion of an oxygen tank aboard that craft.

Heat can also be transferred by conduction, when two objects with different temperatures touch each other and a net amount of thermal energy passes from the "hotter" to the "colder." However, this occurs in only limited scenarios in space, such as the relatively "warm" hand/glove of a spacesuit touching a "cold" object (e.g. a large piece of equipment being manipulated outside the ISS while in Earth's shadow) during EVA.

On Earth heat is commonly transferred by convection currents, e.g. when hot air rises and then falls when it cools. In space or other environments where there is no air or other gas to contribute to this process, convection obviously cannot occur. The limited volumes within spacecraft or the modules of the ISS also greatly limit convection, with fans in the life support system being used to help circulate air.

Toxic substances used in cooling, fuel, and other systems can contaminate the limited volume of air within spacecraft and space stations even when present in very tiny concentrations [1, 4, 18]. The ISS uses liquid ammonia as a coolant. Exposure to tiny amounts of it can irritate the eyes, nose, and throat. Chemicals such as hydrazine, monomethyl hydrazine, and nitrogen tetroxide have been used in propulsion systems but are harmful to humans. Hydrazines in liquid form cause severe burns and penetrate the skin to potentially damage the liver and kidneys as well as produce convulsions or even death. The three U.S. astronauts of the Apollo-Soyuz mission in 1975 were accidentally exposed to nitrogen tetroxide fumes during reentry [19, 20]. One of them briefly lost consciousness before emergency oxygen masks were used. After landing all three were hospitalized for days afterwards due to damage and fluid in the lungs caused by breathing that chemical. Fortunately all of them recovered [19].

A smoldering substance or actual fire can quickly produce dangerous amounts of carbon monoxide. More of that gas is produced by a fire in the microgravity environment of a spacecraft than on Earth because convection of air does not replenish oxygen in the vicinity of a fire. This leads to incomplete burning of material, thus producing larger amounts of carbon monoxide than a similar terrestrial fire would.

Two fires occurred on the Mir space station in the late 1990s [21]. The first, in 1997, resulted after a solid fuel cartridge used to produce oxygen when ignited in a generator instead caught fire and burned out of control for several minutes. Later analysis of air samples showed that this fire produced higher than normal but not dangerous levels of carbon monoxide. The overall danger of the situation was compounded, however, by the fact that 6 people were on board Mir at the time and the fire blocked the path to one of the 2 Soyuz escape capsules, each of which could hold only 3 of them. This would have presented a serious problem for half of the crew if the station had needed to be evacuated [21].

A similar incident occurred in 1998, when paper filters in a device used to remove contaminants from air caught fire. The blaze was quickly contained and produced little smoke. However, subsequent analysis of air using a carbon monoxide sensor showed concentrations of the gas over four times greater than the previous fire in 1997. The carbon monoxide level was high enough to apparently cause one crewmember to develop nausea and a headache. Because carbon monoxide is a colorless, odorless gas that humans cannot sense, it is especially important to monitor its presence via sensors and other means in the closed environment of a spacecraft or space station. Otherwise the first signs of high concentrations of that gas could be when crewmembers develop mild symptoms such as nausea that could progress to seizures, coma, and death.

While formaldehyde is useful for preserving some biological material, it can be hazardous to living humans. Formaldehyde can be released from such sources as experiments in which it is used as a fixative, as well as "outgassing"[5] from materials (e.g. polymeric ones) that still contain it due to its use as part of their manufacturing process. High enough concentrations of formaldehyde in air can cause coughing, wheezing, difficulty breathing, and damage to the lungs [21].

Fortunately formaldehyde can be detected by its distinctive, disagreeable odor at concentrations lower than those that cause injury. Unfortunately there have been incidents in which it was accidentally released in space. These include the Mir-18 mission (1995) when a containment bag leaked paraformaldehyde into the air, and a Shuttle mission in 1991 carrying the Spacelab module for biological studies when a refrigerator overheated and released

---

[5] Gas bubbles of potentially toxic or otherwise harmful chemicals used to make certain materials can be trapped within them during the manufacturing process. Over time significant amounts of these chemicals can be released via "outgassing" into the surrounding air.

formaldehyde from one of its parts. The latter incident resulted in the crew becoming nauseated from the fumes produced but no more serious effects.

Ethylene glycol, an ingredient of antifreeze, has been used in cooling systems such as on Apollo craft and Mir [21]. If inhaled it can cause irritation and pain in the throat, trachea, and the bronchi of the lungs. Exposure to even low concentrations for a prolonged period of time could injure the kidneys. Significant amounts leaked from cooling loops on Mir several times and, because of this chemical's tendency to condense on surfaces while continuing to release dangerous vapors, proved difficult to eliminate from the environment. One such leak in 1997 resulted in three crewmembers developing irritation of their upper airways and skin reactions [19]. The ISS does not use that substance in any of its systems.

Freon is the registered name for various halocarbons that include CFCs (chloroflurocarbons) known to be destructive to the ozone layer. Other types of halocarbons were used to clean hardware prior to Space Shuttle flights. One halocarbon, bromotrifluoromethane, was even used as part of a fire extinguishing system while another, perfluoropropane, is a coolant on the ISS. Although these substances can be toxic at high enough concentrations and with long enough exposure (e.g. in some cases potentially causing liver damage or abnormal heart rhythms), the total amounts present on the ISS are small enough that this is not thought to represent a significant concern [21]. Routine atmospheric analyses are done in the ISS to check for the presence of all the various types of potentially toxic airborne gases just described.

Humans themselves produce carbon dioxide, methane, and other metabolic wastes that must be cleared by environmental systems. Atmospheric levels of carbon dioxide in a spacecraft above about 0.01% of its gases can increase heart and breathing rates, produce headaches, and otherwise disrupt the body's metabolic balance [1]. Also, by one estimate each crewmember on Shuttle flights generated about 1.1 kg of trash per day, with a little over a quarter of that being liquids. Bacterial action on some of this trash (e.g. discarded food and human wastes) can generate new, potentially toxic substances such as sulfur dioxide. Besides its use in coolant systems, ammonia is also generated by human metabolic activity, with production and release augmented by exercise and sweating.

Airborne particles such as lint, dust, shed skin cells manifesting as tiny flakes (e.g. dandruff from the scalp), flecks of paint, tiny bits of food, etc. will all float in microgravity and could be inhaled. Air filters are used on the ISS to help remove them. Also, particles of substances that would otherwise be contained can be accidently released. This occurred on both the Shuttle and the ISS, with materials such as those from waste management systems, animal containment facilities, and etc. finding their way into the atmosphere.

In particular glass is used very carefully due to the risk of contamination by particles of broken glass caused by an accidental breakage—definitely not the kind of thing one would like to inhale.

While certain forms of trauma are more specific to the space environment, such as decompression by exposure to vacuum or at least much lower than normal pressures, astronauts could also experience the same nonfatal but potentially serious ones all of us suffer on Earth. Space travelers could receive accidental cuts, head bumps, punctures, bruises, burns (e.g. those of the electrical variety), muscle strains, etc. that might impair their activities and possibly require treatment.[6] As will be discussed in the next section, if interpersonal relationships among crewmembers were to deteriorate too much there would even be the possibility of a "space brawl" or other deliberate infliction of trauma.

Both the United States and Soviet/Russian space programs have had incidents in which spacefarers either escaped disaster or suffered minor injuries that could have been far worse [19]. In the second suborbital flight of the Mercury program in 1961 astronaut Gus Grissom (later to die in 1967 with two other astronauts in the ground-based Apollo 1 fire) nearly drowned when the hatch to his capsule released prematurely after splashdown. He was rescued but the capsule sank, although it was finally recovered in 1999. The first human to conduct a spacewalk, Alexei Leonov, could have died when his spacesuit expanded so much that he could barely reenter his Voskhod 2 capsule in 1965 [1, 13]. He had to reduce some of the suit's pressure to decrease its size, although at the price of experiencing a case of the bends. The Gemini 8 capsule (1966), crewed by Neil Armstrong and David Scott, went into a fast, uncontrolled spin when a thruster continued firing, with control regained before they lost consciousness due to that rapid rotation.

In 1969 Apollo 12's Saturn V rocket was struck twice by lightning on the launch pad. During splashdown after its return from the Moon a camera broke loose and struck one of the crew on the head, causing a concussion and a cut deep enough to require stitches. Apollo 13 (1970) is well known for the near-catastrophe associated with an electrical short that caused an oxygen tank to explode. One of the three parachutes attached to the Apollo 15 capsule (1971) failed during reentry, a malfunction that would have proven fatal if a second one had also done so.

Some early Soviet flights came close to tragedy when a rocket section almost failed to separate or malfunctioned. These included the first human spaceflight, Vostok 1 (1961), as well as Soyuz 5 (1969) and Soyuz 18a (1975). During the latter mission the rocket's second and third stages did not separate

---

[6] Chapter 5 includes detailed descriptions of these issues.

correctly after launch and the mission was aborted before the Soyuz capsule reached orbit. The cosmonauts on board experienced a maximum g-force of up to about 21 g during their descent, which ended with them landing on a snow-covered slope in the eastern Soviet Union. Their capsule rolled downhill before stopping just short of a nearly 152-m cliff when its parachutes snagged on vegetation. In 1983 leaking fuel caused a fire prior to launch of Soyuz T-10-1, with the crew successfully using their rocket's escape system a few seconds before the craft exploded [22].

Shuttle missions also experienced dangerous events. In 1983 hydrazine fuel leaked out and caused fires in a pair of auxiliary power units as the Shuttle *Columbia* landed. During a 1985 launch of *Challenger* one of the main engines shut down and a second almost did, which would have resulted in the Shuttle not reaching orbit.[7] During the release of a satellite from the payload bay during a 1993 *Discovery* flight both the main and backup explosive release devices detonated instead of just the main ones. The flying debris created when the metal bands holding the satellite were ripped away caused damage to the Shuttle that included puncturing the payload bay bulkhead leading to the crew compartment. Fortunately no injuries occurred.

In 1999 a launch by *Columbia* was marred by the failure of the controllers for two main engines caused by an electrical short, with backup controllers taking over. At the same time a cooling line also ruptured, producing a hydrogen fuel leak that resulted in the Shuttle reaching an orbit slightly lower than planned.

Meanwhile, in the Russian space program a Progress cargo freighter collided with Mir in 1997. This damaged one of the solar arrays used to generate electrical power and punctured one of Mir's modules, causing the station to start depressurizing. The crew was able to seal off the module and repair some damage caused by the collision, but the module itself remained depressurized for the rest of Mir's existence. In 2003 a Soyuz capsule returning to Earth from the ISS was subjected to much greater velocities and its crew to higher g-forces (up to 9 g) than usual. The capsule also landed about 500 km away from its target. After reaching the ground the capsule was dragged about 15 m by its parachute before landing on its side. An American astronaut on board suffered a shoulder injury as a result of those events. Similar high-speed reentries occurred with two other Soyuz capsules in 2007 and 2008. In the latter incident the capsule landed about 475 km from its target and one of the astronauts on board suffered minor injuries to the neck and back.

---

[7] The novel *Shuttle Down* (1981) by G. Harry Stine deals with a Shuttle not reaching orbit and forced to make an emergency landing on Easter Island in the Pacific Ocean.

EVAs on Shuttle and ISS missions have also had accidents. In 1991 an astronaut's spacesuit glove was punctured by a small rod inside it, but this did not result in any significant injury. In 2001 large amounts of ammonia leaked out during a repair of the ISS cooling system, coating an astronaut's helmet and part of his spacesuit with frozen ammonia crystals up to an inch thick. When the ISS left Earth's shadow later during its orbit the heat of the Sun was used to evaporate the ammonia before the astronauts conducting the EVA reentered the ISS. In 2013 an astronaut's helmet started filling with water, requiring discontinuation of the EVA. Before repressurization within the ISS could be completed and the helmet removed there was enough water within it that the astronaut had difficulty seeing and hearing. Fortunately he was not injured.

3.4 Psychological Stress

Besides external hazards such as exposure to vacuum, spacefarers are also at risk from internal dangers created by their own minds. The view of Earth from LEO is spectacular and astronauts at the start of a deep space mission may feel great excitement. However, there are also distinct psychological downsides and risks associated with prolonged space travel.

Early extended missions in space like those on Mir were associated with significant psychosocial stressors. Isolation and loss of social contacts (particularly from family), reduced sensory stimulation, anxiety, boredom, loss of privacy, and overly busy work schedules could all take their toll on the psychological health of spacefarers [1, 23, 24].

Individuals who go into space tend to be self-sufficient and goal-oriented. These traits can be useful for successfully completing a mission. However, such individuals may also either fail to recognize or, out of a sense of "perfectionism," not report psychological problems they may be experiencing due to stress or other factors. Instead they may try to cope on their own, perhaps even indulging in risky behavior to prove (at least to their own satisfaction) that they are "fine."

Recognition of these issues has led to more effective ways to prevent and deal with them. Astronauts undergo extensive psychological testing prior to flight along with continued monitoring during a mission. Crewmembers train together on Earth for a long enough time—often up to 18–48 months for long-duration missions—to hopefully identify any major interpersonal incompatibilities before they reach space. On the ISS they are provided with materials for entertainment[8] and improvements in communication technology help them keep in touch with loved ones back home.

However, new mental stresses will occur during journeys deeper into space, e.g. to Mars. On orbital and lunar missions Earth is still visible as a distinct world and serves as a reminder of home. However, during a voyage to Mars our world will soon shrink to only a bluish point of light, emphasizing the crew's isolation and danger. Unlike missions confined to LEO, the option of evacuating someone who has been injured or shows mental instability back to Earth will be essentially gone. The time required for radio messages to be sent from and to the ship will also become longer and longer. Feelings of isolation could grow into depression.

Early Soviet studies of cosmonauts' behavior showed findings similar to those in individuals confined for long periods to polar science stations or submarines. In the first several months of a mission mood and work performance were generally good, with any irritability or anger generally directed toward "outsiders" like Earth-bound control personnel rather than crewmates. In the next phase fatigue, decreasing motivation, and emotional hypersensitivity and lability became prominent. At that point irritability and frank hostility were more likely to be directed at other crewmembers. In some cases depression and interpersonal conflicts between crewmembers may actually have contributed to early termination of Soviet missions aboard Mir and Salyut space stations.

Generally speaking, using mixed-gender crews has been found to have some advantages compared to those involving only one gender, including overall better social interactions. However, mixed-gender crews on deep space missions could also increase the chances that sexual interactions and conflicts will come into play. Jealousies and development of intimate relationships between crewmembers could interfere with the mission. There have been issues with promiscuity, sexual harassment, and disrupted marriages occurring during long Antarctic missions, with negative effects on overall group functioning. The possibility of such issues occurring in space might make for interesting stories but would be best avoided if possible in real life [23, 24].

Space missions also include medications to treat conditions such as anxiety (e.g. benzodiazepines such as diazepam, lorazepam, or flurazepam), depression (for example, nortryptiline), and actual psychosis (e.g. haloperidol). Such drugs must be used only when necessary. For example, generally well-tolerated ones such as benzodiazepines can be used as aids for sleep or anxiety, while antipsychotic agents must be prescribed very judiciously.

---

[8] Hopefully this entertainment does not include movies such as *Rocketship X-M* (1950), *Marooned* (1969), or *Gravity* (2013), all of which depict highly undesirable things happening to spacefarers.

3.5 Circadian Rhythms and Sleep

Circadian rhythms are biological processes with cycles that recur over a period of approximately 24 hours. The sleep/activity cycle is the classic example, but body temperature, heart rate, blood pressure, and blood levels of cortisol, melatonin, and other substances within the body also follow circadian rhythms [1, 13, 25].

For example, many circadian rhythms, including core body temperature and secretion of growth hormone, are at their minimum around 3–5 a.m. of a person's "local" time. A person's mood, alertness, and ability to perform activities are also lowest at about those times. Conversely, levels of melatonin secretion are roughly opposite to changes in body temperature, highest when the latter is at its least and vice versa. Cortisol levels have a still different pattern, being greatest when someone first wakes up in the early morning with a smaller peak in mid-afternoon, followed by minimum values in early evening.

These rhythms are due to internal "biological clocks" which, in turn, are entrained on cyclical environmental stimuli ("zeitgebers"). The light-dark cycle based on Earth's 24 h rotation with alternating day and night is the most important of these external factors [1, 25]. Even without external light-dark cycles (e.g. someone in a dark cave without a clock), circadian rhythms persist, but with a "free-running" cycle of about 25 hours.

The normal synchronization between a person's internal "clock" and external time cues like light/darkness can be disrupted by rapidly traveling through different time zones on the Earth's surface, producing "desynchronosis."[9] Except on Mars, whose day is only slightly longer than Earth's at 24 h 37 min, space travelers experience light-dark cycles drastically different than on our world. In LEO these cycles are between 80–140 min long, 30–40 % of which is in darkness. Because exposure to bright light is such an important stimulus for "synchronizing" circadian rhythms, those light-dark cycles' effects on orbiting astronauts might be particularly stressful, although this has not been definitely confirmed via formal studies.

Disruption of circadian rhythms by environmental changes can potentially produce insomnia, loss of appetite, fatigue, and psychological stress. These symptoms are not immediately disabling in most people and, in time,

---

[9] Based on my personal experience this term should be used cautiously. Some years ago my wife (also a physician) taught a course on medical terminology at a local university. I substituted for her at several classes when she went on a trip to Europe. Since she was not feeling well right after she returned I taught a last class for her that day too. It being a medical terminology course, I half-jokingly told the students that she was suffering from "a bad case of desynchronosis." The shocked and concerned expressions this information produced on many students' faces made me realize that (a) they were not familiar with the term and (b) they assumed it was some terrible medical condition—possibly infectious, definitely painful, and potentially fatal. I quickly explained to their relief that it was just the fancy name for "jet lag."

resynchronization of circadian rhythms with "local" time occurs. One way to deal with these issues prior to flight is to have astronauts adapt to what will be their "day-night" work-sleep schedules before going into space. On the ISS the overall "time zone" used is Coordinated Universal Time (UTC), the international standard. Astronauts and cosmonauts must adjust to this "artificial" time [1].

Nonetheless, even when other circadian rhythms adapt (at least partially) to conditions in space, the sleep cycle can remain disrupted, with astronauts commonly reporting difficulty sleeping. Other contributing factors to this problem include excitement and stress from being on the mission, noise from equipment and active crewmembers on a different work shift, or uncomfortable surrounding temperatures. Difficulties with sleeping include insomnia and intermittent, poor quality, or even prolonged (up to 12 h) sleep [26].

Problems with sleeping could degrade work performance and alertness during routine tasks or emergency situations. When sleep needs to be delayed caffeine can be used as a mild stimulant. Astronauts often use sedative medications like those described previously to help treat insomnia. However, those drugs can have their own side effects, including continued drowsiness after waking. For this reason non-pharmacological means such as preflight preparation for a particular work-rest schedule and not changing that schedule in-flight are used as primary measures, If these measures are not sufficient, then medications can be used.

On the ISS astronauts have small bunk areas that are lightproof and designed to reduce external noise. They use restraint systems that allow crewmembers to stretch out "supine" relative to the walls of the station and keep them from floating freely during sleep. Despite all these measures, however, astronauts commonly report waking up often during sleep periods, reducing their overall effectiveness in working during the mission. This may also contribute to long-term feelings of fatigue that they can feel during extended missions. Problems with sleep disturbances might be accentuated by the smaller living space that will be available in spacecraft used for long-duration, deep-space missions (e.g. to Mars) compared to that on the significantly larger ISS.

3.6 Risks of Meteoroids and Space Debris

Space is not empty. It contains both natural and human-made material capable of damaging spacecraft and killing astronauts. Meteoroids—particles of stone and iron whizzing through space at an average velocity of about 16–20 km/s as they orbit the Sun—are ubiquitous. There is an estimated 200 kg of meteoroid mass in orbit within 2000 km of Earth's surface at any given time. Most particles are about 0.1 mm or less in diameter, though they can be as large as 1 cm. More rarely, rocks and boulders traverse this region. However, even the tiniest "bullets" can be dangerous due to their high velocities and proportionate amount of kinetic energy.[10] Fortunately space is a very large place and collisions between meteoroids and orbiting spacecraft are infrequent. Still, the possibility of damage to spacefarers and their craft is a real concern [2, 14, 16, 27].

Debris from previous rocket launches and either accidental or deliberate damage to satellites presents a similar hazard. About 2 million kg of human-made objects are estimated to be within 200 km of Earth's surface—a total mass many times greater than "natural" ones like meteoroids. These artificial objects range in size from meters (e.g. old rocket boosters) to tiny flecks of paint. According to a NASA report from 2009, over 19,000 objects larger than 5 cm in LEO and as small as 1 m in geosynchronous orbit (an altitude of about 36,000 km) are currently tracked. The number of objects less than 1 cm in size in LEO is estimated at over 300,000 [27]. The relative velocity of these various pieces of "space junk" is about 10 km/s [14].

Although some of its specific details are dubious or just simply wrong, the 2013 movie *Gravity* depicts an extreme example of how dangerous this might become. It shows wreckage from a missile strike on a satellite in LEO destroying a crewed spacecraft—a possibility uncomfortably close to real life. For example, a Chinese anti-satellite missile test in 2007 reportedly created over 35,000 objects more than 1 cm in size. Some of this debris is thought to have struck and damaged a small Russian satellite in 2013. An active United States communication satellite and a defunct Russian satellite collided in 2009, destroying both of them and produced a large amount of debris. At various times both Space Shuttles and the ISS have fired thrusters to maneuver away from tracked space debris. However, the ISS requires some 30 h to plan and execute such a maneuver to avoid tracked objects.

How long a particular item of space debris remains in orbit is related to factors such as its density, surface area, and altitude. Higher density items will experience less atmospheric drag and would be expected to stay in space longer, as would objects with lower surface areas for the same reason. A derelict satellite in a circular orbit with an altitude between about 200–400 km will likely reenter Earth's atmosphere within a matter of months. One at a higher altitude (400–900 km) could stay in orbit for years to centuries. Greater solar

---

[10] An object's kinetic energy in joules is calculated as $\frac{1}{2}mv^2$, where m is the mass in kilograms and v is velocity in meters/second. Thus, even if an object's mass is very small, if its velocity is very high it will still pack a "wallop." A bullet easily capable of penetrating deep inside a human body may leave the muzzle of a modern rifle moving at about 1.2 km/s but have a mass of only around 0.02 kg.

activity also increases atmospheric drag and accelerates an object's fall from orbit.[11]

Astronauts are most vulnerable to natural and artificial objects during EVA, when they have only a spacesuit and helmet for protection. A meteoroid or debris no more than about 0.35 mm in size and traveling at 10 km/s could potentially penetrate a spacesuit [14]. A particle greater than 1.5 mm in size could produce a "critical leak" as well as direct injury if it penetrates an astronaut's body. The latter could be incapacitating if the object were only 1 mm in size and moving at 10 km/s, and potentially fatal if it were 1.7 mm or larger [6].

American spacesuits are designed to activate a secondary oxygen pack to help maintain suit pressure if the latter falls from its nominal value of 4.3 psi (222 mmHg) to 3.9 psi (202 mmHg). This backup system can sustain pressure for up to about 30 min if the puncture size is less than 4 mm, but it may not be able to compensate long for a larger one. The odds of a 4 mm or greater leak occurring during a 6 h EVA have been estimated at about 1 out of 16,000—very low but potentially catastrophic if it did occur [6].

The ISS itself provides significantly better protection. It uses several layers of a lightweight ceramic fabric to protect key areas such as its habitable compartments. This shielding material "shocks" impacting objects 1 cm or less in diameter (mass about 1.46 g or less) so they are melted and vaporized, thus distributing their energy over a wider area rather than penetrating the ISS.

Overall, the threat of meteoroids and space debris is a very real one that is fair game for science fiction story purposes. However, humans do not have to venture into space to become targets in an interplanetary shooting gallery. Those same meteoroids that could strike orbiting astronauts and craft are much more likely to strike the far larger "bullseye" of Earth's atmosphere. If they enter it they become "meteors," visible in the night sky as brief streaks or occasionally flashes of light. Larger ones reach the ground, where they are termed "meteorites."

The greatest risk to people on the ground, however, is from larger objects—asteroids and comets. One criterion suggested as differentiating a meteoroid from an asteroid is simply size—asteroids have a diameter of about 10 m or more, and meteoroids are smaller than that [28]. Asteroids are mainly located in a belt between Mars and Jupiter, where the largest one, Ceres, with a diameter of about 900 km orbits. The vast majority of asteroids are much

---

[11] This led to the premature demise of the United States' only space station, Skylab. It was launched into orbit in 1973 and inhabited for a total of three missions in that year and 1974. Skylab was expected to stay in orbit until the early 1980s, when it would have been revisited by a Space Shuttle flight and boosted into a higher orbit. Unfortunately solar activity in the late 1970s increased atmospheric drag on the uninhabited station, and it fell from orbit in 1979—2 years before the first Shuttle flight in 1981.

smaller, however, with an estimated tens of millions having diameters less than 1 km.

Asteroids also orbit in locations farther out in the Solar System and, of more concern to us, close to Earth. The latter "near-Earth asteroids" (NEAs)[12] include several classes based on their orbits relative to Earth. They include "Apohele" or "Atira" asteroids located completely within Earth's orbit and that never cross it; "Aten" asteroids with orbits mainly inside Earth's but that can cross it at their greatest distance from the Sun; "Apollos" having orbits that are mainly outside Earth's but that can also cross it; and "Amors" whose orbits are entirely outside of Earth's and (generally) between our world and Mars.

Even relatively "small" asteroids can literally have a big "impact" on Earth. The near-Earth asteroid that exploded over Russia in February 2013 was estimated to be about 20 m in diameter, while the one that blew up over Tunguska in that country in 1908 may have been anywhere between 60–190 m across. The latter might have released as much energy as a 30 megaton bomb when it exploded, while the former had kinetic energy equivalent to "only" about a 500 kiloton bomb. In contrast, the asteroid that devastated the dinosaurs some 65 million years ago may have had a diameter of about 10–15 km and was roughly equivalent to a 1×10^8 megaton bomb.

Comets are thought to originate in either the Kuiper belt, a region extending out about 30 to 50 AU[13] in the same plane that the planets orbit, or the Oort cloud, a spherical region that extends beyond the Kuiper belt up to some 50,000 AU from the Sun. The diameter of a comet's solid "nucleus" can be as great as 30 km, formed of an inner rocky core covered by dust and volatile materials such as frozen water and gases (e.g. carbon dioxide and ammonia)— what has been dubbed a "dirty snowball." As an orbiting comet approaches the Sun it releases both dust and those volatile materials evaporated by Sol's heat in the form of tails. Eventually, after a comet passes by the Sun enough times in its orbit those volatile materials can be exhausted, leaving only the comet's rocky core behind. The latter can still be quite large, however, and is roughly equivalent in potential destructive effects to an asteroid of comparable size.

The Torino scale is one used to assess risks and results if large objects such as asteroids and comets were to strike our world. It runs from 0–10, with lower numbers being better from our standpoint. A score of 0 means that an object has a negligibly small chance of colliding with our planet. A value of 5 means that there is a significant but still not certain chance of collision, but

---

[12] The more general term "near-Earth objects" (NEOs) includes these bodies as well as others in our planet's vicinity, including comets.

[13] An "AU," or "astronomical unit," is Earth's average distance from the Sun, or slightly less than 150,000,000 km.

the object is large enough that it could cause at least regional devastation if it did strike. A score of 10 means that collision is certain and that the object will cause both widespread regional devastation and a global climate catastrophe.

Currently over 10,000 NEOs have been discovered, with a little less than 900 of them being 1 km or greater in diameter [29]. It is estimated that there are roughly 981 NEOs of that latter size, and thus that over 90 % have already been found and can be tracked. However, considering the destructive effects of even the considerably smaller ones cited previously, the chance that a previously unrecognized NEO could wreck havoc at any time remains tiny at any given moment but significant over longer time spans. Science fiction has invoked threats from marauding asteroids and comets both in written works such as Larry Niven and Jerry Pournelle's *Lucifer's Hammer* (1977) and cinematic efforts like *Armageddon* and *Deep Impact* (both 1998).

Another potential menace from space that originated in science fiction before being confirmed in real life is that of "rogue" or "nomad" planets. Fictional examples include the Emperor Ming's world Mongo and the paired Bronson Alpha and Bronson Beta in Philip Wylie and Edwin Balmer's novel *When Worlds Collide* (1933). In real life such solitary planets may have formed in isolation from a star as Jupiter-size or larger worlds. They might also have been ejected from a star's planetary system, perhaps due to gravitational interactions with its other members.

The several isolated planets that have been discovered recently or are theorized to exist are unlikely to be inhabited by alien despots or Lion Men, and space is just too vast for any to be likely to make a beeline towards Earth. However, a large enough one passing close to the Oort cloud could disturb the orbits of the latter's comets enough to send some towards the inner Solar System. This equivalent of a "shotgun" blast of comets could potentially increase the risk of an impact with Earth to worrisome levels. In fact, the closest known rogue planet to Earth (WISE 0855–0714, with an estimated mass somewhere between 3 and 10 times that of Jupiter) is only 7 light-years away. That by itself is a safe distance, but raises the possibility that other yet undiscovered ones might be considerably nearer.

Closer to home, future travelers to the Moon and Mars will be exposed to hazardous materials there. The surfaces of both worlds are mainly covered by rocky material ranging in size from grains of dust to rocks of various sizes. This "regolith" is thought to be about 4–5 m thick in the maria (plains) of the Moon, such as Mare Tranquillitatis (the "Sea of Tranquility") where Apollo 11 landed, and up to roughly 10–15 m in the Moon's highlands. Lunar dust created a nuisance on the Apollo missions by coating spacesuits and equipment. It also has the potential of building up electrostatic charges that, when discharged as sparks, might damage electronic equipment [2].

Such dust could also present a health hazard if astronauts inhale it. Lunar and Martian regolith are considered to be overall not harmful from a chemical standpoint. However, breathing in either variety could physically irritate an astronaut's nasal passages, pharynx, trachea, and lungs. Such dusts could be tracked into a spacecraft (e.g. the Apollo Command Modules) or through the airlock of a Martian habitat on boots and spacesuits after a stroll on the surface. As the Apollo astronauts found, lunar dust in particular is hard and abrasive, with a smell like gunpowder. Certain minerals inhaled on Earth such as by miners, including coal and silica dust, can produce "pneumoconiosis"— damage to the lungs (e.g. "black lung" associated with coal dust) that could produce shortness of breath, chronic cough, and other symptoms. The minerals present in lunar and Martian soil are thought to overall be more "benign" than those that cause pneumoconiosis on Earth, but they are still capable of causing at least short-term respiratory problems.

The higher gravity on Mars compared to the Moon (0.38 g versus 0.17 g) and its very small atmospheric pressure (absent entirely on the Moon) may reduce these risks to some extent by making dust there more likely to "stay in place." On the other hand, unlike on the airless Moon, the thin atmosphere on Mars is able to occasionally stir up great dust storms that can cover virtually the entire planet for weeks at a time.

3.7 Acceleration and Deceleration

Too much acceleration and catastrophic failure of spacecraft can injure or kill. Astronauts in the Mercury program were exposed to accelerations as high as 8 g briefly during launch and up to 11 g with deceleration during reentry. Fighter pilots wearing special suits to maintain blood supply to the brain can remain functional during 9 g turns. However, an unprotected average individual would be able to tolerate no more than about 5 g for a sustained period before losing consciousness [1, 2, 13].

An Apollo capsule returning from space could subject its occupants to as high as a brief 17 g spike as it splashed down [2]. Rocket sled tests in the mid-twentieth century subjected human volunteers to brief accelerations up to 46 g with no long-term adverse effects. By comparison crew on Space Shuttle launches normally experienced much "easier" sustained levels of about 1.2 g during entry to and return from orbit, with brief peak levels up to about 3 g during launch and 2 g when the Shuttle turned in the sky before landing. The landing of a Soyuz capsule is typically associated with a peak acceleration of between 4–7 g. Landings on Mars may be associated with peak levels of up to about 6 g [1]. When relatively high g-forces are involved in launch or reentry it is desirable for the astronaut to be supine so those forces are directed more in the back-to-chest axis rather than head-to-foot to reduce effects on the brain, spine, etc.

But the "design specs" of the human body can be exceeded when things go wrong. The lone cosmonaut aboard the Soyuz 1 capsule in 1967 was killed when its parachutes failed to deploy properly during landing. His craft hit the ground at an estimated 140 km/h followed by an explosion and fire [13]. Seven astronauts were killed aboard the Space Shuttle *Challenger* in 1986 after that craft broke apart during launch when the O-rings of one of its solid rocket boosters failed, allowing pressurized hot gases to leak out and damage nearby structures. This sent the crew cabin plummeting downward to strike the ocean at an estimated speed of about 333 km/hour, with an estimated deceleration of over 200 g. Space Shuttle *Columbia* was destroyed during reentry in 2003, with its seven-member crew dying as their craft broke up.

3.8 The Bottom Line

In real life the dangers inherent to being in space or on the Solar System's other worlds present formidable but potentially solvable challenges to establishing a permanent human presence away from Earth. However, all those perils offer a rich variety of realistic ways writers can put space-based science fiction characters in danger without introducing a single hostile alien. Even in a story set far enough in the future where technology has in some way reduced these risks, a writer must still take these dangers into account if only indirectly by showing how they were controlled or eliminated.

For example, while we can create (if only on a small scale) a livable environment in space, a theoretical option is to modify our bodies to better suit conditions there or on an alien world. The scope of modifications to the human body range from the "tweaks" in human anatomy depicted in Kevin J. Anderson's *Climbing Olympus* (1994) to allow survival on Mars, to the wholesale changes in or replacement of the body in works as disparate as Frederik Pohl's *Man Plus* (1976), Anne McCaffrey's *The Ship Who Sang* (1969), or the previously mentioned Clifford D. Simak story "Desertion." However, while such works show what might happen *if* such modifications were possible, in the here-and-now there is no guarantee that they or anything like them will be feasible. Until or if dramatic advances in genetic engineering, nanotechnology, transferring human consciousness into new biological or artificial bodies, etc. (all subjects for later chapters) make these techniques feasible, we will be "stuck" essentially making do with the vulnerable bodies we have.

Science fantasy (e.g. *Star Wars*) may choose to ignore many space-based dangers unless needed for the plot. "Harder" science fiction may acknowledge those issues but "solve" them by wrapping them in scientific-sounding concepts that actually have little if anything to do with actual science (e.g. the "inertial dampers" used on Starfleet's starships like the *Enterprise* to buffer the effects of rapid acceleration and deceleration).

Conversely, the "hardest" science fiction deals with them head-on in a realistic way. A successful example of such a truly scientific approach occurs in the prologue of Edward M. Lerner's novel *Energized* (2012). That near-future work opens in space, on an asteroid with a diameter of about 2 km that has been diverted into Earth orbit. A huge sunshield has been created to keep the Sun's heat from evaporating the ice on its surface. This presumably renders the effective temperature beneath the shield somewhere around the − 178 °C of the Moon's night side.

In the opening scene two characters wearing spacesuits leave a base and travel out onto that cold surface. But while doing a repair job there one of them murderously removes the two battery packs from the other's suit, cutting off electrical power to its heaters, life support, communications, and other equipment. The victim quickly feels himself freezing, his fingers seem numb, and he loses consciousness within minutes before ultimately dying.

As noted previously, current spacesuits are heavily insulated, retain body heat well, and require a cooling system rather than heaters to keep their occupants at comfortable temperatures. An 80 kg person's basal metabolic rate —the amount of energy in the form of heat generated per unit time at rest—is about 100 watts (100 joules of energy per second). The rate at which this power is radiated from the skin is described by the Stefan-Boltzmann law, $P = A\varepsilon\sigma T^4$, where P is power in watts; A is area in square meters; ε is emissivity (the ratio of energy emitted from an object's surface compared to that of a "black body," which has an emissivity of 1); σ is the Stefan-Boltzmann constant, $5.67 \times 10^{-8}\,\mathrm{Wm^{-2}K^{-4}}$; and T is temperature in Kelvin (K). Human skin has an average emissivity of 0.98, so the power (energy per second) it emits is proportional to its surface area and temperature.

That basal level of 100 watts is created by metabolic activity, and the body must continuously generate it to maintain life. Shivering can temporarily increase the power produced by the body by a maximum of about 100 watts above the basal metabolic rate, to a total of a little over 200 watts. However, the body's immediate fuel stores (e.g. glucose) can be exhausted relatively rapidly to produce this "extra" energy. As we shall see in Chap. 7, if too much heat is lost to a cold environment the body's ability to maintain core (internal) temperature can be overwhelmed.

In the case of a standard spacesuit, that body heat is lost very slowly to the "cold" of space. If this factor alone were involved it would take an astronaut a long time to feel cold and have core body temperature start falling below its nominal value of about 37 °C(98.6 °F). However, other bad things also happen if a spacesuit's electrical power is lost. As part of the breathing process oxygen at an appropriate temperature must be supplied to the astronaut and exhaled carbon dioxide removed. In a standard spacesuit a pair of fans keep these gases circulating. Carbon dioxide is removed by one or more canisters containing lithium hydroxide. They react together to create lithium carbonate and water [16].

If those fans lose electrical power, however, the oxygen supply might cool very quickly. Breathing chilled oxygen will accelerate cooling inside the body, reducing core body temperature and thus producing "hypothermia" much more rapidly than losing overall body heat to space would. If carbon dioxide removal were also impaired due to loss of electrical power to the circulating fans, rapid buildup of that gas within the astronaut's spacesuit might produce symptoms quickly. With mild "hypercapnia" (abnormally increased levels of carbon dioxide in blood), an astronaut could rapidly develop shortness of breath, confusion, headache, and lethargy. Severely increased levels of carbon dioxide could cause feelings of panic, seizures, unconsciousness, and death.

The situation described in the novel could also have other psychological and physiological effects on the astronaut. These could include panic unrelated to metabolic activity or cold, but caused by an "impending sense of doom" associated with the sudden, unexpected life-threatening situation. Under such strong psychological stress the astronaut might perceive he is "freezing" long before his body actually experiences any physical effects due to external heat loss. Under such stressful conditions a person might also have a "vasovagal" response, an autonomic nervous system reflex that could include a drop of heart rate, blood pressure, or both potentially severe enough to cause light-headedness or unconsciousness.[14]

Moreover, the spacesuit described in the novel is significantly different from current models. It is a relatively thin, form-fitting one that is described as not having the heavy insulation or water-cooling system of "standard" pressurized ones, but instead created to provide maximum flexibility and freedom of movement.[15] This suggests that, perhaps as a design condition needed to produce that flexibility, it might require maintaining an astronaut's body temperature, not by retaining body heat and removing it with a cooling system like conventional spacesuits, but by actually letting body heat radiate out to space more rapidly and then resupplying heat via a heater. Such a system might provide the advantage of allowing astronauts to move more freely than a massive pressurized suit at the cost of requiring continuous power from a heating unit

---

[14] A terrestrial example of a vasovagal response is an individual fainting while having blood drawn.
[15] "Skin-tight" spacesuit designs using special materials are now actually being studied—see http://newsoffice.mit.edu/2014/second-skin-spacesuits-0918.

to maintain body temperature. If so, then the astronaut's sensation of rapidly becoming cold when the batteries to the heater were removed might actually be at least partially due to external heat loss. Any one or a combination of the factors just described—rapid loss of body temperature, increased carbon dioxide levels (hypercapnia), a vasovagal response, etc.—could also contribute to him first losing consciousness after a short time and dying soon afterwards.

This detailed analysis of the opening scene in *Energized* illustrates two important points about writing "hard" science fiction. First, there is a reasonable science-based rationale for the events depicted. Second, just enough scientific background information appears on the printed page to make what happens plausible. This particular scene emphasizes *what* happens (the astronaut becomes quickly impaired and dies) rather than bringing the flow of action to a crashing halt with a too detailed (even if scientifically accurate) description of the physics, biology, etc. behind *why* it might be happening as described previously. In short, hard science fiction can include just enough science to keep the story moving, but what it does say (if only very succinctly) is scientifically accurate and possible.

References

1. Clément G. Fundamentals of space medicine. 2nd ed. New York: Springer; 2011.
2. Barratt M. Physical and bioenvironmental aspects of human space flight. In: Barratt M, Pool SL, editors. Principles of clinical medicine for space flight. New York: Springer; 2008. pp. 3–26.
3. Bacal K, Beck G, Barratt M. Hypoxia, hypercarbia, and atmospheric control. In: Barratt M, Pool SL, editors. Principles of clinical medicine for space flight. New York: Springer; 2008. pp. 445–73.
4. Pickard J. The atmosphere and respiration. In: DeHart RL, Davis J, editors. Fundamentals of aerospace medicine. 3rd ed. Philadelphia: Lippincott Williams & Wilkins; 2002. pp. 19–38.
5. Comins NF. The other planets and moons. In: Comins NF, editor. Discovering the essential universe. 4th ed. New York: W. H. Freeman and Company; 2009. pp. 135–99.
6. Norfleet W. Decompression-related disorders: decompression sickness, arterial gas embolism, and ebullism syndrome. In: Barratt M, Pool SL, editors. Principles of clinical medicine for space flight. New York: Springer; 2008. pp. 223–46.
7. Foster P, Butler B. Decompression to altitude: assumptions, experimental evidence, and future directions. J Appl Physiol. 2009;106(2):678–90.
8. Stepanek J. Decompression sickness. In: DeHart RL, Davis J, editors. Fundamentals of aerospace medicine. 3rd ed. Philadelphia: Lippincott Williams & Wilkins; 2002. pp. 67–98.

9. Otto CM. Cardiac masses and potential cardiac "source of embolus." Textbook of Clinical Echocardiography. Philadelphia: Elsevier Saunders; 2004. pp. 407–30.

10. http://imagine.gsfc.nasa.gov/docs/ask_astro/answers/970603.html.

11. Kolesari G, Kindwall E. Survival following accidental decompression to an altitude greater than 74,000 ft (22,555 m). Aviat Space Environ Med. 1982;53(12):1211–4.

12. Holm A, Freedman T, Puskas A. Accidental decompression: a new philosophy for the transports of the 1970s. Aerosp Med. 1970;41(3):277–82.

13. Planel H. Space and life: an introduction to space biology and medicine. Boca Raton: CRC Press; 2004.

14. Clark J. Decompression-related disorders: pressurization systems, barotrauma, and altitude sickness. In: Barratt M, Pool SL, editors. Principles of clinical medicine for space flight. New York: Springer; 2008. pp. 247–71.

15. Altitude Air Pressure Calculator. http://www.altitude.org/air_pressure.php. Accessed 15 April 2015.

16. Locke J. Space environments. In: DeHart RL, Davis J, editors. Fundamentals of aerospace medicine. 3rd ed. Philadelphia: Lippincott Williams & Wilkins; 2002. pp. 245–70.

17. Nunneley S, Adair E. Thermal stress. In: DeHart RL, Davis J, editors. Fundamentals of aerospace medicine. 3rd ed. Philadelphia: Lippincott Williams & Wilkins; 2002. pp. 271–87.

18. Stewart L, Trunkey D, Rebagliati S. Emergency medicine in space. J Emerg Med. 2007;32(1):45–54.

19. Johnston S, Arenare B, Smart K. Medical evacuation and vehicles for transport. In: Barratt M, Pool SL, editors. Principles of clinical medicine for space flight. New York: Springer; 2008. pp. 139–61.

20. Marshburn T. Acute care. In: Barratt M, Pool SL, editors. Principles of clinical medicine for space flight. New York: Springer; 2008. pp. 101–22.

21. James J. Health effects of atmospheric contamination. In: Barratt M, Pool SL, editors. Principles of clinical medicine for space flight. New York: Springer; 2008. pp. 427–43.

22. Hall R, Shayler D. Soyuz: a universal spacecraft. New York: Springer; 2003.

23. Buckey JC. Space physiology. New York: Oxford University Press; 2006.

24. Kanas N, Manzey D. Space psychology and psychiatry. 2nd ed. New York: Springer; 2010.

25. Putcha L, Marshburn T. Fatigue, sleep, and chronotherapy. In: Barratt M, Pool SL, editors. Principles of clinical medicine for space flight. New York: Springer; 2008. pp. 413–25.

26. Buguet A. Sleep under extreme environments: effects of heat and cold exposure, altitude, hyperbaric pressure and microgravity in space. J Neurol Sci. 2007;262(1):145–52.

27. The Threat of Orbital Debris and Protecting NASA Space Assets from Satellite Collisions. 2009. http://www.space.com/3415-china-anti-satellite-test-worrisome-debris-cloud-circles-earth.html. Accessed 15 April 2015.
28. Comins NF. Vagabonds of the solar system. In: Comins NF, editor. Discovering the essential universe. New York: W. H. Freeman and Company; 2009. pp. 203–33.
29. Near Earth Object Program. http://neo.jpl.nasa.gov/stats/. Accessed 15 April 2015.

4

Microgravity and the Human Body

How sweetly did they float upon the wings
Of silence, through the empty-vaulted night,
At every fall smoothing the raven down
Of darkness till it smil'd!

John Milton
Comus

It was the strangest sensation conceivable, floating thus loosely in space,
at first indeed horribly strange, and when the horror passed, not dis-
agreeable at all, exceedingly restful! Indeed the nearest thing in earthly
experience to it that I know is lying on a very thick soft feather bed.

H. G. Wells
The First Men in the Moon

The human body evolved to live optimally at a "normal" gravity of 1 g, equal to a downward acceleration of about 9.81 m/s^2 at Earth's surface. In space our bodies become "weightless" and try to adapt to a new environment of "microgravity" in ways that cause problems both there and when astronauts return to Earth. Science fiction stories and movies often deal with microgravity's effects by either ignoring them or using "artificial gravity" on their spaceships to get around them. The latter can be used realistically, such as having a large space station rotating at an appropriate rate to simulate gravity. Or it can involve a considerably less realistic flip-a-switch gravity field that lets spacefarers walk and move as if they were on Earth.

Objects in low-Earth orbit (LEO) such as the ISS are attracted to our world with an acceleration of about 9.0 m/s^2, only mildly less than what we experience at sea level. The difference is that the downward force Earth exerts on us is counterbalanced by upward forces, such as the ground or floor we stand on, that give us the sensation of weight. In the microgravity environment of space, only the force of gravity itself is present. Anything in orbit is still being attracted to Earth and falling toward it. However, the object's altitude and its motion at a right angle to that attractive force makes it keep missing Earth and remain in "free fall." Spacefarers traveling away from Earth (e.g. en route to Mars) would experience similar conditions.

Gravity is only an attractive force, proportional (based on Isaac Newton's concepts) to the product of the masses of two bodies and inversely proportional to the square of the distance between them. This is expressed in the equation $F = G\dfrac{m_1 m_2}{r^2}$, where F is the gravitational force in newtons, m_1 is the mass of one object in kilograms, m_2 is the mass of a second object in kilograms, r is the distance between them in meters, and G is the gravitational constant (6.67×10^{-11} newtons-meters^2/kg^2) [1].

Albert Einstein's theory of general relativity describes gravity as being due to a curvature in spacetime, the combination of the three "standard" spatial dimensions (e.g. length, width, and height) with time. However, to understand its effects on the human body Newtonian physics is a sufficient approximation. Unfortunately, barring new discoveries in physics showing its existence, classic science fiction staples such as easily controllable gravity levels or anti-gravity fields are not feasible. Instead we will have to deal with the real life effects of microgravity as best we can.

4.1 Space Adaptation Syndrome

Entering microgravity causes immediate and delayed effects on a person's body. These physiological and anatomic changes are collectively described as "Space Adaptation Syndrome" (SAS). About one half to two thirds of astronauts experience symptoms of "space motion sickness" (SMS), a major component of SAS, within minutes to hours of reaching orbit [2–8]. It is associated with problems such as nausea, vomiting, loss of appetite, headache, malaise, sweating, paleness, and dizziness. This is a far cry from Mr. Bedford's description of weightlessness in *The First Men in the Moon* (1901) as being "exceedingly restful."

Up to 73 % of Space Shuttle astronauts flying for the first time developed SMS [4]. It is mildly less likely to occur in astronauts who fly on multiple missions, but if it does the symptoms may be less severe than on the first flight. The chance of developing SMS has not been found to be significantly related to either an astronaut's age or gender. SMS can start as soon as 15 min after entering orbit or as long as 3 days afterwards [9].

Symptoms of SMS worsen with head and body movements and typically last about 24–72 h. They may be accompanied by drowsiness, lethargy, and irritability. These symptoms are similar in many ways to those of motion sickness on Earth, which may occur during travel in a car, aircraft, or on water.

SMS, however, is less likely to be associated with sweating and paleness than terrestrial motion sickness. There is currently no reliable method to predict which astronauts will develop SMS during a first flight and which will not.

Severity of SMS symptoms has been reported to range from mild in 49 % of cases to moderate in 36 % and severe in 15 % [3, 4]. Symptoms are considered "mild" if an astronaut develops only one or a few of them to a mild degree; occur only with head movement; do not interfere with ability to perform mission functions; are associated with no more than one episode of retching or vomiting; and resolve within 36–48 h. "Moderate" symptoms are multiple and persistent, often occurring in cycles in which they become more and then less severe. They may be associated with up to two episodes of vomiting, have minimal impact on astronaut performance, and resolve within 72 h. Symptoms are "severe" if an astronaut develops several that are more pronounced, strongly associated with head movements, include two or more episodes of vomiting, and significantly impact ability to function. They may persist beyond 72 h—in rare cases, as long as 7–10 days.

SMS-related vomiting is especially worrisome. It usually occurs early in the course of SMS, with episodes often separated by several hours and with possible temporary relief of symptoms after one occurs. It may develop without preceding nausea or other warning signs, potentially resulting in vomit floating within the small confines of a spacecraft.

Vomiting within a spacesuit due to SMS would be even more serious and potentially life-threatening, with an astronaut likely to inhale that material into his or her lungs. This could cause both immediate problems with breathing and lead to "aspiration pneumonia" developing within a few days. Because of this risk of sudden vomiting and other symptoms of SMS, astronauts are not scheduled for extravehicular activity (EVA) for at least 3 days after they reach orbit and any symptoms of SMS have resolved.

Susceptibility to SMS does not correspond to a predisposition to the usual type of motion sickness caused by being on a rocking ship or in a rotating chair. The causes of SMS are not entirely established, but the basic problem may be sensory conflict between what an astronaut sees in the space environment and what his or her body's neurovestibular system (the parts of the inner ear and brain that identify the body's orientation and maintain balance) is sensing. The latter system is "wired" based on lifelong experience in 1 g to expect certain relationships between what a person sees or otherwise senses and the individual's body motion and posture [2, 3, 10]. Thus, when someone is standing upright the neurovestibular system is trained to detect that position, the eyes see surrounding objects in their "usual" orientation (e.g. cars with their wheels on the road rather than upside down), and other sensations such as the pressure of one's feet against the ground all combine to give a unified, consistent picture of the surroundings [11].

However, an astronaut entering microgravity has to learn an entirely new set of relationships. As a spacefarer floats inside a craft the neurovestibular system cannot distinguish between up and down as that person's eyes see objects, walls, and other parts of the surroundings change their orientation. This makes it difficult for the individual's brain to distinguish whether the surroundings or the individual is moving, or if he or she is "upright" or "upside down" based on terrestrial concepts that no longer apply. Eventually the astronaut's neurovestibular system and senses adapt to those new relationships, but while that is happening the brain reflexively activates components of the autonomic nervous system[1] and causes the symptoms described previously [12].

Development of SMS may also correlate with how much room astronauts have to move around within their craft after they reach orbit. SMS symptoms were not reported by astronauts in the Mercury or Gemini programs, who remained seated during their entire missions in a cramped capsule and stayed in their spacesuits [3]. Conversely SMS occurred in about one third of Apollo astronauts, who had the "luxury" of a (comparatively) larger Command Module and could remove their spacesuits. As noted previously Shuttle crews had a still higher incidence of SMS. This may have been due to the still greater volume of their craft, and that they routinely removed their spacesuits and moved about performing other activities shortly after reaching orbit [2, 3, 10].

Medications such as dimenhydrinate and meclizine used to treat motion sickness on Earth are not consistently effective for SMS. Another medicine, scopolamine, can be administered orally, by injection, or via a patch. It has also been used prophylactically prior to launch. Scopolamine has been carried on past spaceflights, but it has the disadvantage that using it may delay an astronaut's adaptation to microgravity and be associated with the return of SMS symptoms after it is stopped [2]. Prophylactic use of two medications, promethazine and dextroamphetamine, may also be helpful, particularly in astronauts who have had SMS on previous flights. An injection of promethazine can also improve SMS symptoms after they develop.

Similar issues with the neurovestibular system contribute to occurrence of sensory illusions in about 80 % of astronauts. For example, when an astronaut moves within the spacecraft it may seem that the latter is moving toward him or her rather than vice versa. The inversion illusion—the sensation that one has turned upside down—can also occur shortly after entering microgravity [10, 13–16]

---

[1] As described in Chap. 1, this is the part of our nervous system that works independently of our conscious awareness or direct control.

Astronauts returning to Earth may experience new symptoms as they re-adjust to our world's gravity. "Entry motion sickness" (EMS) can affect them even before they reach the ground during reentry. Symptoms are similar to those of SMS, and as with the latter they are likely to worsen with head movements. EMS is more likely to occur and be more severe after long-duration flights (e.g. months on the ISS) compared to shorter ones, such as Shuttle missions lasting 1–2 weeks.

After returning to Earth astronauts may feel heavy and clumsy, or that they are being pushed to one side while just standing. At night they might feel "levitated" over their beds. Astronauts may even have to relearn how to put their feet back on the ground when rising in the morning without falling out of bed. They may walk with a wide-based gait and have a tendency to fall to one side, or keep walking straight when they try to turn a corner [3, 15]. These problems could persist for weeks, with a typical duration of about 30 days. Astronauts can even experience a "relapse" of some EMS symptoms over that same period, which may take days to resolve.

As part of postflight rehabilitation astronauts are routinely monitored and treated as necessary for 45 days or longer after return to Earth. Measures include checking for "orthostatic" problems (e.g. an excessive fall in blood pressure when moving from a supine to standing position), psychosocial issues, and vestibular function as well as treatments including stretching and aerobic exercise, massage, hydrotherapy, etc [15].

4.2   Cardiovascular Effects

Within minutes of exposure to microgravity about 1–2 L of fluid normally present in the lower body shift to the upper part due to loss of normal gravity-induced "pooling" of blood in the legs when we sit or stand [2, 3, 8, 16–20]. On Earth such a "redistribution" of fluid also occurs when a person moves from a standing to a supine position. However, the amount of fluid involved in doing that is less than the corresponding shift in microgravity. In the latter, this "spreading out" of fluids is associated with distention of the neck veins, facial puffiness, reduction of calf diameter by up to 30 %, and enlargement of organs like the liver and pancreas due to engorgement of their veins. Astronauts may develop nasal congestion, a nasal voice, and headaches. Their legs become thinner and more "birdlike" in appearance.

The body (e.g. volume "sensors" in the heart) perceives this fluid moving to its upper section as being more than it needs. During the first few days in space this "extra" fluid may be lost as urine or move out of blood vessels into a person's tissues and cells, resulting in at least relative dehydration. That reduction in blood volume is thought to be mainly due to fluid shifts within the body rather than loss through the kidneys as urine or "insensible losses" such as perspiration, water vapor in exhaled breath, etc [16]. The astronaut may feel decreased thirst and appetite as the body tries to reach what it thinks is an appropriate level of blood volume, despite the latter being reduced by terrestrial standards. Average reduction of body weight from fluid loss is in the range of about 4%.

Orthostatic reflexes, which compensate for the normal mild drop in blood pressure that occurs when a person first stands up in 1 g from a supine or sitting position, also become impaired. These reflexes cause blood vessels to constrict and heart rate to increase, thus helping the body maintain blood pressure and adequate blood flow to the brain and other organs and tissues.

Impaired orthostatic reflexes and reduced intravascular volume are of little significance while the astronaut is in microgravity. In that environment the body does not need to use those reflexes as in does on Earth, and the lower blood volume an astronaut has there is "physiological." However, when someone returns to our world and experiences the "pull" of gravity again, a liter or more of fluid shifts back from the upper to the lower body. This effect, combined with reduced total blood volume and impaired orthostatic reflexes, can cause heart rate to increase and blood pressure fall enough to decrease blood flow to the brain when the astronaut is sitting or standing. This "orthostatic hypotension" can be severe enough to cause lightheadedness or even unconsciousness. It usually resolves within days with rehydration and increasing activity.

This problem was recognized early in the history of human spaceflight. For example, astronaut Gordon Cooper experienced such symptoms following his 34 h orbital flight in the Mercury *Faith 7* capsule in 1963. After his capsule was retrieved at sea and he stood on the ship's deck his heart rate rose to as high as 188 beats per minute. He became pale, sweaty, and almost passed out, but improved after a short time [2].

The hundreds of astronauts and cosmonauts who have flown since then have varied in how they have been affected on returning to Earth. Many, particularly those on short-duration missions, have reported only mild effects, which may be attributed to various countermeasures developed over the years. The Soviet/Russian space program has used a "Chibas" suit that a cosmonaut wears for several hours at a time [3, 19, 21]. This garment covers the lower body like a pair of trousers and exerts mild negative pressure ("suction") on the legs, simulating the effects of standing by effects such as drawing fluid from the upper body to the legs. Skylab and some Shuttle missions used somewhat similar devices. The rationale for such procedures is to try to reduce loss of orthostatic reflexes by periodically stimulating them.

During reentry and landing an astronaut can wear a suit with functions similar to the G-suit (or, to use a more accurate name, anti-G suit) used by jet pilots [21, 22]. It puts positive pressure on the legs (just the opposite of the Chibas suit) to help prevent blood in the upper body from pooling in the legs at those times. A common method used by returning astronauts to help prevent blood pressure from falling too much is to drink a liter or so of either juice or, alternatively, water with appropriate amounts of electrolytes like salt and also possibly sugars. Doing this just prior to descending to Earth helps to rehydrate them and reduces their chances of developing serious problems with orthostatic issues. This "fluid reloading" technique seems, however, to be more effective for short-duration rather than long-duration flights [3].

How much these problems with orthostatic hypotension might affect astronauts traveling to the Moon or going to Mars is uncertain. It was not reported to be a significant issue during the Apollo flights, with a flight to the Moon lasting only several days before astronauts experienced a lunar gravity of about 0.17 g. However, if someone has been exposed to microgravity for an estimated travel time of about 6 months before reaching the greater gravity (0.38 g) of the Martian surface, symptomatic orthostatic hypotension might occur.

Microgravity has other effects on the heart that are thought to be of little significance. Heart size and mass decrease mildly during long (> 2 weeks) stays in microgravity but recover after return to Earth [2, 23]. Changes in heart rate, blood pressure, measurements of how the heart functions, and occurrence of abnormal heart rhythms have varied widely among different astronauts [17]. However, overall these effects too are small.

Potentially serious abnormal heart rhythms (arrhythmias) have been noted in a few space travelers, but none have been life-threatening. Arrhythmias may be related to stress alone, which produces higher blood levels of catecholamines (e.g. epinephrine, aka "adrenaline") and other chemicals that can "overstimulate" the heart, rather than from microgravity itself or any underlying heart abnormalities. Nonetheless, electrocardiographic monitoring of the heart's electrical activity (including heart rate) is done during EVA and other activities requiring high levels of exercise as a precautionary measure. It was not, however, routinely done during launch or landing in the Shuttle program [2, 17].

4.3 Hematological Effects

The number and mass of red blood cells (erythrocytes) typically decrease slowly to about 10–20 % of preflight values during the first few months in microgravity [3, 17, 24, 25]. This relative anemia may reflect both increased destruction and decreased production of erythrocytes. After return to Earth this abnormality takes several months to recover due to the relatively slow rate of erythrocyte production. Medications such as recombinant human erythropoietin are now available to increase erythrocyte production. They are not routinely used at present but might be considered for very long duration missions such as to Mars. In addition, an increased proportion of abnormally shaped erythrocytes (spherocytes and echinocytes) are seen. This finding does not seem to be functionally important and disappears within several days postflight.

Microgravity is also associated with changes in white blood cells and the immune system [8]. Neutrophils, a type of white blood cell that combats bacterial infections, increase in number on average, perhaps due to stress-induced release of epinephrine, glucocorticoids, and other substances within the body [3, 16]. Conversely, killer T-lymphocytes (another kind of white blood cell) show diminished number and activity, potentially decreasing resistance to viral infections and recognition and destruction of cancer cells. B-lymphocytes, responsible for producing antibodies against bacteria and viruses, and helper T-lymphocyte both show little change in number. The number of eosinophils, which help treat infections from parasites, is decreased [26].

4.4 Musculoskeletal Effects

Relaxed astronauts in microgravity assume a fetal position and lose the normal curve of the thoracolumbar (middle to lower) spine. Height increases by about 3–6 cm, with the spine itself lengthening about 4–7 cm, mainly from decompression of intervertebral disks between the bones (vertebrae) in the spine [27, 28]. This may compress nerve roots as they exit the spine and thus produce back pain. The latter is common during the first few days after entering space. It may also be due to stretching of muscles and ligaments in the lower back. Height returns to normal within about a day postflight as the intervertebral disks get "squished" down again by Earth's gravity [3, 16].

Changes to skeletal muscle and bones represent significant problems with long-term exposure to reduced gravity. In normal gravity they are stimulated to maintain their mass and strength largely by the constant stretching, deformation, and tension caused by activity and constant "pull" of Earth's gravity. It has been estimated that as much as 60 % of the muscles in the human body are used to resist gravity [3]. With that stress removed in microgravity, the weight-bearing muscles of the legs and back lose much of their mass (atrophy) Not surprisingly, muscles in the upper extremities, which are normally far less stressed by terrestrial gravity, show at most mild losses [2, 16, 29, 30]. Besides losing overall mass, muscle fibers do not contract as effectively in microgravity, and they are weakened and more easily damaged [31].

To help counteract these effects astronauts undergo extensive, rigorous exercise training preflight to help build up strength and endurance. This can include sustained aerobic exercise (e.g. jogging) to more than 50 % of a person's maximal level, total workouts of up to 1–1.5 h at least 3 times a week, and playing sports such as racquetball or baseball to improve agility [2]. Nonetheless, soon after entering microgravity some deconditioning still occurs. EVA in particular requires great physical effort, and the reduced exercise ability associated with muscle mass loss and other factors makes such strenuous activities more difficult [3].

Many methods have been used to deal with microgravity-induced losses in muscle mass and strength. The Soviet space program used electrical stimulation of muscles on some early flights. Muscle stimulation is still used on the ISS for research purposes [3]. On long-duration missions, including those on Salyut space stations, cosmonauts wore a "Penguin" suit fitted with elastic bands, straps, and buckles that offered resistance to bending and stretching. The idea was that the wearer would put mild stress on muscles during movement, thus providing resistive exercise during normal activities [3, 20, 21].

However, the main method used to reduce muscle mass loss and deconditioning in space is a vigorous exercise program [2, 3]. During long-duration missions astronauts perform both isotonic (aerobic) exercise (e.g. using a treadmill with the astronaut kept on it using devices such as bungee cords, or an exercise cycle) and isometric exercise (weightlifting-type exercises against a resistance). On the ISS the total amount of time devoted to exercise is about 2.5 h in a day. This includes the time needed to prepare for it (e.g. changing into appropriate clothing) and cleaning up afterwards, with about 1.5 h actually spent exercising [3]. About 4–6 such sessions are performed each week. Typically most of the session is devoted to isotonic exercise, with resistive exercises having a smaller but important role.

However, despite vigorous exercise programs cosmonauts on Mir for over a year showed a 20 % decrease in leg circumference associated with muscle atrophy. Work capacity of calf muscles decreased fourfold. Total muscle mass stabilized in 6 months at around 80–85 % of preflight value [2, 3, 29]. Typical current exercise programs show similar results.

After returning to Earth astronauts usually experience muscle weakness and soreness, particularly in the lower back and lower extremities [8]. These symptoms may take months to improve despite exercise. With time and activity, muscle mass can return to preflight levels. Magnetic resonance imaging (see Chap. 8) can be used to quantitatively assess muscle mass postflight and evaluate its recovery [2].

Reduced gravity also has serious effects on bones. The skeleton provides support for the body, protects internal organs and the blood cell-forming system in bone marrow, and is a reservoir for calcium. The latter is essential for muscle contraction, regulation of heart rhythm and contraction, blood clotting, and neural and hormonal functions. Microgravity causes demineralization of bone, decreased skeletal mass, and calcium loss. On the 84-day Skylab-4 mission urinary excretion of calcium increased over the first month, then stabilized at a level high enough to increase the risk of developing kidney stones. Fecal excretion of calcium increased linearly without leveling off, consistent with continuing loss of calcium from the skeleton [32].

This bone loss may be due to many factors. Bones in the lower back, pelvis, and legs are particularly affected by the loss of normal gravity-induced stresses on them. Activity of osteoclasts (cells that reabsorb bone) may increase, while that of osteoblasts (cells that form bone matrix) decreases. The net effect is more brittle bones and an increased risk of fractures, especially on return to Earth [3, 29, 32–37]. These problems are similar to those seen in osteoporosis, a disease that is particularly common in older women and is also associated with decreased bone mass, with annual losses as high as 1–5 % in postmenopausal women [2, 38].

In addition to maintaining muscle mass and conditioning, exercise programs in space help reduce loss of bone mass. With earlier exercise regimens the degree of loss in bone mass averaged about 15–20 % in missions of 6 months or more, similar to levels of muscle mass loss. On long duration missions overall bone loss has averaged about 1–2 % per month [3, 39]. As expected, weight-bearing bones like the calcaneus (heel bone) and other bones in the legs showed a greater average percentage of loss, while bones in the upper extremities demonstrated little or no loss. Bone mass loss has been found to average about 1.0–1.5 % per month in the hip and lower back (lumbar spine). After returning to Earth's gravity bone mass increases. However, it may not return to a normal level in the bones most affected or overall even years later [3, 29, 40].

These levels of bone mass loss are still better than the 30–40 % losses in both muscle and bone masses found in chronically bedridden individuals on Earth who cannot exercise. Studies using subjects placed at bed rest for prolonged periods have shown loss of bone mass from normally weight-bearing bones, just as is seen in space [33]. However, under worst-case circumstances, such as an astronaut injured too severely to exercise and too far away (e.g. en route to Mars) for timely return to Earth, levels of bone loss could potentially reach 50 % or greater. A reduction of bone density to less than 60 % of normal markedly increases the risk of fractures. Moreover, it is not known how effective vigorous exercise will be for missions lasting much more than a year in microgravity, since the longest time any person has been in space continuously is currently 438 days. Astronauts living in the greater gravity of the Moon (0.17 g) or Mars (0.38 g) would presumably experience smaller muscle and bone losses, but how much less is unknown.

Astronauts have their skeletal status checked before and after flight using a test called "dual-energy x-ray absorptiometry." It involves directing a pair of x-ray beams with different energies at a person's bones. Bone mineral density can then be measured by determining how much each x-ray beam is absorbed by those bones. Both active and retired astronauts are also followed every year with bone density studies.

New methods to reduce bone loss have been successful on the ISS. The Advanced Resistive Exercise Device (ARED) arrived there in late 2008 [3, 41–46]. Unlike the older Intermittent Resistive Exercise Device (IRED), ARED allows astronauts to use higher levels of resistive exercise. It also specifically targets bones that are most susceptible to loss in mass and density, such as those in the lower extremities. ARED uses vacuum cylinders and inertial flywheels to simulate in microgravity the constant mass and inertia of free weights used for exercise on Earth.

Recent reports indicate that use of ARED is associated with a degree of bone loss much lower than that previously seen, with overall bone mass remaining close to preflight levels in both men and women. In some astronauts this significant reduction in bone loss may have been due to using both ARED and taking medications of a class called "bisphosphonates" (also called "diphosphonates") that can be prescribed to prevent osteoporosis here on Earth. However, since the use of resistive exercises in space has been associated with development of mild back pain and musculoskeletal injuries, including possible intervertebral disk herniation, ARED might also potentially increase the risk for developing these problems [44, 47].

ARED might also be helpful for reducing the usual microgravity-induced loss in muscle mass and conditioning. Thus, as an alternative to the "standard" exercise regimen on the ISS that emphasizes aerobic exercise on a treadmill or stationary cycle, astronauts can now follow one involving at least 30 min of that type of exercise daily as well as 90 min using ARED 3–6 days per week. Unfortunately ARED is too large and massive for any craft going to the Moon or Mars in the foreseeable future. Nonetheless, its success in dramatically reducing bone loss indicates that this is a manageable problem.

Besides exercise programs, prophylactic medical measures would also have to be considered for long-duration missions such as to Mars. These might include not sending astronauts with conditions such as increased urinary excretion of calcium (thus, among other things, predisposing them to formation of kidney stones), a prior history of kidney stones, or known osteoporosis. Diets low in salt and high in both calcium as well as proteins derived from vegetable rather than animal sources can also promote bone health [2]. Supplemental Vitamin D may also be useful. During a long-duration mission periodic monitoring of urinary excretion and blood levels of calcium might also help guide the use of exercise and other countermeasures.

While the immediate focus on reducing bone and muscle loss is on the risks of fractures and other problems during long-duration missions, skeletal loss in particular remains an important consideration regarding an astronaut's health after return to Earth. That goal is to reduce future risks of fractures, osteoporosis, and other long-term adverse effects. None of these problems represent absolute contraindications to long stays in space, but they do need to be addressed to maximize astronauts' health during and after flight.

4.5 Effects on Vision

A newly recognized hazard of microgravity is that it can be associated with long-term changes in vision. One study surveyed 300 astronauts and found that 29 % of those on short-term missions and 60 % on long-term ones reported problems with near- or far-sightedness (primarily the latter). In some cases these visual changes have persisted for years after flight, and they may be permanent [3, 48]. Determining the reasons for and management of these problems is an active area of research.

Potential causes for these problems include direct damage to the retina (the light-sensitive tissue at the back of the eye), increased pressure within the eyes, and mildly increased pressure within the head ("intracranial pressure") that then produces swelling ("papilledema") where the optic nerve enters each eye. Some of these effects may be due to factors such as microgravity-induced fluid shifts from the lower to upper body, mildly increased levels of carbon dioxide in the spacecraft's atmosphere affecting arterial blood flow within the brain, or perhaps even alterations in metabolism involving vitamins such as B_{12} [49]. Resistive exercise, such as with ARED, might also increase the risk of developing impaired vision due to an increase in blood pressure and perhaps intracranial pressure [46].

Reduced outflow of venous blood from the head may also contribute to increased intracranial pressure and secondary effects on the eyes. In 1 g the internal jugular veins (described in Chap. 2) are primarily responsible for venous drainage from the brain when a person is supine, while the paired vertebral veins (running through the back of the neck) are mainly involved when an individual is upright. In microgravity, with the body no longer being "upright" as defined by a gravitational gradient, the internal jugular veins may assume more importance for this venous drainage and the vertebral arteries less. However, when the internal jugular veins become larger as more blood tries to pass through them they may be partially compressed by muscles in the neck. This partial obstruction to venous outflow could then contribute to decreased outflow of both cerebrospinal fluid from the brain and clear fluid (aqueous humor) present in the front of the eye, ultimately leading to increased pressures within both the brain and the eyes [48].

4.6 Gastrointestinal Effects

Microgravity's effects on the esophagus, stomach, intestines, and others organs of the gastrointestinal system are, except for the vomiting associated with SMS, overall relatively mild. For example, astronauts tend to have reduced appetites independent of SMS, partly due to decreased taste sensation associated with mild swelling and other effects on the tongue. On Earth swallowing is somewhat easier when a person is in a standing or sitting position instead of being supine, with microgravity in some ways simulating the latter.

Movement of air out of the gastrointestinal tract, either via eructation (the fancy medical term for belching) or passing of flatus, is more likely to occur in an uncontrolled fashion in microgravity. Astronauts may experience "heartburn" due to acid in the stomach refluxing into the esophagus or "water brash" as a burning irritation if stomach acid and swallowed saliva make it all the way up to the throat and mouth. Another unglamorous body function, constipation, may be increased in microgravity. Emptying and motility of the stomach are slowed, and the liver and pancreas become mildly engorged due to the fluid shifts described previously. However, these are minor effects and, overall, digestion of food seems at most little affected by being in space.

Microgravity also makes release and handling of solid waste matter more challenging. The ISS has two toilets that, like the ones on the Space Shuttle and Mir, use negative pressure to help remove feces from the immediate area for processing (e.g. by vacuum drying), storage, and eventual disposal (such as by bringing it to Earth). These toilets include foot and waist restraints as well as handholds. Astronauts and cosmonauts on prior spacecraft either had to "hold it" if the mission was short enough or employ "containment systems" that involved direct application of colostomy-type bags and similar low-tech equipment to the vicinity of the anus. Prior to launch, bowel cleansing regimens and diets designed to reduce production of feces (e.g. ones low in fiber and milk products) can be employed as ancillary measures.

4.7 Genitourinary and Endocrine Effects

Kidney ("renal") function itself seems to be little affected by microgravity. However, what the kidneys do to help maintain fluid and electrolyte balances is affected by changes elsewhere in the astronaut's body, including the reduced blood volume and increased calcium release from bone associated with microgravity. Both of those latter factors are associated with an increased risk of developing kidney stones.

Stones can form in a kidney and stay there (nephrolithiasis) or travel down the tube leading to the bladder, the ureter. Even a stone as small as 2–3 mm in diameter could block off the ureter at its junction with the bladder. This "urolithiasis" typically causes incapacitating pain, nausea, and vomiting.[2] If this obstruction involved the right ureter, it could produce pain in the right lower quadrant of the abdomen and raise the possibility of acute appendicitis. A common method of assessing for the presence of a ureteral stone and its effects (e.g. leakage of urine near the obstructed kidney) is a computed tomography (CT) scan of the abdomen—a test not available on the ISS, much less a mission to the Moon or Mars. A standard x-ray is much less sensitive. An ultrasound scan may represent a good alternative to a CT scan [50]. In any event, while the ISS does not have an x-ray machine, it does have an ultrasound device [3].

So far there has been at least one probable incident of a ureteral stone occurring in space, in a cosmonaut on a long-duration mission, along with several that have been reported postflight after short-duration missions. However, in the context of a Mars mission, an obstructing stone that was too large to pass into the bladder on its own might ultimately cause a urinary tract infection (UTI). In a worst case scenario the latter could, without adequate medical or other treatment, lead to "sepsis," a serious complications associated with a fall in blood pressure, damage to multiple body organs, and possibly death.

Urine collection in microgravity is not as challenging as solid waste management but can still be problematic. A male astronaut can urinate into the funnel-shaped end of a hose connected to a space toilet system that uses negative pressure to pull urine into it. Less sophisticated techniques include diapers (still routinely used as part of EVAs) and, like those used in Gemini flights, a tube and bag connected to a male astronaut by a condom-like attachment (a system that is not a viable option for female astronauts). The generic option is to use a space toilet in a conventional sitting position.

---

[2] Unfortunately I can attest to the development and severity of these symptoms with urolithiasis based on personal experience.

Some hormones produced by the body's endocrine system show mild changes while in microgravity. There may be a small reduction in blood levels of some catecholamines such as norepinephrine, while others (e.g. epinephrine) remain essentially unchanged. Levels of free T_3, a thyroid hormone, may decrease while that of cortisol and aldosterone, produced by the adrenal glands, can increase [37, 51]. A mild degree of insulin resistance, a condition associated with type 2 diabetes mellitus, can develop during flight. Fortunately, however, all these effects appear to be of little if any significance and resolve after return from space [9].

4.8 Pulmonary Effects

Microgravity has only mild effects on the lungs [3, 9]. When a person stands upright in Earth's gravity, blood flow to the upper (apical) regions of the lungs is less than at their bases. This leads to a "ventilation-perfusion mismatch" at each lung's apex, where there is an "excess" of air relative to the amount of blood available for gas exchange (adding oxygen and removing carbon dioxide) [52]. In microgravity blood flow is distributed more "evenly" throughout the lungs, thus increasing the efficiency of gas exchange at each apex. However, microgravity's other effects include that the amount of air moved in or out of the lungs during a normal breath, the "tidal volume," is mildly decreased, the rate of breathing may mildly increase, and total volume of air in the lungs can fall slightly. The net result of all these changes is that microgravity normally does not significantly affect overall respiratory function.

The respiratory tract may also be affected indirectly by floating particulate matter in the spacecraft. Inhaling tiny particles in the form of dust or those released by various materials can, for example, irritate the throat and cause coughing. This can be worsened by activities such as exercise that increase the rate and depth of breathing. This problem is reduced by using food and other substances with a low potential to release particles, as well as environmental control measures such as forced air circulation and air filters.

4.9 Dealing with Microgravity

Real-world techniques for dealing with microgravity involve either better adapting the human body to it or using external forces to simulate some fraction of Earth's gravity. The former technique currently involves the "simple" countermeasures described previously, including exercise programs, appropriate diets, and medications to help treat symptoms and physiological effects of microgravity. However, these methods deal only indirectly with microgravity and at best partially alleviate some of its adverse effects.

More advanced techniques such as genetic engineering might be used to directly change the human body itself to make it "healthier" and function better in microgravity. The potential downside of doing that is, unless they could later be reversed, any major modifications might make it more difficult to work in "normal" gravity after returning to Earth. An excellent science fiction example of this concept occurs in Lois McMaster Bujold's novel *Falling Free* (1988). It describes "Quaddies," humans with bodies genetically engineered to have their legs replaced by a second pair of functional arms. This modification allows them to work better in microgravity, where legs are of limited usefulness and arms are much more important for performing activities.[3]

The other alternative is to effectively "bring gravity" with us into space by providing "artificial" gravity. Current spacecraft already transplant essential components of a terrestrial environment (e.g. a breathable atmosphere and comfortable temperatures) into space. Since gravity is a force associated with mass (e.g. a person's body) and the acceleration acting on it, techniques that produce acceleration could recreate the effects of gravity.

Astronauts experience "linear acceleration" during launch and reentry. As described in Chap. 3 the effective g-forces acting on them can be much greater than that produced on Earth's surface. The occupants of a spacecraft that was constantly increasing its velocity in a straight line would also feel a "force" pushing them back in the opposite direction from that of the craft's travel. This is the same sensation of being pushed deeper into the seat that a passenger in a car experiences when the driver presses down hard on the accelerator pedal.

However, there are serious limitations with using this technique as a constant means of "artificial" gravity. Something would have to provide the force and energy (the latter being related to the work performed on the spacecraft, which is the force acting on it multiplied by the distance it moves) needed to maintain at least a constant acceleration. On-board chemical fuel is obviously finite, and other energy sources on the craft—e.g. nuclear fission, fusion, or (as in *Star Trek*'s starships) matter-antimatter reactions—have varying degrees of efficiency regarding creation of such energy as well as significant technical issues with safely implementing them.

For example, matter-antimatter reactions result in 100 % conversion of matter into energy. However, there are formidable challenges to creating and storing enough antimatter to provide a spacecraft's fuel so that it produces

---

[3] However, as will be described in Chap. 12, there are significant genetic, anatomic, and other issues with actually making this change.

only desired effects (controlled power and acceleration for a starship) rather than undesirable ones (uncontrolled release of energy that makes the spacecraft explode or producing enough radiation to harm the crew). While the principles of physics involved are well established, the rate-limiting factor for implementing these "exotic" fuels is the technical complexity involved that, overall, far exceeds present-day capabilities and plans.

"Living off the land" by using sources of energy in space itself—e.g. the pressure produced by the Sun's light itself against a solar sail, solar panels, or even collecting hydrogen in space itself for fusion engines as envisioned by the Bussard ramjet concept—is another possibility. However, any techniques that use sunlight as an ongoing source of energy to produce acceleration would be less effective at greater distances from the Sun. Moreover, the degree of acceleration provided by "solar" techniques is overall very small. It would be enough to provide a craft's propulsion over a long period of time, with the rate of change in its velocity dependent on factors such as its mass, the size of the solar sail, etc. However, the acceleration would still be far too low to produce artificial gravity at a significant fraction of 1 g for any crew onboard. A ground- or space-based laser could be used to provide additional energy, particularly in the outer Solar System and beyond, but obviously has its own technical challenges. Likewise using a Bussard ramjet has issues with gathering and utilizing enough hydrogen in space to serve as fuel for its (still hypothetical) fusion engines.

Even if (and hopefully when) such methods become feasible and practical, the uniform sensation of artificial gravity they would produce would be based on the spacecraft traveling with constant acceleration and in only one direction. Change the degree of acceleration (e.g. decreasing it as the craft approaches its destination, such as a distant star system) or the ship's direction (which would require another appropriate force to act on the craft), and the strength and direction of an artificial gravity field would also change. This would obviously have to be done in a controlled fashion to prevent injury to the craft's occupants, such as from dangerously rapid deceleration (spacecraft are not typically equipped with internal airbags) or being flung against a wall.

Also, a craft orbiting the Earth or a space habitat in a location such as a Lagrange point[4] is not, in fact, moving overall in a straight line. Instead of linear acceleration, producing "artificial gravity" on it would require rotating the ship or habitat, which would thus have an "angular velocity"—its speed of rotation and instantaneous direction of travel, the latter constantly changing

---

[4] A Lagrange point is one of five positions in space (L_1 through L_5) where, in oversimplified terms, the gravitational attraction of two bodies (e.g. the Sun and the Earth) effectively "balance" each other. An object such as a space habitat or satellite could thus "stay put" in that position in space. The actual concepts and effects are more complex than this, but this will serve as a very rough approximation of the idea.

in an overall circular path. This is associated with the outward "centrifugal force" that anyone on a fast-rotating merry-go-round or similar amusement park ride experiences. Once enough force is applied to an object in space, it will continue to rotate unless another force acts on it (e.g. friction associated with atmospheric drag, retro rockets, etc.).

There are three basic ways of producing artificial gravity via rotation [53]. The most obvious is to rotate the entire craft or structure, a method proposed as early as the first decade of the twentieth century by the visionary Russian "rocket scientist" Konstantin Tsiolkovsky [16, 54]. A popularized version of this technique was the "wheel in space" proposed by Wernher von Braun and Willy Ley in the early 1950s, subsequently depicted in varying guises in such visual media as the television series *Men into Space* (1959), the *Doctor Who* serial "The Wheel in Space" (1968), and cinematic works such as *Conquest of Space* (1955) and *2001: A Space Odyssey* (1968).

A second technique, potentially applicable for a trip to Mars, involves connecting the crewed part of the craft to another section of the original rocket having similar mass via a tether. The two sections would then rotate around their common center of mass, located along the tether. This would, in effect, put the crew inside a (Trojan?) horse of an interplanetary merry-go-round. In fact, during the Gemini 11 flight in 1966 the crewed capsule rendezvoused with an Agena rocket booster and, as part of an EVA, attached a 30-meter tether between the two craft. After the Gemini capsule fired its thrusters the two vessels were set in slow rotation around their mutual center of gravity along the tether, generating a minimal amount of artificial gravity (about 0.0005 g) [3, 53].

A third method would require a larger craft. The crewed part of *Discovery One* in *2001: A Space Odyssey* includes a continuously rotating centrifuge that produces a low degree of artificial gravity. A short-term method for providing artificial gravity is a small centrifuge or other type of rotating device within a craft that can subject individual astronauts to loads up to 1 g or even more. Such a rotating machine was used as part of Neurolab experiments on a Space Shuttle flight in 1998, with subjects subjected to forces equivalent to up to 1 g for as long as 7 min at a time [54]. The ISS does not have facilities for conducting long-term centrifuge studies on human subjects.

The primary limiting factors shared by all three of these methods include how fast the system must spin to produce a desired degree of effective artificial gravity and how far the crew must be from the center of rotation. Put more technically, these factors are respectively the system's angular velocity and radius from the system's center of gravity (approximately the same as its center of mass) to a particular point on a hypothetical or real (e.g. the wall of a cylinder) circle surrounding it. A "space wheel" is circular but is also three-dimensional in a doughnut-shaped configuration called a "torus." Depending on how "thick" the doughnut portion (the part outside the hole) is, the actual radius/distance from its center of gravity will vary depending on where a crewmember is located within that structure.

The degree of "artificial gravity" produced at a particular point in a rotating structure is proportional to the square of its angular velocity (how fast it is rotating) and directly proportional to its distance from the center of gravity. The von Braun "wheel in the sky" was conceived as having a radius from its approximate center of gravity to its outer edge of approximately 125 ft. (about 38 m). Producing 1 g at that location would have required a rotation rate of a little over 5 rotations per minute (rpm). Much larger structures, such as the proposed cylindrical Kalpana space habitat with an outermost radius of about 250 m, would only need to be rotated at about 2 rpm [55]. Conversely, a cylindrical spacecraft with a radius of 4 m would require a rotation rate of about 15 rpm to produce 1 g at its wall.

Both physical and neurovestibular factors effectively limit generation of artificial gravity at 1 g as a viable option to very large structures, e.g. the Kalpana space habitat or, in science fiction, literally astronomically sized ones such as depicted in Larry Niven's *Ringworld* (1970). Because the degree of inward centripetal force that generates the "gravity" in a rotating structure is directly proportional to the distance (radius) from the center of gravity, moving closer to the latter will produce a lower perceived level of gravity. Thus, an astronaut positioned horizontally with feet "standing" on the outer wall of the rotating craft and head pointed toward the center of gravity will have a higher level of "gravity" at the former and lower one at the latter.

For a large enough structure this gradient may be very small, e.g. about 2 % for a theoretical astronaut about 2 m in height in a very large structure with a radius of 100 m. This gradient is small enough that the astronaut would be unlikely to perceive it. However, for a more realistic radius (based on current technology) of 10 meters the gradient increases to 20 %, which could be noticeable [53]. The gravitational gradient would increase quickly in spacecraft with still smaller radii.

Running along the interior rim of a circular rotating spacecraft would also affect the degree of artificial gravity. Doing this in the direction of the rotation would increase the latter, while jogging in the opposite direction would decrease the level of artificial gravity.

Besides the centripetal force in a rotating structure that gives the sensation of gravity to someone at rest, another factor is the "Coriolis force." If an astronaut in a rotating structure were to try to move straight outward from a point closer to the center of gravity to one farther out, the Coriolis force acts perpendicular to that individual. The net effect is that the astronaut would feel "pushed" to one side in the direction of rotation of the structure. Conversely, if the astronaut were to move straight toward the center of gravity the sideways "push" would be opposite to the direction of the structure's rotation. The magnitude of this Coriolis force is directly proportional to the angular velocity of the rotating structure and the velocity of movement outward or inward along a line extending out from the center of gravity [53].

A person can tolerate and adapt to low levels of continuous rotation, head-to-foot gravity gradients, and Coriolis force. However, past certain limits movement becomes problematic, and the inner ear's neurovestibular system, responsible for our sense of balance, is affected enough to produce symptoms emulating motion sickness such as dizziness, nausea, vomiting, etc. With a rotation rate of several times per minute or more, when an astronaut simply turns his or her head stationary objects would appear to rotate and continue to do so even after head movement ceased [53].

Likewise, an astronaut standing on the outer wall of a rotating spacecraft and lifting an object "up" toward the center of gravity could, if the gravity gradient were high enough (e.g. fast rotation rate and/or relatively small radius of the craft) feel it become "lighter," potentially making the object more difficult to handle. Only movement parallel to the axis of rotation (think of a vertical line passing through the center of a doughnut perpendicular to the circle formed by the "dough" part) would be free of these effects. Otherwise, although they might be able to adapt to some degree, astronauts in a rotating ISS module (typical radius of about 2 m) or spacecraft of comparable diameter would be constantly experiencing these disorienting effects if the rate of rotation were enough to produce a significant amount of artificial gravity.

It is thought that a constant rotation of 2 rpm would be well tolerated. However, as noted previously it would take a structure of Kalpana size to achieve a maximum of about 1 g at its outer wall with that rate. For the roughly 2 m radius of an average ISS module, a rotation rate of about 15 rpm would produce a maximum of about 0.5 g at its wall—and this value would decrease rapidly with minimal movement or positioning relative to the center of gravity. The maximum dimension (about 108 m) of the ISS includes its linear array of inhabited modules. If that string of modules were to be put into rotation end-to-end around the overall center of gravity (radius of 54 m), a 2 rpm rotation would still produce at most about 0.25 g at each end [53]. However, if this were done an astronaut moving from an end of the ISS through modules closer to its center would experience the gravitational gradients and Coriolis effects described previously.

The minimum level of sustained (or perhaps intermittent) artificial gravity needed to significantly reduce the most serious effects (e.g. loss of bone and muscle mass) of microgravity on the human body is also uncertain. Current thought is that, while a lunar-level gravity of about 0.17 g might be helpful for movement inside a large enough structure, it would probably still be associated with significant bone and muscle loss, fluid shifts, etc. A Martian-level gravity of about 0.38 g would be better, but how much so is unknown.

Besides the potentially beneficial and adverse effects of artificial gravity on the human body, there are also significant technical issues with employing it. These include problems with structural integrity (e.g. what if the tether connecting a crewed craft with a "counterweight" snapped) and the greater challenges of activities such as EVA on a rotating rather than merely orbiting ISS-sized structure [54].

In short, using "artificial gravity" via rotation in smaller spacecraft and structures may be a case of trading one set of potential problems for another. In science fiction stories a writer can decide that everything goes well—the tether doesn't break, humanity or an alien race creates a habitat large enough to minimize the aforementioned problems with gravity gradients, etc. Unfortunately, in real life the problems and risks associated with artificial gravity have to be balanced carefully against its benefits.

Finally, several other methods could also (at least theoretically) be used to produce or at least emulate gravity in space. Since gravitational force is proportional to mass, using a whole Earth-mass planet as a spaceship—perhaps Earth itself, as in Stanley Schmidt's *Lifeboat Earth* (1978),[5] or alien worlds in the *Fleet of Worlds* series[6] by Larry Niven and Edward M. Lerner—would provide 1 g of "all-natural" gravity and plenty of room for characters, colonists, etc. However, providing the means of propulsion for such a large "craft" using known principles of physics instead of the imaginative but speculative ones in the aforementioned novels would be problematic.

Extremely dense matter, such as that contained in a white dwarf star, neutron star, or a black hole of suitable mass, might also (at least in theory) be a means of providing gravity on a starship. One might envision such a highly advanced craft constructed as a series of two or more concentric hollow spheres around a solid, inner sphere of white dwarf or neutron star material. The innermost hollow sphere would thus become analogous to the surface of Earth, while the outer sphere could represent this synthetic world's "sky." Such a ship would be reminiscent of the hollowed-out asteroid/generation ship depicted in the *Star Trek* episode "For the World is Hollow and I Have Touched the Sky" (1968) or, more distantly, to the *Star Wars* series' Death Stars, but with its artificial gravity provided by at least tenuously "realistic"

---

[5] This work is a sequel to Dr. Schmidt's novel *The Sins of the Fathers* (1976), which provided the background for why Earth had to be made mobile.
[6] This set of five novels was first published between 2007 and 2012.

means rather than the intrinsically unrealistic (by known scientific principles) flip-a-switch variety.

Unfortunately this concept presents myriad practical issues. First and foremost is the obvious difficulty of acquiring enough white dwarf or neutron star material to provide gravity on such a craft. It would also require capabilities far beyond current human ones to place enough metal and other material in space to fabricate the spherical portions designed for human habitation. The amount of white dwarf and neutron star material needed to create 1 g at the "surface" of this hypothetical craft can also be estimated. A white dwarf is the remnant of a star that, while it was on the "main sequence" and using hydrogen as its fuel for fusion at its core, originally had a mass of about 0.8–8.0 that of our Sun.[7] The total mass of a white dwarf can be up to about 1.4 of a "solar mass"[8] (the "Chandrasekhar limit"), although one of 0.6 is most typical [56].

Creating a gravitational force of 1 g (an acceleration of roughly 9.81 m/s²) at the "floor" of our spherical starship could (in theory) be done by using a sufficient mass of equally spherical white dwarf material placed an appropriate distance away from that floor. By Newtonian physics one body (m) would accelerate toward another (M) due to gravity by the relationship $a_g = \dfrac{GM}{r^2}$, where a_g is the gravitational acceleration ("attraction") in m/sec², G is the gravitational constant 6.67×10^{-11} m³/kg sec², and r is the distance in meters between the two objects [1]. By the "shell theorem" a spherical body would act as if all its mass were concentrated at its center, so r would be the distance from its center to the object being attracted to it. Thus, plugging in Earth's mass of about 5.98×10^{24} kg and its radius of roughly 6.363×10^6 m into that equation gives an approximate value for a_g at its surface of around 9.81 m/sec².[9]

A white dwarf has a typical density (mass divided by volume) of 1×10^9 kg per cubic meter (kg/m³), compared to Earth's average (but by no means homogenous) density of 5.52×10^3 kg/m³ [56, 57]. Using that value for a white dwarf's density, the aforementioned values for Earth's mass and radius, and the equation for the volume V of a sphere ($V = \frac{4}{3}r^3$), a white dwarf with Earth's mass would have a calculated radius of a little less than 113 km. Thus, if a "ball" of white dwarf material with this mass were suspended at the exact center of a sphere whose surface was Earth's average radius (about 6.363×10^6 m) from the white dwarf's center, a person on the outside of that sphere would

---

[7] Our Sun itself will become a white dwarf in roughly six-and-a-half billion years.

[8] One solar mass is the mass of our Sun, or about 1.99×10^{30} kg.

[9] It should be noted, however, that Earth is not quite spherical, with a slightly larger radius at its equator than at the poles; its mass is not uniformly distributed; and its rotation also adds very slightly to the downward force experienced by a surface dweller. All these factors also contribute, if only very mildly, to the net force that "pulls" us toward the ground.

experience "normal gravity" with a "downward" acceleration of 9.81 m/sec². An intrepid astronaut could (theoretically) walk along the outer surface of that human-made sphere as easily as we walk across the floor of a room.

Some difficulties with this scenario include first harvesting that substantial amount of white dwarf material, carving it into an exactly spherical shape (any large "bumps" or "pits" would lead to significant variations in gravitational acceleration over them), and keeping its position fixed within that rather large outer sphere, whose radius would be the same as Earth's. However, since a_g is related not only to the mass of the white dwarf material but also to a person's distance r from its center, the size of our spaceship could be made much smaller by using a white dwarf sphere with far less mass and having the equally spherical craft's "floor" line the former's surface. Using the above equations, a sphere of white dwarf material with a radius of 35 m and associated mass of 1.796×10^{14} kg would create a gravitational acceleration of 9.78 m/sec² (minimally less than 1 g) at such a floor covering its surface.

As with "artificial gravity" produced by a rotating structure, the astronaut standing upright would experience a head-to-foot gradient, with gravitational acceleration at the head of about 8.75 m/sec²—a relatively mild difference of about 11 % between the "top" and "bottom" of the body. Thus, if (and a very great "if" it is!) this rather large amount of white dwarf material (total volume just under 180,000 m³) could be obtained, in theory it could provide artificial gravity for a starship. This would still, however, leave significant problems regarding how to actually propel that substantial mass between stars, cooling the white dwarf material down to a manageable level (the typical surface temperature in "full-sized" white dwarfs ranges from between about 4000–40,000 K—much more than enough to give our intrepid astronaut "walking" on one a "hot foot"), protecting the spaceship's inhabitants from high-energy electromagnetic radiation produced by the white dwarf, and carving it into a sphere [56].

What this "thought experiment" of using white dwarf material for artificial gravity illustrates is that there can be a huge gap between what is theoretically possible and the practical problems involved with actually doing it. Similar calculations using material from a neutron star (the remnants of stars with masses between about 8–25 times that of our Sun, with a typical density of around 1×10^{17} kg/m³) show that it would be even less suitable as the central "walk on" core of a starship, and the inner surface of the latter would have to be a considerable distance away from the very small but very massive neutron star material kept in place at its center. Thus, in spite of all of its own technical challenges, generating artificial gravity via rotation of a large enough craft or habitat is much "simpler."

Magnetism is also a potential attracting force that could be used to simulate gravity [53]. Water and other chemicals within the human body are weakly "diamagnetic" and can be attracted by a sufficiently strong magnetic field. Small, low-mass animals such as mice and frogs can be "levitated" in powerful magnetic fields that effectively negate the attractive force of gravity. However, doing this requires extremely strong fields produced by massive and power-hungry devices, including use of cryogenic superconducting materials, to generate them by current technology. The health effects of long-term exposure to those fields on humans are also uncertain. Also, such magnetic fields would much more powerfully attract strongly magnetic materials such as "ferromagnetic" metals, as well as interfere with electronic and other devices affected by magnetism (e.g. standard computer hard drives). This would obviously limit use of many materials on the craft.

A technique employed in early science fiction movies such as *Destination Moon* (1950) involves using "magnetic boots" to keep spacefarers' feet attached to their craft. This could indeed help prevent an astronaut from floating freely in microgravity under certain circumstances. Unfortunately the latter have extremely limited applicability to current spaceflight technology. For magnetic boots to keep an astronaut anchored to the walls of a spacecraft, the latter must have a sufficiently large mass of ferromagnetic material (e.g. certain types of stainless steel) for the magnets to attract them. Likewise those magnets in the boots must be powerful, able to keep a perhaps 80 kg or so human body from pulling away from its "anchor" against the craft.

While ISS modules and spacecraft contain some stainless steel, even if the latter were magnetic (not all stainless steel is) they also contain larger proportions of nonmagnetic materials such as aluminum and carbon fiber. These provide good structural strength and, because they are not as dense as iron, have less mass than a given volume of the latter. Current rocket technology can lift only limited amounts of mass into orbit and beyond for a given set of engines and quantity of fuel. Thus, if less of this finite mass needs to be devoted to the rocket components themselves, more can be allocated to crew, cargo, and equipment. Aluminum, titanium, and other metals "lighter" than iron have a clear advantage regarding that issue. Unfortunately, however, magnetic boots do not work on them.

Also, even if a person were wearing sufficiently strong magnetic boots pressed against the interior of a spacecraft's steel hull, any movement an astronaut makes with legs or upper body could have unhelpful results. For example, one or both boots could inadvertently come free, negating their primary potential benefit and, if both boots come free, possibly sending an astronaut into an unexpected trajectory toward a wall. Likewise, if the latter happened to a spacewalking astronaut treading across the ship's hull, that unfortunate individual could go sailing off into the void. Conversely, if the boots provided too strong an attachment the astronaut might accidentally "bob" too far forward, backwards, or sideways, thus putting perhaps injurious strains on the ankles.

The strength of the magnets would also have to be "fine-tuned" so that their attraction to the ship's hull was strong enough to keep them attached to it when needed but not so strong that the astronaut could not voluntarily lift them off or slide them along the "floor" when needed. This would in turn depend partly on the distance between the boot magnets and the spacecraft's wall—and if the latter were circular instead of flat, the boots would make only as much direct contact with it as the curve of its hull allowed.

Moreover, magnetic boots would provide little if any protection against the physiological effects on the human body in microgravity, such as fluid shifts or loss of bone and muscle mass. An astronaut might do a slight amount of extra exercise when moving the boots across or (hopefully one at a time) away from the ship's hull, but the net beneficial effect of this is questionable. Also, the volume inside current spacecraft such as a Soyuz capsule is very limited, with astronauts and cosmonauts spending much of their time secured in their seats. Likewise the ISS has foot restraints in its modules for when its occupants must secure themselves to a single spot to perform an experiment or other duty. In short, the major "advantage" of magnetic boots is that it allowed some early space-based movies and TV shows made on Earth to at least acknowledge the presence of microgravity without having to employ obvious "special effects" such as marionette-like rope systems to simulate it.

4.10 The Bottom Line

The adverse effects of microgravity on the human body represent significant but not insurmountable challenges to maintaining an astronaut's health during short- and especially long-term missions. SMS, EMS, as well as bone and muscle loss can now be at least adequately managed during current missions to LEO. Long-term effects of microgravity on astronauts' health (e.g. some degree of permanent bone loss and visual impairment), however, remain significant areas of concern. There are greater risks associated with long-duration missions to Mars and other destinations far from Earth, such as high rates of bone and muscle loss if an injured or ill astronaut were unable to exercise enough to counteract them. However, although all these are significant issues, none of them preclude a reasonably healthy and successful trip to and from Mars.

Unlike NASA and other space agencies, science fiction writers have many choices for dealing with microgravity in space-based stories. Even if they prove impossible in real life, "flip-a-switch" artificial gravity generators are, with plot-driven exceptions (e.g. an early scene aboard a Klingon ship in *Star Trek VI: The Undiscovered Country*), spectacularly effective and reliable in science fiction. Authors may choose to surround this scientifically "soft" element with more realistic "hard" ones to compensate for it, or simply employ it as a well-established convention.

Likewise, many of the less savory aspects of the human body's functions (particularly of the excretory variety) in microgravity can by tacit assent go unmentioned or be simply ignored in stories. Certainly a writer can, for true realism, mention the nasal congestion, nausea, bloating, facial puffiness, and other effects of SMS and fluid shifts. However, for story purposes this is not necessary.

Science fiction writers also have the luxury of introducing spacecraft and habitats large enough that the effects of microgravity can, as described previously, be reduced dramatically by rotating them at a slow rate. Letting technologically advanced aliens do that puts the onus of solving the enormous practical problems with constructing such artifacts on their shoulders (assuming they have shoulders) rather than on humanity's. Or our distant descendants can do it using techniques (e.g. nanobots—see Chap. 11) yet to be discovered and, more importantly, paid for with their money rather than ours. Starting the story *after* all the tremendous resources and work need to make those craft and habitats is done certainly simplifies the author's job.

References

1. Walker J. Gravitation. In: Walker J, editor. Halliday & Resnick. Fundamentals of physics. 9th ed. Hoboken: Wiley; 2011. pp. 330–58.
2. Buckey JC. Space physiology. New York: Oxford University Press; 2006.
3. Clément G. Fundamentals of space medicine. 2nd ed. New York: Springer; 2011.
4. Ortega H, Harm D. Space and entry motion sickness. In: Barratt M, Pool SL, editors. Principles of clinical medicine for space flight. New York: Springer; 2008. pp. 211–22.
5. Reschke M, Harm D, Parker D, Sandoz G, Homick J, Vancerploeg J. Neurophysiologic aspects: space motion sickness. In: Nicogossian AE, Huntoon CL, Pool SL, editors. Space physiology and medicine. 3rd ed. Malvern: Lea & Febiger; 1993. pp. 228–60.
6. Stewart L, Trunkey D, Rebagliati S. Emergency medicine in space. J Emerg Med. 2007;32(1):45–54.
7. Thornton WE, Bonato F. Space motion sickness and motion sickness: symptoms and etiology. Aviat Space Environ Med. 2013;84(7):716–21.

8. Williams D, Kuipers A, Mukai C, Thirsk R. Acclimation during space flight: effects on human physiology. Can Med Assoc J. 2009;180:1317–23.

9. Baker E, Barratt M, Wear M. Human response to space flight. In: Barratt M, Pool SL, editors. Principles of clinical medicine for space flight. New York: Springer; 2008. pp. 27–57.

10. Reschke M, Bloomberg J, Paloski W, Harm D, Parker D. Neurophysiologic aspects: sensory and sensory-motor function. In: Nicogossian AE, Huntoon CL, Pool SL, editors. Space physiology and medicine. 3rd ed. Malvern: Lea & Febiger; 1993. pp. 261–85.

11. Oman C, Howard I, Smith T, Beall A, Natapoff A, Zacher J, et al. The role of visual cues in microgravity spatial orientation. http://ntrs.nasa.gov/archive/nasa/casi.ntrs.nasa.gov/20030068201.pdf.

12. Oman C. Human visual orientation in weightlessness. Levels of perception. New York: Springer; 2003. pp. 375–98.

13. Clark J, Bacal K. Neurologic concerns. In: Barratt M, Pool SL, editors. Principles of clinical medicine for space flight. New York: Springer; 2008. pp. 361–80.

14. Parmet A, Gillingham K. Spatial orientation. In: DeHart RL, Davis JR, editors. Fundamentals of aerospace medicine. 3rd ed. Philadelphia: Lippincott Williams & Wilkins; 2002. pp. 184–244.

15. Payne MW, Williams DR, Trudel G. Space flight rehabilitation. Am J Phys Med Rehabil. 2007;86(7):583–91.

16. Planel H. Space and Life. An introduction to space biology and medicine. Boca Raton: CRC Press; 2004.

17. Hamilton D. Cardiovascular disorders. In: Barratt M, Pool SL, editors. Principles of clinical medicine for space flight. New York: Springer; 2008. pp. 317–59.

18. Convertino V. Consequences of cardiovascular adaptation to spaceflight: implications for the use of phamacological countermeasures. Gravit Space Biol. 2005;18:59–70.

19. Antonutto G, Clément G, Ferretti G, Linnarsson D, Traon A, Di Prampero P. Physiological targets of artificial gravity: the cardiovascular system. In: Clément G, Bukley A, editors. Artificial gravity. New York: Microcosm Press and Springer; 2007. pp. 137–62.

20. Aubert A, Beckers F, Verheyden B. Cardiovascular function and basics of physiology in microgravity. Acta Cardiologica. 2005;60:129–51.

21. Nicogossian AE, Sawin C, Grigoriev AI. Countermeasures to space deconditioning. In: Nicogossian AE, Huntoon CL, Pool SL, editors. Space medicine and physiology. Malvern: Lea & Febiger; 1993. pp. 447–67.

22. Burton R, Whinnery J. Biodynamics: sustained acceleration. In: DeHart RL, Davis JR, editors. Fundamentals of aerospace medicine. 3rd ed. Philadelphia: Lippincott Williams & Wilkins; 2002. pp. 122–53.

23. Perhonen M, Franco F, Lane L, Buckey J, Blomqvist C, Zerwekh J, et al. Cardiac atrophy after bed rest and spaceflight. J Appl Physiol. 2001;91:645–53.

24. Kirkpatrick A, Campbell M, Jones J, Broderick T, Ball C, McBeth P, et al. Extraterrestrial hemorrhage control: terrestrial developments in technique, technology, and philosophy with applicability to traumatic hemorrhage control in long-duration spaceflight. J Am College Surg. 2005;200(1):64–76.

25. Planel H. Cosmic radiation. Space and life. An introduction to space biology and medicine. Boca Raton: CRC Press; 2004. pp. 121–40.

26. Huntoon CL, Whitson P, Sams C. Hematologic and immunologic functions. In: Nicogossian AE, Huntoon CL, Pool SL, editors. Space physiology and medicine. 3rd ed. Malvern: Lea & Febiger; 1993. pp. 351–62.

27. Kalb R, Solomon D. Space exploration, Mars, and the nervous system. Arch Neurol. 2007;64(April):485–90.

28. Sayson JV, Hargens AR. Pathophysiology of low back pain during exposure to microgravity. Aviat Space Environ Med. 2008;79(4):365–73.

29. Shackelford L. Musculoskeletal response to space flight. In: Barratt M, Pool SL, editors. Principles of clinical medicine for space flight. New York: Springer; 2008. pp. 293–306.

30. Jaweed M. Muscle structure and function. In: Nicogossian AE, Huntoon CL, Pool SL, editors. Space physiology and medicine. 3rd ed. Malvern: Lea & Febiger; 1993. pp. 317–26.

31. Narici M, Zange J, Di Prampero P. Physiological targets of artificial gravity: the neuromuscular system. In: Clément G, Bukley A, editors. Artificial gravity. New York: Microcosm Press and Springer; 2007. pp. 163–90.

32. Schneider B, LeBlanc A, Taggart L. Bone and mineral metabolism. In: Nicogossian AE, Huntoon CL, Pool SL, editors. Space physiology and medicine. 3rd ed. Malvern: Lea & Febiger; 1993. pp. 327–33.

33. LeBlanc A, Spector E, Evans H, Sibonga J. Skeletal responses to space flight and the bed rest analog: a review. J Musculoskelt Neuronal Interact. 2007;7:33–47.

34. Cavanagh P, Licata A, Rice A. Exercise and pharmacological countermeasures for bone loss during long-duration space flight. Gravit Space Biol. 2005;18(June):39–58.

35. Tamma R, Colaianni G, Camerino C, Di Benedetto A, Greco G, Strippoli M, et al. Microgravity during spaceflight directly affects in vitro osteoclastogenesis and bone resorption. FASEB J. 2009;23(8):2549–54.

36. Orwoll E, Adler R, Amin S, Binkley N, Lewiecki E, Petak S, et al. Skeletal health in long-duration astronauts: nature, assessment, and management recommendations from the NASA bone summit. JMBR. 2013;28(June):1243–55.

37. Strollo F. Chapter 4. Hormonal changes in humans during spaceflight. Advances in Space Biology and Medicine. 1999; 7:99–129.

38. Finkelstein JS, Brockwell SE, Mehta V, Greendale GA, Sowers MR, Ettinger B, et al. Bone mineral density changes during the menopause transition in a multi-ethnic cohort of women. J Clin Endocrinol Metab. 2008;93(3):861–8.

39. Canan J. Health effects of human spaceflight. Aerosp Am. 2013(September):24–30.

40. Qin Y. Challenges to the musculoskeleton during a journey to Mars: assessment and counter measures. J Cosmol. 2010;12:3778–80.
41. Advanced Resistive Exercise Device. 2013. http://www.nasa.gov/mission_pages/station/research/experiments/1001_prt.html. Accessed 15 April 2015.
42. Lang T. Bone Loss in Long-Duration Spaceflight: Measurements and Countermeasures. 2013. http://astronautical.org/sites/default/files/issrdc/2013/issrdc_2013-07-16-0945_lang.pdf. Accessed 15 April 2015.
43. Smith SM, Heer MA, Shackelford LC, Sibonga JD, Ploutz-Snyder L, Zwart SR. Benefits for bone from resistance exercise and nutrition in long-duration spaceflight: Evidence from biochemistry and densitometry. J Bone Miner Res. 2012;27(9):1896–906.
44. LeBlanc A, Matsumoto T, Jones J, Shapiro J, Lang T, Shackelford L, et al. Bisphosphonates as a supplement to exercise to protect bone during long-duration spaceflight. Osteoporos Int. 2013;24(7):2105–14.
45. Nagaraja M, Jo H. The role of mechanical stimulation in recovery of bone loss—high versus low magnitude and frequency of force. Life. 2014;4(2):117–30.
46. Hargens AR, Bhattacharya R, Schneider SM. Space physiology VI: exercise, artificial gravity, and countermeasure development for prolonged space flight. Eur J Appl Physiol. 2013;113(9):2183–92.
47. Scheuring RA, Mathers CH, Jones JA, Wear ML. Musculoskeletal injuries and minor trauma in space: incidence and injury mechanisms in U.S. astronauts. Aviat Space Environ Med. 2009;80(2):117–24.
48. Wiener TC. Space obstructive syndrome: intracranial hypertension, intraocular pressure, and papilledema in space. Aviat Space Environ Med. 2012;83(1):64–6.
49. Zwart S, Gibson C, Mader T, Ericson K, Ploutz-Snyder R, Heer M, et al. Vision changes after spaceflight are related to alterations in folate- and vitamin B-12-dependent one-carbon metabolism. J Nutr. 2012;142:427–31.
50. Smith-Bindman R, Aubin C, Bailitz J, Bengiamin RN, Camargo CA, Corbo J, et al. Ultrasonography versus computed tomography for suspected nephrolithiasis. N Engl J Med. 2014;371(12):1100–10.
51. Tou J. Models to study gravitational biology of mammalian reproduction. Biol Reprod. 2002;67(6):1681–7.
52. Pickard J. The atmosphere and respiration. In: DeHart RL, Davis JR, editors. Fundamentals of aerospace medicine. 3rd ed. Philadelphia: Lippincott Williams & Wilkins; 2002. pp. 19–38.
53. Bukley A, Paloski W, Clément G. Physics of artificial gravity. In: Clément G, Bukley A, editors. Artificial gravity. New York: Microcosm Press and Springer; 2007. pp. 33–58.
54. Clément G, Bukley A, Paloski W. History of artificial gravity. In: Clément G, Bukley A, editors. Artificial gravity. New York: Microcosm Press and Springer; 2007. pp. 59–93.

55. Globus A, Arora N, Bajoria A, Straut J. The Kalpana One orbital space settlement revised. http://alglobus.net/NASAwork/papers/2007KalpanaOne.pdf. Accessed 15 April 2015.
56. Comins NF. The deaths of stars. Discovering the essential universe. 4th ed. New York: W. H. Freeman and Company; 2009. pp. 309–50.
57. Comins NF. Gravitation and the motion of the planets. Discovering the essential universe. 4th ed. New York: W. H. Freeman and Company; 2009. pp. 27–60.

5
Space Medicine: Paging Dr. McCoy

Life is short, the art long, opportunity fleeting, experiment treacherous, judgment difficult.

Hippocrates
Aphorisms

A Diagnostician had the most important job in the hospital, Conway thought, as he donned radiation armor and readied his patients for the preliminary examination.… For a Diagnostician to look at a patient that patient had to be unique, hopeless and at least three-quarters dead. When one did take charge of a case, though, the patient was as good as cured—they achieved miracles with monotonous regularity.

James White
Hospital Station (1962)

Practicing space medicine can be easy in science fiction. As illustrated early in *The Empire Strikes Back* (1980), injuries inflicted by hungry wampas respond well to merely immersing the victim in a bacta tank. As noted in Chap. 2, the coffin-like autodoc machines depicted in *Ringworld* (1970) and other works by Larry Niven provide the most complex medical care with little aid from flesh-and-blood medical personnel. The glossy high-tech equipment and spacious facilities located on large Federation starships such as the *Enterprise* in its various incarnations put to shame modern-day emergency rooms and surgical suites. Future humans might be genetically engineered to be more resilient to damage and heal more rapidly, or treated by injecting molecular-sized nanobots (See Chap. 11) that immediately go to work detecting and repairing body cells and organs. Physicians and nurses could routinely meet the challenges of managing the medical needs of exotic aliens such as those depicted in James White's "Sector General" stories and novels.

But providing medical care in space now and in the foreseeable future is not easy and involves many limitations. The most obvious is that present-day space travelers are unavoidably cut off from the full infrastructure of medications, medical personnel, supplies, and equipment available on Earth. Calling "911" will not bring an ambulance speeding to them to transport an ailing astronaut to an emergency room that, while not as sophisticated as Dr. McCoy's, is still much more capable of providing care than any current or near-future space-based medical facilities. Instead they will, at least initially, have to make do with whatever is available on the International Space Station (ISS) or their spacecraft—and hope that it is enough to deal with medical emergencies.

5.1 Medical Care in Space

The finite resources currently available in space make it impossible to treat every potential injury and illness [1]. Medical care is thus focused on dealing with those that are most likely to occur and most treatable given the limited types and amounts of medications, diagnostic equipment, and on-site expertise available. Selecting what supplies to send into space is also constrained by considerations of mass and size. Only a restricted amount of total mass can be launched at one time, and there is only limited space available for them onboard the ISS and, even more so, current and near-future spacecraft. A NASA report estimated that it would require 40 pounds of Earth launch mass (the rocket itself, propellant, etc.) to send 1 lb of medical equipment to Mars [2]. As desirable as it might be to have a computed tomography (CT) scanner or magnetic resonance imaging (MRI) machine available to acquire images of the body's internal structures, these devices are currently far too massive and large to be practical for current space-based medicine [3].

As described in previous chapters, microgravity and other aspects of the space environment also present unique problems regarding maintaining health (e.g. bone and muscle loss) and potential types of injury (risk of rapid decompression, excessive acceleration or deceleration, etc.) In addition, anything that could hurt or make an astronaut sick on Earth could also do so in space.

Table 5.1 lists injuries and illnesses that have been reported to occur during space missions. None have been life-threatening, although some could interfere (and have done so) with astronauts' ability to work effectively. Superficial abrasions, bruises, and minor lacerations (cuts) are common on spacecraft, particularly ones involving the hands [4]. Fortunately, so far no astronaut has received a laceration so serious as to require surgical repair. The spacesuits used for extravehicular activity (EVA) may produce chafing of various (and sensitive) parts of the body, and the gloves in those suits can damage fingernails directly or cause bleeding beneath them.

Table 5.1 Illnesses, Injuries, and other medical issues reported during space missions

Space adaptation syndrome, particularly space motion sickness, with symptoms that could include nausea, vomiting, headache, paleness, dizziness, sweating, and drowsiness
Loss of appetite associated with space adaptation syndrome or independent of it
Insomnia and other sleep disturbances
Nasal and sinus congestion
Headache
Subungual hematoma (bleeding beneath fingernails or toenails)
Skin irritation, infections (boils), and rashes
Burns
Lacerations (cuts), bruises, abrasions (scrapes) of the skin
Upper respiratory infection (e.g. a cold)
Fungal infection
Nosebleeds
Pneumonitis (inflammation of the lungs)
Urinary tract infections and stones
Prostatitis
Constipation and diarrhea
Muscle strains and back pain
Decompression sickness and barotrauma
Abrasions and foreign bodies involving the eyes
Decreased visual acuity

As noted in Chap. 4, changes to the back occurring in microgravity (e.g. stretching of ligaments in the vertebral column) commonly produce varying degrees of pain. Astronauts experience spasms mainly in the lower back that can be intense and require medications to treat. These pains are typically worst during the first few days in space and then improve. Muscle strains involving the shoulders, forearms, anterior abdomen, and lower back commonly occur due to various activities such as EVA, or the need to maintain uncomfortable postures for long periods while working at a laboratory bench. Astronauts may feel tingling or "electric" sensations in the thighs and legs, possibly due to irritation involving peripheral nerves as they exit the spine.

In microgravity astronauts do not have to worry about injuries to muscles and bones due to falling. However, it is common for them to handle and move (e.g. during EVA) equipment and objects with great enough mass and hence potential momentum (the product of mass and velocity) that they could potentially produce fractures, crush injuries, etc. For example, if an astronaut's hand or leg were in the way of a massive object moving toward the wall of a module it might indeed cause a significant injury. Even in the 0.17 g of the Moon an astronaut on the Apollo 17 mission suffered a back strain while drilling for a core sample of lunar material. Astronauts have even sustained muscle and other soft tissue damage when one of the stretched elastic straps used to secure items or stabilize one exercising on a treadmill unexpectedly snapped loose and struck a vulnerable body part [4].

Mild strains involving the ligaments of the hands, knees, and ankles are common in space. Those strains are typically associated with movement during EVA or exercising on a treadmill. Astronauts with previous injuries to the knees or other joints may have increased susceptibility to these kind of space-based strains. An astronaut with a sprained ankle will not need to avoid "putting weight" on it as on Earth. However, if the injury is so severe that treadmill or other exercise using the injured limb cannot be resumed quickly, there is a risk that the microgravity-induced loss of muscle mass and bone density described in Chap. 4 will occur at a greater rate and become more severe.

Astronauts can also develop health issues that, while perhaps not dramatic enough for a thrill-packed science fiction tale, can interfere with their ability to work. As noted in the previous chapter microgravity might predispose an individual to develop gastroesophageal reflux and heartburn. Mild gastritis (inflammation or irritation of the stomach's lining) may also occur, with symptoms that can include sensations of burning, bloating, or discomfort in the upper abdomen as well as nausea, vomiting, or loss of appetite. Such complaints are reported frequently by astronauts [5]. However, because these problems are also common in a terrestrial setting it is difficult to determine whether microgravity itself was responsible for them or if other factors, such as stress contributing to gastritis, were involved.

Likewise unheroic conditions such as constipation and, less frequently, diarrhea can occur in space. The former may be related in part to decreased large intestine activity in microgravity and mild dehydration if an astronaut does not drink enough fluids. Constipation is typically most likely to occur in the first days after reaching orbit. It can be managed with good hydration, use of standard drug store-type medications that stimulate bowel activity, soften stools, or provide bulk in the form of fiber, or—if necessary—careful use of an enema.

Diarrhea is an especially messy and unpleasant event in microgravity. It might be related to ingested foods or, though less likely due to the thorough medical examinations and quarantine procedures astronauts undergo prior to flight, an infectious process such as viral gastroenteritis. Episodes of diarrhea are generally self-limited, and symptomatic relief can (as with constipation) be provided using over-the-counter level medications.

Astronauts and cosmonauts can also develop abdominal pain. Causes range from innocuous ones such as simple gas to potentially life-threatening events like appendicitis. A member of the first crew of cosmonauts aboard the Soviet Salyut 7 space station in 1982 experienced the sudden onset of lower abdominal pain that initially could not be distinguished from appendicitis. Fortunately it resolved, and in retrospect was thought to have been most likely due to urolithiasis (see Chap. 4). While the latter is a painful and significant problem, overall it is less serious than appendicitis. This type of pain could also be associated with diverticulitis (inflammation or infection of a small "pouch" coming off the colon) or, in a female astronaut, a ruptured ovarian cyst or (in an unlikely but not impossible scenario, as described in Chap. 10) an ectopic pregnancy[1] [6].

Besides the short- and long-term problems with vision in microgravity described in Chap. 4, astronauts can experience temporary or permanent impairment of hearing [1, 7–9]. This may be associated with consistently high ambient sound levels on the ISS and other craft due to noisy equipment such as fans used for air circulation as well as other components of the environmental control system. However, other factors such as a higher than terrestrial-standard percentage of carbon dioxide in the local atmosphere or fluid shifts to the upper body might also contribute.

Other problems reported during spaceflight range from annoying ones such as sinusitis and toothaches to potentially serious ones such as abnormal heart rhythms. For example, a cosmonaut on the Salyut 6 space station in 1978 developed a severe toothache that required treatment with pain medications for 2 weeks [1]. Another cosmonaut performing EVA during a mission aboard Mir in 1987 developed a rapid abnormal heart rhythm ("tachyarrhythmia") that was considered serious enough to result in his being returned to Earth early [7]. Another incident of a crewmember on Mir developing a significant (though fortunately asymptomatic) tachyarrhythmia occurred in 1995.

Table 5.2 summarizes common types of medications stocked on the ISS [10, 11]. Similar medications have been used on prior space missions [12]. Both NASA and the Russian Aviation and Space Agency provide their own sets of medical kits, supplies, and other equipment. These include a mixture of drugs that could be found in a home's bathroom medicine cabinet or first-aid kit along with others that would require a prescription or administration by medical personnel. All these drugs and supplies have one thing in

---

[1] In an ectopic pregnancy a fertilized egg does not reach a woman's uterus but begins to divide and grow elsewhere. This occurs most commonly in a fallopian tube, which connects an ovary to the uterus, but also on occasion in the ovary or abdominal cavity. It nearly always results in death of the embryo and could be life-threatening for the woman—for example, if a complication such as rupture of the fallopian tube containing the growing embryo and resulting hemorrhage occurs.

Table 5.2 Types of medications commonly carried on current space missions

Class of medication	Indication	Examples
Analgesics and non-steroidal anti-inflammatory agents	Pain relief, reduction of inflammation	Acetaminophen (Tylenol) Acetylsalicylic acid (aspirin) Meperidine (Demerol, IM or IV) Ibuprofen (Motrin) Diclofenac Prednisone Pyridium (for urethral discomfort) Hydrocodone
Antiemetics	Nausea, vomiting	Promethazine (Phenergan) Procholorperazine (Compazine) suppositories
Antihistamines and H_1 blockers	Nasal congestion, mild allergic reactions	Diphenhydramine (Benadryl) Loratadine (Claratin)
Central nervous system stimulants	Increase alertness	Dextroamphetamine (Dexedrine)
Cardiovascular agents	Angina pectoris, arrhythmias, congestive heart failure	Trinitroglycerin (sublingual and patches) Lidocaine (IV) Atropine (IV) Metoprolol Adenosine (IV) Epinephrine (IV) Furosemide (IV) Dopamine (IV) Bretylium (IV) Propranolol Verapamil (IV)
Antibiotics	Treatment of bacterial infections	Ampicillin Ciprofloxacin Metronidazole (Flagyl) Cefadroxil Azithromycin
Sedative-hypnotics	Insomnia, anxiety	Zolpidem (Ambien) Flurazepam (Diflucan) Temazepam (Restoril) Diazepam (IV)
Antidepressant and antipsychotic agents	Psychiatric problems	Haloperidol (Haldol, oral and injectable) Nortriptyline Luoxetine (Prozac)

Table 5.2 (continued)

Class of medication	Indication	Examples
Gastrointestinal agents	Diarrhea, constipation, heartburn, hemorrhoids	Diphenoxylate/atropine (Lomotil) Bisacodyl (Dulcolax, oral and suppositories) Loperamide (Imodium) Mylanta DS Milk of magnesia Pepto Bismol Simethicone Omeprazole Hydrocortisone suppositories
Dermatological agents	Skin rashes and infections	Kenalog 0.1 % cream Neosporin ointment Mycolog cream
Ophthamologic and otic agents	Eye and ear irritation and infections	Cyclopentolate ophthalmic solution Corticosporin otic solution Topical anesthetics and antibiotics

Medications are taken orally unless specified as being given IM or IV, or used in the form of a suppository, cream, ointment, or solution
IM intramuscular. *IV* intravenous

common—once they are used they cannot be replaced except by resupply from Earth. This means that they must be used judiciously, especially during any future missions to the Moon or Mars.

Medications may be absorbed, metabolized, and excreted by the body differently in microgravity than on Earth. Fluid shifts, decreases in blood volume and muscle mass, and even possible changes in how enzymes in the body function may alter how quickly medications become effective and how long their effects last [12, 13].

Medical supplies and equipment on the ISS are similar to those used by terrestrial paramedics. These include syringes; needles for drawing blood, starting intravenous lines, and administering medications; various types of bandages, dressings, pads, and swabs; suture material; scalpels; surgical masks and gloves; a pulse oximeter for measuring oxygen saturation in the blood; and a stethoscope and blood pressure cuff. Besides specifically medical items like scalpels, hemostats, tongue depressors, a thermometer, ophthalmoscope, otoscope, etc. the ISS also carries mundane but useful ones such as dental floss and lip balm. It also has machines for monitoring and recording the heart's electrical activity, a defibrillator, and basic equipment to help an astronaut who is short of breath by providing oxygen and other means.

An ultrasound machine is also available for taking images of internal organs such as the heart as part of studies on how the human body responds to microgravity. Current versions of that device may weigh as little as 2.5 kg or less, a mass-to-benefit ratio that falls within a reasonable range for lofting one into orbit or beyond [3]. However, if needed that machine could also be used to check for serious injuries such as a pneumothorax (see Chap. 2) and bleeding within the abdomen [11, 14]. Its use in that context would be similar to how such devices are used on Earth for initial assessment of some types of internal injuries via a FAST ("*Focused Assessment with Sonography in Trauma*") scan of the abdomen and chest [15, 16]. Because neither the ISS nor any near-future spacecraft will have an x-ray machine, if required that ultrasound device might also be pressed into service for nonstandard but helpful purposes such as checking for subtle fractures involving the feet and ankles [4].

On the ISS two crewmembers are assigned as Crew Medical Officers (CMOs), the same number used on Space Shuttle flights. CMOs who are not physicians (and most are not, due to the relative scarcity of astronaut-physicians) receive paramedic-level training in diagnosing and treating medical and dental problems. This typically includes 34 h of medical training provided at least 6 months before launch, with additional refresher courses taken during long-duration missions [7]. Advice is also available from a ground-based Crew Surgeon, Deputy Crew Surgeon, and other medical personnel.

In space it may be better for physician-astronauts to lean toward being "generalists," able to treat the widest gamut of the most likely illnesses and injuries with the limited medical resources available, than "specialists." For example, a trauma surgeon would be better qualified to treat certain injuries better than a non-surgeon physician or a non-physician CMO. However, without having the right "tools" such as a fully staffed operating suite available, the trauma surgeon might, under some circumstances, be able to do no more than a physician with less high-level training.

Likewise, having specialized medical equipment and other hardware on a spacecraft that a non-physician or even a physician without specific training could not use would be a waste of limited onboard space. Modern communications systems also allow use of "telemedicine," with terrestrial physicians able to monitor telemetry information (e.g. heart rate, heart rhythm, blood pressure, etc.) via radio communications and observe astronaut-patients in real time while advising CMOs in space [17, 18]. Nonetheless, it seems prudent to include a physician in the crew of any mission designed to stay far from Earth for a long time (e.g. one to Mars) [7]. It has been suggested that the CMO on such a mission should have at least 2–3 years of surgical training due to the importance of having those particular skills available if needed to treat crewmates [19].

Severity of medical problems on the ISS could range from minimal to potentially fatal. In a standard classification system Class I medical events are considered medically minor and do not impact the mission [20].

Class II events are more significant and further classified based on their seriousness. A Class IIa problem is considered manageable with resources on the ISS and unlikely to require shortening the overall mission or bringing the affected individual back to Earth for definitive treatment. Class IIb issues can be initially dealt with on the ISS but may require early (but not emergency) return of the astronaut. Class IIc problems are serious enough that, if the patient cannot be stabilized and successfully treated on the ISS, emergency evacuation might be necessary.

A Class III event cannot be adequately managed on the ISS and requires bringing the astronaut-patient back to Earth on an emergency basis for care. Two Soyuz capsules, each capable of holding three crewmembers, are always docked at the ISS. Since the maximum crew size on the ISS is no more than 6, if an individual requires evacuation for medical reasons up to 2 crewmates could accompany the patient back to Earth.

However, due to the Soyuz capsule's small size an alternative is to send only the patient and a crewmate while the craft's third seat would be used to place any needed medical support equipment [7]. Unfortunately this could leave only one three-seat capsule at the ISS for what could be 4 remaining crewmembers if an emergency evacuation of the station were required later. Should that occur, drawing straws or some other method to decide who stays would have obvious dramatic possibilities in a science fiction story, but it is clearly a situation best avoided if possible in real life.

CMOs on the ISS or spacecraft in LEO have the option of "stabilize and transport," with an injured or ill crewmember treated initially in space and then, when it can be safely done, returned to Earth. My story "Hearts in Darkness" concerns a cardiologist on the ISS forced to weigh the risks of keeping a crewmate who had suffered a myocardial infarction (MI, or "heart attack") onboard or deciding his patient was stable enough to send back to Earth for definitive care.[2] This kind of clinical decision has no perfect answer but only choices that, based on the best assessment possible, have greater or lower risks. Unfortunately, even if the overall odds favor one approach over another, any of the available options could still have a bad outcome for an individual patient.[3]

One concern relevant to any injured individual returning to Earth is that blood pressure could drop dangerously due to the fluid shifts and impaired orthostatic reflexes described in Chap. 4. Measures such as rehydrating the

---

[2] *Analog Science Fiction and Fact* March 2002.
[3] While the risk of a spacefarer having an MI is very low, particularly in view of the extensive health evaluations each undergoes, it is not zero. One cosmonaut had a major MI at the age of 49, just 2 years after his last trip into space [5]. It is conceivable that, with bad timing, that serious medical event could have occurred while he was on a mission.

patient and maintaining a supine position during reentry might be helpful. However, with certain problems such as significant blood loss that could not be adequately replaced or an MI, the patient might still be vulnerable enough to a fall in blood pressure (with associated increase in heart rate) during and immediately after return to Earth to cause his or her condition to deteriorate further. This might result in development of shock and damage to internal organs (including the brain) due to their receiving an inadequate supply of blood and oxygen [11]. The effects of increased g forces alone during reentry could also contribute to compromising someone with a serious injury or illness. Such factors will need to be taken into account on a case-by-case basis when making a decision whether to evacuate the individual or not [19–21].

Both telemedicine and evacuation become increasingly problematic on missions beyond LEO, where the finite speed of radio waves (roughly 300,000 km/s in space) becomes an issue. Communication delays of about 1 s each way between a ground-based physician and a CMO on the Moon would be a minimal impediment. However, it would take an average of roughly 20 min after a CMO on Mars asks a desperate question concerning a severely injured crewmate before a physician on Earth will hear it, and a minimum of another 20 min before a reply arrives. That is plenty of time for bad things to happen. While the time needed to evacuate an astronaut from LEO is measured in a minimum of hours, it would be several days from the Moon and months from Mars.

The sad truth is that an astronaut might die from injuries or other life-threatening conditions that could have been successfully treated on Earth. My story "The Last Temptation of Katerina Savitskaya"[4] deals with that scenario in the context of a mission to Mars, particularly the psychological effects of such a "no-win" situation on a surviving crewmate. Like those of physics, the "cold equations" of medical practice cannot be ignored, and no matter how hard we try to work around them they will have the final say.

5.2 Preventive Care

The maxim that prevention is better than cure is even more true in space than on Earth [7]. Astronauts and astronaut candidates are rigorously screened to make sure that only the most fit and healthy leave our world.[5] These screening measures include obtaining comprehensive medical histories, performing

---

[4] *Analog Science Fiction and Fact* September 2008.
[5] Current standards in the U.S. program also include requirements not specifically related to health. For example, to "fit" in current craft and, in some cases, spacesuits, astronaut candidates must typically have a height between 152 cm and 191 cm (just under 5 feet tall to 6 feet 3 inches tall).

Table 5.3 Medical evaluation of astronaut candidates and astronauts

Medical history and exams	Cardiopulmonary tests	Laboratory and other tests
Physical examination, including a rectal examination and proctosigmoidoscopy	Blood pressure	Blood workup includes chemistries, blood counts, etc
Women also receive a pelvic examination and Pap smear	Pulmonary function tests	Urinalysis
Musculoskeletal examination, including assessment of muscle mass and bone	Exercise treadmill test	Measurement of 24-h urinary excretion of calcium
Complete eye and ear examination, including audiometry (a "hearing test"), tests of visual acuity, color and depth perception, and tonometry to check for glaucoma (increased pressure in the eyes)	Echocardiogram	Bone densitometry studies Stool checked for blood and parasites
	12-lead electrocardiogram and 24-h Holter monitor (for continuous recording of heart rhythm)	Drug screen
		PPD skin test (for tuberculosis)
Dental examination, including x-rays		Screen for sexually transmitted diseases
Complete neurological examination, including an electroencephalogram		Pregnancy test (premenopausal women) •
Psychiatric interview and psychological tests (includes Minnesota Multiphasic Personality Inventory II and Multidimensional Aptitude Battery)		X-rays of chest, sinuses, and skull. Abdominal ultrasound. Mammograms and pelvic ultrasound in women

Special attention is given to ensure that astronauts are as healthy as possible before going into space. This includes pre-screening them for existing diseases as well as conditions such as increased risk of developing kidney/ureteral stones that could be significantly worsened due to the physiological effects of microgravity and other conditions in space

periodic physical and dental examinations, and ordering diagnostic tests to discover potential problems before they manifest themselves during a mission. Current astronauts must also undergo a thorough annual examination. Those scheduled for long-duration missions (e.g. months or longer, as on the ISS) undergo multiple exams starting about 180 days prior to launch and for at least a month after returning to Earth.

Major procedures used to evaluate astronauts and astronaut candidates are summarized in Table 5.3 [4]. Most are ones that would be routinely done on us stay-at-home adults depending on our age and gender. Others, like genetic screening for cancer risk or checking for increased urinary excretion of calcium that could predispose to kidney stone formation, go beyond this (at least to some degree) to help identify asymptomatic health issues that could manifest themselves for the first time after an astronaut reaches space. For example, exercise tests assess work capacity; screen for presence of coronary artery disease, which could cause chest pain, myocardial infarction, and life-threatening arrhythmias; and evaluate heart rate and blood pressure responses to stress.

Likewise a female astronaut found to have mutations in the *BRCA1* and *BRCA2* genes that are known to carry a significantly higher risk for developing breast or ovarian cancer might not be a good candidate for a long-duration mission to Mars. This risk was illustrated by an incident involving a physician at an Antarctic station in 1999 who found a lump in her breast [1, 22]. Due to severe winter weather conditions precluding any planes landing there for many weeks, she could not be immediately evacuated for evaluation. She and other personnel had to perform a biopsy on the mass, process it to make a stained slide with its cells, and transmit images of that sample to outside medical experts. When it was determined that the mass was cancerous, chemotherapy medications were airdropped to the station for initial treatment before the weather improved enough that she could be evacuated for definitive care.

Although that physician went on to have a full recovery, the prognosis for a female astronaut on Mars who developed such a health issue would be much more grim. As we will see in the next chapter, radiation exposure in space would also significantly increase the risk of later developing cancer in an astronaut who was already predisposed based on genetic markers (including ones associated with colorectal cancer, a deadly form of skin cancer called "melanoma," and other types) or even a strong family history of cancer [1]. Other conditions in space, such as impairment of immune function in microgravity and exposure to toxins on the spacecraft, could also increase overall risk of developing cancer [22]. These dangers are less for an astronaut training for a mission of short duration or one in LEO where timely evacuation is feasible, but they would remain significant.

Diagnostic tests used for screening and ongoing medical evaluations of astronauts prior to flight are typically noninvasive and carry minimal or no risk. Echocardiography uses ultrasound to detect abnormalities of the heart's size, valves, and function. Pressure changes and fluid shifts in microgravity increase the risk of sinus problems, and x-rays of those areas can help identify individuals with increased susceptibility. Abdominal ultrasound can detect gallstones, which could cause cholecystitis (inflammation of the gallbladder). Electroencephalograms measure brain wave activity and detect abnormalities that might indicate an increased risk of seizures.

Some health problems in space could be due to "bad timing" alone and probably would have occurred even if the individual had stayed on Earth. Others, such as new kidney stones, could be at least partly due to changes from microgravity and other factors in the space environment. In either case the goal is to try to prevent trouble before it happens [23].

There are also questions regarding how aggressive such preventative measures must be. For example, should an astronaut traveling to Mars have a prophylactic appendectomy, trading the very small but certain risk of having surgery on Earth for the smaller chance of developing potentially catastrophic appendicitis while in space? Assuming a worst-case scenario in which an unfortunate astronaut developed appendicitis far from Earth, treatment with intravenous antibiotics alone could be effective in some cases. An ultrasound device could be used to identify an associated abscess and guide placement of a catheter via a small incision to drain it [11]. On the other hand, those antibiotics will be in very limited supply, and if draining the abscess is ineffective further treatment options (such as removing a ruptured appendix) are much more problematic.

Likewise an exercise treadmill test can usually detect blockages in coronary arteries ("coronary artery disease," or CAD) severe enough to significantly restrict blood flow. However, this and similar tests are less likely to identify milder blockages from atherosclerotic plaques that, as noted in Chap. 2, could unexpectedly rupture and form a blood clot that suddenly blocks a coronary artery, causing an MI. An exercise test itself carries only minimal risk. However, diagnostic procedures that are more sensitive for detecting CAD carry very small but higher risks. For example, having either coronary angiography or a CT scan of the coronary arteries will expose a patient to x-ray radiation. Both tests also require intravascular injections of a "contrast agent," a material that can cause uncommon but serious side effects such as kidney damage or death. Coronary angiography is also considered an "invasive" procedure.[6] Another test involving x-ray radiation, electron-beam computed tomography

---

[6] Coronary angiography requires passage of a long thin tube (catheter) into an artery in the arm or groin that is then threaded to the heart. The contrast agent injected into the coronary arteries shows up on a fluoroscopy machine. Locations in a coronary artery where its flow is "pinched" and reduced indicate the presence of blockages. Risks of coronary angiography are low but include causing an MI, stroke, or death.

(EBCT), evaluates the presence of calcium in the coronary arteries, which may be a marker for CAD, particularly in younger individuals. However, its overall value for predicting future cardiac events such as MI requires further clarification [24].

Too aggressive screening with medical tests can cause its own problems. Many tests have varying probabilities of producing a "false positive" result—that is, the test indicates an abnormality is present when it really is not. Such a test result might require further evaluation with more costly, complex, and riskier tests to determine whether it really is a "true positive" (the abnormality really is there) or a false positive, and perhaps could even erroneously cause an astronaut to be disqualified from going into space. Procedures such as whole-body CT and MRI scans can also find abnormalities of uncertain significance that may actually be benign. Such findings could include innocuous masses such as cysts, very small benign tumors, minor variants from "normal" anatomy, and other incidental findings often lumped together under the medical slang term "incidentalomas" [1]. However, because it may not be possible to determine that they are innocuous by their appearance alone on such tests, once these abnormalities are found additional testing or even surgical procedures may be needed to prove that they really are not something more serious instead.

Certain health problems, even if stable, carry sufficient risk that they might disqualify astronauts from flight. Someone with known CAD carries an increased risk of MI even if that condition is well controlled on medication. High blood pressure can increase the risk of stroke and other health issues, and so astronauts must have blood pressures within an acceptable range to qualify for spaceflight. For example, an elevated blood pressure of 140/90 mmHg could disqualify an astronaut from flight. Diabetes mellitus is also a major risk factor for CAD and other significant health problems [1]. Chronic health conditions such as significant inflammatory bowel disease that might, in its normal course, require hospitalizations for treatment could also preclude selection as an active astronaut [2].

Some pre-existing medical conditions might be acceptable under some circumstances. Visual acuity standards for understandably higher for those piloting a spacecraft compared to those who are only along for the ride. Individuals with chronic conditions that require regular dosing of medications (e.g. thyroid replacement therapy) might be fit for short duration missions. However, they might not qualify for longer ones to the Moon or Mars, where they would suffer significant health consequences if their medications were lost and could not be replaced soon enough. Likewise, those with health problems that require periodic monitoring (such as checking blood glucose) to assess their need for further therapy may also be excluded, particularly for long-duration and distant missions. On the other hand, prospective spacefarers who would be disqualified for long-duration missions, such as those with a history of kidney stones, might be cleared for short-duration ones (defined as less than 30 days) [25].

Astronauts also have several physical examinations done within days of their flight to make sure nothing new has developed since they were last checked. Crewmembers are kept in restricted quarters for about a week prior to flight to reduce the chance that they will acquire an infectious disease (e.g. the common cold) from non-flying personnel before their launch.

5.3 Exercise and Nutrition

As described in Chap. 4, astronauts routinely perform both isotonic (aerobic) and isometric (resistive) exercise to maintain overall fitness and reduce bone and muscle loss in microgravity. Maintaining overall strength and conditioning as much as possible is especially important for astronauts who perform EVA. The Orlan spacesuit used in the Russian space program can have a total mass of about 70 kg, and the extravehicular maneuvering unit (EMU) used by American astronauts has one of about 130 kg [26]. Even in microgravity a great deal of muscular effort is needed to move and maneuver such masses due to their intrinsic inertia.

Besides exercising, astronauts should also eat a balanced diet with enough kilocalories[7] to maintain body mass and overall health [1, 27]. Typical dietary requirements in space are in the range of 2500–3000 kcal/day, with even more kilocalories/day required if energy expenditure increases, e.g. with strenuous EVA. Many factors in the space environment may, however, suppress appetite and associated food intake. These include astronauts feeling warmer due to reduced body heat dissipation, mildly increased levels of exhaled carbon dioxide compared to terrestrial standards, and their vigorous exercise programs.

Some of the healthiest foods, such as fresh fruit and vegetables, are not available in space. Conversely, high-calorie foods containing fat and simple sugars that have less intrinsic nutritional value take up the least space for the calories they provide. High-fiber foods that would be desirable from a general health standpoint are also bulkier and generate more solid waste (feces)—disadvantages in the space environment. The preprocessed foods that astronauts typically eat thus represent a balance between the healthiest varieties and those that provide more calories at the cost of some overall nutritional value.

---

[7] Note that what is commonly called a "calorie" with food is really a "kilocalorie," or one thousand calories.

Astronauts should ideally consume certain minerals and nutrients such as salt and protein within relatively narrow ranges. For example, too much salt could increase urinary excretion of calcium and cause an increased risk of kidney stones. Our bodies produce vitamin D with exposure to sunlight. However, markedly reduced exposure to the Sun on the ISS and spacecraft means that vitamin can need to be provided as an oral supplement. Certain other vitamins (e.g. vitamin C) acting as antioxidants (see Chap. 9) can be provided in pill form. Animal studies indicate that some of them might provide at least mild protection against the cancer-causing effects of increased radiation levels in space. Sugars can also contribute to tooth decay, so limiting their intake is important for reducing the chance of dental caries ("cavities") and toothaches in space, where no neighborhood dentist is available.

Foods on the ISS may be refrigerated, frozen, dried (e.g. fruit), and preserved via such methods as "thermostabilization" (a fancy way of saying heating them) or irradiation to destroy bacteria and other causes of spoilage [7]. Fluid shifts to the upper body, with attendant swelling of the tongue and nasal congestion, may contribute to decreasing the taste and smell of foods and make eating less enjoyable. Adding condiments and spices to food may help this. Drinking fluids is typically done via plastic bags with straws, while solid foods are in prepackaged units that are individualized for each crewmember. Care must be taken while eating to prevent food from floating away, and any utensils used are magnetized and placed on a metal surface for the same reason when they are not being used. Overall, although the mechanics of eating and drinking are more challenging in microgravity, spacefarers typically learn the appropriate skills quickly.

5.4 Infectious Diseases

Infections present a major health risk in space. As described previously, vigorous screening and isolation of astronauts prior to going into space helps reduce this hazard. However, in the closed environment of a spacecraft or a habitat such as the ISS they will still be exposed to bacteria, fungi, and viruses from many different sources [28].

Concentrations of airborne bacteria and fungi are decreased but not eliminated by systems that filter particles in the air on space vessels. The upper limit for microorganisms on the ISS is 1000 colony-forming units/m^3 of air, which is slightly higher than the standard used for U.S. office buildings. Another source of infection is water that condenses on various surfaces, particularly near waste collection systems [4]. Samples of water from those latter areas have shown significant amounts of various microorganisms that, should an astronaut suffer a cut near those systems, could result in an infected wound.

Drinking water is another possible source of contamination. On Space Shuttle missions, water came from stored supplies and fuel cells. Water produced by the latter was first disinfected by passing it through an iodinated resin. That system reduced levels of bacterial contamination to terrestrial standards. Recycling of contaminated water derived from air, urine, and wash water is needed on the ISS and long missions away from Earth. Bacteria and viruses are destroyed as part of the recycling process on the ISS. After processing, the number of microbes in its water is much smaller than typical drinking water on our world.

Preflight processing via thermostabilization or other methods reduces levels of aerobic bacteria (those that need oxygen, compared to "anaerobic" ones that don't and, in some cases, can actually be killed by it) in food to no more than 10^4 per g. Food is also tested specifically for the absence of bacteria that commonly cause food poisoning. Because they can harbor microbes transmittable to crewmembers, experimental animals are also tested for disease-causing microorganisms ("pathogens") before launch. Fortunately plants are not a significant source of microorganisms that can infect humans. Payload items, especially microbial experiments, could also cause contamination.

However, the most important source of potentially dangerous microorganisms in space is the astronauts themselves. Even a "healthy" individual has about 10^{12} (one followed by twelve zeros) bacteria on the skin, 10^{10} in the mouth, and 10^{14} in the gastrointestinal tract, particularly the intestines [29]. Fortunately the vast majority of these bacteria are "commensal"—that is, they do not cause disease or do so only after an area on the host is injured (e.g. a cut on the skin). As noted in Chap. 1, some are even "symbiotic," producing substances like Vitamin K (needed for blood clotting) beneficial to the host. These "good" bacteria also protect against infection by more virulent microorganisms by competing with them for the same nutrients, producing chemicals that kill them, and providing chronic, low-level stimulation of the body's immune system so it is continuously "primed" to fight infections.

Microorganisms are continuously shed from skin, mucus membranes (e.g. the tissue inside the nose and mouth), and the gastrointestinal and respiratory tracts. Sneezing, coughing, and even talking release microbe-laden aerosols into the confined environment of a spacecraft, ready to be inhaled by other crewmembers. On Earth these tiny droplets settle to the floor within seconds or minutes. In microgravity they remain suspended until colliding with some surface such as a crewmember's skin or mucus membranes, where they can potentially cause an infection. This risk can be made worse if skin and other tissues are either too dry or too moist due to humidity levels on their craft, with either condition potentially predisposing to infections. Urine and especially feces are other major sources of contamination. Methods used to manage them are described in Chap. 4 [7].

Maintaining personal hygiene to reduce risk of infection and for psychosocial reasons is difficult in microgravity. Washing hands, face, hair, and especially showering require water that is in limited supply and may need to be recycled, along with problems of removing soap, microorganisms, and other contaminants. Containment and collection of released water is also a major problem, since sprayed water will "float" in spherical drops throughout the cabin. Even the ISS does not have a shower,[8] with its crew having to use methods such as wet wipes, cloths, a small water spray, and soap squeezed from toothpaste-type tubes or released from a dispenser. A fan in the special compartment used for cleaning helps draw any excess water into a drain leading to a storage tank [7]. With no laundry facilities onboard, clothing is worn for a certain amount of time, then bagged and stored after being replaced by new clothes.

The space environment may also make disease-causing bacteria more likely to cause infections and more resistant to a wider range of antibiotics. How much of this might be due to the effects of mutation from radiation, microgravity itself, or other factors remains uncertain. A person's immune system function may also be at least mildly depressed in microgravity, increasing vulnerability to infection. These issues, along with ones like the limited availability of antibiotics, their possible decreased effectiveness due to reduced absorption by the body in microgravity, and more rapid degradation (a shorter "shelf life") in space could add up to a significant problem for crewmembers during long-duration missions.

Various types of "extremophile" bacteria and other microorganisms can also live under conditions that, in some ways, are even harsher than those typical in space. For example, *Deinococcus radiodurans* is one of a small number of bacteria that can survive acute radiation levels well over a thousand times greater than would kill a human being. Fortunately these bacteria, while they can cause food to spoil, do not cause disease in humans.

It has also been reported that *Streptococcus mitis*, a type of bacteria that can be harmful to us, was recovered from a camera on the Surveyor 3 lunar probe retrieved about 31 months after its landing by astronauts on Apollo 12, whose own craft landed near the probe [7]. If so, this means that microorganism was

---

[8] The U.S.'s Skylab space station (launched in 1973) had a "shower" consisting of a tubular cloth bag an astronaut could get inside and then squirt warm water and apply liquid soap onto his body. Unfortunately this technique resulted in significant amounts of fluids escaping into the craft that then had to be vacuumed up to prevent it getting into equipment or being inhaled by the crew [5].

able to survive such conditions as vacuum, high levels of radiation, lack of water, large temperature swings from far below the freezing point of water to well above it, etc. This illustrates how much "hardier" the bacteria and other microorganisms we carry into space are than us as well as how difficult it may be to protect astronauts from them [30].

Some microorganisms, particularly certain types of molds and fungi, can also "infect" and corrode equipment on the ISS and other craft. Crewmembers on the ISS routinely swab surfaces and check the air and water for microbial contaminants. All these risks to personnel and the spacecraft itself emphasize the need to keep their environment as "germ-free" as possible, at least to the level where risks of infection and damage are reduced to the lowest levels attainable.

Despite all precautionary measures, however, astronauts have developed infections in space. Most have been minor infections involving the skin, throat, and eyes that were acquired after reaching orbit or beyond. However, in spite of quarantine procedures 1 of the 3 astronauts on the Apollo 7 orbital mission in 1968 developed a cold after reaching space. Not surprisingly his 2 crewmates subsequently caught it. Although they successfully completed their nearly 11-day mission, those illnesses made their work considerably more difficult. That event did, however, help lead to more stringent guidelines to prevent preflight infections. In current practice astronauts are quarantined for about a week prior to flight to reduce the risks of acquiring a viral or other infection from Earthbound individuals [7].

More serious infections have also occurred. An astronaut on the ill-fated Apollo 13 flight developed a significant UTI that was at least partly related to dehydration resulting from the limited water supply available after the craft's accident. Other astronauts and cosmonauts have also developed UTIs as well as infections of the prostate gland (prostatitis). In fact, the cosmonauts aboard the Salyut 7 space station returned to Earth prematurely after only 56 days of a planned 216-day mission due to one of them developing prostatitis and a severe UTI [7, 20]. Problems with urinary retention—an inability to pass urine—have also been reported [31]. This could not only cause an astronaut discomfort, perhaps even requiring passing a catheter into the urethra to drain the bladder, but also predispose to development of a UTI.

5.5 First Aid and Surgery in Space

No one has needed to have surgery in space—yet. But if an astronaut required immediate treatment or timely medical evacuation to Earth was not an option (e.g. on a mission far beyond LEO), someday an operation might need to be done there. Telemedicine with appropriate specialists back on our world and medical computer programs designed to assist with diagnosis and treatment might help guide a CMO forced to do a procedure under less than ideal circumstances. However, factors such as the biological and physical effects of microgravity as well as the limited diagnostic tests and other medical resources available in space could all complicate care [7].

As described in Chap. 4, microgravity causes circulating blood volume and red blood cell mass to fall. This results in at least a relative state of mild anemia and dehydration, thus decreasing a person's reserves in case of severe bleeding [1, 15, 24, 32, 33]. Options for providing blood transfusions, including from crewmates (assuming they had compatible blood types—see Chap. 13) would obviously be limited.

In addition, fluid shifts to the upper body in microgravity could make it more likely that head trauma would cause swelling and damage to the brain. Slow or inadequate healing of bone fractures might also occur due to microgravity-induced loss of calcium and bone mass as well as lack of the normal gravitational stress that promotes bone growth [7]. In fact, animal studies have demonstrated impaired healing of fractures in microgravity [34]. Postoperative wound healing could be complicated by increased susceptibility to infection by microorganisms that, as noted previously, may more virulent and resistant to antibiotics than on Earth [7, 35].

Some standard techniques used to diagnose particular medical problems will not work in microgravity. On Earth someone suspected of having blood or other fluid in the chest due to injury, pneumonia, etc., can be turned on one side before taking a chest x-ray to look for gravity-induced "layering out" of fluid. This diagnostically helpful layering will not occur in microgravity. Likewise an x-ray taken in a sitting position will not show air rising to the upper part of the abdomen after rupture of the stomach or intestines. Air-fluid levels often seen in terrestrial gravity when there is obstruction of the small intestines would also not be seen on x-rays taken of the abdomen. This could make it more difficult to decide whether surgery might be needed to prevent life-threatening intestinal damage. These issues are currently moot since even the ISS does not have an x-ray machine. However, when one eventually does make it to orbit or beyond these effects would become relevant.

On Earth initial management of a patient with hypotension (low blood pressure due to dehydration, blood loss, or other causes) includes placing the individual on his or her back and raising the legs higher than the head. Gravity would then make blood in the legs go to the upper body. This would help provide more blood and the oxygen in it to the brain both directly and by making the heart pump more blood with each beat. Swelling from a fracture or serious strain to a joint in an upper or lower extremity could be reduced by elevating it. Likewise, someone who has swelling of the brain ("cerebral edema," such as might be caused by head trauma) or too much fluid in the lungs from a dysfunctional heart (congestive heart failure) may benefit from being placed head upright. This can reduce the relative "excess" of blood and fluid in those areas.

None of those quick, simple maneuvers would work in microgravity. While medications and other ways can be used to deal with these kinds of medical emergencies, they may take longer to implement than those other measures—time that could mean the difference between life and death. Also, due to fluid shifts and other effects of microgravity astronauts most likely already have mildly increased "intracranial" (within the skull) pressures. Thus, a particular degree of head injury could result in higher (perhaps dangerously so) intracranial pressures in microgravity than in a terrestrial environment.

Techniques for treating minor cuts, bruises, muscle sprains, and burns in microgravity are similar to first-aid methods on Earth. For example, a superficial cut could be cleaned with an antiseptic agent (e.g. via an alcohol swab) and an adhesive bandage applied. A cut that may have foreign (and potentially infectious) material in it could be carefully irrigated by using a syringe to squirt a small amount of sterile saline into the wound. A gauze pad placed over it and the tip of the syringe would help absorb any fluid splashing out as part of that irrigation process. A topical antibiotic ointment could also be administered either prophylactically or if a wound were to show subsequent signs of infection.

A strained finger or even a simple, nondisplaced fracture of any digit can be treated with a splint, wrapping it against an adjacent finger or toe, or other standard means. Bigger splints are also available on the ISS for injuries to larger joints such as ankles.

As noted in Chap. 4, fractures of long bones (the humerus, radius, or ulna in the upper extremities or femur, tibia, or fibula in the lower ones) could occur under certain circumstances in microgravity. That risk is increased due to bone mineral loss. Fractures might be more likely to occur during strenuous activities, such as EVA. But even assuming CMOs have the capability to diagnose (in the absence of an x-ray machine) and treat such fractures, there are problems with the availability and use of materials to treat them.

For example, open surgical procedures for certain severe types of fractures require operative capabilities and stabilizing/reparative hardware not available on the ISS—much less near-future spacecraft or on Mars. Even creating a plaster cast by mixing plaster of Paris (assuming that powdery material, a form of calcium sulfate called "gypsum," was actually available) with water would be challenging in microgravity. Casts created from fiberglass-coated bandages might also be used and have the advantage of being lighter than comparable plaster casts. However, fiberglass can also produce significant amounts of out-gassing[9] of harmful materials, which could potentially limit its use [19].

Some types of fractures are normally stabilized by applying traction. On Earth this typically involves providing tension by using weights. Systems using springs or elastic bands might provide enough tension to substitute for weights in microgravity.

The various types of burns discussed in Chap. 2 can also occur. First-degree and, if they are not extensive, second-degree burns can be treated using current capabilities in space. Treating a first-degree burn may include cleaning the area, cold compresses, oral pain medication, and perhaps a topical antibiotic. A protective dressing might also be used. Localized second-degree burns can be managed similarly to first-degree burns, though with more aggressive attention to reducing the risk of infection via cleaning the area and using topical antibiotics.

Second-degree burns and especially third-degree burns covering extensive areas of the body can be associated with dramatically increased risk of infection as well as fluid losses due to disruption of skin integrity. Definitive treatment of third-degree burns may require operative procedures such as placement of skin flaps—capabilities that will not be available away from Earth for a long time to come.

Astronauts have developed first-degree burns in the form of sunburns due to harmful ultraviolet (UV) light from unfiltered sunlight.[10] Sunlight passing through spacecraft windows that do not block UV rays can produce such first-degree burns after only seconds of exposure [5]. Windows on the ISS have glass panes designed to prevent harmful levels of UV light from passing through them. One crewmember received second-degree burns on the forearm during the 1997 fire on Mir described in Chap. 3. Fortunately that injury was not great enough to require evacuation to Earth for definitive treatment.

Minor eye injuries are common in space. An astronaut's eyes can be irritated or even suffer painful abrasions due to the unavoidable presence of tiny floating particles and low levels of noxious gases in the confined environment of spacecraft. If the taut elastic cords used to secure astronauts during treadmill exercise suddenly come loose a cord might whip up and strike someone's eye—an injury that has happened at various times during missions [5, 36].

Evaluating eye injuries in space is, except for the absence of a slit lamp for a more detailed examination, similar to methods on Earth. The examination might include using a magnifying lens to look at the eye, checking it both externally and internally with an ophthalmoscope, and use of orange fluorescein

---

[9] See Footnote 5 in Chap. 3 for a review of outgassing.
[10] The dangers of ultraviolet light will be described more fully in Chap. 6.

strips that make an abrasion of the cornea (the clear tissue on the front of the eye) "light up" when a small penlight equipped with a "cobalt blue" filter is shone on it.

Treating a suspected foreign body in an eye on Earth can include irrigating the eye with a spray of water or saline. However, in microgravity a different technique can be used. Since free water forms floating spheres there, a small aqueous globe can be created in front of the affected astronaut's eye. That individual can let it touch the eye and then blink, thus providing a flushing action. An alternative method is to use a bag containing a small amount of water that is then squeezed out gently into the eye. Yet another technique involves having the injured individual wear goggles holding a small amount of water between one lens and the eye [7, 36].

Various standard types of eye patches and medications (e.g. topical analgesics and antibiotics, particularly in the form of ointments) are also available. Besides causing sunburn in space, excessive UV light from unfiltered sunlight can quickly produce inflammation ("keratitis") of the cornea. Also, while not an acute concern, exposure to various aspects of the space environment such as UV light, radiation, etc. can increase an astronaut's long-term risk of developing cataracts, an increase in opacity ("clouding") of the lens in an eye that impairs vision [37].

Another potential risk is that an astronaut could have a mild allergic reaction or even a life-threatening "anaphylactic" or "anaphylactoid" one while in space. These are caused by the body's immune and inflammatory system responding in a self-harming manner to a foreign substance, which might include a certain food or medication. Allergic reactions might be mild and self-limited, e.g. developing a rash, or much more serious. For example, anaphylactic and anaphylactoid reactions[11] can cause symptoms and signs such as a dangerous drop in blood pressure, constriction of bronchi in the lungs leading to severe difficulty breathing, swelling of the face and skin, and vomiting. The effects on the lungs, cardiovascular system, and brain (due to impaired oxygen supply resulting from these cardiopulmonary insults) can prove fatal even with appropriate treatment.

Prospective astronauts can be prescreened for known allergies and tested for their response to common types of "allergens." Certain medications on board the ISS could be used to treat such reactions, ranging from oral diphenhydramine to treat hives, to subcutaneous or intramuscular injections of epinephrine via a spring-loaded auto-injector for severe ones.

---

[11] In an anaphylactic reaction a person has been previously exposed to a particular foreign substance ("antigen") that may have produced little or no response, but experiences a severe reaction on being exposed to it again. Conversely, an anaphylactoid reaction occurs by a different mechanism, and the person may have that same kind of severe reaction after first exposure to an offending substance.

Research on doing various procedures and operations in microgravity has been almost exclusively limited to those done on mannequins or animals either in space or during short periods of microgravity when a plane goes into a steep, rapid descent as part of "parabolic flight" on Earth. These have included performance of cardiopulmonary resuscitation (CPR), placing IV and urinary catheters, suturing wounds, splinting and casting limbs, and various simple types of abdominal surgery [7, 11]. No actual operation has been done on astronauts in space, and only very basic surgical procedures have been attempted during parabolic flights [15, 16].

With bleeding following a serious injury or during an operation, venous blood forms hemispheres of increasing size instead of oozing [19, 38]. This effect actually makes control of venous bleeding easier than in a terrestrial environment. Simple measures such as direct compression of a bleeding area, application of appropriate dressings, or using a tourniquet on a limb would still work. On the other hand, the force and volume of bleeding from veins increases in microgravity, perhaps because their walls are less compressed than "normal" [15]. During an operation blood would also adhere to surgical instruments and gloves in microgravity, making them increasingly difficult to use.

A spurting artery produces blood droplets suspended in the air, which contaminate the immediate area and could be inhaled by anyone nearby if they are not contained. This might be done by putting a clear tent- or box-like arrangement secured over the operative field, with the surgeon and assistants putting their arms through porthole-like openings in it to reach the patient. "Gloveboxes" using a similar technique are routinely employed on the ISS for various types of activities such as performing experiments, and the methods used with them could be adapted for performing surgery [7]. A system using a clear dome filled with a sterile saline solution has also been described. It would be placed in watertight contact with the patient's skin and have equally watertight ports for inserting instruments and the surgeon's gloved arms [39].

In space, gravity would also not hold temporary wound dressings in place, although surface tension and blood clotting might be enough to do this. Even mundane tasks like collecting and measuring body fluids (e.g. urine and blood) are difficult, since they will not flow "downward" or settle to the bottom of a container.

The usual gravity-assisted technique of raising an intravenous (IV) fluid bag higher than the patient to make fluid flow into a vein will not work in microgravity. This problem can be addressed by pumping up a pressure bag around the plastic IV fluid bag to squeeze its contents into the patient. An even simpler method would be to squeeze the IV fluid bag by hand. That latter technique would not work well, however, to provide a controlled, continuous infusion of a medication.

Irrigating solutions such as sterile saline commonly used to cleanse minor wounds and during major surgery on Earth would be more difficult to control due to the risk of splashing. Careful application and, if needed, suction of these fluids would be required. Since such fluids would, like all other medical supplies, be available in limited quantities, filtered water might also be used as an alternative [4]. During an operation on the abdomen, viscera like the intestines could float out of the body, making the procedure more complicated and increasing the risk of infection [16].

Maintaining sterility during an operation in space is challenging. Suture material would float, making wound closing and keeping that material sterile more difficult. Normally a packaged surgical gown is "shaken out" by the person who will wear it, taking special care that it does not touch a nonsterile surface. In microgravity a second person would have to secure the edge of the gown to prevent it from floating free and being contaminated. The operating area could also be contaminated by airborne particles that, in a terrestrial operating room, would settle to the floor.

Even putting on surgical gloves might be difficult—trying to remove and put on the first one could make the still-packaged second glove float away. Sterile drapes used to cover the patient would need to be fastened down to prevent them from drifting off. Surgical instruments such as scalpels and hemostats would have to be secured when not being used. Methods for doing this include using magnetized implements or pads, Velcro attachments, or keeping them secure in lightweight sterile pouches near the surgical field. Sharp items such as needles and scalpels could be kept in place by sticking them into secured Styrofoam blocks [16]. Washing and disinfecting surgical equipment will require special equipment, but at least on space habitats and structures as large as the ISS it should be possible to have them available.

The patient, surgeon, and assistants would also need to be restrained to prevent them from floating away during an operation. This could be done by means such as placing restraining bands on the patient and having the surgeon use dedicated footholds attached to the interior surface of the craft/habitat.

Even providing adequate lighting for a major operation might be difficult on current craft [7]. The bright lights typical of an operating room on Earth are necessary so that the surgeon can directly see what is going on beneath the skin (e.g. a bleeding blood vessel) and deal with it appropriately. Available alternatives in space (e.g. a bright flashlight held by another crewmate, or a strap with a light worn on the surgeon's head) may be less effective.

Providing anesthesia would also be more challenging. Local anesthesia, such as injecting lidocaine before suturing a deep laceration, would be little affected by microgravity. Likewise sedation and anesthesia with short-acting intravenous medications such as midazolam or etomidate along with an analgesic (e.g. fentanyl) should, if needed for procedures such as setting a wrist fracture, be reasonably well tolerated [11].

Deeper levels of intravenous sedation might be feasible if appropriate ventilatory support were present, since the medications used to provide it typically decrease a person's respiration ("breathing") [40]. Such support might, however, have to be limited to using basic devices such as a mask-bag connected to an oxygen supply. It could also include insertion of an endotracheal (ET) tube into a person's trachea, just before that "breathing tube" divides into branches for each lung.

The ISS carries a small supply of ET tubes as well as a small ventilator. However, use of the latter typically requires 100 % oxygen, and the "expired" gas released from the patient-ventilator system would have an oxygen concentration not such less that that. Unless that oxygen-rich gas could be vented or diluted with nitrogen (with either option reducing the station's stores of those gases), it could increase the overall percentage of oxygen in the surrounding hazard and create a fire hazard that, while not at an Apollo 1 level (see Chap. 3), might still be significant.

Spinal and epidural anesthesia, general anesthesia using inhaled gases, and nerve blocks would present problems with effectiveness and/or safety in space [11]. Spinal and epidural anesthesia both require injecting appropriate agents into one of several spaces around the spinal cord. The epidural variety also involves placing a catheter for continuous administration of medications.

Spinal anesthesia requires an injection into the lower spine and is used for operations involving the lower body (roughly the mid-abdomen on down). Epidural anesthesia can potentially be applied anywhere along the spine. However, the safety and effectiveness of spinal anesthesia is dependent on gravity for making sure that only the desired parts of the body are affected, and its use in microgravity would not be feasible [19].

Conversely, epidural anesthesia and local nerve blocks could potentially be effective in space. However, they both require specialized medical training to perform, and currently planned spacecraft or the ISS may well not have anyone on board trained to do them. General anesthesia induced using standard inhaled gases would, in the (by surgical suite standards) very confined and poorly ventilated spaces of even the ISS, likely result in those gases reaching and affecting the surgeon and other attending medical personnel as well as the patient.[12] In short, any level of anesthesia above the local or intravenous

---

[12] Needless to say it is not considered good medical practice for a surgeon to take a nap while in the middle of performing an operation. While such slumber might refresh the doctor it would be unlikely to do the patient any good.

variety has serious issues regarding how well it will work and how safe it would be for all concerned [16].

Robotic surgery might also be feasible. The device could perhaps even be operated remotely by a surgeon on Earth if the patient is in LEO or at least no farther than the Moon [41, 42]. The size and mass of current standard robotic systems preclude this as an option for major present-day operations in space. Smaller devices for less complex types of surgery such as an appendectomy are, however, being evaluated [43].

When feasible, "minimally invasive" techniques such as "laparoscopic" surgery might also be used. This technique is used for procedures such as gallbladder surgery and requires only limited incisions to introduce instruments and a small camera system. Laparoscopes are thin, tubular instruments that allow transmission of images from the end inside the patient via a camera to an external video monitor. While they clearly require appropriate surgical skills to use, under some circumstances they might be employed to assess injuries within the abdomen and perhaps perform procedures to control bleeding. However, the resources needed to perform even laparoscopic surgery in space are not currently available on the ISS or any planned future missions, and for now it remains only a theoretical option.

Overall, the current assessment is that, while surgery would be more challenging in many ways and performed more slowly in microgravity, it would still be feasible to some degree under appropriate conditions.

Although it is hoped that they will never be needed, the prospects of successfully performing some less complex but potentially life-saving procedures in microgravity are better. Placing a chest tube for treatment of a pneumothorax (see Chap. 2), performing an emergency tracheostomy, or doing CPR and advanced cardiac life support (ACLS) could all be done with available resources on the ISS and (less certainly) on a large enough spacecraft sent to Mars [7].

As with doing surgery, CPR requires that both the patient and the rescuer be appropriately restrained. Otherwise the first attempt at chest compressions would send both hurling away from each other courtesy of Newton's Third Law of Motion, expressed as "For every action there is an equal and opposite reaction." A rescuer providing chest compressions might also potentially tire out easily due to muscular deconditioning in microgravity, so having crewmates frequently change who does the compressions would be even more essential than it is on Earth. An alternative would be to use a mechanical device such as an air-powered "thumper" to provide chest compressions to a secured patient instead.

Besides chest compressions and respiratory support, ACLS procedures also include use of a defibrillator and specific medications. Those resources are available on the ISS. A small defibrillator should also be included on a spaceship bound for Mars. However, if a rescuer pressed a defibrillator's paddles down on a victim's chest with enough pressure to be effective for shocking the heart, the same issues with action-reaction during chest compressions would apply. Fortunately it would take only a few more seconds to place two adhesive pads instead of paddles on the victim's chest before using the defibrillator.

5.6 The Bottom Line

Science fiction writers have many more options available for providing medical care in space and alien worlds than present-day medical professionals. We can imagine yet-undeveloped "wonder drugs" of the future or products of advanced alien technology that work far more rapidly and better than current ones. Easy-to-use devices such as Dr. McCoy's small all-purpose medical tricorder or those contained in C. M. Kornbluth's "The Little Black Bag" (1950) could be employed in an extraterrestrial environment to work marvelous cures.[13] All the myriad ways the human body could be modified, repaired, enhanced, and improved using biological and other methods via nanotechnology, genetic engineering, cybernetics, etc. are also fair game for science fiction. Although, as later chapters will describe, those techniques are in their infancy in real life, their capabilities are much vaster in our imaginations.

For example, it would be very helpful to have something similar to the "autodoc" mentioned at the beginning of this chapter to use either in space or on a planetary surface. Simply place someone with serious injuries inside it, dial in what is wrong with the unfortunate individual or, better yet, let the machine have full diagnostic capabilities to figure that out, then wait for the autodoc to make those repairs just as we now wait for food to heat up in a microwave oven. Even using it more as a "semiautodoc," with a flesh-and-blood physician controlling its mechanical/robotic parts like a compact one-person surgical suite (as depicted in Larry Niven's 1980 work *The Patchwork Girl*), would be extremely helpful. Such a machine would be particularly useful in a spacecraft that would otherwise have a very limited number of trained medical personnel or conventional equipment to provide care for the most seriously sick and injured.

However, before that device puts surgeons and other medical practitioners into unemployment lines, many practical issues would have to be addressed.

---

[13] Unless, in Dr. McCoy's case, the patient is suffering from "redshirtosis." The mortality associated with that condition is very high despite all his advanced skills and technology.

The repair process could require a patient with major, life-threatening injuries to be adequately and safely sedated for hours or even days while being treated. Medications used for such purposes typically reduce a person's drive to breathe as part of their sedating effects. Those powerful enough to produce prolonged and deep (e.g. general) anesthesia require placing an ET tube and, in the context of a completely automated device, using a ventilator to assist or even take over the patient's own breathing.

Unfortunately a sedated and intubated individual cannot perform a variety of normal, healthy functions. These include the inability to swallow saliva, or to cough and clear secretions from the lungs. Failure to do those things can help lead to development of pneumonia as well as "plugging" of bronchial tubes by mucus and other materials, impairing the patient's ability to exchange oxygen and carbon dioxide. In an intensive care unit (ICU) this can be done to some degree by periodically suctioning the mouth and through the endotracheal tube using catheters. How an autodoc might do this, however, is uncertain.

There would also have to be continual monitoring of all vital signs, measurement of blood oxygen and carbon dioxide levels, adjustment of ventilator settings (including how often it "delivers a breath" by pushing humidified and oxygenated air into the lungs, the volume of air given with each breath, and many other parameters) based on the patient's respiratory condition at a given time. In theory a sufficiently well-programmed and "smart" machine could do these things as part of a complex set of feedback loops. In practice there would remain many points of failure, such as what might happen if an endotracheal tube became dislodged or plugged with mucus. Placing or replacing an endotracheal tube can at times be challenging for even the most skilled medical personnel. It requires proper positioning of the patient's neck and visualization of the vocal cords with a lighted laryngoscope to make sure the tube goes into the trachea (where it belongs) rather than the adjacent esophagus (where it does not). While it might one day be possible to build a machine that could do this (e.g. a medical droid in the *Star Wars* movies would presumably have that capability), incorporating it into an autodoc not much larger than a coffin would be challenging.

Besides requiring help with pulmonary functions, our deeply sedated patient would also be unable to move. While this might seem a trivial observation, from a medical standpoint it presents significant issues. If a terrestrial autodoc's treatments last for many hours or days, a patient lying solely supine during that time would be at risk of developing "pressure sores" or "decubitus ulcers" on various parts of the body (e.g. the back, buttocks, and heels in contact with whatever material is lining the bottom of the autodoc). These sores and ulcers are related to the skin in those areas not receiving as much blood as they should due to blood vessels and other tissues being compressed and "sandwiched" between the bottom of the autodoc and the mass of the rest of the patient's body. Over time this can cause large areas of skin to become inflamed, infected, and even die.

On Earth, measures used to help prevent development of these sores and ulcers in bedridden patients with reduced levels of consciousness include turning them periodically onto one or the other side, using special mattresses as well as protectors for heels and other vulnerable body areas, and even powered pressure-redistributing devices. Sores can also be produced by friction, shearing, and other forces, as when body parts such as the knees or ankles rub against each other. Even lack of continued "lubrication" of the eyes by tears in a person who cannot blink can abrade and damage their delicate corneas.

Here too it would be challenging for an autodoc to provide all these important functions. Merely turning the patient in a device able to partially rotate may require securing the individual to it, and runs the risk of the straps, other cushioning materials, or even the person's legs pressing against each other causing pressure or friction injuries. "Floating" the patient above the bottom of the autodoc by using continuous streams of air would present the obvious difficulty of making sure that all parts of the body, with their different masses, were "levitated" relatively equally by these streams, as well as requiring a continuous power supply.

In microgravity these considerations might seem less of a problem, since the patient would no longer be pressing down against the autodoc's bottom. However, that individual would instead need to be firmly secured so as to avoid continuously bumping against the sides of the autodoc, and any materials used to do this or to simply cover the patient (e.g. the equivalent of a hospital gown) could still, over time, potentially scrape against skin enough to cause injury. Moreover, being in microgravity would accelerate the bone and muscle loss that a bedridden terrestrial individual experiences due to inability to exercise.

There are also serious issues with providing hydration, nutrition, and waste disposal for the person in an autodoc. Intravenous fluids can be provided via a peripheral vein or a central one, such as a subclavian vein (located beneath the clavicle, or "collarbone"), or an internal jugular vein in the neck. The catheters themselves can, however, be sources of infection. It is also unclear how the autodoc would, without human assistance, change IV bags or the equivalent as well as provide a variety of different mixtures of sterile saline concentrations, electrolytes such as potassium, etc. based on the blood tests it will also presumably be capable of obtaining and analyzing.[14]

---

[14] This problem might also apply to any blood transfusions needed by an injured individual, including that the blood be of a compatible type. However, perhaps by the time an autodoc is feasible a "one type

Since the sedated individual in the autodoc also will not be eating, if the stay in it is for more than a few days adequate nutrition would also need to be provided. Without such support a person's body will start using its own long-term stores of energy. The latter include carbohydrates, such as obtained from glycogen stores in the liver, which are quickly exhausted; fat stores; and proteins (e.g. in muscles). A severely ill or injured individual in particular needs adequate amounts of nutrients to assist in the healing process and to resist/fight infection.

These nutrients—carbohydrates, lipids, proteins, vitamins, electrolytes, etc.—must be given in adequate amounts and absorbed by the person's body. A feeding tube can be passed into the stomach or upper small intestine via the nose or mouth, or more invasively by placing a tube through the skin directly into the stomach or the jejunum (the middle section of the small intestine). However, in a sedated patient material delivered into the stomach could "reflux" back into the esophagus. A small amount of this food might even find its way into the lungs despite the trachea leading to them being mostly sealed off by an endotracheal tube with an inflated cuff at its end. Also, the ailing individual's gastrointestinal tract may not be working properly, leading to problems such as the food delivered not being adequately digested and absorbed.

Another alternative is to give all nutrients completely intravenously—"total parenteral nutrition." This method can actually be used to provide nutrition indefinitely. However, it carries both acute and long-term risks such as infections or blood clots forming on the long catheter used to deliver it that, in turn, could potentially break off and travel to the lungs (a "pulmonary embolism").[15]

The autodoc would also need the means to collect urine and fecal material. The former can be done in both men and women by inserting a catheter directly into the bladder via the urethra or, less commonly, directly into the bladder. In men an external condom catheter can also be used. Here too, as with endotracheal intubation, having the autodoc itself place a catheter would be problematic, and might be better done for it by a human assistant. However, if the autodoc were to otherwise operate completely autonomously it would also need to collect all the urine created, dispose of it, and as needed analyze (e.g. for blood cells, proteins, glucose, bacteria, etc.) and culture it if a UTI (a common problem in this setting) is suspected.

Collecting and disposing of fecal matter would, as in real life, be even more challenging. Based on the principle of "what goes in must come out," whatever solid wastes were in the person at the time of the injury/illness that required being placed in an autodoc will eventually be expelled. Those wastes would

---

replenishes all" form of artificial blood might also be available for at least emergency purposes.
[15] See Chap. 7 for more about these issues.

then need to be removed and the patient cleaned in a very timely fashion after they make their exit. The reason for the latter is not merely aesthetic. Fecal material is highly enriched in bacteria that could cause infections involving nearby skin, the urinary tract, etc. A rectal tube can sometimes be useful in patients with diarrhea to help reduce soiling. However, a rectal tube has much more limited applicability if feces are solid. The latter, if not allowed to exit the rectum or if they become impacted there, would need to be removed by other means, such as digital disimpaction,[16] enemas, etc. that, again, might prove difficult for an autodoc.

All the challenges of creating an autodoc described so far have only involved supporting the injured or sick patient rather than actually *doing* anything to treat and cure anyone. As noted in Chap. 2, the human body can be burned, bleed internally and externally, penetrated by various objects, damaged by blunt trauma, and otherwise physically insulted in many different ways. The skull and bones can be fractured, joints dislocated, skin and internal organs lacerated, bruised, and torn, limbs partially or completely amputated, fleshy protuberances such as the nose, ear, and (in males) the external genitalia ripped off. Such injuries can range anywhere from mild to disabling to life-threatening to quickly fatal. The medical skills and resources needed to deal with them extend from first aid to major surgery with, in the severely injured, potentially many months of rehabilitation and recuperation that may still lead to varying degrees of permanent disability. Likewise, myriad diseases and events such as a myocardial infarction, severe infection, etc. can also be serious enough to require major medical interventions.

No matter how much medical technology and treatments advances in the future, if the human body remains essentially unchanged its own intrinsic vulnerability to illness and injury will limit what can be done to protect and heal it. For an autodoc or any other single device, no matter how sophisticated, to be able to treat all or even part of the vast spectrum of illnesses and injuries our bodies are vulnerable to would require a dazzling array of diagnostic and therapeutic capabilities.

Thus, for example, while performing different types of major surgery have some aspects in common (e.g. need for anesthesia, monitoring of vital signs, infusion of fluids and sometimes blood, etc.) they can differ markedly in details, including the particular surgical skills and techniques required. Likewise, some of the instruments and ancillary items (e.g. hardware such as pins, rods, nails, screws, etc. for dealing with serious fractures) are very specific for particular types of operations, with no "one size fits all."

---

[16] Yes, this does mean putting a (hopefully gloved) finger into the rectum and literally scooping out any solid waste encountered there.

Thinking beyond current medical technology, as will be seen in later chapters we are currently a long way from realizing the possible benefits of medical nanotechnology, stem cell therapy, etc. and uncertain how far these "futuristic" methods can even be developed. Even using smaller versions of diagnostic equipment such as x-ray machines, CT scanners, far more sophisticated versions of robotic devices for performing surgery, etc., and trying to compress the current or much better capabilities of a modern well-staffed hospital into a single machine would be, to say the least, challenging. This would be the case no matter how well the autodoc was programmed to know how to do surgery, administer medications, etc.—if a treatment required far more equipment, tools, drugs, etc. than it had available to employ, all that "textbook" knowledge and skill it possessed could not be applied.

The purpose of this analysis is not to "prove" that an autodoc is impossible. Such a device or other SF creations with similar functions (e.g. bacta tanks) are not only extremely useful for story purposes but machines with only some of its imagined capabilities could improve overall healthcare in real life. Instead it is meant to show that there are major challenges to bringing such SF concepts into actual medical practice as opposed to presenting them as a fait accompli in a story or other work. Perhaps, instead of having an "automatic doctor" that is (at least relatively speaking) a "jack-of-all-trades" capable of doing a wide range of care on its own or with limited human assistance, a "semiautodoc" might be more practical. This could involve a division of labor, with medically trained humans (or other sentient beings) performing certain functions such as providing the machine with the physical medications, fluids, etc. it would "decide" to administer, as well as less desirable but necessary ones like cleaning up a patient's messy bowel movement.[17] Likewise a flesh-and-blood surgeon or other specialist could work together with the machine as a team to perform an operation better than either could alone.

In short, SF stories placed in the present and near future will have to take into account all the many issues with medical care in space described in this chapter. Going much beyond those capabilities requires either locations such as large space habitats or spaceships, city-sized colonies on Mars, the Moon, large asteroids, etc., well-developed worlds outside the Solar System that have extensive medical infrastructures, or invoking sophisticated but speculative future advances in nanotechnology and other fields. But even if it cannot provide the exact details of how to achieve those things, SF can give us goals we can aim to reach.

---

[17] The latter task is one that nurses and other medical personnel would, however, gladly delegate if possible to an autodoc.

References

1. Buckey JC. Space physiology. Oxford: Oxford University Press; 2006.
2. Hamilton D, Smart K, Melton S, Polk J, Johnson-Throop K. Autonomous medical care for exploration class space missions. 2007. http://ntrs.nasa.gov/archive/nasa/casi.ntrs.nasa.gov/20070032039.pdf. Accessed 15 April 2015.
3. Sargsyan A. Medical imaging. In: Barratt M, Pool SL, editors. Principles of clinical medicine for space flight. Berlin: Springer; 2008. p. 181–207.
4. Gray G, Johnston S. Medical evaluations and standards. In: Barratt M, Pool SL, editors. Principles of clinical medicine for space flight. Berlin: Springer; 2008. p. 59–67.
5. Marshburn T. Acute care. In: Barratt M, Pool SL, editors. Principles of clinical medicine for space flight. Berlin: Springer; 2008. p. 101–22.
6. Hamilton D, Scheuring R, Jones J. Right lower quadrant abdominal pain in a female crewmember on the International Space Station. Aviat Sp Environ Med. 2007;78(Suppl. 4):A89–98.
7. Clément G. Fundamentals of space medicine. 2nd ed. Berlin: Springer; 2011.
8. Clark J, Allen C. Acoustics issues. In: Barratt M, Pool SL, editors. Principles of clinical medicine for space flight. Berlin: Springer; 2008. p. 521–33.
9. Smith S, Nixon C. Vibration, noise, and communication. In: De Hart RL, Davis JR, editors. Fundamentals of aerospace medicine. 3rd ed. Philadelphia: Lippincott Williams & Wilkins; 2002. p. 154–83.
10. Taddeo T, Armstrong C. Spaceflight medical systems. In: Barratt M, Pool SL, editors. Principles of clinical medicine for space flight. Berlin: Springer; 2008. p. 69–100.
11. Stewart L, Trunkey D, Rebagliati S. Emergency medicine in space. J Emerg Med. 2007;32(1):45–54.
12. Putcha L, Pool SL, Cintrón N. Pharmacology. In: Nicogossian AE, Huntoon CL, Pool SL, editors. Space physiology and medicine. 3rd ed. Philadelphia: Lea & Febiger; 1993. p. 435–46.
13. Graebe A, Schuck E, Lensing P, Putcha L, Derendorf H. Physiological, pharmacokinetic, and pharmacodynamic changes in space. J Clin Pharmacol. 2004;44(8):837–53.
14. Kirkpatrick A, Jones J, Sargsyan A, Hamilton D, Melton S, Beck G, et al. Trauma sonography for use in microgravity. Aviat Sp Environ Med. 2007;78(Suppl. 4):A38–42.
15. Kirkpatrick A, Campbell M, Jones J, Broderick T, Ball C, McBeth P, et al. Extraterrestrial hemorrhage control: terrestrial developments in technique, technology, and philosophy with applicability to traumatic hemorrhage control in long-duration spaceflight. J Am Coll Surg. 2005;200(1):64–76.
16. Drudi L, Ball C, Kirkpatrick A, Saary J, Grenon S. Surgery in space: where are we at now? Acta Astronautica. 2102;79:61–6.
17. Simmons S, Hamilton D, McDonald PV. Telemedicine. In: Barratt M, Pool SL, editors. Principles of clinical medicine for space flight. Berlin: Springer; 2008. p. 163–79.

18. Cermack M. Monitoring and telemedicine support in remote environments and in human space flight. Br J Anaesth. 2006;97(1):107–14.
19. Campbell M, Billica R. Surgical capabilities. In: Barratt M, Pool SL, editors. Principles of clinical medicine for space flight. Berlin: Springer; 2008. p. 123–37.
20. Johnston S, Arenare B, Smart K. Medical evacuation and vehicles for transport. In: Barratt M, Pool SL, editors. Principles of clinical medicine for space flight. Berlin: Springer; 2008. p. 139–61.
21. Stepaniak P, Hamilton G, Olson J, Gilmore S, Stizza D, Beck B. Physiologic effects of simulated +Gx orbital reentry in primate models of hemorrhagic shock. Aviat Sp Environ Med. 2007;78(Suppl. 4):A14–25.
22. Barr Y, Bacal K, Jones J, Hamilton D. Breast cancer and spaceflight: risk and management. Aviat Sp Environ Med. 2007;78(Suppl. 4):A26–37.
23. Pietrrzyk R, Jones J, Sams C, Whitson P. Renal stone formation among astronauts. Aviat Sp Environ Med. 2007;78(Suppl. 4):A9–13.
24. Hamilton D. Cardiovascular disorders. In: Barratt M, Pool SL, editors. Principles of clinical medicine for space flight. Berlin: Springer; 2008. p. 317–59.
25. Payne MW, Williams DR, Trudel G. Space flight rehabilitation. Am J Phys Med Rehabil. 2007;86(7):583–91.
26. Locke J. Space environments. In: De Hart RL, Davis JR, editors. Fundamentals of aerospace medicine. 3rd ed. Philadelphia: Lippincott Williams & Wilkins; 2002. p. 245–70.
27. Smith S, Lane H. Spaceflight metabolism and nutritional support. In: Barratt M, Pool SL, editors. Principles of clinical medicine for space flight. Berlin: Springer; 2008. p. 559–76.
28. Horneck G, Klaus DM, Mancinelli RL. Space microbiology. Microbiol Mol Biol Rev. 2010;74(1):121–56.
29. Pierson D. Microbiology. In: Nicogossian AE, Huntoon CL, Pool SL, editors. Space physiology and medicine. 3rd ed. Philadelphia: Lea & Febiger; 1993. p. 157–66.
30. Larios-Sanz M, Kourentzi K, Warmflash D, Jones J, Pierson D, Willson R, et al. 16S rRNA beacons for bacterial monitoring during human space missions. Aviat Sp Environ Med. 2007;78(Suppl. 4):A43–7.
31. Stepaniak P, Ramchandani S, Jones J. Acute urinary retention among astronauts. Aviat Sp Environ Med. 2007;78(Suppl. 4):A5–8.
32. Huntoon CL, Whitson P, Sams C. Hematologic and immunologic functions. In: Nicogossian AE, Huntoon CL, Pool SL, editors. Space physiology and medicine. 3rd ed. Philadelphia: Lea & Febiger; 1993. p. 351–62.
33. De Santo N, Cirillo M, Kirsch K, Coreale G, Drummer C, Frassl W, et al. Anemia and erythropoietin in space flights. Seminars Nephrol. 2005;25(6):379–87.
34. Morey-Holton E, Hill E, Souza K. Animals and spaceflight: from survival to understanding. J Musculoskelet Neuronal Interact. 2007;7:17–25.
35. Farahani R, DiPietro L. Microgravity and the implications for wound healing. Int Wound J. 2008;5(4):552–61.

36. Manuel F, Mader T. Ophthalmologic concerns. In: Barratt M, Pool SL, editors. Principles of clinical medicine for space flight. Berlin: Springer; 2008. p. 535–44.

37. Jones J, McCarten M, Manuel K, Djojonegoro B, Murray J, Feiversen A, et al. Cataract formation mechanisms and risk in aviation and space crews. Aviat Sp Environ Med. 2007;78(Suppl. 4):A56–66.

38. Campbell M, Billica R, Johnston S. Surgical bleeding in microgravity. Surg Gynecol Obstet. 1993;177:121–5.

39. Rosen R. Blood in zero gravity: NASA tries to prepare for surgery in space. 2012. http://www.theatlantic.com/technology/archive/2012/10/blood-in-zero-gravity-nasa-tries-to-prepare-for-surgery-in-space/263171/. Accessed 15 April 2015.

40. Agnew J, Fibuch E, Hubbard J. Anesthesia during and after exposure to microgravity. Aviat Sp Environ Med. 2004;75(7):571–80.

41. Haidegger T, Sanor J, Benyo Z. Surgery in space: the future of robotic telesurgery. Surg Endosc. 2011;25:681–90.

42. Espiner T. NASA's robonaut 2 scrubs up for space surgery. 2014. http://www.bbc.com/news/technology-26872884. Accessed 15 April 2015.

43. Rutkin A. Mini robot space surgeon to climb inside astronauts. 2014. http://www.newscientist.com/article/dn25341-mini-robot-space-surgeon-to-climb-inside-astronauts.html#.U4fElRb7CF4. Accessed 15 April 2015.

6

Danger! Radiation!

Dr. Leonard McCoy: "Are you out of your Vulcan mind? No human can tolerate the radiation that's in there!"
Captain Spock: "As you are so fond of observing, doctor, I am not human."

Star Trek II: The Wrath of Khan (1982)

"Well, some mighty queer specimens came out of the radioactive-affected areas around the bomb targets. Funny things happened to the germ plasm. Most of 'em died out; they couldn't reproduce; but you'll still find a few creatures in sanitariums—two heads, you know."

Henry Kuttner and C. L. Moore, writing as "Lewis Padgett"
"The Piper's Son" (1945)

"Radiation" in science fiction can be used to kill, injure, or simply put characters at risk. Many stories and novels written during the first decades after Hiroshima, such as Walter M. Miller, Jr.'s novel *A Canticle for Leibowitz* (1960), depicted the deadly effects of radiation on humans and other life in the apocalyptic scenario of nuclear war. Even earlier, Robert A. Heinlein (using the pseudonym "Anson MacDonald") in his 1940 story "Solution Unsatisfactory" depicted planes dropping radioactive dust on hostile cities to rapidly wipe out entire populations. Merely mentioning the word "radiation" in a science fiction work puts the reader on alert that dramatic and potentially terrible things could (and probably will) happen.

Although less common in modern science fiction, radiation has also been used as a virtually magical means to create monstrous threats to humanity or, at the other end of the scale, to help it develop new plot-furthering abilities. Science fiction movies in the 1950s taught us that the right kind of radiation exposure could make ants (*Them!*), tarantulas (*Tarantula*), grasshoppers (*Beginning of the End*), and even humans (*The Amazing Colossal Man*) grow to enormous sizes. In written science fiction the "mutant" offspring of post-World War III survivors might develop telepathy (such as in the Kuttner and

Moore series of "Baldy" short stories quoted previously), enhanced intelligence (e.g. the 1953 novel *Children of the Atom* by Wilmar H. Shiras), or other abilities rendering them superior to us mediocre members of *Homo sapiens*. Radiation could also either accelerate human evolution, as in Edmond Hamilton's 1931 "The Man Who Evolved," or create dramatic atavisms such as those presented in Hollis Alpert's 1960 short story "The Simian Problem."

The problem with how radiation is depicted in science fiction is not that, in the right circumstances, it will not harm people or produce mutations in progeny. It certainly can. Instead it is the details—the actual acute and chronic damage/changes radiation exposure can cause, how long it takes for symptoms to develop or death to occur, and what kind of real effects it has on individuals and their children—that can be wildly exaggerated or simply wrong. And at the other end of the spectrum, often the risk of radiation exposure is downplayed or goes unmentioned in science fiction situations where it is indeed a real threat.

6.1 What is Radiation?

The way the word "radiation" is commonly used is only one specific variety of how physicists define the term. Its general definition is simply "the emission and propagation of energy; also, the emitted energy itself" [1]. This energy can be carried in "acoustic" waves—that is, sound waves with frequencies (see below) ranging from far below those our ears can hear to those much higher (ultrasound). Thus, technically whenever you or I speak to other people we are bombarding them with radiation (though hopefully, unless a deranged dentist has just put a radioactive filling in one of our teeth, only of the acoustic variety).

"Particulate" radiation, as the name indicates, consists of atomic or subatomic particles with high velocities and energy. These include alpha particles (helium atoms stripped of their electrons, leaving only their nuclei), beta particles (free electrons or positrons),[1] and other parts of atoms such as free protons and neutrons. This definition also includes nuclei heavier (having more protons and neutrons) than helium, such as high-energy particles of beryllium, boron, and other elements originating from outer space that are part of "cosmic rays" (more on this later).

---

[1] A positron is a form of antimatter identical to an electron except for having a positive charge rather than the negative one of an electron. Isaac Asimov's classic stories notwithstanding, no one has found a way yet to use positrons to help create advanced robotic brains.

Table 6.1 The electromagnetic spectrum

Name	Frequencies (Hz)	Wavelengths (meters)
Gamma rays	$>3 \times 10^{19}$	$<1 \times 10^{-11}$
X-rays	3×10^{17}–3×10^{19}	1×10^{-8}–1×10^{-11}
Ultraviolet light	7.5×10^{14}–3×10^{17}	4×10^{-7}–1×10^{-8}
Visible light	4.3×10^{14}–7.5×10^{14}	7×10^{-7}–4×10^{-7}
Infrared light	3×10^{12}–4.3×10^{14}	1×10^{-3}–7×10^{-7}
Microwaves	3×10^{9}–3×10^{12}	1×10^{-1}–1×10^{-3}
Radio waves	$<3 \times 10^{9}$	$>1 \times 10^{-1}$

Any type of electromagnetic energy is also a form of radiation. Electromagnetic waves consist of oscillating electrical and magnetic fields traveling together through space [2]. In a vacuum these waves travel at their maximum velocity of slightly less than 3×10^8 m/s and move at a minimally slower velocity through air. The number of oscillating cycles an electromagnetic wave (or any other kind of wave, such as sound waves) completes each second is its "frequency," measured in hertz (Hz, with one Hz = one cycle per second). The distance the electromagnetic wave travels during each cycle is its "wavelength," measured in meters. There is an inverse relationship between frequency and wavelength, described by

$$f = \frac{c}{\lambda}$$

where f is the frequency in hertz, c is the aforementioned speed of electromagnetic energy in a vacuum (commonly called the "speed of light"), and λ is the wavelength in meters.

Table 6.1 shows the frequencies and corresponding wavelengths of different regions of the "electromagnetic spectrum." Radios, televisions, and cell phones use "radio waves"—electromagnetic radiation having the longest wavelengths. For example, in the United States a standard AM (amplitude modulation) radio receives signals in a band of frequencies ranging from 535–1705 kHz (1 kHz = 1000 Hz), with slightly different ranges in other countries. The range of frequencies used by cell phones is much higher than that, in certain bands located between about 806 MHz (1 MHz = 1,000,000 Hz) and 2690 MHz.

Electromagnetic waves in the next higher band of frequencies are called "microwaves." They are used for radar and, of course, microwave ovens. Although we cannot see infrared light, in the next higher range of frequencies, we can sense it as heat. Our eyes are sensitive to only a tiny segment of the electromagnetic spectrum, the part we commonly call "light," with wavelengths slightly shorter and frequencies just greater than the infrared range.

Specialized cells called cones in the retina, a membrane lining the back of the eye, perceive electromagnetic radiation with wavelengths between 700 and 400 nm (one nanometer, or nm, $= 1 \times 10^{-9}$ m) and associated frequencies of 4.3×10^{14}–7.5×10^{14} Hz as "colors." What we see as "red" corresponds to the longest wavelengths (about 650–700 nm) and lowest frequencies in that range. Other colors—orange, yellow, green, blue, indigo, and violet, along with their intermediate shades—represent progressively shorter wavelengths and higher frequencies.

Ultraviolet (UV) light produced by the Sun or artificial sources is divided into three basic ranges of wavelengths shorter than visible light—UVA (<400–315 nm), UVB (<315–280 nm), and UVC (<280–100 nm) [3]. X-rays and gamma rays, with still shorter wavelengths and higher frequencies, are used for (among other things) obtaining images of internal body organs and treating certain diseases such as cancer.

Max Planck (1858–1947), Albert Einstein (1879–1955), and other twentieth century physicists found that electromagnetic energy can, in different settings, act either as a wave or a particle. When it acts like the latter these "discrete," massless "bundles" of energy are called "photons." The energy of a photon is proportional to its frequency in the relationship

$$E = hf$$

where E is energy in joules, h is Planck's constant (about 6.625×10^{-34} J-s), and f is the frequency in hertz. The energy of a photon is typically measured in a unit called an "electron volt" (eV), defined in technical terms as the kinetic energy gained by an electron that is accelerated through an electrical potential of one volt. One electron volt is equal to about 1.602×10^{-19} J.

6.2 Biological Effects of Radiation Exposure

In case you have any doubts, being bitten by a radioactive spider will not give you the power to climb walls. Exposure to high levels of gamma radiation will not make you intermittently turn into an unjolly green giant with anger management issues. And despite what is depicted in the 1936 Bela Lugosi-Boris Karloff movie *The Invisible Ray*, it will also not make a radiation-exposed character glow in the dark, threaten the innocent with instant death from his radioactive touch, and ultimately vaporize in a blinding burst of light.

What is commonly called radiation refers to a very specific type—"ionizing radiation." Both high-energy particles (e.g. alpha and beta particles produced by the decay of uranium and other radioactive elements) and electromagnetic radiation whose photons have sufficient energy can use that energy to strip electrons away from atoms in biological systems (e.g. human beings) [4]. This "ionizes" the atoms and molecules of important chemicals and structures within our cells. UVB, UVC, x-rays, and gamma rays all have photons with enough energy to be ionizing.

However, radio waves, microwaves, infrared, and visible light, and UVA do not have enough energy to ionize atoms. The most these forms of "non-ionizing" radiation can do is to increase the thermal energy ("heat") of atoms and molecules within food, water, and other materials. Thus the heating element glowing red in a conventional oven generates enough infrared radiation for cooking. A microwave oven can generate microwaves with sufficient power to heat a cup of water to make tea—a capability that also explains why it is a bad idea to put your hand in that appliance while it is running lest the water in its cells and blood boil too. UVA has enough energy to tan skin or, with prolonged exposure, cause sunburn.

Ionizing radiation can directly disrupt chemical bonds and otherwise damage proteins, DNA, and other parts of cells so that they cannot function properly or even be destroyed. This form of radiation also causes indirect damage by producing highly reactive substances called "free radicals." In living organisms important free radicals include atoms and molecules consisting of oxygen alone, such as superoxide (O_2^-), or of oxygen and hydrogen (e.g. the hydroxyl free radical) [4, 5]. Oxygen and hydrogen atoms are, of course, part of water (H_2O)—a ubiquitous and necessary component of our bodies both within and outside cells.

The ultimate effects of radiation exposure in an individual are based on many factors. These include the type of radiation (electromagnetic or particulate) received; how many and which body cells are affected; whether a person's whole body or only certain areas and organs are exposed; the total dose of ionizing radiation delivered; whether a particular total dose is given all at once over a short time (e.g. minutes or hours) or in separate, individually much smaller doses over a long time (such as months or years); and how well or even if the exposed person's body can repair radiation-induced damage to cells.

If the injury to cells caused by the direct and indirect effects of radiation is severe enough the cells will die. Less severe damage may cause cells to not function properly. They might also fail to reproduce, thus further depleting the body's supply of healthy cells. In particular a radiation-induced change or "mutation" to the cell's DNA that is not enough to kill it could, if the cell does divide, result in creation of new defective copies. These "malignant" cells might grow out of control, potentially replacing normal cells or growing into tumors that injure and destroy surrounding healthy tissue. This process is called "cancer."

Ionizing radiation can wreck havoc by causing either "germ cell" or "somatic cell" mutations. Germ cells are the ones adults possess that are essential for sexual reproduction, e.g. sperm cells in males and egg cells (oocytes and ova) in females. Somatic cells include all the other types of developed cells that make up the human body's organs and tissues.

Some somatic cells, such as the various kinds that ultimately produce red and white blood cells, may undergo many divisions throughout a person's lifetime. These and other cells that undergo repeated and often rapid division (e.g. those in hair follicles or the lining of the intestines) are the most sensitive to radiation-induced damage and possible mutation. However, if not too many of these cells are destroyed by radiation a person's body could replace them. Other somatic cells, such as ones in the heart and brain, do not divide further after they reach their final stage of development. They are more resistant to damage from radiation. But if these cells are damaged or destroyed, the human body has only a very limited ability to repair or replace them.

This point deserves stressing. Exposing children and adults to high enough levels of radiation can only damage or kill their somatic cells. It will not confer any kind of superpower (Sorry, Fantastic Four!) or transform anyone into a bloodthirsty reptilian mutant as in the 1959 movie *The Hideous Sun Demon*. Ionizing radiation can only make healthy somatic cells unhealthy in conventional ways, by causing cancer or other abnormalities and injuries described in standard medical textbooks.

The situation is slightly different in the case of radiation-induced changes to germ cells. Sperm cells are among the most sensitive to damage from radiation. They are produced continuously during a healthy male adult's lifetime, at a typical rate of several hundred million per day. Starting from "scratch," it takes roughly several months for mature sperm cells (spermatozoa) to develop. How long they live after that depends on where they wind up, but that lifespan is usually measured only in days.

This high "turnover" rate contributes to sperm cells' sensitivity to radiation damage. However, it also means that if a man receives a radiation dose sufficient to kill some of them (more on this later) he might, over a period of months, eventually produce more to get his sperm count back to normal levels both quantitatively and qualitatively.

Egg cells are less susceptible to radiation damage and death than sperm cells. One reason is that egg cells are somewhat better shielded. They are located in a woman's two ovaries, almond-shaped organs typically several centimeters in length within her lower abdomen. The sperm cells stored in a man's two testes have only the thin skin of the surrounding scrotum to protect them. Conversely both the ovaries and the egg cells within them are located inside the woman's body, where overlying layers of skin and fat help protect them.

Women produce all the immature egg cells (oocytes) they will ever have even before they are born, with the number of egg cells continually falling after birth. Thus they do not undergo the ongoing creation via cell division and development that sperm cells do in adult men. As noted previously, replicating cells, like those that produce sperm cells, white blood cells, etc., are more sensitive to radiation-induced damage than ones like egg cells, heart muscle cells (myocardiocytes), or other cells that are no longer dividing and either are (or are nearly) mature. Egg cells fall into the latter category, thus contributing to their relative resistance to radiation. However, unlike sperm cells, if the radiation dose is enough to destroy egg cells, a woman cannot replace them.

A dose of radiation may be enough to cause a mutation in a sperm or egg cell's DNA, but not sufficient to destroy the cell or render it incapable of being involved in successful fertilization. If such a radiation-damaged sperm and/or egg cell participates in fertilization, the altered DNA will be incorporated into the resulting offspring. The genetic mutations in such progeny may be severe enough that death occurs even before birth. Babies that do survive may have problems with development of their brains and other parts of their bodies, as well as an increased lifetime risk of cancer. A radiation-induced genetic mutation in offspring's DNA that involves changing the coding of as little as a single amino acid differently than "normal" can be enough to cause health problems (e.g. sickle cell trait and anemia).

There is a tiny chance that a radiation-induced mutation could be beneficial rather than harmful. Such mutations occurring over very long time spans in large populations may have contributed to human evolution. However, every "improvement" was counterbalanced by many, many more "failures." If a beneficial radiation-induced mutation occurs at all it will most likely be something subtle, such as a slight change in an enzyme that improves its function. Unfortunately, even such a hypothetical "good" mutation might be accompanied by others that are deleterious.

Moreover, intelligence, longevity, and other complex human characteristics are influenced by many genes as well as myriad environmental and other factors.[2] The odds of radiation-induced changes in germ cells improving rather than impairing such traits in the children produced may not be zero but it is vanishingly remote. For all these reasons (and in some cases others), far from developing psychic powers, super intelligence, two heads, the ability to fly, etc., real-life "mutants" would be expected to be unhealthy in various purely conventional ways.

---

[2] Chapter 12 will go into this in much more detail.

6.3 Radiation Dosage and the Human Body

Radiation exposure is measured in rads ("radiation absorbed dose") or grays (Gy, with 1 gray = 100 rads) [6, 7]. A rad is defined as the dose of radiation that deposits 0.01 J of energy into one kilogram of matter. The specific effect of radiation on human tissues and organs is measured in rems ("roentgen equivalent man"), or sieverts (Sv, with 1 sievert = 100 rems). One rem is equal to one rad multiplied by a "quality factor" (QF). This QF is 1 for electromagnetic radiation, beta particles, and lower-energy protons; 10 for neutrons; and 20 for alpha particles. Therefore, for electromagnetic radiation 1 rad always equals one rem. Neutrons and alpha particles impart more destructive energy as they pass through our tissues than the other types of radiation do, hence their higher QFs. Thus, for a higher-energy neutron one rad is equal to 1×10 (the QF), or 10 rem. This QF applies primarily to the long-term risks of radiation exposure (e.g. development of cancer) and its cumulative effects rather than the short-term problems caused by a single large dose of radiation.

Our annual exposure to background radiation here on Earth is in a typical range of 0.1–0.3 rad [8, 9]. This is equivalent to 0.1–0.3 rem, or 1–3 mSv (mSv = millisievert, one-thousandth of a sievert). While some of this background radiation comes from extraterrestrial sources such as cosmic rays, most of it originates from natural and artificial terrestrial sources. For example, radon is a radioactive element produced by the decay of tiny amounts of uranium and other naturally occurring radioactive elements present in some building materials. Radon is produced as a gas that can be inhaled by individuals living in those dwellings. Medical diagnostic tests may also contribute to a person's annual radiation exposure. A chest x-ray delivers about 0.01 rad. For comparison, the safety limit for whole-body exposure in radiation workers is 5 rem per year [10].

Table 6.2 summarizes the effects of increasing acute doses of ionizing radiation to the entire body and the delayed damage they can produce [5, 7, 11–13]. A whole-body acute dose of between roughly 20 and 50 rad delivered over minutes to hours will at most make counts of white blood cells, platelets, and sperm cells fall for a short time, beginning about one day after exposure. It will not produce symptoms or cause death, although it will increase the long-term risk of cancer.

An exposure between 50 and 150 rad will damage cells in the bone marrow, skin, and those lining the stomach and intestines, producing the "prodromal syndrome." Within hours of receiving that dose the person may develop redness of the skin, fever, nausea, vomiting, weakness, cramps, and diarrhea. These symptoms typically resolve within 2 days, and the survival rate is virtually 100 %.

Table 6.2 Acute and delayed effects of whole-body radiation

Radiation dose (rad)	Primary cells injured	Symptoms and signs	Latency	Outcome
Acute effects				
~20–50	White blood cells and platelets	None. Mild, transient decrease in white cell and platelet counts. Fall in sperm cell counts	1 day	Survival—100 %
50–150 (Prodromal syndrome)	Skin, white blood cells, gastrointestinal tract	Redness of skin, fever, nausea, vomiting, weakness, cramps, diarrhea	Hours	Resolution of symptoms within 48 h. Survival— nearly 100 %
150–400 (Hematopoietic syndrome)	Bone marrow	Prodromal syndrome, plus anemia, skin rash, mouth ulcerations, bleeding, infections. Marked fall in blood cells and platelets. Hair loss at higher doses	3–4 weeks	Three week asymptomatic period after prodromal syndrome, then anemia etc. Hair loss delayed for 2–3 weeks. Death rate 5–50 %. May need bone marrow transplant
400–800 (Hematopoietic/gastrointestinal syndrome)	Bone marrow and gastrointestinal tract	Prodromal and hematopoietic syndromes but more severe, with vomiting and diarrhea	1 day	Death rate 50–90 % within 6 weeks even with intensive treatment
800–2000 (Gastrointestinal syndrome)	Small intestine	Prodromal syndrome, plus bloody diarrhea, loss of appetite, infection of the intestine and septic shock	4–6 days	Death rate approaching 100 % at lower doses in this range and 100 % at higher ones, up to 2–2.5 weeks after exposure

Table 6.2   (continued)

Radiation dose (rad)	Primary cells injured	Symptoms and signs	Latency	Outcome
>2000 (Central Nervous System syndrome)	Brain	Prodromal syndrome, plus headache, apathy, tremors, unsteady gait, convulsions, coma	Minutes to hours	Death rate 100 %, usually within several days to a week
Delayed effects				
No threshold for cancer risk. Increasing acute and lifetime doses associated with increasing risk	White blood cells, thyroid, breast, lung, bone, skin, etc. for cancer. Lens of the eye. Sperm cells and ova	Variable for cancer. Cataracts. Infertility. Genetic defects in offspring	Variable. Months to years	Latent period varies from 4 to 10 years (leukemia) to up to 25 years (thyroid and breast cancers)

Note that these syndromes may overlap with each or occur sequentially

A dose between 150 and 400 rad will produce the same (though probably more severe) initial symptoms as the prodromal syndrome along with more serious and potentially fatal ones later. The cells in the bone marrow that produce red and white blood cells, as well as those that create the platelets the body needs to make blood clot, will be severely damaged. The person will be asymptomatic after the prodromal syndrome ends but then develop the "hematopoietic syndrome" about 3–4 weeks after the exposure. At that time there will be a marked drop in blood cell counts, resulting in anemia, bleeding, and an increased risk of infection. Even with the best treatment, potentially including a bone marrow transplant, mortality ranges from between about 5 % at the lower end of this radiation dose range to 50 % at its upper end.

At the higher doses in this range and at greater doses the victim's hair will eventually fall out due to damage to hair follicles. However, this will not occur until about 2–3 weeks after the radiation exposure. The person's hair should eventually grow back, unless for some reason a very high dose of radiation was directed at the scalp but the body as a whole received a sublethal dose. Thus, any scene depicted in print, television shows, or movies that shows a character pulling out his or her hair or merely having it fall out shortly after receiving a dose of radiation at least this high is not accurate.

An acute dose between 50 to several hundred rad also temporarily reduces fertility in men by damaging and destroying sperm cells. However, nonfatal doses higher than 400 rad could make a man sterile. Similar doses could also kill enough of a woman's oocytes to render her sterile too.

Individuals exposed acutely to 400–800 rad will develop the "hematopoietic-gastrointestinal" syndrome. They experience the symptoms of the prodromal syndrome but much more severely than at lower doses, with incapacitating vomiting, diarrhea, and dehydration. If they survive long enough they will develop a worse case of the hematopoietic syndrome within a few weeks. The mortality rate even with optimal medical care is as high as 90 % within 6 weeks.

An acute dose between 800 and 2000 rad will, except for a very small chance for survival at the lowest end of that range with intensive treatment, eventually prove fatal. Besides all the symptoms previously described, within 4–6 days after exposure the person will then develop the "gastrointestinal syndrome" with bloody diarrhea, loss of appetite, and infections caused by bacteria released from the damaged intestines, eventually culminating in septic shock. Death typically occurs within about 2 weeks.

Doses greater than 2000 rad are uniformly fatal and produce the "central nervous system syndrome." Symptoms such as headache and tremors can develop within hours after exposure. These are followed by more severe problems such as unsteady gait, apathy, difficulty thinking, convulsions, and eventually coma. Death occurs within several days to about a week—too short a time for development of effects such as hair loss or the hematopoietic syndrome to occur.

The risk of developing cancer increases linearly with both increasing levels of a nonfatal acute dose and the total amount of radiation received in one's lifetime. For example, although this is definitely not recommended, if a person were to receive one rad per day for a year (equal to a total annual dose of about 365 rad) that individual would not develop any of the symptoms this same dose delivered over an hour would cause. However, giving this total dose in daily exposures of 1 rad would definitely increase the chances of developing cancer years later. The typical time, or "latency period," it takes for different types of cancer to develop after acute radiation exposure varies. Leukemia, a form of cancer involving white blood cells, may manifest within 4–10 years. Thyroid and, in women, breast cancer may take up to 25 years. Individuals can also develop cataracts, a "clouding" of the lens in each eye that impairs vision.

6.4 Realistic Settings for Radiation in Science Fiction

Outside of comic books and B-movies, radiation can figure as a plot element in more realistic science fiction scenarios. The 1998 novel *Lethal Exposure* by Kevin J. Anderson and Doug Beason opens with a character being murdered by being exposed to a lethally high dose of radiation. Over the handful of days it takes him to die he is depicted as enduring the vomiting, diarrhea, dehydration, and symptoms of the central nervous syndrome such as confusion, apathy, and ultimately coma and death that a real-life victim would.[3] He lives only long enough to help others in the novel find out who killed him and why.

On the other hand, Spock's death scene in *Star Trek II: The Wrath of Khan* (1982) is moving and dramatic—but, unless his half-Vulcan physiology is dramatically more sensitive to radiation than the standard human variety, not realistic. Even an acute exposure of a few thousand rad would take many minutes if not hours to start affecting him and around several days to kill him. By that time, the damage to his brain would be far too great for him to utter any noble last words.

An area where science fiction and (unfortunately) real life overlap is depictions of the aftermath of a nuclear explosion or reactor accident. Victims in the vicinity of an exploding atomic bomb but far enough away to survive its initial heat and blast will also be exposed to radiation created by the bomb's detonation. Fissionable uranium and new radioactive elements created by the atomic explosion itself are inherently unstable, breaking down over time into ultimately more stable elements. The time it takes a radioactive element to change ("decay") into another radioactive isotope[4] of the same or different element, or into a nonradioactive isotope or other element (e.g. lead), is measured by its half-life—that is, how long it takes half the radioactive element's atoms to decay.

The half-lives of radioactive materials used in nuclear weapons or created by their detonation vary greatly. Uranium-235 and plutonium-239, used individually in atomic bombs and as the trigger to produce the required temperatures to start a fusion reaction in hydrogen bombs, have half-lives of 703.8 million and 24,100 years, respectively.

A nuclear explosion will produce high levels of both electromagnetic radiation (gamma and x-rays) and particulate radiation (e.g. alpha and beta particles as well as neutrons). It will also produce new types of radioactive elements with half-lives much shorter than the original uranium or plutonium. Strontium-94 has a half-life of only 75 s, emitting most of the total beta particles produced by its decay as a "burst" over a period of minutes. Others have half-lives measured in days (strontium-89, 50.5 days; iodine-131, 8 days; barium-140, 12.8 days) to decades (cesium-137, 30.17 years; strontium-90, 28.8 years). These latter and similar radioactive isotopes are dispersed as "fallout"—a mixture of dust, vaporized water, and these radioisotopes in the

---

[3] I provided medical advice to the authors regarding these effects of radiation, and their depiction is indeed realistic.

[4] As noted in Chap. 1, isotopes are varieties of an element that have the same number of protons but different numbers of neutrons in their nuclei.

form of fine particles shot up into the upper atmosphere after a nuclear blast that then "rains" back down to the ground [14–18].

The radiation produced by this fallout is primarily in the form of beta particles, with some radioactive elements also producing gamma rays. When outside a person's body beta particles can usually be stopped by shielding equivalent to only a few millimeters of aluminum. However, if radioactive material is breathed in or ingested in the form of food or water the beta particles have little distance to travel to damage internal tissues and cause much greater injury.

Some commonly produced radioisotopes are particularly likely to cause this type of damage inside the body. Drinking water can be contaminated by cesium-137, which dissolves easily in it. Ingested iodine-131 is primarily taken up by the thyroid gland, whose healthy cells are then subjected to high, localized levels of damaging beta particles. Strontium-90 settles into soil and is taken up by plants growing there. Individuals eating those plants or the products of animals that have eaten them (e.g. milk from cows) will selectively concentrate it in their bones, thus injuring them.

Nuclear power plant accidents release many of the same or similar radioisotopes into the atmosphere, perhaps as part of steam or, in the case of the reactor that exploded at Chernobyl in 1985, the uranium-235 fuel itself. The latter decays by production of alpha particles, whose ability to penetrate material is so low that they can be stopped by a sheet of paper. Here too, however, once inhaled (e.g. by the firefighters who breathed in smoke produced by the fire and explosion at Chernobyl) or otherwise entering the body, they are particularly damaging (remember that "quality factor" of 20 mentioned previously?) It is estimated that radiation levels near the exposed reactor core at Chernobyl may have been as high as 30,000 rad/h—enough to produce an ultimately fatal dose within minutes.

Other major, well-publicized nuclear power plant accidents such as at the Three Mile Island plant in 1979 or the damage caused to reactors at Fukushima in Japan made by an earthquake and tsunami in 2011 released enough radioactive material within the facilities themselves to potentially deliver a fatal dose within a short time. Fortunately the levels of radioactivity released into the environment outside those plants' immediate areas were far too low to cause acute injuries. However, the amounts released were high enough to potentially at least mildly increase the long-term risk of individuals in nearby areas (particularly children) developing leukemia, breast, thyroid, or other forms of cancer, either directly or by ingesting contaminated food or plants.

Even before the first nuclear reactor was created in late 1942, science fiction stories were already attempting to realistically depict both the potential benefits and risks of nuclear power. Notable examples include Heinlein's "Blowups Happen" (1940) and Lester del Rey's "Nerves," published in mid-1942. Although many details of such stories turned out to be wrong, their attempts to present tales as scientifically accurate as possible based on what was known at that time is still a worthy goal today.

6.5 Radiation in Space

From both a science fiction and real world standpoint radiation poses a particularly high risk in an extraterrestrial environment. Radiation levels in space are significantly higher than on Earth. Our Sun is a major source of both electromagnetic and particulate radiation. It constantly ejects a stream of about 1.3×10^{36} particles per second—mainly protons and electrons, with much smaller amounts of helium and heavier nuclei. This "solar wind" permeates the entire Solar System. The kinetic energies[5] of the particles in the solar wind are relatively low, generally in the range of a few kilo-electron volts (keV, with 1 keV = 1000 eV) [6, 19].

The Sun also intermittently produces solar flares—emissions of mainly electromagnetic radiation such as x-rays and gamma rays, along with some high-energy protons and electrons. There are also bursts of between 1×10^{11} and 5×10^{13} kg of particles released over a short time—"coronal mass ejections" (CME). Worse still are "solar particle events" (SPE) in which the number and intensity of high-energy particles shot into space are particularly high [20]. SPEs contain protons and electrons that can have energies in the range of tens to hundreds of mega-electron volts (MeV, with one MeV = one million electron volts) [5].

The number and severity of solar flares and CMEs correlates roughly with the Sun's sunspot cycle. Sunspots are areas where the Sun's magnetic field protrudes to its "surface." Their number rises and falls over approximately 11-year cycles. When they are at their maximum there can be up to about 1100 solar flares and CMEs per year.

Another source of radiation in space is the previously mentioned cosmic rays. The "ray" part of the name is a misnomer, coined before the discovery that they are actually very high-energy particles rather than, like x-rays or gamma rays, part of the electromagnetic spectrum. Individual protons (the nuclei of the most common form of hydrogen) make up about 90 % of cosmic ray particles. The remainder consist mainly of alpha particles with a smattering of nuclei of heavier elements, electrons, positrons, and other subatomic particles [5, 6, 21, 22].

---

[5] Kinetic energy is the energy associated with movement of a particle or other object.

Cosmic rays can have so much energy—up to several giga-electron volts (GeV, with 1 GeV = 1 billion eV) or even more—that they travel at speeds that are a significant fraction of c. While nothing composed of matter can actually reach that speed, cosmic rays with the greatest energy can achieve more than 99 % of c. The Sun can produce protons and other particles that qualify as cosmic rays based on their energies and speeds. However, these solar cosmic rays are about a million times less intense than the particles in the solar wind [19].

Cosmic rays with the highest energy, and thus capable of individually causing the greatest damage to cells, originate outside the Solar System ("extrasolar" cosmic rays). They come from elsewhere in our Milky Way Galaxy or from intergalactic space, originating in sources such as exploding stars (supernovae). Unlike particulate radiation from the Sun, whose levels can increase markedly above the average ones produced by the solar wind due to CMEs and other events, the level of extrasolar cosmic rays remains fairly steady. There is an inverse relationship between solar activity and the intensity of extrasolar cosmic rays within our Solar System—that is, when one increases the other tends to decrease.

Fortunately for us, Earth's surface is well protected from all this space radiation by its atmosphere and magnetic field [22, 23]. The former, particularly its ozone layer, shields us from the harmful UVB and UVC, x-rays, and gamma rays produced by the Sun. Earth's magnetic field traps electrons, protons, and other charged solar particles in the Van Allen belts. The latter were discovered by the United States' first two satellites, Explorer 1 and Explorer 3, in 1958. They consist primarily of an outer belt located between around 13,000–60,000 km above sea level and an inner belt averaging about 1000–10,000 km in height. The outer belt holds mainly trapped electrons with typical energies of 0.1–10 MeV. The inner belt consists primarily of protons with energies as great as about 100 MeV [23].

Astronauts in low-Earth orbit (LEO) and especially those venturing into interplanetary space (e.g. the Apollo missions and future journeys to Mars) lack much or all of this protection. Individuals on the International Space Station (ISS), orbiting at an average altitude of a little over 400 km, are no longer effectively shielded by Earth's atmosphere. The metals and other materials composing the ISS itself provide some shielding from electromagnetic radiation such as x-rays and gamma rays, but little from cosmic rays. Fortunately astronauts are still protected from charged solar particles (mainly electrons and protons) by the Van Allen belts, except for one area over the South Atlantic where the inner belt dips down as low as about 200 km.

The net result is that individuals on the ISS receive roughly one to two hundred times the radiation exposure per year that we here on Earth do. This works out to up to approximately 20 rem per year, compared to the previously mentioned 0.1–0.3 rem per year for us on the ground. SPEs and other solar activity can add significantly to these totals. For example, cosmonauts on the Mir space station in 1989 received an "extra" 6–7 rem during an SPE that lasted nearly 6 days. The Space Shuttle *Atlantis* was docked at Mir when this occurred, and astronauts had to stay within the interior of the craft where shielding was greatest.

In fact, because of such hazards astronauts are classified as radiation workers. NASA guidelines limit the total amount of radiation they are allowed to receive during a mission and cumulatively over their careers. The latter is based on how much radiation would be expected to increase an astronaut's risk of developing cancer by 3 % compared to the age-and gender-matched general population. For example, a 30-year-old male astronaut has a career limit of 62 rem and a woman astronaut the same age 47 rem. Limits are lower in women due to their additional risk of developing breast cancer and longer average life expectancy compared to men. At age 55 career limits are 147 rem for men and 112 rem for women. These values are higher than in younger astronauts because older ones are more likely to succumb to other natural causes before developing radiation-induced cancers than younger ones are [5, 13].

A major concern regarding future space missions is that the radiation risk in interplanetary space or on the surface of the Moon is substantially greater than in LEO. Away from our world the best protection astronauts have is the spacecraft itself or whatever other limited materials are available. In the case of that event involving Mir in 1989, if astronauts had been on a deep-space mission or on the Moon at that time their estimated exposure would have been enough to cause a 10 % chance of death.

Far worse was an SPE that occurred mid-way between the Apollo 16 and 17 missions, in August 1972. The estimated radiation exposure for an astronaut outside the Command Module of the Apollo craft would have been as high as 241 rem/h—more than enough (as shown in Table 6.2) to cause symptoms and a significant chance of death. Even within the protection of the Command Module total exposure could have been at least 350 rem.

However, should a writer decide to threaten astronauts in orbit or either en route or on the Moon with a solar flare, CME, or SPE (such as during the fictional Apollo 18 mission in James A. Michener's 1982 novel *Space*), several factors must be taken into account. For any of these solar events to present a significant hazard they must be directed toward Earth. If one occurs, for example, on the side of the Sun facing away from our planet we will not have to worry about it. Measured at its equator the Sun takes about 25 days to make one rotation. Given that the duration of flares and CMEs is typically no more than a few hours, as long as our planet is not in these solar events' "crosshairs" astronauts won't be significantly affected.

Also, the time required for radiation produced by these solar events to reach Earth varies widely. X-rays and gamma rays, traveling at c, will traverse the average distance between the Sun and Earth of about 150 million km in a little over 8 min. Protons and electrons released in the flare may also be traveling at near that speed and so will arrive not long afterwards. However, the total amount of radiation exposure to orbiting astronauts caused by a solar flare, while of concern for increasing their long-term risk of cancer, will almost certainly not be great enough to cause acute injury if they are, for example, protected within their craft.

The situation is both better and worse with CMEs and especially SPEs. The high-energy particles produced in these solar events travel at speeds ranging from about 600 km/s to, rarely, as high as 3000 km/s. A little math shows that even the latter will take about 14 h and the former nearly 3 days to reach Earth. Given that there is now an extensive network of satellites monitoring solar activity, a dangerous CME or SPE should be detected soon enough to warn astronauts it is coming and to seek protection. However, because the overall radiation levels produced by these two types can be much higher than those produced by solar flares, the amount of protection available to the astronauts may not be enough to prevent acute (and potentially fatal) injury.

As noted previously, the number and severity of solar flares and CMEs show a rough correlation with the sunspot cycle, giving a very general idea of when the risk from these events will be highest. The potentially most dangerous type of solar event, SPEs, occur much less frequently than flares or CMEs. However, SPEs with the highest energies and amounts of particles can occur at any time, independent of the sunspot cycle. Such SPEs can last as long as several days to a week. Thus, while there may be time to warn astronauts an SPE is coming, if the latter is a particularly bad one the astronauts may not have the resources to adequately protect themselves.

Such radiation exposure is an even greater concern beyond the vicinity of Earth and its Moon, particularly during the long period when astronauts are traveling to their destinations. During the 253 days it took the Curiosity rover to reach Mars it found higher overall levels of radiation in interplanetary space than previously reported. Based on Curiosity's measurements it was estimated that astronauts would receive an average dose of about 66 rem during a round-trip voyage to Mars using current propulsion systems [24]. To this must be added radiation exposure while they are on the surface of Mars, which has only a minimal ozone layer, an atmospheric pressure just less than 1 % of Earth's at sea level, and no global magnetic field like our Van Allen belts for protection.

The radiation hazards of other Solar System real estate depicted in science fiction can, however, be far worse. Stories such as Isaac Asimov's "Christmas on Ganymede" (1942) and movies as disparate as *Fire Maidens of Outer Space* (1956), *2001: A Space Odyssey* (1968), and *Europa Report* (2013) involve travel to the vicinity of Jupiter. Unfortunately for human visitation Jupiter's magnetic field is far stronger than Earth's, and its equivalent of Van Allen belts holds many times greater levels of particulate radiation. All four of its largest moons—Ganymede, Callisto, Europa, and Io—orbit within Jupiter's magnetic field and are thus continuously bathed by that radiation. At Europa the exposure rate may be as high as about 30,000 rad per day. This is enough radiation to provide an astronaut in a spacesuit with a dose that, within a matter of minutes, would be sufficient to cause symptoms such as vomiting soon afterwards and, in well less than an hour, ultimately be a 100 % fatal one [25].

Radiation also poses serious problems in other science fictional scenarios. A starship moving at a significant fraction of the speed of light would also be struck by particles of interstellar gas (mainly hydrogen) and dust. These could have a speed relative to the ship at least as great as the latter, so these protons and other particles would effectively act as particulate radiation. While the density (mass per volume) of these particles is low, ranging from as little as about 100 to as much as 1×10^{17} hydrogen atoms or molecules per cubic meter, a ship traveling at a velocity that is a significant percentage (~ 10 % or more) of *c* would still encounter enough of them to represent a hazard [26].

Most of the stars in our Galaxy are smaller, longer-lived, and cooler than our Sun [27]. Any potentially habitable planets around them would also need to be closer to those stars. Unfortunately these type K and M stars are particularly predisposed (for at least a significant part of their lifetimes) to produce flares that bathe any planets in the vicinity with damaging ultraviolet radiation. This could not only reduce the chances of complex life developing there or injure any visiting astronauts, but also help strip the atmosphere from those planets [28].

Very high radiation levels are also present close to a neutron star, black hole, or regions of space with nearby supernovae.[6] Our Galaxy is thought to have a black hole at its center with an estimated mass in the range of 4 million times that of our Sun, and so its vicinity would be particularly unhealthy for visiting space travelers. Some other galaxies have "active galactic nuclei" that emit huge amounts of electromagnetic energy, including x-rays and gamma rays. Active galactic nuclei are thought to be due to matter "accreting" near a massive black hole in the galaxy's center.

---

[6] Such neutron stars and black holes are the extremely dense remnants of massive stars that have exploded as Type II supernovae. Individual black holes can combine into a single one having many millions of times the mass of our Sun.

Catastrophic astronomical events within our Galaxy could bombard Earth with dangerous levels of radiation. A Type Ia supernova occurs when a white dwarf star in a binary system with a giant companion explodes [29]. It releases both electromagnetic radiation (e.g. x-rays and gamma rays) over a period of months, as well as a longer and larger burst of particulate radiation in the form of cosmic rays [30]. Gamma-ray bursts (GRBs) are associated with extremely high-energy explosions, primarily in distant galaxies but occasionally in our own. They occur as single rather than recurrent events, producing a beam of gamma rays that typically lasts from seconds to minutes,[31, 32] although one recently observed in a distant galaxy lasted about 20 h [33]. It has been suggested that a GRB might be caused by the explosion of an extremely massive star (a "hypernova") or produced by merging black holes [29, 33].

The radiation produced by either a Type Ia supernova occurring within tens of light-years[7] of Earth (or perhaps even thousands of light-years away if it sends radiation more directly at us), or a powerful GRB occurring in our Galaxy and beamed toward us, could wreck havoc on our world. While the immediate effect of this radiation to humans on Earth might be relatively mild, our protecting atmosphere could be significantly damaged.

For example, the ozone layer protecting us from UVB and UVC from the Sun could be destroyed and take many years to recover, thus bombarding both terrestrial plant and animal life (including us) with dangerous levels of ionizing radiation. Besides significantly increasing our cancer risk in the long run, humans would be more immediately affected by loss of the plants and animals we use for food. Large-scale destruction of plants using photosynthesis to remove carbon dioxide from our atmosphere and release oxygen might, due to increases in carbon dioxide levels, accelerate global warming.

Also, chemical interactions between the oxygen and nitrogen in our atmosphere caused by a supernova explosion or GRB would create compounds such as nitric acid, thus producing acid rain, and nitrogen dioxide. The latter absorbs visible light and so might actually cause global cooling instead. Due to their potential global effects GRBs and nearby supernova explosions have been suggested as possible causes or contributors to prior mass extinctions in our world's history, such as the Ordovician mass extinction about 440 million years ago [34].

Fortunately the odds of any such cosmic catastrophe occurring close enough in our celestial neighborhood to cause significant damage to Earth are extremely low. Such events are estimated to occur on average about once every several hundred million years. Currently there are no known stars thought likely to become a supernova anytime soon that are close enough to Earth to

---

[7] A light-year is the distance light travels in a vacuum in one year, or about 9.461×10^{12} km.

represent a major danger. GRBs represent a minimal but more unpredictable threat. It is not inconceivable that our present-day or near-future world could prove singularly unlucky and be affected by a GRB, thus making it a potential plot device in a disaster story.

6.6 Protecting Against Radiation

Astronauts can be shielded from space-based radiation with widely varying levels of effectiveness and complexity. Spacesuits offer relatively little protection. Cosmic rays pass easily through them. Even particles with far lower energy, such as a 0.5 MeV electron or a 25 MeV proton, can do it—and particles with these and higher energies are common in space. Current spacecraft offer suitable protection against electrons and protons with similar energy levels, but not against much higher ones.

Lead shielding stops high-energy electromagnetic radiation and lower energy particulate radiation. However, it poses the obvious problem of having much greater density (mass divided by volume) than metals such as aluminum that are typically used for spacecraft, and thus is more difficult and costly to send into space. Also, when cosmic rays and other very high-energy particles pass through lead and similar "massive" metals, they can interact with this "shield's" atoms and actually produce more radiation exposure to unfortunate astronauts than they would have done by passing through "lighter" metals or water.

Another option is to use a barrier of water or certain types of plastic around at least one section of a spacecraft, which could serve as a "safe room" if the Sun puts out more radiation than normal (e.g. during a CME or SPE). Generating an electrostatic and/or magnetic field around the ship or a base on the Moon or Mars to repel charged particles like electrons or protons could also be used. However, this would not be effective against electromagnetic radiation and of very limited value for very high-energy charged particles like most cosmic rays. Power requirements and the size of such a system might also be prohibitive based on current (though not necessarily future) technology.

Another option on the Moon or Mars is to use "native" materials. For example, the Moon's regolith (the fine pulverized rock covering its surface) could be put into the equivalent of sandbags to serve as barriers for a habitat on its surface, or the habitat could be at least partly buried "underground." The same method could be used on Mars.

For example, the relatively low-energy particles in the solar wind penetrate only nanometers into lunar regolith. Solar particles with much higher energies of about 1 MeV have an intensity about a million times less than average ones in the solar wind. They pierce the regolith to a depth of only about a few tens of microns (one micron = one millionth of a meter). Galactic cosmic rays are about one hundred times less abundant than these 1 MeV particles from the Sun. Those with energy of 1 GeV can penetrate a meter into the Moon's regolith, and their energy may be considerably greater than that. Thus, while not a perfect means to shield against radiation, using regolith can substantially reduce radiation exposure [19].

These same considerations also apply to shielding Earth-based characters from radiation. In addition there are some medications and nutrients that may at least mildly protect individuals from the damaging effects of radiation [11]. The former include drugs such as amifostine and N-acetyl-L-cysteine, both of which might be helpful for reducing the adverse effects of receiving a high acute dose of radiation. Naturally occurring substances such as cysteine and glutathione as well as antioxidant nutrients contained in foods, such as vitamins C, A, and E may also reduce risk. Individuals exposed to radioactive I-131, which can be released during nuclear reactor accidents, can be treated with prophylactic potassium iodide to block uptake of I-131 into the thyroid gland.

6.7 The Bottom Line

Science fiction stories dealing with radiation can take several approaches. The first is to deal with it in a fully realistic manner, using the actual effects of radiation as described previously. Taking the information in this chapter into account would certainly "harden" a science fiction tale.

This same information could also be used as only as the factual background for dealing with radiation-related issues. Heinlein's short story "The Long Watch" (1949) gives little detail about the specific symptoms experienced by an unfortunate character waiting to die from exposure to radioactive material, but it does not make any glaring errors either. Or a futuristic story could include genetically altered humans who have at least a mild degree of enhanced resistance to radiation. An intermediate step would be to identify individuals who have greater natural resistance to radiation than average. These methods should be used with caution, however, since a dramatic reduction in human sensitivity to radiation could involve drastic changes to our bodies and how they work.

An "anti-radiation" pill far more effective than current ones, or using nano-technology to either protect our bodies against or heal them from the effects of ionizing radiation, are other possibilities. Here again, however, unless the story involves reasonably far-future "magical" (in Arthur C. Clarke's sense) advanced technology, the details should not contradict the known effects of radiation exposure.

Another option for the (hopefully) not too far future is to have characters live in a large space habitat. Having a great mass of material between characters and space would, except for a relatively minor contribution from cosmic rays, keep them reasonably safe from radiation [35].

A different but, at least for story purposes, potential approach is to ignore or downplay at least some radiation hazards and just get on with the story. An author can simply mention that particular "real world" issues have been dealt with or "solved," such as by creation of better shield generators, anti-radiation medications, or other methods described previously. Writing a serious scientific paper on the danger of space radiation requires a direct and detailed approach on how to deal with it. Fortunately writing a story dealing with that hazard does not have to be as specific in the "solutions" it describes.

Of course, a writer armed with all this scientific knowledge could still ignore everything in this chapter and write a science fantasy story in which radiation *does* produce superhuman mutants or other unrealistic effects. Hopefully all but the very youngest readers will know such things are impossible. And if the idea is part of a good story, readers may still be willing to go along for the ride.

References

1. Radiation. In: Parker SB, editor. Concise encylopedia of science & technology. New York: McGraw-Hill; 1997.
2. Walker J. Electromagnetic waves. Halliday & Resnick. Fundamentals of physics. 9th edn. Hoboken: Wiley, Inc.; 2011.
3. ISO 21348 Definitions of Solar Irradiance Spectral Categories. ftp://ftp.ngdc. noaa.gov/STP/SOLAR_DATA/SolarOnlineTemp/AIAA/SampleImages/ISO_ DIS_21348_(E)_table1.pdf. Accessed 15 April 2015.
4. Scala RJ. Biological effects of ionizing radiation. In: Early PJ, Sodee DB, editors. Principles and practice of nuclear medicine. 2nd edn. St. Louis: Mosby-Year Book, Inc.; 1995. pp. 118–30.
5. Committee on the Evaluation of Radiation Shielding for Space Exploration NRC. Managing space radiation risk in the new era of space exploration. 2008. http://www.nap.edu/catalog.php?record_id=12045. Accessed 15 April 2015.
6. Planel H. Cosmic radiation. Space and life. An introduction to space biology and medicine. Boca Raton: CRC Press; 2004. pp. 121–40.
7. Stratmann HG, Nordley GD. Biological hazards and medical care in space (Part two). Analog Sci Fict Fact. 1998; 118(5):55–71.
8. Sources and Effects of Ionizing Radiation. Vol. 1. 2008. http://www.unscear.org/ docs/reports/2008/09-86753_Report_2008_GA_Report_corr2.pdf. Accessed 15 April 2015.
9. Sources and Effects of Ionizing Radiation. Vol. 1. Scientific Annex. 2008. http:// www.unscear.org/docs/reports/2008/09-86753_Report_2008_Annex_B.pdf. Accessed 15 April 2015.

10. Introduction to Ionizing Radiation. 2013. https://www.osha.gov/SLTC/radiationionizing/introtoionizing/ionizinghandout.html. Accessed 15 April 2015.
11. Buckey JC. Space physiology. New York: Oxford University Press; 2006.
12. Chapter 9. Radiation and radiobiology. In: Nicogossian AE, Huntoon CL, Pool SL, editors. Space physiology and medicine. 3rd edn. Malvern: Lea & Febiger; 1994.
13. Mcphee JC, Charles JB. Human health and performance risks of space exploration missions. 2009. http://ston.jsc.nasa.gov/collections/trs/_techrep/SP-2009-3405.pdf. Accessed 15 April 2015.
14. Chapter 2. Fallout from nuclear weapons. 2005. http://www.cdc.gov/nceh/radiation/fallout/default.htm. Accessed 15 April 2015.
15. Chapter 3. Estimation of doses from fallout. 2005. http://www.cdc.gov/nceh/radiation/fallout/default.htm. Accessed 15 April 2015.
16. Chapter 4. Potential health consequences from exposure of the United States population to radioactive fallout. 2005. http://www.cdc.gov/nceh/radiation/fallout/default.htm. Accessed 15 April 2015.
17. Nuclear detonation: weapons, improvised nuclear devices–radiation emergency medical management. 2013. http://www.remm.nlm.gov/nuclearexplosion.htm. Accessed 15 April 2015.
18. Planning guidance for response to a nuclear detonation. 2010. http://www.epa.gov/rpdweb00/docs/er/planning-guidance-for-response-to-nuclear-detonation-2-edition-final.pdf. Accessed 15 April 2015.
19. The terrestrial planets and their satellites. In: Encrenaz T, Bibring J-P, Blanc M, Barucci M-A, Roques F, Zarka P, eds. The solar system. 3rd edn. Berlin: Springer; 2010.
20. Committee on Solar and Space Physics and Committee on Solar-Terrestrial Research NRC. Radiation and the International Space Station: recommendations to reduce risk. 2000. http://www.nap.edu/openbook.php?record_id=9725. Accessed 15 April 2015.
21. Cosmic Rays. McGraw-Hill encyclopedia of astronomy. 2nd edn. New York: McGraw-Hill, Inc.; 1993.
22. Moldwin M. An introduction to space weather. New York: Cambridge University Press; 2008.
23. Stratmann HG. Space weather: the latest forecast. Analog Sci Fict Fact. 2012; 132(5):20–30.
24. Zeitlin C, Hassler DM, Cucinotta FA, Ehresmann B, Wimmer-Schweingruber RF, Brinza DE, et al. Measurements of energetic particle radiation in transit to Mars on the Mars Science Laboratory. Science. 2013;340(6136):1080–4.
25. Jun I, Garrett HB. Comparison of high-energy trapped particle environments at the Earth and Jupiter. 2005. http://trs-new.jpl.nasa.gov/dspace/bitstream/2014/37958/1/04-0913.pdf. Accessed 15 April 2015.
26. Chapter 5. The formation of stars. In: Green SF, Jones MH, editors. An introduction to the sun and stars. New York: Cambridge University Press; 2004.

27. Comins NF. Characterizing stars. Discovering the essential universe. 4th edn. New York: W. H. Freeman and Company; 2009; pp. 257–77.

28. Cuntz M, Guinan EF, Kurucz RL. Biological damage due to photospheric, chromospheric and flare radiation in the environments of main-sequence stars. 2010. http://arxiv.org/pdf/0911.1982v1.pdf. Accessed 15 April 2015.

29. Comins NF. The deaths of stars. Discovering the essential universe. 4th edn. New York: W. H. Freeman and Company; 2009;309–50.

30. Ellis J, Schramm DN. Could a nearby supernova explosion have caused a mass extinction? Proc Natl Acad Sci U S A. 1995;92:235–8.

31. Meszaros P. Gamma-ray bursts. 2006. http://arxiv.org/pdf/astro-ph/0605208v5.pdf. Accessed 15 April 2015.

32. Zhang B, Meszaros P. Gamma-ray bursts: progress, problems & prospects. 2008. http://arxiv.org/pdf/astro-ph/0311321.pdf. Accessed 15 April 2015.

33. Ackermann M, Ajello A, Asano K, Atwood W, Axelsson M, Baldini L, et al. Fermi-LAT observations of the gamma-ray burst GRB 130427 A. Science. 2014;343(6166):42–7.

34. Melott AL, Lieberman BS, Laird CM, Martin LD, Medvedev MV, Thomas BC, et al. Did a gamma-ray burst initiate the late Ordovician mass extinction? 2004. http://arxiv.org/pdf/astro-ph/0309415.pdf. Accessed 15 April 2015.

35. Globus A, Arora N, Bajoria A, Straut J. The Kalpana One orbital space settlement. Revised. http://alglobus.net/NASAwork/papers/2007KalpanaOne.pdf.

7
Suspended Animation: Putting Characters on Ice

He had visualized his dead body enclosed in a rocket flying off into the illimitable maw of space. He would remain in perfect preservation, while on earth millions of generations of mankind would live and die, their bodies to molder into the dust of the forgotten past.

Neil R. Jones
"The Jameson Satellite" (1931)

I, Anthony Rogers, am, so far as I know, the only man alive whose normal span of eighty-one years of life has been spread over a period of 573 years. To be precise, I lived the first twenty-nine years of my life between 1898 and 1927; the other fifty-two since 2419. The gap between these two, a period of nearly five hundred years, I spent in a state of suspended animation, free from the ravages of katabolic processes, and without any apparent effect on my physical or mental faculties.

Philip Francis Nowlan
Armageddon 2419 A.D. (1928)

Suspended animation—the process of greatly slowing or stopping the body's metabolic rate so a person does not age appreciably—has many uses in science fiction. Early stories such as Edward Bellamy's *Looking Backward: 2000 – 1887* (1887), H.G. Wells's *When the Sleeper Wakes* (1899), the debut of Buck Rogers (cited above), and Laurence Manning's *The Man Who Awoke* (1933) employed mysterious drugs, hypnosis, gases, or freezing to put characters into prolonged "sleep" for one-way time travel to the distant future. More recent science fiction, including Frederik Pohl's novel *The Age of the Pussyfoot* (1969) and Larry Niven's "The Defenseless Dead" (1973), have explored the social and political dimensions of dealing with the frozen dead-but-not-dead vividly described by the term used in those works, "corpsicle."

Other potential uses for suspended animation could be to stabilize a critically injured individual (e.g. on a battlefield) before transporting the victim to a medical facility for definitive care or, extending that idea, preserving someone with a currently fatal disease for years until medical science finds a cure. Suspended animation has been a pivotal plot element in television shows such as *Red Dwarf, Lost in Space,* and episodes of the original versions of *The Twilight Zone* (e.g. 1961's "The Rip van Winkle Caper" and 1964's "The Long Watch") and *Star Trek* ("Space Seed," from 1967). It has also played a secondary role in movies such as *2001: A Space Odyssey* (1968), *Dark Star* (1974), *Alien* (1979) and its sequel *Aliens* (1986), and *Avatar* (2009).

In some of those works and in literary science fiction, suspended animation is used to facilitate either interplanetary travel or much longer interstellar journeys. For example, in A. E. van Vogt's 1944 story "Far Centaurus" a spaceship's occupants use the "Eternity drug" during their centuries-long voyage. Putting astronauts into suspended animation would have the important practical benefit of greatly reducing the food, energy, and recycling requirements such a "sleeper ship" would need to support active versus "suspended" passengers for months or years. More importantly it prevents those characters from dying of old age during the decades or more of ship time it might take them to reach a distant star.

But can a person really be put into a state virtually indistinguishable from death—with no breathing, heartbeat, or even detectable cellular activity—and then revived days, months, or years later, none the worse for wear? While this is far beyond current human capabilities some terrestrial creatures can already do it to varying degrees—and it is possible that someday we too might, at least to some degree, mimic their abilities [1, 2].

7.1 Suspended Animation and Hibernation in Animals

Conditions of extreme cold, lack of water, and reduced oxygen cause animals ranging in complexity from bacteria to mammals to drastically reduce or even stop metabolic activity. Bacterial spores have been revived after being trapped in amber, ice, and salt crystals for many thousands to millions of years [3–5]. Tiny animals such as brine shrimp cysts, tardigrades, and nematodes can be completely dried out, with no measurable metabolism, then brought "back to life" when water is restored [6]. Intertidal marine invertebrates like barnacles and mollusks experience repeated freezing and thawing, as well as lack of oxygen, during winter as tides rise and fall [7]. Spiders, ticks, and many insects can survive prolonged temperatures of $-50\,°C$ or even lower. Land snails are capable of "estivation"—a state of dormancy induced by drastically reduced availability of water. They can remain dormant in their shells for several years.

Certain cold-blooded vertebrates have similar abilities. Spadefoot toads and lungfish may estivate for years. Goldfish can do without oxygen for up to 16 h at a temperature of 20 °C and for a week at 5 °C. Crucian carp and fresh water turtles spend up to 4 months submerged in water just above the freezing level, with a metabolic rate about 10 % of normal and without breathing. Adult females of the wood frog *R. sylvatica* can typically survive 13 days at a body temperature of about – 2.5 °C, with no respiration, heartbeat, or blood flow and with 65 % of their bodies converted to ice. Members of that species living in the coldest climes have developed even greater protective mechanisms, such as accumulating high levels of cold-protecting chemicals in brain tissue. For example, Alaskan frogs have been reported to survive freezing to body temperatures as low as – 16 °C and endure a 2-month bout of being frozen at – 4 °C [8, 9].

At least 6 of the 34 orders of mammals contain species that "hibernate," typically but not always during certain times of the year [10–12]. With one partial exception all are small animals, such as the murine opossum, European hedgehog, brown bats, and some types of lemurs, squirrels, mice, and hamsters [8, 13].

Most mammals that hibernate do it during cold winter months and/or under conditions of food shortage. Hibernation is triggered primarily by a fall in environmental temperature—perhaps by the thermoregulatory center ("thermostat") in the warm-blooded animal's brain resetting itself to a lower temperature. Other factors such as decreased daylight and food or, in some hibernators, internal biorhythms may also contribute. Some mammals, such as hamsters, do not hibernate if they are kept continuously in a warm environment. Others (e.g. squirrels) show torpor ("sluggishness") [14] during winter months even if they are kept warm.

At environmental temperatures of 0 °C (the freezing point of water) or less the metabolic rates of those other mammals fall to as little as 1–3 % of normal. Body temperature, usually 35–38 °C, decreases to 2–6 °C. The arctic ground squirrel's body temperature can even drop below freezing to as low as about – 3 °C [14, 15]. Dwarf and mouse lemurs, primates native to the island of Madagascar, do not hibernate but instead estivate, becoming dormant primarily under dry rather than cold conditions.

The black bear is the only large mammal that "hibernates"—and its capability is much less than these smaller animals, enough so there it may not be considered "true" hibernation [10]. A black bear's body temperature falls only mildly from normal at about 38 °C to a range of 31–35 °C, metabolic rate drops as low as 30 % of normal, and heart rate goes down from its typical 40 beats per minute to 10 beats per minute [8].

Unlike cold-blooded vertebrates such as frogs, a hibernating mammal's breathing and heartbeat slow but do not stop. A ground squirrel's heart rate decreases from its usual 200–300 beats per minute to 3–10 per minute as body temperature falls from 37° to a range of 4–7 °C. The squirrel's breathing rate drops from 100 to 150 breaths per minute to only 1–2 per minute [11, 16]. Electroencephalographic (brain wave) studies indicate the hibernating state resembles deep sleep, without the rapid eye movement (REM) phase close to a waking level when dreaming occurs [11, 17, 18].

Entering a state approaching suspended animation has survival benefit for animals periodically exposed to cold and lack of oxygen. For mammals, maintaining normal body temperature during winter would require eating much more at a time when food is scarce or mobilizing their bodies' energy stores, particularly from fat [10]. To do that a mouse would have to add an amount of fat three times its body's normal weight. Metabolic rate would also have to increase several times greater than normal to make up for body heat lost to the cold environment.

By hibernating, ground squirrels use about 12 % of the energy they would need if they stayed active. The rate at which biochemical processes occur falls two- to threefold with each 10 °C reduction in body temperature. This accounts primarily for the thirtyfold or more decrease in metabolic rate when the hibernating animal's body temperature falls to just above 0 °C. At this much lower metabolic rate its fat stores are adequate to meet its nutritional needs.

To return to normal body temperature and "wake up," hibernating animals use stored fat ("brown adipose tissue") to generate the heat needed for internal rewarming via specific biochemical reactions. However, the net energy saved by hibernating is much greater than what is needed for rewarming. This is despite the fact that mammals do not hibernate continuously during winter. Instead they have repeated cycles of prolonged hibernation alternating with short periods of rewarming and wakefulness.

Hibernation may also reduce an animal's aging rate. In one study hamsters that hibernated poorly when exposed to cold had a life expectancy of 727 days, compared to 1093 days in those who hibernated well. The average number of active, non-hibernating days was 687 for the poor hibernators and 809 days for good hibernators—a nearly 18 % increase in effective lifespan [10].

7.2 Suspended Animation and Humans

The basic reason why some fish, amphibians, mammals, etc. can go into suspended animation or hibernate and we humans cannot is that, unlike ours, their biochemistry and anatomy are adapted to do this.

Freezing an animal's or a person's body produces ice crystals both inside and outside its cells. This is nearly uniformly lethal when it occurs inside cells, mainly by damaging subcellular structures (e.g. the nucleus and mitochondria) and forming "compartments" within the cells' fluid (cytoplasm). When extracellular water freezes it disrupts connections between cells and draws unfrozen water from them. This leaves the cells dehydrated and greatly increases the concentrations of ions and other solutes. This in turn can cause irreversible denaturing of proteins (think of an egg after it has been hardboiled) and other permanent injuries. Cells are virtually certain to die when ice content exceeds 65% of total tissue or body water. Some cell components are damaged well below that level [19].

Animals capable of suspended animation and hibernation have evolved bodies and metabolic processes optimized to do it. Freezing must be confined to extracellular fluid spaces, cell shrinkage curbed, cell membranes and proteins protected. The animals' cells must remain viable despite the marked reduction of oxygen and nutrients produced by freezing and diminished blood circulation, respiration, etc.

Different species do this in different ways. Non-mammals make their cell membranes more resistant to low temperatures by changing the lipids (a group of organic compounds that includes fats) in them and producing protective substances (e.g. trehalose and proline). The amounts and varieties of enzymes and proteins within cells also change to improve cold tolerance. Special proteins outside the cell induce slow freezing and prevent formation of large ice crystals. This regulates passage of water from intracellular to extracellular areas and minimizes cell damage.

For example, in freeze-resistant species of frogs the bulk of their frozen blood and other internal fluids forms in large body cavities (e.g. the abdomen) rather than their organs, thus reducing damage. In these and other non-mammals low-molecular weight carbohydrates like glycerol, sorbitol, and glucose act as "cryoprotectants"—essentially "antifreeze"—inside cells [8]. These chemicals accumulate in high concentrations, reducing ice formation and stabilizing proteins against denaturation. For example, the wood frogs described previously use chemicals such as glucose and urea, while certain tree frogs and the Siberian salamander employ glycerol [7].

Lowering body temperature (and hence metabolic rate) drastically reduces the amounts of carbohydrates, proteins, and fat needed to maintain life. However, even the frozen or hibernating animal must use some of these "fuels" to survive, and make do with reserves already stored in its body. These animals use biochemical pathways for generating energy that are more efficient than we humans employ. Their body organs are also much more tolerant to lack of oxygen than ours are.

Cell metabolism involves processes that generate adenosine triphosphate (ATP), the primary molecule providing energy for cell functions, and those that use ATP. As described in Chap. 1, reactions that use oxygen to create

ATP (aerobic metabolism) are much more efficient and produce many more molecules of ATP than those that do not use it (anaerobic metabolism). In mammals, one molecule of glucose (a sugar and carbohydrate) could produce up to a net number of 36 molecules of ATP via biochemical pathways that use oxygen. This process generates waste products of water and carbon dioxide, both of which are relatively easy to clear from the body.

When sufficient oxygen is not available, we and other animals must use anaerobic metabolism instead. However, in humans this process produces only a net of 2 molecules of ATP per glucose molecule. Anaerobic metabolism also produces lactate as a waste product, an acid that is more difficult to eliminate than the water and carbon dioxide created by aerobic metabolism.

For example, a turtle's biochemistry is much better adapted than ours to handle the reduced energy and increased lactate production associated with anaerobic metabolism. Its brain can tolerate conditions for months that would irreversibly damage our more complex ones within minutes. Goldfish use a metabolic pathway whereby lactate is converted to ethanol and carbon dioxide, which are excreted through its gills. Marine mollusks use a more efficient form of anaerobic metabolism than humans do. It generates up to 7 ATP molecules from a molecule of glycogen (the storage form of glucose) compared to the 2 ATP molecules that we do to provide energy when enough oxygen is not available [6].

Human physiology allows none of these options. The human brain in particular is extremely dependent on having enough oxygen available to create the ATP and other energy sources it needs to function. Though the brain constitutes only about 2 % of an adult's weight it receives 13 % of the body's total blood flow (with the oxygen and nutrients that blood carries) and consumes 20 % of its oxygen. To mimic the extreme abilities of a turtle or goldfish would require wholesale changes in our anatomy, biochemistry, and metabolic processes.

Mammals capable of hibernating use physiological adaptations less drastic than fish or amphibians but still considerably different than any *H. sapiens* can muster. The exact mechanisms for many of these adaptations, such as hibernators' ability to maintain ATP production and cell function at low temperatures that would kill the cells of non-hibernators like us, are still uncertain. One important difference is their metabolic response to reduced oxygen availability. Humans and other non-hibernating mammals use primarily carbohydrates at such times—an only modestly efficient and rapidly depleted source of energy. Hibernators suppress carbohydrate metabolism and use their fat stores instead. The latter act as a more efficient reserve of food due to fat generating a little more than twice the amount of energy per gram than carbohydrates do.

At present the substances that trigger and sustain hibernation in mammals are not well understood. Hibernation induction trigger (HIT) is a hypothetical chemical (or set of chemicals) that may be produced by some winter-hibernating mammals [14, 17, 20, 21]. Its existence has been suggested by research in which blood from a hibernating animal (e.g. a ground squirrel) was reportedly able to induce hibernation after being injected into another of the same species that was not hibernating. A specific chemical in the blood that does this has not been identified. If it does exist, it has been suggested that it may be an opiate, a substance that stimulates the same or similar cell receptors in the brain as opium and morphine do. A recent report indicated that administering drugs stimulating adenosine receptors in arctic ground squirrels induced torpor in those animals [15]. Metabolic hormones such as FGF21 have also been found to be involved in inducing torpor [16]. Neurotransmitters such as serotonin may also play a role [11].

As noted previously only small mammals, up to the size of a marmot (a type of ground squirrel weighing up to about 5 kg), can truly hibernate. Larger mammals, like us humans, had less physiological need to evolve this ability. As an animal's body mass increases its basal metabolic rate also increases, but not as rapidly. For example, a tenfold increase in body mass produces only a fivefold rise in metabolic rate. Thus, instead of the huge fat stores (300 % greater than its usual body weight) a mouse would need to maintain normal body temperature through the winter if it did not hibernate, a human-sized animal would require increasing body mass by a more practical 20 %. Small mammals also require relatively little energy to rewarm their bodies. Larger ones would require much more energy to "heat up" to normal body temperature again, thus decreasing the net energy-saving benefits of hibernation [10].

7.3 How the Human Body Responds to Cold

To summarize, the reptiles, fish, turtles, frogs etc. described previously and, to a lesser degree, hibernating mammals have bodies well adapted to tolerating cold. Those adaptations include optimizing the balance between energy production and utilization (particularly supply and demand for ATP), protecting against adverse effects of reduced body temperature (e.g. blood clotting), modifying cell functions and, in a hibernating mammal, markedly lowering its body's temperature set-point (its internal "thermostat").

Compared to those animals, humans have a pitifully limited ability to respond to cold [22, 23]. Despite what is portrayed in Edgar Rice Burroughs's story "The Resurrection of Jimber-Jaw" (1937) or how Captain America managed to remain young long after World War II ended, after being frozen in a block of ice for decades or millennia you will *not* revive as good as new once you are defrosted.[1] Likewise, as we will see later, even using futuristic medical technology the prospects for astronaut Frank Poole's ultimately surviving his attempted murder by HAL (described in Sir Arthur C. Clarke's 1997 novel *3001: The Final Odyssey*) seem equally bleak. Instead, in any of these cases you would just be a cold corpse.

At room temperature (about 22 °C) an unprotected person can maintain a normal core (internal) temperature of about 36.4–37.5 °C quite well. However, if the effective surrounding temperature drops much below this level, and especially if it reaches 0 °C or lower, we need clothing, gloves, etc. to conserve body warmth. Without them we run the risk of developing "hypothermia," defined as a core body temperature that has fallen to 35 °C or less.

Exposure to cold activates the thermoregulatory center in the human brain's anterior hypothalamus. The latter structure activates compensatory reflexes that try to maintain the temperature of critical organs (e.g. brain and heart)—even if that means sacrificing body parts less immediately essential to life, such as fingers and toes. Blood vessels constrict, particularly in the skin, arms, and legs. While this reduces blood flow and heat loss in those areas it also makes them more vulnerable to injury, such as frostbite. The "hunting phenomenon" also occurs within minutes of exposure to cold. This reflex makes constricted blood vessels dilate temporarily, allowing blood to flow within fingers and toes, and then "clamp down" again in irregular cycles. How much this actually protects hands and feet from cold is, however, uncertain.

Shivering, defined technically as involuntary contraction of muscles in the body's trunk, increases heat production. However, shivering does this at the expense of increasing the expenditure of energy and nutritional stores. Heart rate, blood pressure, and breathing rate all increase. Being cold makes a person uncomfortable and eventually agitated.

Vasoconstriction and shivering intensify if the core body temperature falls to the threshold of hypothermia at 35 °C. At that temperature metabolic rate increases to as much as six times normal due to the individual trying to generate enough heat to keep internal organs "warm." Mild hypothermia, a core temperature between 32 and 35 °C, increases urine output due to changes in kidney function, and blood sugar levels rise for a time. Brain function also decreases, with the person potentially becoming confused and apathetic.

With moderate hypothermia (core temperature between 28 and 32 °C) the body's ability to resist further falls in temperature becomes increasingly impaired. The victim alternately shivers violently and stiffens. Breathing rate is

---

[1] Even if Otzi, the Iceman, had not been in poor health or met with probable foul play he would not have been alive after his thaw in 1991. http://www.iceman.it/en/node/233.

initially high but later diminishes as core temperature continues to go down. Oxygen consumption and heart rate are usually decreased by 50 % of normal at 28 °C [22]. Metabolic rate drops below normal, declining about 6 % for each additional 1 °C fall in core temperature. Confusion worsens and the person may be either unresponsive or combative. Fluid losses cause dehydration. Fingers, toes, and lips turn blue. Serious but not necessarily lethal arrhythmias (abnormal slow or fast heart rhythms) such as atrial fibrillation, in which the two upper chambers (atria) of the heart "quiver" rather than contract, may develop.

A core temperature between 20 and 28 °C represents severe hypothermia. At about 25 °C shivering stops. The person becomes increasingly irrational as core temperature drops within this range. Up to half of people with moderate and severe hypothermia have paradoxical undressing, with removal of their clothes accelerating the fall of body temperature. Ultimately the person becomes unresponsive, stiff, and no longer responds to painful stimuli. Heart rate, breathing rate, and blood pressure all decrease further. Metabolic rate drops to 50 % or less of normal. Ventricular fibrillation, a lethal arrhythmia in which the two primary pumping chambers of the heart, the right and left ventricles, quiver instead of contract, may occur. It is the usual terminal event in severe hypothermia.

At 20 °C or less (profound hypothermia) the person appears gray, does not breathe, and looks dead. In some individuals this death-like state, with no measurable heart rate or blood pressure, may occur at the lower end of the severe hypothermia range (about 20–24 °C) [22]. The brain shows no electrical activity. Ventricular fibrillation becomes increasingly likely as core temperature drops and ultimately the heart stops.

But there is some chance for recovery even in cases of marked hypothermia. The reduced metabolic rate and oxygen requirements that help protect hibernating mammals from low body temperature also, to a limited extent, protect humans. Blood flow to the brain decreases 6–7 % for every 1 °C fall in core temperature. However, as noted previously metabolic activity also falls at about the same rate, thus decreasing the brain's energy requirements and delaying permanent damage.

The net effect is that, although the risk of brain injury and death is still significant, recovery may still be possible even in extreme situations. For example, individuals immersed in very cold water for up to a littler longer than an hour can, with treatment, recover despite severe hypothermia. The current record for lowest core body temperature in a victim who survived profound hypothermia is 13.7 °C [24]. This individual was immersed in near-freezing water for about 80 min and required 9 h of resuscitation efforts. She recovered with no clinical brain damage. A child about 2.5 years old survived submersion in cold water for 66 min, with core temperature falling to 19 °C, without neurological impairment [22]. A woman had full neurological recovery following medically induced hypothermia to a core temperature of 9 °C for an hour as part of an experimental treatment method for cancer [25]. Thus, individuals with hypothermia should not be considered dead until they have been rewarmed to near normal temperatures and aggressive resuscitation efforts have not worked.

Rewarming a person with hypothermia can be "passive," such as bringing the victim into a room-temperature environment, using an insulated sleeping bag as covering, or having a rescuer provide body heat. "Active" rewarming may involve external heating pads, putting the person into warm water, or using forced air convection heaters. Treatment of severe and profound hypothermia includes infusing heated (40–45 °C) saline intravenously and into the victim's abdomen as well as administering warmed oxygen. Core temperature should be monitored continuously (e.g. with a thermistor probe inserted into the rectum). A cardiopulmonary bypass (heart-lung) machine or other methods to stabilize the victim's cardiovascular status and facilitate rewarming, such as extracorporeal membrane oxygenation (ECMO), can also be used to treat severe hypothermia.[2]

"Cardiopulmonary resuscitation" (CPR) can be used if no breathing or pulse is present. However, merely jostling a severely hypothermic person, such as during CPR or moving him or her into a warmer environment, might cause the heart to go into ventricular fibrillation. Standard treatments for that arrhythmia such as electrical shocks to the chest and medications probably will not work until the victim is sufficiently rewarmed.

Although a hypothermic individual requires less oxygen due to reduced metabolic rate, there is also less oxygen than normal available to supply those needs. As body temperature falls, hemoglobin in red blood cells gives up oxygen less readily. At 12 °C it does not give up oxygen at all. The small amount of oxygen normally dissolved in blood can be used by the brain and rest of the body for a short time [26]. However, that oxygen is not enough to prevent increased production of harmful substances that can damage the brain and other internal organs during and after rewarming as part of "reperfusion" (restoration of blood supply) injuries.

Once oxygen levels are restored, the metabolic activity in surviving cells increases again. Unfortunately this also allows those harmful chemicals produced during "hypoxia" (low oxygen levels in body tissues) to act. Those

---

[2] Although the exact ways they do it are somewhat different, machines used for cardiopulmonary bypass and ECMO temporarily assist or take over the functions of the heart and lungs to provide gas exchange of oxygen and carbon dioxide as well as blood circulation. In a patient with marked hypothermia they can also help rewarm "chilled" blood in that individual's body.

chemicals injure cell components directly (e.g. by causing swelling and rupture of cells); produce oxygen free radicals (see Chap. 6) that help damage cell proteins, lipids, and DNA; and stimulate substances that cause inflammation (for example, cytokines) and blood clotting (e.g. thromboxane A_2), leading to more internal damage [27]. The end results can include coma, seizures, brain dysfunction ranging from memory loss to a persistent vegetative state, and brain death.

Thus, even if the person survives hypothermia itself without significant permanent injury, body cells can still be damaged during and after rewarming. Vital organs like the brain and heart may be injured directly, or indirectly by blood clots (e.g. a stroke). Life-threatening infections and pulmonary edema (fluid in the lungs) can occur after even the best available treatment [28].

7.4 Therapeutic Hypothermia

Hypothermia is not always a bad thing [29, 30]. In certain medical situations, under very controlled conditions, intentionally cooling a person may help protect the brain, heart, and other vital organs. "Therapeutic hypothermia" may entail cooling the patient's whole body or individual organs, e.g. packing the head with ice bags to "chill" the brain, or infusing cooled fluids intravenously or even directly into the heart. Some survivors of a cardiac arrest are now routinely treated with mild hypothermia [31, 32]. Such a patient's core body temperature is typically reduced to 32–34 °C for 12–24 h using methods such as chilled intravenous (IV) fluids, cooling blankets, and ice bags to reduce risk of brain and other organ damage after resuscitation [27, 33]. The patient is also given medications for sedation and to reduce shivering. A ventilator supports breathing. Such techniques also have potentially useful applications for treating trauma, severe acute blood loss, and stroke [15].

Other methods using hypothermia can also be employed during brain surgery, treating head trauma or other conditions producing increased pressure in or around the brain, and as part of surgery on the heart and the large artery arising from it, the aorta. The prime rationale for using hypothermia in these situations is to prevent brain and other organ damage due to the body's inability to supply adequate oxygenated blood to them during an operation. These measures may involve stopping the patient's heart, supporting circulation using cardiopulmonary bypass, and reducing core temperature to as low as 14 °C and perhaps even less. One classification scheme used for surgical procedures defines "mild" therapeutic hypothermia as a core body temperature between 28.1 and 34 °C, "moderate" as 20.1 and 28 °C, "deep" as 14.1 and 20 °C, and "profound" as 14 °C or less [34].

Inducing increasing levels of therapeutic hypothermia significantly slows the metabolic rate but still falls far short of full suspended animation. Even with these techniques the "safe" limit for keeping a person at deeper levels of therapeutic hypothermia is only about one hour. After that, the complications described previously (damage to body organs, blood clots, etc.) become increasingly likely.

Research using dogs suggests that the safe limit for higher levels of hypothermic "suspended animation" might be pushed from one to several hours. In one method the dog is cooled to less than 10 °C. Its blood is drained and replaced with a maintenance solution similar to the fluid inside cells. This solution reduces risk of blood clots forming during both cooling and, before it is replaced by blood again, rewarming [35, 36]. Dogs have recovered completely after being in that state of profound hypothermia for about 2 h.

A NASA study will reportedly assess a potential method for inducing and maintaining a torpor-like state involving mild hypothermia in astronauts en route to Mars [37, 38]. As noted previously, the rationale for doing this includes reducing the amount of "consumables" (e.g. food, water, oxygen, clothing, etc.), decreasing the need for "living space" and equipment (such as a food galley, exercise equipment, and so on) needed by a crew of "active" rather than sleeping astronauts, and perhaps alleviating some of the potential adverse psychological issues/stressors during long-duration missions described in Chap. 3. The mass and volume of crew "living space" saved by doing this might then be used instead for radiation shielding, employing launch vehicles with lower lifting capacity, etc. On the other hand, astronauts would still age at the same rate as they do on Earth.

Proposed methods for inducing mild hypothermia include special IV catheters designed for infusing chilled fluids (e.g. a saline solution at 4 °C) or a "transnasal cooling system." [32, 37, 39] The goal would be to slowly (at the rate of about 1 °F/h) reduce an astronaut's core body temperature to a range of about 89–93 °F (32 to 34 °C). The transnasal cooling system involves placing a catheter (a thin hollow tube) about 10 cm long into the nose. A special cooling device would then deliver a mixture of oxygen and a liquid coolant into the nasal cavity. Oxygen is typically given at a flow rate of 40–50 L/min [39, 40]. The coolant becomes a mist that, as it evaporates and removes heat, cools the nasal cavity to as low as about 2 °C. Tympanic membrane (eardrum) temperature can be monitored to estimate the average temperature of the brain. Core body temperature would be evaluated using various methods, such as via a catheter in a vein or by probes placed into the esophagus, bladder, or rectum [39, 40].

In this protocol cooled astronauts would also be given standard IV medications (e.g. as a continuous infusion) such as midazolam, fentanyl, propofol, and sufentanil for sedation [40]. They might also require specific muscle relaxants, such as pancuronium, to reduce shivering [32]. Such medications are typically given in some combination (e.g. propofol and sufentanil).

In terrestrial patients treated with mild hypothermia, cooling and sedation is generally maintained for up to several days (typically about 24 h). In the NASA proposal, an individual astronaut might remain in a sedated-hypothermic state for up to 14 days at a time [38]. He or she would be rewarmed at a rate between 1 and 4°F/h, over a period ranging from 2 to 8 h, with use of methods such as warming pads and then remain active for several days before being sedated/cooled again for another up to 2-week period. At any one time only a single astronaut might be active to watch over the rest of the "sleeping" crew and be responsible for their routine care, dealing with any complications or emergencies, as well as general mission-related tasks (e.g. communicating with Earth).

The sedated/hypothermic crewmembers would presumably have their heart rates and rhythms, blood pressures, oxygen saturation, etc. continuously monitored via leads and other means similarly to how this is done for patients in a terrestrial intensive care unit. Their needs for water and nutrition would be provided intravenously. Urine would be collected by (presumably) an internal or external catheter system.

Proposed measures to reduce overall bone and muscle loss in microgravity include administering bisphosphonates, low-level electrical stimulation of muscles, and perhaps some level of artificial gravity (see Chap. 4 for more regarding these methods). Robot manipulator arms are also envisioned to help manipulate cooled crew as needed to manage their care. Astronauts would also need to be adequately secured while "sleeping" in either microgravity or a low level of artificial gravity to prevent them from floating away inside their craft. A recycling system for oxygen and water is also envisioned.

The methods described in this proposal [37] for providing mild hypothermia, sedation, nutrition, etc. are all based on currently available procedures. However, implementing them to put astronauts into a torpor-like state for prolonged periods in space also carries risks that cannot be adequately quantitated at present, particularly in the timeframe and context of a trip to Mars. The maximum duration any individual is said to have been successfully kept in such a cooled/sedated state at one time is about 14 days [37]. However, this proposed method for astronauts would involve many repeated cycles of prolonged torpor and wakening in the space environment—conditions whose safety and efficacy are well in excess of current knowledge.

There are also many known medical issues and risks associated with these procedures, and there may well be others that are not known yet. The intranasal cooling system described previously requires both oxygen and coolant. It would need to be determined how much of each would be brought as fixed supplies or recycled in some way to keep astronauts hypothermic. Oxygen use with this method would also need to be quantitatively compared to its use if crew were awake and active. The intranasal cooling system also has been found to carry risks of its own, including epistaxis (nosebleeds), nasal whitening, damage to nasal tissue, and gas bubbles under the skin around the eyes ("periorbital emphysema") [39, 40].

While the reported rates of such complications are very low, if this technique were used on a sufficient number of crewmembers who underwent the cooling-waking procedure multiple times during their trip, the chances of at least one occurring might increase significantly. On the other hand, keeping environmental temperatures in the module where the cooled/unconscious crew is located to below normal comfort levels but above freezing could help reduce the need to maintain cooling using the intranasal system, chilled IV fluids, cooling pads, etc.

The proposal envisions placing 2 IV lines for long-term use in the astronauts to administer sedation, medications, nutrition, etc. as well as obtaining blood samples. They will presumably be a type of PICC ("peripherally inserted central catheter") line or similar type of catheter [41, 42]. One of these long, thin, sterile tubes may be inserted into a relatively large vein in an arm (e.g. at the "antecubital fossa," the "front" of the crease in a person's elbow). A small ultrasound device can be used to locate a suitable vein. A PICC line or another type of catheter (e.g. a "Hickman catheter") can also be "tunneled" under the skin just beneath a clavicle ("collarbone") and inserted into an internal jugular or other nearby vein.

On Earth fluoroscopy can be used if necessary to help safely advance the catheter within the vein. A PICC or other type of central IV line[3] must be long enough that its tip can be positioned properly, typically near the junction between the superior vena cava and the right atrium. A chest x-ray may be used to confirm its tip is at the right location. However, not even the ISS carries a fluoroscope or x-ray machine, and these capabilities are also not expected to be present on near-future craft headed for Mars. Alternate techniques, such as a portable system that uses the heart's electrical (electrocardiographic, or ECG) pattern for guidance, would be more feasible in space.

A PICC line or similar catheter (including a tunneled one) could also be inserted into a vein in a lower extremity (usually the femoral vein, near the "crease" between the thigh and pelvis). This site of insertion is used less commonly than one in the upper body, however, for reasons that include a potentially greater rate of infection and the risk associated with a thrombus ("blood clot") forming in the femoral and other lower body veins. Nonetheless, the method proposed for cooling/sedating astronauts does suggest its use for one of the IV lines [37].

---

[3] A "central" IV line is one located at least partly in a vein inside the chest. It may extend into the right side of the heart itself (e.g. the right atrium).

These 2 IV lines would presumably be placed in astronauts before launch. On Earth a PICC and any kind of tunneled line can remain in a person's body for weeks or months [43]. However, given enough time one of these IV lines could become occluded due to a thrombus developing in it. Injecting a "blood thinning" medication into the IV catheter can reduce the risk of this happening and also help "clear" it should a thrombus occur. Nonetheless, this type of IV line can develop a thrombus that cannot be removed by this method, potentially requiring that it be replaced while en route to Mars. Doing this while maintaining sterility would be challenging in microgravity or low artificial gravity even by a crewmember trained in and having adequate supplies/equipment for performing the procedure.

Thrombi can also form around the IV line that could occlude the vein it is in. One or more such blood clots could even break off and "embolize" to the lungs, resulting in a pulmonary embolism (see Chap. 2) with effects ranging from clinically insignificant to life-threatening depending on the size of the clot(s). Inflammation ("thrombophlebitis") of the vein might accompany the clot. A line can also break ("fracture") and need to be replaced. It is unclear from the proposal how long each individual central line would be expected to remain in an astronaut's body (e.g. only during the flight to Mars, and perhaps with placement of new ones prior to return to Earth). However, the risk of one or more such complications occurring increases the longer each line is in an astronaut's body.[4]

Moreover, astronauts would also be at increased risk for developing significant infections due to the presence of the indwelling IV lines, from the IV fluids they receive if not kept sterile during handling (e.g. when connecting them to a line), the need to repeatedly obtain blood samples for testing, the catheter-based urine collection system, their sedated state, etc. These risks range from local skin infections to potentially much more serious ones such as a urinary tract infection, or aspiration pneumonia caused by inadvertent passage of saliva and other material from the mouth, throat, or stomach into the lungs. Evaluating for a possible pneumonia typically includes obtaining a chest x-ray, but as noted previously this is not expected to be feasible in space for a long time to come. Bacteremia (bacteria in the blood) can also result from any of those or other sources of infection and could lead to life-threatening sepsis and shock.

Treatment of a serious infection might require removal and replacement of one or both central lines; analyzing and culturing blood, urine, and sputum; then using antibiotics. However, microgravity and perhaps even low levels of artificial gravity may be associated with at least mild depression of astronauts'

---

[4] A blood pressure cuff should also not be used on an arm with a PICC line due to it increasing the risk of the line clotting, the vein being damaged, etc.

immune systems, thus potentially rendering them more susceptible to acquiring infections and making it more difficult to treat them. Moreover, not all microorganisms may be sensitive to available antibiotics. As noted in Chap. 5, the space environment itself might help make microorganisms "hardier" and more difficult to treat. Hypothermia itself may also increase susceptibility to infections [44].

Because "sleeping" astronauts will not be able to eat for up to about 2 weeks at a time, they would receive "total parenteral nutrition" (TPN)[45]. This is a well-established method that includes IV infusion of maintenance fluids as well as carbohydrates (e.g. glucose), electrolytes (such as sodium and potassium), vitamins, trace minerals, amino acids (for protein production), and lipids. Water requirements would be in the range of about 2.5 to 3 L per 24 h. Blood would also need to be sampled at times to make sure that myriad blood chemistries (e.g. levels of glucose and electrolytes, liver function tests, etc.) were all acceptable. The amounts and contents of the TPN constituents would then need to be adjusted as needed by the conscious (and hopefully conscientious) caregiver. Terrestrial patients receiving TPN are weighed periodically to evaluate whether they are receiving too much or too little fluids and nutrients. Alternative means to do this would be necessary in space.

Medical expertise will also be limited on-site to one or at most a small number of conscious crewmembers, with advice from Earth-based personnel being significantly delayed as the craft approaches Mars. Thus, if only one crewmember is usually awake at a time, and each takes turns at that solitary watch, presumably all would need to have the training and experience to handle a medical or other emergency (e.g. malfunction of the module's life support system, a fire, a meteoroid puncture of the craft with slow loss of pressure, etc.) This sole "watcher" would also need to sleep conventionally during part of his or her lonely shift. During that time the safety of those other astronauts would have to be entrusted entirely to their monitoring equipment and both its prompt recognition of a problem (e.g. low oxygen saturation, abnormal heart rate, etc.) and ability to send an alarm to the napping one on duty.

If a medical emergency were to occur, one or more fellow astronauts (particularly those with more medical expertise than the conscious caregiver) could be wakened to provide help. However, this process may take several hours to accomplish safely, and the awakened individual (including one who is a medical professional) might have enough residual effects from the hypothermia and sedation to impede that crewmember from functioning at the needed level for an uncertain time afterwards. Moreover, if a sedated/cooled astronaut required immediate care (e.g. due to a cardiopulmonary arrest), a solitary conscious crewmate/caregiver would have no extra help at all.

Thus, whether due to an accident, medical emergency, etc., while "asleep" astronauts will, even without a misguided descendent of the HAL 9000 aboard, be very vulnerable to injury and death. While hopefully only a theoretical risk, they would also be at the mercy of the one or small group of individuals awake and any serious psychological issues they might develop in space, including deliberately injuring their helpless crewmembers.[5]

A less paranoid but still worrisome danger is that an astronaut who is sedated and receiving analgesics would not feel the pain of an occluding ureteral stone, an inflamed gallbladder, or other serious event. Thus, the conscious crewmember on duty may not be able to recognize anything is wrong (e.g. when the affected astronaut develops a fever, rapid heart rate, a fall in blood pressure, etc.) until the problem is far advanced, making management more difficult and, perhaps, survival less likely.

Also, as emphasized in Chap. 5 spacefarers en route to the Red Planet will be able to carry only a limited number of medications and other irreplaceable supplies. Thus, for example, it will be necessary to carry an adequate amount of dressings, sterile items such as suture material, etc. needed to maintain and, if needed, replace their central lines.

Astronauts will not be taking food by mouth for most of their trip to and from Mars, thus saving most of the mass of foodstuff/liquids that would otherwise be needed. However, TPN and IV solutions have their own mass (e.g. a 1 L bag of an IV fluid has, by definition, a mass of 1 kg from the fluid itself, not counting the additional mass of the plastic bag holding it). Thus, if each "sleeping" astronaut receives about 3 L of IV fluids per 24 h, the total mass of those fluids would be a minimum of at least 3 kg. The astronaut on watch would also be engaged frequently with handling and changing IV fluid bags.

Special equipment (e.g. intranasal cooling systems along with a supply of coolant, devices for monitoring the ECG, heart rate, etc.) will also be required. The infusion rate of TPN fluids would need to be both closely monitored and regulated to avoid adverse effects. These could include fluid overload, electrolyte imbalances (e.g. potassium, calcium, trace elements, etc.), glucose levels that are too low or high (with the potential need to administer insulin if blood glucose rises too high), and other abnormalities in blood chemistries such as the elevated levels of triglycerides that can occur with lipid solutions used for TPN [42, 46]. Moreover, although TPN can theoretically be used indefinitely, it also carries other specific risks such as liver dysfunction as well as development of "sludge" within the gallbladder and an increased risk of gallstones [47, 48].

---

[5] To quote the satiric question raised by the ancient Roman writer Juvenal, "But who is to guard the guards themselves?"

Also, although its overall risk would be anticipated to be low, significant blood loss could occur due to the presence or replacement of central lines. For example, depending on the site and method used, placing a new PICC or similar line could potentially result in the catheter going through the wall of a vein in the chest or other area not amenable to external compression. As noted in Chap. 4, astronauts in microgravity are already at risk of developing mild anemia. This could be exacerbated by the repeated blood sampling needed to monitor their blood chemistries and other tests while they are "sleeping." Use of agents such as erythropoietin that stimulate red blood cell production could potentially help compensate for this. However, their use would add an additional level of complexity and need for supplies.

Decisions would also have to be made regarding how to best protect a sleeping astronaut's airway and respiratory function. As noted in Chap. 5, there are both pros and cons to intubating a sedated individual. Simpler methods such as various types of "oral airway" that just go into the mouth could be used. However, they would not offer as much protection against development of aspiration pneumonia as an endotracheal tube would. Moreover, ancillary measures short of intubation (e.g. use of an oxygen mask) would also be less effective if a sleeping astronaut's respirations became too depressed. This is a significant and ongoing risk with the types of medications (e.g. propofol, fentanyl, etc.) used to continuously maintain sedation and analgesia. These medications have at best a narrow window between doses used to keep a patient unconscious with an adequate respiratory drive and those that would depress the latter too much. Any use of a paralyzing agent such as pancuronium would require that intubation be done or at the very least readily available. Additional medications to reverse the effects of agents such as midazolam, fentanyl, pancuronium, etc. would also need to be available if needed.

The level of mild hypothermia envisioned would also have significant effects on how medications are distributed within, used, metabolized, and cleared by the body [49, 50]. In general, during hypothermia sedatives and other drugs would be expected to act longer and require lower dosages than those used in someone with a normal body temperature to avoid toxic effects (e.g. respiratory depression). These effects would be due to a combination of factors, including the body's overall reduced metabolic rate during hypothermia, as well as decreased rates of metabolism of drugs by the liver and their clearance by the kidneys. Dosages of drugs would have to be carefully titrated and their effects monitored especially carefully during induction of hypothermia and rewarming, when metabolic rates are changing significantly over relatively short periods of time.

Hypothermia also causes shifts of important electrolytes such as potassium, magnesium, and phosphate into cells [44, 51]. This leads to lower levels of these electrolytes in the blood, which if severe enough could cause problems such as arrhythmias. During rewarming the opposite effect occurs, with electrolytes shifting out of cells back into the bloodstream. This could result in problems such as "hyperkalemia" (an above normal blood level of potassium), which can also cause arrhythmias.

Chapter 5 also described some of the problems associated with keeping an individual immobilized for days at a time, including breakdown of skin, dryness of the eyes, etc. Another risk is development of a "deep vein thrombosis" (DVT). In terrestrial gravity immobility increases the possibility that a blood clot (thrombus) will develop in a vein, usually but not exclusively in a person's lower extremity. Here too, if the thrombus "breaks off" and travels through the venous system to the lungs it can cause a pulmonary embolism. Alternatively, in individuals with a patent foramen ovale (see Chap. 3) it can result in a "paradoxical embolism" by traveling from the right to the left atrium and into the body's systemic arteries. Once there, it can travel to the brain and cause a stroke, or to arteries in the rest of the body and cause damage elsewhere. The risk of a DVT occurring in microgravity or a low level of artificial gravity is uncertain, and could be reduced by using various standard methods and medications (e.g. "blood thinners"). However, if it did occur the results could be disastrous.

The proposed torpor-inducing technique would also have several at least potentially manageable gastrointestinal issues. Prior to induction of sedation/hypothermia it would be highly desirable for astronauts to have intestines as free as possible of fecal material. This could be accomplished by using medications and techniques routinely used prior to colonoscopy, diets low in bulk-producing fiber, etc. prior to launch. However, if individual astronauts are to be reawakened for several days at a time during their trip to Mars, decisions would have to be made about whether they should simply continue their TPN, or eat/drink during that time. If they are allowed to eat, the types of food they ingest should hopefully be optimized and perhaps others means used to reduce the amount of fecal material they would pass when they are again cooled and sedated. On the other hand, the possible adverse effects on symbiotic gut bacteria and food absorption when astronauts do start eating again (e.g. when they reach Mars) after receiving predominately or exclusively TPN for up to 6 months en route would also have to be considered.

The effectiveness of proposed countermeasures to reduce musculoskeletal loss is also unclear. As noted in Chap. 4, while the combination of bisphosphonates with use of the Advanced Resistive Exercise Device (ARED) was found to be effective for helping preserve bone mass, how effective bisphosphonates alone, without resistive exercises, would be for the duration of inactivity envisioned is unknown. Likewise, the effectiveness of long-term electrical stimulation of muscles in micro- or reduced gravity to maintain muscle mass and function is uncertain. Besides technical issues with creating artificial gravity, it is unknown what level would be required to adequately reduce musculoskeletal losses even in combination with these other potential countermeasures.

In short, having astronauts spend most of their time "sleeping" en route to Mars has clear benefits. They include reduced mass requirements for launch, more efficient use of resources sent into space, and reduction of psychological stressors. However, it also has many tradeoffs, risks, and complexities whose full extent and degree are currently unknown. First and foremost it will have to be established that humans can, over the very extended periods envisioned, be safely, reliably, and repeatedly put into the cooled/sedated state proposed and rewarmed without significant impairment of their health.

Assuming that studies on Earth show that this method is feasible, the many specific hazards of doing this in space—the risk of infections and metabolic problems, limited supplies and medical expertise, etc.—would also have to be assessed. For example, astronauts on the ISS might undergo this process and be carefully evaluated before it is used on those going to Mars. Even if all study results were favorable, however, and appropriate steps taken to reduce the odds of significant complications (e.g. a catheter-related infection or thrombosis, excessive loss of musculoskeletal mass, etc.), there will be some types and degrees of risks associated with this method that would not be present with astronauts having a "normal" sleep-wake cycle during their journey.

Ultimately, based on the best data available after thorough study, the overall risks versus benefits of having astronauts either "awake" or sleeping during their mission to Mars will have to be carefully weighed. For example, even if the chance of an astronaut developing brain damage or dying (e.g. from septic shock) with the cooling/sedating protocol is very low, is it still worth that risk given its catastrophic results if it did occur? Here too, as in many medical situations, there may be no clear right or wrong decision, but one ultimately has to be made based on the best available assessment of risks and benefits. Only time and further testing will help clarify how feasible this particular method is for use either here on Earth or in space.

7.5 Cells and Cold

The focus so far has been on the enormous difficulties of putting a whole person into suspended animation. However, for individual body cells and tissues the picture is more optimistic.

Successful "cryopreservation" of human sperm cells in liquid nitrogen (temperature about $-196\,°C$) was first reported over 60 years ago. Since then comparable techniques for cryopreserving other human cells—oocytes (immature egg cells), stem cells (precursors of mature cells), embryos, and tissues like skin, bone, and corneas—have been developed. Similar cells from cattle, mice, and other animals can also be preserved.

Before being cooled to subfreezing temperatures cells are exposed to cryo-protectants such as glycerol, dimethyl sulphoxide (DMSO), sucrose, or ethylene glycol. Ice formation may be induced in the extracellular space by "seeding" it with ice or a chilled needle. This gradual, controlled freezing of extracellular fluid helps prevent the sudden production of large, rigid ice crystals that would otherwise occur and damage cells.

At subfreezing temperatures above about $-130\,°C$ ice crystals can still grow and injure cells. To prevent this, long-term storage of cells requires using liquid nitrogen to reach temperatures well below that level. Although most cryopreservation methods use gradual cooling, an "ultrarapid" protocol for embryos involving initial exposure to high concentrations of cryoprotectants followed by immediate immersion in liquid nitrogen has been described [52].

Depending on the type of cell and protocol performed, when they are ready to be used cryopreserved cells may be "thawed" to room temperature as gradually as $8\,°C/min$, or as rapidly as $100–300\,°C/min$. Unfortunately cells are exposed to the same (if not worse) risks of ice crystals forming and other damage occurring during rewarming as they are during the original cooling. Some will inevitably be destroyed during either part of the cooling-rewarming process.

Sperm cells seem to be the "hardiest" ones for cryopreservation, in bulk if not individually. A small cryopreserved sample of them will contain many hundreds of millions, so even if a significant fraction do not survive the preservation process many others will. Current opinion is that an appropriately large specimen of sperm cells can be stored indefinitely for later warming and successful fertilization, with the current record for this being about 21 years [53]. Cryopreserved mature egg cells may not be as successful for later fertilization, but oocytes are now commonly used for this purpose [54].

The upper "safe" limit for storing human embryos is not known, but a successful implantation and birth has been reported after nearly 20 years of cryopreservation [55]. The rate of births following intrauterine transfer of thawed human embryos produced by in vitro fertilization appears to be similar to those when "fresh" (non-frozen) embryos were used. Thus, based on present-day capabilities, for science fiction purposes a colony starship bringing cryopreserved sperm cells, egg cells, and embryos to populate a new world is reasonable, particularly if artificial uteruses (see Chap. 10) to bring them to birth are developed in the meantime.

But while isolated or suspensions of cells as well as "simple" tissues can be safely cryopreserved, there are no proven techniques yet for freezing and banking large, complex human organs like livers or hearts for transplantation [56]. These organs are especially vulnerable to injury during the freezing process. It is much more difficult to ensure that appropriate amounts of cryoprotectant are reaching the tightly bound cells of a large three-dimensional object like the liver than with individual, suspended sperm and other types of cells. If too little cryoprotectant is used the cells are damaged during freezing. Use too much of it and those cells are damaged by the toxic effects of the cryoprotectant itself. Large body organs also contain different kinds of cells and require an intact system of blood vessels to function properly. The ideal cooling and thawing rates to minimize destruction of those various types of cells while also preventing rupture of fine blood vessels (capillaries) may be significantly different.

At present, large organs destined for transplantation can be preserved for only a matter of hours after cooling [57]. A heart should ideally be transplanted within 4 h after it is removed from the donor and a lung within 6–8 h. A liver or pancreas should be transplanted within 12 h and a kidney within about 18 h. Preservation techniques involve perfusing organs with cold solutions to get their temperatures down to about 4 °C. In all cases organs should be transplanted as soon as possible after cooling.

The 1987 novel *The Legacy of Heorot* by Larry Niven, Jerry Pournelle, and Steven Barnes present a future in which problems with preserving a person's brain from damage due to ice crystals during cold-induced suspended animation have been partially but not completely solved. Some colonists on an alien world who reached it by being placed in suspended animation for the trip are thought to suffer from "Hibernation Instability" due to at least some ice crystals forming in their brains as part of that process. The neurological issues thought to be associated are said to include "…major memory losses, impairment of motor skills, mood swings and clinical personality disorders…" Use of this concept in the novel demonstrates that the authors were indeed cognizant of this real-life risk. Thus, although they describe capabilities far beyond current ones, their story is strengthened by this recognition of the actual biological issues involved.

7.6 Making Your Characters Chill Out

But despite these encouraging words about cryopreserving human cells and, perhaps someday soon, whole organs for long periods of time, we have still not yet exhausted all the issues with putting someone's whole body into suspended animation.

A typical way to do this in science fiction—putting the character in a special chamber to presumably flash-freeze (or freeze dry) them, or even have them exposed suddenly to extremely low temperatures in deep space (as in Sir Arthur C. Clarke's 1997 novel *3001: The Final Odyssey*)—would merely result in massive destruction of cells and body parts/organs with, as usual, the brain being the most vulnerable to damage. In addition to all the metabolic and physiological issues described previously, the human body is a large, complex three-dimensional structure whose wide variety of tissues could simply not be "frozen" so quickly, particularly if the goal is to successfully rewarm it later to a healthy state.

As mentioned in Chap. 1, the number of cells in the human body has been estimated to be at least around 3.72×10^{13} [58]. If the low temperatures someone's body were subjected to were originating outside it there would be a significant differential in time between cooling the person's skin and his or her internal organs. The cooling rates of different organs would also vary. Meanwhile a person's various metabolic processes—particularly those that provide blood, oxygen, and nutrients to the brain—would not be slowing at the same rate in different parts of the body.

And as described previously, as its temperature decreases blood tends to clot and releases oxygen less readily. Thus, before internal body temperature reached anywhere near the freezing point it is extremely likely that internal organs—especially the brain—would be deprived of oxygen long before their metabolic rates had slowed enough to require a commensurate amount less of it, resulting in irreversible damage. Draining a person's blood and replacing it with cryoprotectant and other fluids could help reduce this problem. However, having this described or shown when a character goes into suspended animation in a science fiction work is definitely not the norm.

As if things were not difficult enough, a few degrees before its freezing point water expands. This too can damage cells, especially if that expansion happens rapidly. Once at and below the freezing point of water, major damage to cells will continue to occur due to formation of ice crystals within and outside them. Our bodies are also filled with many extracellular water-containing liquids besides blood, such as bile, the cerebrospinal fluid bathing the brain and spinal cord, urine, etc. Having a person's blood literally freeze in his or her veins, arteries, and especially the tiny capillaries will also damage them. Similar if not worse damage would occur to the brain as the cerebrospinal fluid within and surrounding it cools and gives that unfortunate individual a literal case of brain freeze.

Even under the most controlled conditions, of necessity requiring both internal (e.g. infusion of chilled fluids) and external cooling and reversal of that process during rewarming, successful whole-body suspended animation would be tremendously difficult. But despite this pessimistic assessment regarding its practical feasibility, in theory the cells in a living human body could be preserved by a process called "vitrification." The concept behind vitrification is to cool the water in the body and its cells in such a way that it does not actually freeze—that is, turn from a liquid to a solid, i.e. ice.

Cooling produces slowing not only of biochemical (e.g. enzymatic and protein-synthesizing) activities but also the translational motion of water, cryoprotectants, and other molecules in a liquid state. Cooling water and other fluids below their freezing points strongly predisposes them to freezing. However, under certain conditions water can remain liquid down to a temperature of about $-41\,°C$, becoming "supercooled" [59]. With further cooling ice crystals will form until, at a low enough temperature, water becomes a "glass"—a liquid too cold to flow or freeze, in an overall arrested state of motion. At that point the viscosity of water and other fluids is so high that both metabolism and molecular motions in cells cease, with cellular components such as nuclei staying in their "correct" anatomical places. Such "vitrified" cells are truly in suspended animation.

Unlike freezing, successful vitrification itself should not injure cells. However, as with other cooling methods cells would still be at risk of freezing and injury during rewarming. There is also the very practical difficulty of cooling water and other fluids to the "glass transition temperature"—the point at which vitrification occurs and there is no longer a danger of freezing—before a significant number and size of damaging ice crystals form. There is even some controversy about what the actual glass transition temperature for water is, ranging as low as about $-100\,°C$ to what is thought to be a more likely value of about 136 K, [60, 61] or $-137\,°C$.

Both freezing and vitrification are successfully used as methods to cryopreserve suspensions of individual cells, such as sperm cells. In theory vitrification of large organs would be an excellent way to preserve them indefinitely for transplantation when needed [56]. Even now research is being conducted on possibly doing this with kidneys based on prior successful animal experiments [62].

However, as with other cooling methods, it is likely that some organs will be more challenging than others to successfully vitrify and rewarm. Factors such as the overall size and number of cells in the organ (e.g. small organs such as ovaries might be "easier" than larger ones like livers), the possible differing responses of various types of its cells to cryoprotectants and the overall process of vitrification, and how well cells can be perfused and protected during these procedures will all come into play.

Another important consideration for preserving organs via vitrification or other hypothermic techniques is how much "reserve" an organ has after the cooling-rewarming process—that is, how many cells it could afford to have die or not function immediately for the transplant to still succeed. For example, neither loss of a relatively small number of cells in a cryopreserved kidney nor a delay of days before it starts to function well would be major impediments, since renal (kidney) function can be supported in the meantime by dialysis and other means. On the other hand, a transplanted heart should ideally start working well immediately without a prolonged recovery phase, although modern devices such as ventricular assist devices (see Chap. 14) might be used to help "tide it over" until it can.

It should come as no surprise that successfully inducing and reversing suspended animation via vitrification for an entire human body would be at least as challenging (if not more) as the less intense cooling methods described previously and for all the same reasons. If vitrification were attempted it would have to combine many techniques, all of them requiring a great deal of time, preparation, and coordination.

The process might begin with procedures as mundane as inserting a catheter to drain a person's bladder of urine, as well as cleansing the stomach, small intestines, and colon of material that does not need to be preserved. As described previously regarding dog experiments, blood would have to be drained while somehow maintaining adequate oxygen and nutrients to the brain and other organs during the whole cooling process. Cooling of both the outside and inside of the body would need to synchronized, cryoprotectants would have to be administered in finely tuned doses and rates—the list of real-world difficulties goes on and on.

Perhaps, while all the bugs for successfully vitrifying and reviving whole humans are being worked out, methods could be developed to induce a state in humans mimicking hibernation by non-cryogenic means. For example, if the chemical or chemicals that help induce and regulate hibernation in ground squirrels, hamsters, and other mammals could be identified, either they or new drugs derived from them might be applicable to humans too. At present, however, this possibility remains highly speculative.

There also might be other ways to induce hibernation in humans. Hydrogen sulfide, the gas that gives rotten eggs their distinctive odor, has recently been found to induce a hibernation-like state in mice [63, 64]. Besides smelling bad, hydrogen sulfide in high atmospheric concentrations is toxic and inhaling enough of it can be fatal. However, in one study mice were exposed to a very low concentration (80 parts/million) of this gas for 6 h. During that time they gradually reduced their oxygen consumption and metabolic rate to about 10 % of normal. Their core body temperature fell to only 15 °C in an environmental temperature of 13 °C, and their breathing rate decreased from 120 to less than 10 breaths per minute. The mice were said to have experienced no long-term health effects from this method simulating hibernation. It should be noted, however, that among many other differences between them, some (but not all types) of mice can hibernate as part of their normal capabilities while humans cannot. Thus, extrapolating effects that work in mice to what works in us members of *H. sapiens* is also uncertain.

Human trials using hydrogen sulfide or similar chemicals have been proposed [65] but are not currently being conducted. Moreover, studies using hydrogen sulfide in other mammals, such as pigs and sheep, had little or no effect on inducing torpor [14]. Thus, the value of using hydrogen sulfide for this in larger animals (e.g. humans) remains very uncertain.

A recent science fiction novel (2012), *Bowl of Heaven* by Gregory Benford and Larry Niven, seems to pay homage to this research. Its opening chapter describes a character reviving from suspended animation remembering how "they had given him the stinky sulfur gas…" before entering that state. This "throwaway" phrase suggests that the process used to induce suspended animation might have included hydrogen sulfide, which is (to put it mildly) very "stinky" indeed. Further research will be needed to determine whether that gas might be used as at least an adjunct to other methods (the novel also uses the phrase "cold sleep" for the process). However, whether or not hydrogen sulfide does turn out to be useful, this phrase suggests that the authors were cognizant of this research and included an element of real, cutting-edge science into their work, thereby strengthening it.

Finally, several IV medications can depress the brain's metabolic functions. Barbiturates and newer agents like etomidate are used for this purpose during general anesthesia. However, these too can only be administered for relatively short times. Outside of science fiction there is still no medication, gas, or any other substance that can start a person on the road to safely reaching the twenty-fifth century.

7.7 Prospects for Reviving the Frozen

Despite the odds against them a small number of officially dead individuals have elected to be stored in liquid nitrogen in the hope of being revived someday. Works such as Robert Ettinger's book *The Prospects of Immortality* (1962) helped bring this idea of "freezing for the future" to public attention. This use of "cryonics" assumes (or at least hopes) that the means to revive them will eventually be developed, along with other advanced medical capabilities such as finding a cure for whatever killed them in the first place or perhaps even rejuvenate their cells and make those people "young" again. It also assumes that someone in the future will actually want to revive them and that their brains were not irreparably damaged before being chilled and preserved.

Legally a person must already be clinically dead (no respiration or heartbeat, no response to deep pain, etc.—see Chap. 13 for more on this) before the cryopreservation process can be started. Though techniques for preparing bodies for long-term cryopreservation and storage vary in details, basic procedures are the same [66]. In the "ideal" situation, immediately after clinical death the newly deceased's body is injected with heparin (to reduce blood clotting) and various protective medications to try to reduce cell injury. Machines for performing CPR, including a "thumper" for chest compressions and a respirator, are used. The body is cooled with ice bags, a cooling blanket, and/or other means to a little above the freezing point of water. As soon as possible the patient is connected to a heart-lung machine. Blood is drained and replaced by a cryoprotectant. After this procedure is completed the CPR machine is stopped.

The body is then cooled further in two stages. First is a rapid phase over several hours using liquid nitrogen vapors. Its aim is to cool the body as quickly as possible to a little beyond the glass transition temperature to try to achieve vitrification of body cells. One protocol uses a target temperature for this first phase of about $-125\,°C$ (although, as noted previously, the glass transition temperature of water may actually be as low as $-137\,°C$). This rapid cooling is followed by slower cooling over about 2 weeks to the temperature of liquid nitrogen ($-196\,°C$). The body is then stored in a "cryostat"—a generally cylindrical storage unit made of metal (similar to a thermos bottle) or fiberglass. There the person is immersed in liquid nitrogen, suspended vertically in a head-downward position. This position is used so that the feet and not the head will thaw first in case of a cooling problem, such as a prolonged power outage to the cooling system.

Unfortunately, despite all these preparations and even with brain and other tissues not showing gross signs of significant injury, based on current concepts it is virtually certain that irreparable brain damage occurred long before this preservation process was complete. Starting within minutes after clinical death, with each subsequent passing second before the body is cooled to the temperature of liquid nitrogen, more and more cells—including those in the brain—are being injured and dying. Perhaps future medical research will prove that assessment wrong, but until it does the prospects of meaningful revival of these preserved individuals appears to be very dubious at best. Starting cryopreservation while the person is still living might reduce this damage but also raises ethical and legal issues.

In some cases, instead of the whole body only the head of a person is preserved—what has been called "neuropreservation." This procedure has potential advantages, including that cooling just the brain could proceed more rapidly and more effectively than that of the entire body, and the head would take up less space in liquid nitrogen during long-term storage.

The greatest disadvantage of this latter technique is illustrated by the 1962 movie *The Brain That Wouldn't Die*. However, the possibility that a successful human head transplant could be performed may not be as remote as it might seem. Head transplants involving dogs and monkeys were performed over 40 years ago and, at least in the very short term, were successful [67, 68]. A significant limitation of the technique used in those experiments was that the spinal cord of the "recipient" was not attached to the "donor's" head, so the latter could not move its new body.

A recent paper outlined possible procedures for the first human head transplantation, including attachment of the spinal cord, various individual nerves, and other tissues (e.g. the trachea and esophagus, the "tubes" connecting the throat to the lungs and stomach, respectively) from the "donated" head to the "host" body [69]. Such procedures would, at least in theory, allow normal function of the whole new head-body system. Nonetheless the surgical challenges to successfully doing all this remain formidable. Even if the operation were successful there would still be the serious long-term risk of the new body rejecting the foreign tissue of the donated head. Such rejection is a problem with more conventional transplantation procedures and might conceivably be managed to varying degrees with current anti-rejection medications. How well these or future medications would prevent rejection in this situation without causing unacceptable side effects is uncertain. Managing rejection if it did occur would also obviously present a difficult dilemma.

Although not directly related to the idea of suspended animation per se, the subject of head transplantation suggests another science fiction staple—brain transplantation and the "brain-in-a-jar." For example, the former figures prominently in Robert A. Heinlein's 1970 novel *I Will Fear No Evil*, in which the brain of a male character is transplanted into the body of a beautiful young woman. It can be argued that, despite its enormous difficulties, a head transplant would be significantly "simpler" than a brain transplant in terms of both keeping the latter alive during the process as well significantly increasing the already great complexities of making all the intricate neurological, vascular, and other connections to the host body (e.g. the "cranial" nerves connecting the donated brain to the host's eyes, ears, etc.). Also, Heinlein's plot device of having the dead woman's consciousness remain despite her brain being replaced by a man's, however interesting it might be from a psychological and metaphysical perspective, would (to say the least) be very difficult to explain by any known biological process.

Likewise the idea promulgated by such as works as Curt Siodmak's 1942 novel *Donovan's Brain*, the *Outer Limits* television episode "The Brain of Colonel Barham" (1965), and other works that an isolated brain-in-a-jar could gain superhuman powers such as telepathy, grow larger and develop greatly enhanced intelligence, etc. emphasize "fiction" over "fact." Nonetheless, in theory an isolated brain could be kept alive and functioning if life-sustaining oxygen, nutrients, etc. were supplied to it and waste materials (such as carbon dioxide) removed via its blood vessels. Connections might even be made directly to its visual, auditory, speech and other specialized areas so that it could see, hear, and otherwise communicate with the external world. How long such an isolated brain could survive both physically and psychologically in such a state is a different matter. However, those kinds of questions are well suited for presentation in science fiction.

7.8 Methods Equivalent to Suspended Animation

For interstellar journeys suspended animation is only one method of getting characters to an exoplanet well within the time frame of a normal human lifespan. Albert Einstein's theory of special relativity includes the concept that an observer in one frame of reference will measure the passage of time in a second frame of reference moving relative to his or her own differently than another observer within that second one [70, 71]. Put less technically, to a person on Earth (one frame of reference) time would appear to pass more slowly on a spaceship traveling near the speed of light (the second frame of reference) compared to that recorded by a terrestrial clock. Conversely, for astronauts on that spaceship time on Earth would seem to go by more quickly than that recorded on their craft.

This "time dilation" effect is related to the fact that the speed of light, c, is a universal constant, measured as the same by all observers wherever they are. Time dilation is mathematically described by this equation:

$$\Delta t = \frac{\Delta t'}{\sqrt{1 - \dfrac{v^2}{c^2}}}$$

Here $\Delta t'$ is the "proper time," the interval of time measured by an observer in a frame of reference moving (e.g. in a spaceship) at speed v relative to a second frame of reference (e.g. Earth). Similarly Δt is the corresponding time interval that would be measured in that second frame of reference. The velocities reached during terrestrial modes of human transportation such as walking, traveling in cars, trains, or jets, etc. are miniscule (in the range of a few to several thousand kilometers per hour) compared to c (about 300,000 km/s). At these conventional speeds the fraction v^2/c^2 in the above equation is nearly zero. Thus, the time Δt measured by a person standing still on a city sidewalk is actually longer than $\Delta t'$, the time measured by someone driving past in the street nearby at 40 km/h. However, passage of a second of time for the person at rest on the sidewalk would be less than a quadrillionth of a second longer— a difference far too tiny for either individual to notice.

Conversely, as v starts to approach the value of c time dilation becomes more and more of a factor. If v is one-tenth of the speed of light ($0.1c$) for a speeding spacecraft, a little math shows that for every second measured by a clock on the ship ($\Delta t'$) a terrestrial clock (Δt) would show passage of 1.005 s. When v gets very close to c this effect becomes increasingly significant. For each second recorded on a craft moving at $0.95c$ the corresponding time interval measured on Earth would be 3.2 s. Cranking up the ship's speed to $0.999c$ increases the ratio to one second on the ship corresponding to about 22.4 s of terrestrial time. Expressing the latter relationship in years means that, for every birthday that an astronaut on the spacecraft has, someone on Earth would celebrate a little over 22 of them.

As mentioned in Chap. 6, a light-year is the distance travels in one year. The closest star system to Earth is the Alpha Centauri system, a set of three stars all a little over 4 light-years away. Besides time dilation, traveling at speeds close to light also effectively "contracts" the outside universe, making distances appear shorter to anyone traveling near c. Thus, to an intrepid astronaut traveling at $0.999c$ all the way from Earth to Epsilon Eridani, a star similar to our Sun at a distance of about 10.5 light-years (in our frame of reference), the distance to that alien star would seem about 10.5/22 or just under 0.5 light-year away, based on how long it would take to get there measured by clocks on the spaceship.

In short, time dilation based on "relativistic" speeds—that is, motion at speeds close to that of c—is a well-established part of the known laws of physics. It would "shorten" the passage of time for a space traveler relative to the terrestrial rate enough that it would effectively, though not literally, be equivalent to suspended animation.

Unfortunately there are daunting (though not insurmountable) technical issues involved with propelling an occupied spaceship to relativistic speeds. Even with the use of matter-antimatter engines such as in *Star Trek* the energy requirements to do this are also enormous, particularly because the ship would markedly increase its mass (again, relative to a observer on Earth) as its velocity gets very close to c, and thus the force needed to accelerate it.

In addition it would take a significant amount of time to accelerate the ship to a high fraction of c in the first place. Energy and time would also need to be expended to decelerate it so that it can go into orbit around an alien world rather than just zip right by it and its parent star at a speed near c. Thus, the time dilation examples given previously are "best case" scenarios, and it would actually take significantly more subjective time on the spaceship before astronauts arrive at their destination. Other difficulties too numerous to mention here also come into play.

Nonetheless, such pesky details as how to do it have not stopped writers from using near-light speed travel as "hard" science fiction. An extreme example of this occurs in Poul Anderson's novel *Tau Zero* (1970), whose spacefaring characters travel so far and fast they live to see the end and rebirth of the universe itself. More speculative alternatives include using ideas borrowed from classic science fantasy such as a "hyperdrive"; postulating that a future science will prove us wrong about the speed of light being the ultimate speed limit for physical objects; or assuming that the latter is true but finding loopholes (e.g. wormholes, "shortcuts" between one part of the universe and another) to work around it.

For example, while light itself can move no faster than c and no physical object can move at that speed in what Einstein called "spacetime"—in the very simplest terms, the universe we live in, with three spatial dimensions and one of time—spacetime itself can expand faster than c. In fact, current cosmology postulates this happened a tiny fraction of a second after the "Big Bang" itself created our universe roughly 13.8 billion years ago during an extremely brief but important "inflationary phase." If, for example, a localized area of spacetime enclosing a spaceship could be artificially manipulated to expand faster than c then the ship and its occupants could also travel at an effectively "superluminal" speed, even though c would still have its conventional value within that "bubble."

Here too a myriad of details would need to be worked out before (or if) this could become a practical means of space travel. However, this concept and others used by some science fiction writers that "expand" on known scientific principles reflect an underlying knowledge of established physics before manipulating its laws or extrapolating beyond them. For example, Dr. Stanley Schmidt, former editor of *Analog Science Fiction and Fact*, uses a novel way to achieve velocities greater than c in his novels *The Sins of the Fathers* (1976) and its sequel *Lifeboat Earth* (1978) that incorporates actual science.[6] His method involves the "Rao-Chang drive" that allows physical objects which, in our "reality," can approach but never reach or exceed the speed of light to "tunnel"

---

[6] The fact that Dr. Schmidt has a PhD in physics helps explain why he is able to successfully do this.

to another in which supraluminal ones can be achieved, without actually accelerating to *c* itself.

In a still more speculative mode, others authors have used ideas for (essentially) producing "true" suspended animation borrowed from physics instead of biology. One postulated property of a "stasis field" would be that time does not pass within such a field. Thus, anyone or anything in "stasis" would remain unchanged for very long times, putting them effectively into suspended animation during that time. Vernor Vinge coined the term "bobble" for the spherical stasis fields with potentially very long lifespans depicted in his novels *The Peace War* (1984) and *Marooned in Realtime* (1986), portraying the serious implications of their use. In one of my own stories, "To Him Who Waits,"[7] I employed a stasis field for the considerably less serious purpose of helping my main character, stuck at a future spaceport, pass the time while waiting for the long-delayed flight to his homeworld to arrive.

As a perhaps even less scientifically plausible method, it is never explained in either *The Empire Strikes Back* (1980) or *Return of the Jedi* (1983) what carbonite is and how it can put Han Solo into suspended animation. However, we certainly see what it does, and based on Han's reaction to the process and his appearance while and after being encased in carbonite it appears to be an effective but unpleasant technique.

Finally, biologically based methods that do not use hypothermia as their primary technique could also be envisioned. Allen Steele's 2002 novel *Coyote* presents colonists on a starship traveling up to $0.2c$ using a "biostasis" technique that does not involve "standard" cryogenic methods for suspended animation. Instead their naked and hairless bodies float in closed containers within a gelatinous material, with an oxygen mask covering the lower face and (presumably IV) plastic tubes in their arms. They are kept in a coma-like state by means of drugs and prevented from aging by use of "homeostatic stem-cell regeneration, telomerase enzyme therapy, and nanotechnical repair of vital organs" rather than primarily by slowing metabolic processes via hypothermia and vitrification. This represents an imaginative combination of different methods to induce the functional equivalent of suspended animation.

Later chapters will describe how stem cells, telomerase, and medical nanotechnology all still face formidable challenges as means to maintain or restore health. As noted in Chap. 5 and earlier in this chapter, many other measures (e.g. for maintaining respiration, avoiding infections from IV lines, providing nutrition, etc.) besides those depicted in the novel would also need to be used to keep a sedated individual healthy. While "cushioning" a person's body in a gelatinous substance might help in some ways to prevent the pressure sores described

---

[7] *Analog Science Fiction and Fact*, December 1999.

in that latter chapter, it might also raise questions such as if it might macerate the skin, how it would prevent skin infections, possible issues with waste and other disposal (including urine, solid waste, and even perspiration), etc.

Fortunately, for purposes of the narrative none of these additional "realistic" issues need to be addressed. Using the concepts alone of biostasis via stem cells, medical nanotechnology, etc. is more than enough to add sufficient plausibility for story purposes, with the author creating an imaginative means to use "suspended animation" in it based on elements of actual medical science.

7.9 The Bottom Line

While theoretically not impossible, keeping a person in suspended animation for days, months, or years carries many daunting challenges. As with concepts that clearly violate established scientific laws but are useful for plots, a writer can simply present suspended animation as a given while being silent or extremely terse about how it is actually done. A magical "stasis beam" or medical treatment that somehow slows down biological activity at the molecular level can also be used as an "explanation" that really just describes "what" it does and ignores "how" it does it.

To add a dollop of "hard" science to a story the writer might describe in brief, general terms that a vitrification process was used and some of the problems mentioned previously that would need to be solved to make it successful. This at least lets readers know that the author is aware of these issues—and makes the former aware of them too if they do not know them already. Likewise, mentioning far-future but not impossible techniques using nanotechnology to aid the process, extensive genetic engineering to make the human body more conducive to suspended animation, and similar allusions to the science involved can add to the story's plausibility.

For story purposes it might actually be better to leave some real-life details unmentioned. For example, *2001: A Space Odyssey* depicts three crewmembers "sleeping" within coffin-like chambers. Monitors showing their markedly reduced heart rates and breathing rates indicate that they are in hibernation, at a level similar to the ground squirrels and other small mammals described previously (at least until HAL lowers those rates to a terminal zero). The presumptive reason why they are hibernating is, as noted previously, plausible—to reduce consumption of limited oxygen and food supplies.

However, even assuming that this could be done (more likely than true suspended animation but still difficult), as mentioned previously, in real life this might cause more problems than it solves. The same issues described in Chap. 5 that would make creating an "autodoc" difficult, including a "hibernating" individual's susceptibility to pressure sores, infections, hydration, nutrition, etc. would also apply with that method. Thus, it may be prudent for even a "hard" science fiction work to allude to such difficulties but not dwell on them in needless (again, at least for story purposes) detail.

For example, Charles E. Gannon takes this reasonable approach in *Fire with Fire* (2013). That novel employs suspended animation both for one-way travel into the future and space travel. Its descriptions of the process, including blood exchange, colonic cleansing, and use of catheters and IV lines depict the difficulties with initiating and maintaining it with an effective level of verisimilitude.

In short, putting characters into suspended animation or an equivalent variation of it for very long periods of time will have to ignore the known science involved, only allude to the methods used, or postulate great advances over current knowledge and medical care. Although these techniques may have at best a questionable basis in our "real" world, they are fair game as ways to write entertaining and perhaps even informative science fiction.

References

1. Stratmann HG. Suspended animation: the cold facts. Analog Sci Fict Fact. 2000; 120(4) 43–55.
2. Bellamy R, Safar P, Tisherman SA, Basford R, Bruttig SP. Suspended animation for delayed resuscitation. Crit Care Med. 1996;24:S24–47.
3. Vreeland RH, Rosenzweig WD, Powers DW. Isolation of a 250 million-year-old halotolerant bacterium from a primary salt crystal. Nature. 2000;407:897–900.
4. Cano RJ, Borucki MK. Revival and identification of bacterial spores in 25- to 40 million-year-old dominican amber. Science. 1995;268:1060–4.
5. Murray AE, Kenig F, Fritsen CH, McKay CP, Cawley KM, Edwards R. Microbial life at – 13 °C in the brine of an ice-sealed Antarctic lake. Proc Natl Acad Sci U S A. 2012;109:20626–31.
6. Storey KB, Storey JM. Metabolic rate depression and biochemical adaptation in anaerobiosis, hibernation and estivation. Q Rev Biol. 1990;65:145–74.
7. Storey KB. Biochemistry of natural freeze tolerance in animals: molecular adaptations and applications to cryopreservation. Biochem Cell Biol. 1990:687–98.
8. Zancanaro C, Biggiogera M, Malatesta M. Mammalian hibernation: relevance to a possible human hypometabolic state. 2004. http://www.esa.int/gsp/ACT/doc/ARI/ARI Study Report/ACT-RPT-BIO-ARI-036501-Morpheus-Verona.pdf. Accessed 15 April 2015.
9. Costanzo JP, do Amaral MCF, Rosendale A, Lee RE Jr. Hibernation physiology, freezing adaptation and extreme freeze tolerance in a northern population of the wood frog. J Exp Biol. 2013;216:3461–73.

10. Nedergaard J, Cannon B. Mammalian hibernation. Phil Trans R Soc Lond. 1990;326:669–86.
11. Kilduff TS, Krilowicz B, Milsom WK, Trachsel L, Wang LCH. Sleep and mammalian hibernation: homologous adaptations and homologous processes? Sleep. 1993;16:372–86.
12. Kaciuba-Uscilko H, Greenleaf JE. Acclimatization to cold in humans. 1989. http://ntrs.nasa.gov/archive/nasa/casi.ntrs.nasa.gov/19890013690_1989013690.pdf. Accessed 15 April 2015.
13. Woodland Jumping Mouse. 2014. http://www.arkive.org/woodland-jumping-mouse/napaeozapus-insignis/. Accessed 15 April 2015.
14. Bouma HR, Verhaag EM, Otis JP, Heldmaier G, Swoap SJ, Strijkstra AM, et al. Induction of torpor: mimicking natural metabolic suppression for biomedical applications. J Cell Physiol. 2012;227:1285–90.
15. Jinka TR, Toien O, Drew KL. Season primes the brain in an arctic hibernator to facilitate entrance into torpor mediated by adenosine A(1) receptors. J Neurosci. 2011;31(30):10752–8.
16. Nelson BT, Ding X, Boney-Montoya J, Gerard RD, Kliewer SA, Andrews MT. Metabolic hormone FGF21 is induced in ground squirrels during hibernation but its overexpression is not sufficient to cause torpor. PLoS One. 2013;8:e53574. Accessed 6 May 2015.
17. Bolling SF, Tramontini NL, Kilgore KS, Su T-P, Oeltgen PR, Harlow HH. Use of "natural" hibernation induction triggers for myocardial perfusion. Ann Thorac Surg. 1997;64:623–7.
18. Berger RJ. Cooling down to hibernate: sleep and hibernation constitute a physiological continuum of energy conservation. Neurosci Lett. 1993;154:213–6.
19. Storey KB. Organic solutes in freezing tolerance. Comp Biochem Physiol. 1997;117A(3):319–26.
20. Vybiral S, Jansky L. Hibernation triggers and cryogens: do they play a role in hibernation? Comp Biochem Physiol. 1997;118A:1125–33.
21. Jinka TR, Duffy LK. Ethical considerations in hibernation research. Lab Animal. 2013;42:248–52.
22. Brown DJ, Brugger H, Boyd J, Paal P. Accidental hypothermia. N Engl J Med. 2012;367(20):1930–8.
23. Giesbrecht GG, Bristow GK. Recent advances in hypothermia research. Ann N Y Acad Sci. 1997;813:663–75.
24. Gilbert M, Busund R, Skagseth A, Nilsen P, Solbø J. Resuscitation from accidental hypothermia of 13.7 °C with circulatory arrest. Lancet. 2000;355:375–6.
25. Niazi SA, Lewis FJ. Profound hypothermia in man. Report of a case. Ann Surg. 1958;147:264–6.
26. Dexter F, Kern FH, Hindman BJ, Greeley WJ. The brain uses mostly dissolved oxygen during profoundly hypothermic cardiopulmonary bypass. Ann Thorac Surg. 1997;63:1725–9.

27. Neumar RW, Nolan JP, Adrie C, Aibiki M, Berg RA, Bottiger BW, et al. Post-cardiac arrest syndrome: epidemiology, pathophysiology, treatment, and prognostication. Circulation. 2008;118(23):2452–83.

28. Vaagenes P, Ginsberg M, Ebmeyer U, Ernster L, Fischer M. Cerebral resuscitation from cardiac arrest: pathophysiologic mechanisms. Crit Care Med. 1996;24:S57–68.

29. Kochanek PM. Bakken lecture: the brain, the heart, and therapeutic hypothermia. Cleve Clin J Med. 2009;76(Suppl 2):S8–12.

30. Marion DW, Leonov Y, Ginsberg M, Katz LM, Kochanek PM, Lechleuthner A. Resuscitative hypothermia. Crit Care Med. 1996;24:S81–9.

31. Arrich J, Holzer M, Herkner H, Müllner M. Hypothermia for neuroprotection in adults after cardiopulmonary arrest. 2010. http://www.ucdenver.edu/academics/colleges/medicalschool/departments/medicine/intmed/imrp/CURRICULUM/Documents/ArrichJ--HypothermiapCPR2010.pdf. Accessed 15 April 2015.

32. Kim F, Nichol G, Maynard C, Hallstrom A, Kudenchuk PJ, Rea T, et al. Effect of prehospital induction of mild hypothermia on survival and neurological status among adults with cardiac arrest: a randomized clinical trial. JAMA. 2014;311(1):45–52.

33. Peberdy MA, Callaway CW, Neumar RW, Geocadin RG, Zimmerman JL, Donnino M, et al. Part 9: post-cardiac arrest care: 2010 American Heart Association guidelines for cardiopulmonary resuscitation and emergency cardiovascular care. Circulation. 2010;122(18 Suppl 3):S768–86.

34. Tian DH, Wan B, Bannon PG, Misfeld M, Lemaire SA, Kazui T, et al. A meta-analysis of deep hypothermic circulatory arrest versus moderate hypothermic circulatory arrest with selective antegrade cerebral perfusion. Ann Cardiothorac Surg. 2013;2(2):148–58.

35. Safar PJ, Tisherman SA. Suspended animation for delayed resuscitation. Curr Opin Anesthesiol. 2002;15:203–10.

36. Behringer W, Safar P, Wu X, Kentner R, Radovsky A, Kochanek PM, et al. Survival without brain damage after clinical death of 60–120 min in dogs using suspended animation by profound hypothermia. Crit Care Med. 2003;31(5):1523–31.

37. Torpor inducing transfer habitat for human stasis to Mars. 2014. http://www.nasa.gov/sites/default/files/files/NIAC_Torpor_Habitat_for_Human_Stasis.pdf. Accessed 6 May 2015.

38. Brumfield B. Sleeper spaceship could carry first humans to Mars in hibernation state. 2014. http://www.cnn.com/2014/10/07/tech/innovation/mars-hibernation-flight/. Accessed 15 April 2015.

39. Castren M, Nordberg P, Svensson L, Taccone F, Vincent JL, Desruelles D, et al. Intra-arrest transnasal evaporative cooling: a randomized, prehospital, multicenter study (PRINCE: Pre-ROSC IntraNasal Cooling Effectiveness). Circulation. 2010;122(7):729–36.

40. Busch HJ, Eichwede F, Fodisch M, Taccone FS, Wobker G, Schwab T, et al. Safety and feasibility of nasopharyngeal evaporative cooling in the emergency department setting in survivors of cardiac arrest. Resuscitation. 2010;81(8):943–9.

41. Johansson E, Hammarskjold F, Lundberg D, Arnlind MH. Advantages and disadvantages of peripherally inserted central venous catheters (PICC) compared to other central venous lines: a systematic review of the literature. Acta Oncol. 2013;52(5):886–92.

42. Olveira G, Tapia MJ, Ocon J, Cabrejas-Gomez C, Ballesteros-Pomar MD, Vidal-Casariego A, et al. Parenteral nutrition-associated hyperglycemia in non-critically ill inpatients increases the risk of in-hospital mortality (multicenter study). Diabetes Care. 2013;36(5):1061–6.

43. Christensen LD, Rasmussen HH, Vinter-Jensen L. Peripherally inserted central catheter for use in home parenteral nutrition: a 4-year follow-up study. JPEN J Parenter Enteral Nutr. 2013;38(8):1003–6.

44. Varon J, Marik PE, Einav S. Therapeutic hypothermia: a state-of-the-art emergency medicine perspective. Am J Emerg Med. 2012;30(5):800–10.

45. Boullata JI, Gilbert K, Sacks G, Labossiere RJ, Crill C, Goday P, et al. A.S.P.E.N. clinical guidelines: parenteral nutrition ordering, order review, compounding, labeling, and dispensing. JPEN J Parenter Enteral Nutr. 2014;38(3):334–77.

46. Berlana D, Barraquer A, Sabin P, Chicharro L, Perez A, Puiggros C, et al. Impact of parenteral nutrition standardization on costs and quality in adult patients. Nutr Hosp. 2014;30(2):351–8.

47. Raman M, Allard JP. Parenteral nutrition related hepato-biliary disease in adults. Appl Physiol Nutr Metab. 2007;32(4):646–54.

48. Baudet S, Medina C, Vilaseca J, Guarner L, Sureda D, Andreu J, et al. Effect of short-term octreotide therapy and total parenteral nutrition on the development of biliary sludge and lithiasis. Hepatogastroenterology. 2002;49:609–12.

49. Anderson KB, Poloyac SM. Therapeutic hypothermia: implications on drug therapy; 2013. http://www.intechopen.com/books/therapeutic-hypothermia-in-brain-injury/therapeutic-hypothermia-implications-on-drug-therapy. Accessed 15 April 2015.

50. van den Broek MH, Groenendaal F, Egberts AG, Rademaker CA. Effects of hypothermia on pharmacokinetics and pharmacodynamics. Clin Pharmacokinet. 2010;49(5):277–94.

51. Manola A, Geronilla GG, Kallur KR, Slim H, Lundbye J. The impact of therapeutic hypothermia on serum potassium. J Am Coll Cardiol. 2012;59(13s1):E62.

52. Van Steirteghem A, Van den Abbeel E, Camus M, Devroey P. Cryopreservation of human embryos. Bailliere's Clin Obstet Gynaecol. 1992;6:313–25.

53. Horne G, Atkinson AD, Pease EH, Logue JP, Brison DR, Lieberman BA. Live birth with sperm cryopreserved for 21 years prior to cancer treatment: case report. Hum Reprod. 2004;19(6):1448–9.

54. Edgar DH, Gook DA. A critical appraisal of cryopreservation (slow cooling versus vitrification) of human oocytes and embryos. Hum Reprod Update. 2012;18(5):536–54.

55. Dowling-Lacey D, Mayer JF, Jones E, Bocca S. Live birth from a frozen–thawed pronuclear stage embryo almost 20 years after its cryopreservation. Fertil Steril. 2011;95:1120.

56. Fahy GM, Wowk B, Wu J. Cryopreservation of complex systems: the missing link in the regenerative medicine supply chain. Rejuvenation Res. 2006;9:279–91.

57. Watson CJ, Dark JH. Organ transplantation: historical perspective and current practice. Br J Anaesth. 2012;108 (Suppl 1):i29–42.

58. Bianconi E, Piovesan A, Facchin F, Beraudi A, Casadei R, Frabetti F, et al. An estimation of the number of cells in the human body. Ann Hum Biol. 2013;40(6):463–71.

59. Moore EB, Molinero V. Structural transformation in supercooled water controls the crystallization rate of ice. Nature. 2011;479(7374):506–8.

60. Velikov V, Borick S, Angell CA. The glass transition of water, based on hyper-quenching experiments. Science. 2001;294(5550):2335–8.

61. Capaccioli S, Ngai KL. Resolving the controversy on the glass transition temperature of water? J Chem Phys. 2011;135:104504–1.

62. Fahy GM, Wowk B, Pagotan R, Chang A, Phan J, Thomson B, et al. Physical and biological aspects of renal vitrification. Organogenesis. 2009;5(3):167–75.

63. Blackstone E, Morrison M, Roth MB. H_2S induces a suspended animation-like state in mice. Science. 2005;308:518.

64. Li RQ, McKinstry AR, Moore JT, Caltagarone BM, Eckenhoff MF, Eckenhoff RG, et al. Is hydrogen sulfide-induced suspended animation general anesthesia? J Pharmacol Exp Ther. 2012;341(3):735–42.

65. Mike Roth on mice and men and suspended animation. 2011. http://www.wired.com/business/2010/02/mark-roth-on-mice-and-men/. Accessed 15 April 2015.

66. Alcor Procedures. http://www.alcor.org/procedures.html.

67. White RJ, Albin M, Yashon D. Neuropathological investigation of the transplanted canine brain. Transplant Proc. 1969;1:259–61.

68. White RJ, Wolin LR, Masspust LC, Taslitz N, Verdura J. Primate cephalic transplantation: neurogenic separation, vascular association. Transplant Proc. 1971;3:602–4.

69. Canavero S. HEAVEN: the head anastomosis venture project outline for the first human head transplantation with spinal linkage (GEMINI). Surg Neurol Int. 2013;4(Suppl 1):S335–42.

70. Tipler PA, Llewellyn RA. Chapters 1 and 2. Relativity I and II. In: Tipler PA, Llewellyn RA, editors. Modern physics. 5th edn. New York: W. H. Freeman and Company; 2008.

71. Brehm JJ, Mullin WJ. Introduction to the structure of matter. New York: Wiley; 1989.

8

Telepathy, Using the Force, and Other Paranormal Abilities

It is one thing for the human mind to extract from the phenomena of nature the laws which it has itself put into them; it may be a far harder thing to extract laws over which it has no control. It is even possible that laws which have not their origin in the mind may be irrational, and we can never succeed in formulating them.

Sir Arthur Stanley Eddington
Space, Time, and Gravitation (1920)

Use the Force, Luke.

Obi-Wan Kenobi
Star Wars: A New Hope (1977)

Characters with paranormal mental abilities are a staple of science fiction. The term "psi powers" has been used to describe such more-than-human talents as reading or controlling another person's mind, using thoughts alone to move objects or instantaneously transport them (including one's own body) from one place to another, and seeing into the future [1]. Similarly the use of electronics or other technologies to produce such effects or augment a person's natural capabilities has been dubbed "psionics."

Such abilities may initially be latent and developed in the course of a story, such as in the 1968 movie *The Power* when a character learns of his psychic powers only when attacked by another with similar ones. They may be present at birth, literally so as depicted in Jerome Bixby's "It's a *Good* Life" (1953), which describes a newborn with superhuman abilities causing havoc and death. Psi powers may be limited to isolated individuals as in the 1981 movie *Scanners*, the shared ability of whole human societies (e.g. Alfred Bester's 1956 work *The Stars My Destination*, where self-teleportation is common), or possessed by alien races. They may develop in future stages of human evolution as portrayed in John W. Campbell's "Forgetfulness" (1937), be caused by radiation-induced mutations as in Henry Kuttner and C. L. Moore's "Baldy" stories, or created by scientific means (e.g. the Lens in E. E. "Doc" Smith's "Lensman" series).

The common thread linking these myriad abilities is the idea that our brains, either alone or with technological assistance, are capable of performing actions beyond our current understanding of physics and human biology. Science fiction describing psi powers typically concentrates on asking the question "*If* this were possible, what would happen?" while glossing over or ignoring scientific explanations for them. Readers and viewers can wonder what they would do if they possessed such superhuman abilities, or how they and others might interact with someone who had them.

Psi powers represent ways of mentally manipulating and sensing the present (and perhaps future) state of the matter and energy that constitute our everyday world that go beyond our normal capabilities. Trying to understand how they might do that and if such abilities are really possible involves first seeing what we know our brains are capable of doing. Similarly, to see how paranormal abilities apparently violate some laws of physics we must recognize what those laws are.

8.1 . Telepathy

In its most general sense telepathy involves the direct transfer of thoughts from one being to another. This may be one-way, with a telepath being able to read another person's thoughts but not send his or her own to someone else. Alternatively individuals with telepathy might be able to both send their thoughts to non-telepaths and receive the latter's. Or two-way telepathy may be a skill possessed by multiple telepaths.

Human beings already have many ways to share their thoughts and feelings. Talking, reading, watching television and movies, etc. are everyday ways of doing this. Even simple gestures like a hug or kiss can do this without any verbal content. Modern devices like cell phones and computers let us communicate what we are thinking at speeds limited to that of light itself.

However, in theory telepathy could eliminate all these "middlemen" methods and let us commune with each other without the ambiguity, imprecision, and perhaps (!) even dishonesty that the spoken and printed word may have. Telepathic aliens might even consider its apparent lack in humans a proof that *H. sapiens* is not really an intelligent species, as in one of my stories,

"Neighborhood Watch,"[1] or at least a major impediment to developing a tech-nological civilization as suggested in another, "Achromamorph's Burden."[2] Robert A. Heinlein's *Time for the Stars* (1956) presents both a significant limi-tation to telepathy (about 10 % of identical twins can read each other's minds, or in some cases, those of very close relatives, but not anyone else's) and an enormous advantage (telepathic communication is instantaneous even over a span of many light-years).

But, as various science fiction tales point out, telepathy could be a mixed blessing. Someone acquiring that ability may learn how greatly other peoples' thoughts differ from their spoken words, as in the 1961 *Twilight Zone* episode "A Penny for Your Thoughts." If telepaths are a tiny minority in a society they might be treated with suspicion and persecuted by "normal" humans, as in Kuttner and Moore's "Baldy" stories or A. E. van Vogt's 1946 novel *Slan*. Non-telepaths might fear that a telepath could violate their mental privacy or be jealous of the advantage that ability might give in personal, job-related, and other relationships. They might fear that an unscrupulous telepath could use the information gleaned from others' minds to manipulate, blackmail, or otherwise injure them. Even worse is the possibility that someone could not only passively read another's thoughts but even alter them and control the other person's mind, like the Mule in Isaac Asimov's *Foundation* series.

The notion of telepathy is a simple one when looked at holistically. We all know what it feels like to "think," in terms of an experience of forming words and images in our minds. Likewise we are familiar with what it is like to speak and listen to someone else as part of a conversation. Combining those conceptions in various ways leads to the idea of directly communicating with another's mind.

However, the concept that thoughts can be received from and perhaps sent directly into another person's mind at least partly presupposes that we know what thoughts are. Our thinking processes involve electrochemical actions by specialized cells, neurons, within our brains. An adult's brain contains up to an estimated 100 billion neurons, although a recent estimate suggests it may be closer to 86 billion. Estimates of the number of neurons specifically as-sociated with the human brain's cerebral cortex range between about 16 and 26 billion [2, 3].

Neurons typically have four basic anatomic components [4]. The "soma" constitutes the main body of the neuron, containing its nucleus, mitochon-dria, and other essential organelles. Multiple "dendrites" are extensions of the neuron. They project from the soma and repeatedly branch out like the limbs

---

[1] *Analog Science Fiction and Fact*, January/February 2013.
[2] *Analog Science Fiction and Fact*, February 2000.

of a tree to form a complex network. Dendrites receive incoming electrical impulses from other neurons and send them to the soma.

A single "axon" (aka a "nerve fiber") typically projects from the soma. Not all neurons have an axon, but those that do have only one. The axon transmits electrical impulses away from the soma to other neurons or other types of cells (e.g. those in muscles). Like dendrites, an axon can branch many times and make connections with many other cells. Axons in a human adult can be up to a little over a meter long.

The fourth basic part of a neuron, the "axon terminal," is located at the end of an axon's branch. It forms part of a functional unit called a "synapse" by which electrical impulses are transmitted from the soma to other cells. The axon terminal represents the "sending" end of these electrical impulses. In one type of synapse an axon terminal releases chemicals called "neurotransmitters" packaged in tiny bubble-like vesicles. These move across the "synaptic cleft," a gap between the axon terminal of one neuron and the corresponding end of a second neuron's dendrite or another type of cell (e.g. a muscle cell). The "post-synaptic" area of the target cell has receptors on its membrane that "absorb" the neurotransmitters, which in turn stimulate or suppress the cell's electrical activity. In some synapses electrical activity can be sent directly from one cell to another without these kinds of chemical intermediaries.

Thoughts are formed and memories stored in a complex system of intricately connected neurons within our brains, supported by other types of cells such as glial cells. Each neuron connects to and modulates the activity of as many as thousands of other neurons and in turn is affected by many others via their synaptic connections. The number of synapses per neuron in the human brain is not known exactly. Estimates range between 1000 and 30,000 synapses for each neuron in the human cortex, for a total (using that estimated number of 16 billion cortical neurons cited previously) of 16–480 trillion [3].

Synapses in the human brain allow neurons to communicate and process information in the form of patterns of electrical activity between and within them. Ongoing research continues to reveal more details about how this is done, including the possibility that dendrites may not only be passive "wires" for carrying electrical impulses but also involved in the "computational" actions of neurons [5].

One theoretical mechanism for telepathy is that it could involve detecting the electrical activity within the brain associated with a person's thoughts and interpreting the information coded by it. Modern technology has already taken the tiniest first steps toward doing this. The brain's overall electrical activity can be measured by "electroencephalography" (EEG)[6]. This technique uses many small electrodes positioned over a person's scalp to measure the summed impulses produced by large groups of neurons within the brain. These hundreds or thousands of neurons produce "waves" of electrical activity that the electrodes detect and send to an amplifier system. Those patterns of electrical impulses can then be viewed on a screen or printed as a continuous readout.

Unfortunately an EEG machine provides only the broadest picture of electrical activity in various parts of the brain. There is no current analog to the helmet that "Doc" Brown wears to (not very successfully) receive electrical impulses from Marty McFly's brain using a wired suction cup and read the latter's thoughts in the movie *Back to the Future* (1985). However, more sophisticated devices can recognize what electromagnetic and metabolic activity may, at least in the most general sense, correspond with our thoughts.

While EEG measures electric currents produced by the brain's neurons, magnetoencephalography (MEG) detects changes in the magnetic fields associated with those currents [7]. Magnetic fields generated by the brain are even weaker than its electric fields, or even the ambient magnetic "noise" produced by electronic and other powered devices and wires present in a typical house or city. MEG must be done in a room shielded from outside magnetic fields, and it requires a large machine with sensitive detectors arranged around the subject's head. This technique has been used to assess what parts of the brain are associated with language processing and which are activated when a person sees another make hand motions or facial gestures.

Electrocorticography (ECoG) detects electrical activity using an electrode array placed inside the skull directly over a part of the brain. Recent research suggests that electrical signals obtained via this method from the temporal lobe could be "coded" into individual vowels and consonants. This raises the possibility that a person could silently "speak" by thinking words whose electrical analogs would then be decoded into words and sentences [8]. However, this capability has not yet been fully demonstrated, and the need to implant the electrical array inside the skull (with its attendant risks) would tend to restrict this technique to individuals with neurological impairments that render them unable to communicate by conventional means.

Other methods create actual images of the brain that show which areas are most active when a person receives some outside stimulation (e.g. seeing an object), is asked to think of something, or performs an activity. These include functional magnetic resonance imaging (fMRI), functional near-infrared spectroscopy (fNIRS), positron emission tomography (PET), and single-photon emission computed tomography (SPECT) [9, 10]. These techniques evaluate brain activity indirectly via changes in its metabolic activity and blood flow and can produce color-coded images showing active parts of the brain.

For example, when an area of the brain becomes more active, blood flow to it increases to meet increased metabolic demands. The "fresh" blood delivered has a higher saturation of oxygen than that present before the area became more active. A method such as fMRI, which assesses differences in magnetization between blood containing higher and lower saturations of oxygen, can detect this local change in blood flow and thereby identify an active region within the brain.

The machine used for fMRI contains a powerful electromagnet and is large, heavy, and stationary. The room where fMRI is done must not contain any material that could be attracted by the electromagnet and requires the person being studied to be immobile during scanning. The equipment used to perform fNRIS is much more portable. It includes a band-like arrangement of sensors worn against the scalp that allows a person to move and perform activities during the study. Like fMRI, fNRIS detects regions in the brain with greater and lower oxygen saturation in blood and thus differences in activity. While fNRIS cannot localize these areas as well as fMRI, its greater ease of use is a major advantage for assessing brain function [9].

PET typically involves injection of substances the brain uses for metabolism, such as glucose, coupled to short-lived radioactive elements. As they decay these radioactive "tracers" emit positrons (the positively charged, antimatter equivalent of an electron) that are detected by the PET scanner. Tracers can also be used to assess neurotransmitter activity in the brain, which helps provide information on how different areas function. SPECT involves injection of longer-lived radioactive materials whose location in the brain corresponds with regions of increased cerebral blood flow and activity.

However, the best these methods can do is tell us that a person is thinking, or at least what areas of the brain become more active with thinking, but not specifically *what* is being thought. Telepathy can, at a trivial level, be likened to the relation between a radio receiver and transmitter. In this analogy a telepath would "receive" the thoughts "transmitted" by another person's mind, presumably if both were tuned to the same frequency. Similarly, communication between two telepaths could be compared to a pair of individuals too far apart to speak directly to each other chatting on their cell phones, only without the cell phones.

Unfortunately this rudimentary radio analogy for telepathy breaks down when it comes to the details. The electromagnetic energy the human brain generates differs from the alternating current used for radio waves, and it is not modulated—that is, continuously altered to contain information corresponding to our thoughts—in the way that radio waves are. Like other cells, neurons can both maintain and change the electric potential (voltage) across their cell membranes. They do this by regulating the concentrations of different ions (see Chap. 1), particularly those of sodium and potassium, between the inside and outside of the cell. A difference between the net charge within the cell and the charge produced by ions in its immediate environment creates a voltage potential (gradient) across its membrane, somewhat similar to how the chemicals in a battery create one between its positive and negative poles.

Normally a neuron maintains an electrical potential of about -70 millivolts (mV) across its membrane, its "resting potential." By actively pumping or otherwise regulating the flow of ions in or out across its cell membrane, a neuron can change its electric potential and the electric current it produces. If this gradient reaches a "threshold" level, around -55 mV, it generates an "action potential." This is a "pulse" of electricity and its associated current produced by a sudden change in the electric potential from an initially negative value towards a positive one, reaching a peak of up to about $+100$ mV. This positive value rapidly falls back toward the resting, negative electrical potential due to further movements of ions across the neuron's cell membrane. The electric current produced by these events can be transmitted from one neuron to the many others connected to it, thus stimulating them too.

The electric pulses produced by action potentials are "all or nothing" individual events. They do not occur unless the neuron is "depolarized" to its threshold level, and the time before another action potential is generated within a neuron will vary. In this way they represent discrete, "digital" events. However, a neuron's electrical activities are also "analog" in that its resting potential can vary continuously over a range of voltages below the threshold level. Thus, a neuron whose resting potential has changed to -60 mV due to interactions with its environment will be more likely to reach its threshold potential of about -55 mV and so produce an action potential than one whose resting potential is greater, e.g. -70 mV.

In contrast, radio waves are cyclical events, with their voltage and current continuously rising and falling. Audio and video information can be included in them by analog methods, such as varying (modulating) the amplitude or frequency of the radio wave ("AM" and "FM," respectively.) Alternatively radio waves can also transmit data digitally by including specific groups and patterns of pulses that a receiver then decodes.

Thus, the neurons in our brains produce, encode, transmit, and decode the electrical activity and voltage states corresponding to our thoughts differently than standard techniques employed for radio waves. Moreover, while the power used by the brain has been estimated at about 20 W, only a tiny fraction of that amount is actually used to generate electricity[11]. The net effect is that the brain is a very weak transmitter of electromagnetic energy. There is also no objective evidence that humans can consciously detect the presence of electric and magnetic fields with our brains[12]. We are constantly exposed to the electric and magnetic fields propagated by electric wiring and devices and even the magnetic field produced by the Earth without being aware of their presence, and their power far exceeds that generated by individual brains.

Also, the process of thinking involves not only changes in electrical potential and current involving our neurons as they interact with each other, but also how many are in a relatively stable "rest" state with similar resting potentials at a given time. At such times no information involving those neurons is being "transmitted" within the brain.

In short, any mechanism for telepathy involving electricity and magnetism presupposes that the electromagnetic fields produced by one brain can actually be detected by another brain; that there actually is information encoded within those fields representing thoughts; and that the receiving brain can interpret that information. Based on the known biological, biochemical, and physical principles I have described and others, electromagnetic energy does not appear to be a suitable medium for telepathy (at least in humans).

Thus, if telepathy using our brains alone did exist it would require other mechanisms that are either currently unknown or highly theoretical. One possible "candidate" is based on quantum mechanics. The latter has been invoked as a way to help explain human consciousness[3] and, by extension, our thoughts [13]. For example, quantum entanglement is an experimentally confirmed phenomenon in which two or more particles interact in such a way that their quantum states (e.g. the spins or positions of two electrons or the quantum states of photons) are "linked" to each other [14]. Put less technically, if the quantum state of one particle changes then that of the other will too, and it can do so at an apparent speed faster than light. This phenomenon has also been demonstrated under certain circumstances with macroscopic diamonds [15].

By making a tremendous leap beyond the atomic scale to that of everyday existence, this principle could be invoked as a way that the atoms and molecules of one person's brain might "connect" to that of another to allow telepathy. However, current thought is that, for all practical purposes (e.g. under extreme conditions incompatible with human life, such as very high or low temperatures or energy states), such specific quantum effects do not apply beyond the atomic and subatomic level. Moreover, "entanglement" by itself cannot actually be used to transfer external information.[4] Thus, barring a radical rethinking of current ideas on the subject, quantum mechanics seems no more promising than the "classical" physics of electromagnetism as a means for telepathy.

Conversely, science fiction can postulate the existence of energy fields or properties of matter other than the ones we know exist that would allow telepathy to occur. A human or alien brain might manipulate a "psionic field,"

---

[3] This subject will be discussed more fully in Chap. 14.
[4] However, as we shall see later, another concept that involves quantum entanglement, "quantum teleportation," could be used to manipulate quantum states at a distance under some circumstances.

perform a "Vulcan mind meld," or "use the Force" to receive, send, and manipulate thoughts. Unfortunately just giving a name to these things and describing or showing what they do does not allow their existence to be experimentally tested, quantified, or prove that they exist. For now at least, the idea of telepathy as a natural ability of humans is not only unconfirmed but also lacks a potentially demonstrable rather than purely hypothetical mechanism.

However, the situation may be more optimistic regarding aliens with innate telepathic abilities. One possibility is that an extraterrestrial could develop the biological equivalent of a simple radio transceiver within the equivalent of its cerebral cortex. Also, some animals on Earth, primarily aquatic ones such as sharks, certain types of fish, eels, and dolphins as well as a few species that spend at least some time on land (e.g. bees, some varieties of echidna, and platypuses) have a sense that humans apparently lack—"electroreception" [16, 17].

The bodies of creatures with electroreception have receptors that detect electric fields produced near them by other living things (e.g. potential prey or predators). This "passive electroreception" involves just sensing that electrical energy, which has a very low frequency in the tens of hertz (Hz).[5] A smaller number of species can actually produce their own electric fields—"active electroreception"—using special organs derived from certain types of muscle and nerve cells. Changes or distortions in those self-generated fields help them evaluate their environment, including the presence of other living things (friendly or otherwise) around them. Animals using active electroreception may produce pulses of electrical energy or even create electromagnetic waves at frequencies up to thousands of Hz. These frequencies can even be changed (modulated), raising the possibility that such electric fields could be used to transmit and receive more complex information.

Intelligent aliens would have to possess a much more sophisticated version of electroreception for any but the most rudimentary "telepathic" communication. In terrestrial animals the range of electroreception is also very short—typically less than a meter. This is due to other animals (with rare exceptions such as electric eels) generating only extremely weak electrical fields, therefore requiring that they must be close by for detection. Producing electrical fields also requires energy and potentially, as in the case of eels, specialized cells. Electroreception is also more effective in water instead of air, since the former is a much better conductor of electricity.

The picture that emerges from this is of a sentient, perhaps amphibious alien that can generate sufficiently powerful electric fields and detect them well enough to effectively enable it to communicate telepathically with others of its kind. The possibility such communication could only be short-range,

---

[5] 1 Hz = 1 cycle/s.

within distances we humans use when talking to each other directly, suggests that the close proximity and physical contact Mr. Spock needs to perform the aforementioned Vulcan mind meld may indeed by necessary.

This discussion about the feasibility of telepathy has, overall, assumed it to be a "natural" process in either unmodified human beings or perhaps those altered via biological means (e.g. genetic engineering, or a "mutant" produced by radiation or evolution), rather than employing electronics or similar physical means to produce the same effect. Future technology might allow at least parts of the brains of two people to be linked, though not necessarily enough to share actual thoughts. A recent report described experiments using two rats whose minds were "linked" together using implanted cortical microelectrode arrays [18]. Motor and sensory information from one rat's brain was transmitted to the second's brain.

In one experiment the first rat was trained to identify a visual stimulus (a light going on) and then press one of two levers, receiving a reward of water if it pressed the correct lever. The second rat, also in a cage with two levers and this reward system but no on-off light system, had its brain "wired" to receive electrical activity from the first one's. It was found that, when the first rat successfully pressed the correct lever after its light went on, the second rat, despite having no light to indicate when to do it, nonetheless pressed the correct lever a significantly high percent of the time to receive its own water reward. Similar experiments were also consistent with the second rat's brain being "linked" to the first's regarding both motor and sensory information.

As impressive as this is, however, this procedure requires implantation of microarrays over specific parts of the brain—an invasive procedure; sensory and motor information could be shared only between two rats who both had the microarrays implanted (in this case, one-way, although in principle two-way communication might be possible); and such a technique would, at least in its current capabilities, be unable to transmit/share complex information such as abstract thoughts and speech. Besides requiring considerable advances to qualify as "telepathy," this technique raises the specter of true "mind control" if electrical impulses from one person's brain could be used to control another's.

Hal Clement's story "Impediment" (1942) cautions about another potential problem with telepathy. In it telepathic aliens learn with great effort to interpret the thought patterns of a human character tricked into cooperating with them, expecting that they will then be able to read the minds of all other Earthlings. They discover to their dismay that each individual human has a unique pattern or "language" of mental processes. Thus, although they have learned how to read one person's mind, they would have to repeat the same long painstaking method on other cooperative individual humans before they could read their minds too.

8.2 Telekinesis

Moving objects using only the unaided power of one's mind would be a useful skill. Besides bending spoons this ability would be useful for extracting an X-Wing fighter from a swamp as in *The Empire Strikes Back* (1980), moving dice to win at gambling in the 1961 *Twilight Zone* episode "The Prime Mover," or taking revenge on high school bullies as in the film *Carrie* (1976). Comic book characters (e.g. Jean Grey of the X-Men) may use it for good or ill. Jack Vance's novella "Telek" (1952) depicts a society in which this talent can be learned from other people without being born a "mutant." Aliens might also teach it and other paranormal powers to previously "normal" humans, including castaways such as Valentine Michael Smith in Robert A. Heinlein's *Stranger in a Strange Land* (1961) and Charlie Evans in the *Star Trek* episode "Charlie X" (1966).

Besides manipulating external objects, telekinesis can also encompass levitation—using that power to make one's own body float or fly, a power effectively possessed by Superman and many other superheroes as well as used in science fiction works such as Isaac Asimov's story "Belief" (1953). A broader definition of telekinesis includes the ability to manipulate matter at the atomic or molecular level. For example Element Lad of DC Comics' Legion of Super-Heroes can transmute one element into another, while his fellow member Chemical King could speed up or decrease the rate of chemical reactions. The ability to mentally change the kinetic energy of atomic-level particles would also enable someone to dramatically change an object (or person's) temperature, which could in turn manifest as being able to start fires— "pyrokinesis"—as in Stephen King's 1980 novel *Firestarter* or C. L. Cottrell's "Danger! Child at Large" (1959).

Unfortunately, as helpful as these abilities might be they violate basic principles of physics and human biology. As described in the previous section on telepathy our brain cells do generate electromagnetic waves and energy. However they are both far too weak to move macroscopic objects or even be used to consciously cause changes at the atomic level.

Some aspects of telekinesis are superficially analogous to two real-world phenomena, magnetism and gravity. Both seem to involve "actions at a distance" by "invisible forces," with objects moving in the absence of direct contact by other objects such as our hands. A material that has been temporarily or permanently magnetized can either attract or repel others capable of being magnetized. Magnets have a "north" and a "south" pole, with magnetic field lines moving away from the north pole and towards the south. Interactions between their magnetic fields make "opposite" poles (north and south) attract each other and "like" poles (two north or two south poles) repel one another.

A very strong magnetic field can also affect atoms and molecules in a living organism's body [19]. This is particularly the case with hydrogen atoms and water molecules, both of which can act as very tiny, weak magnets. This principle is used for fMRI and the more general use of this technique as a medical imaging test, magnetic resonance imaging (MRI). These tests employ a powerful magnetic field to act on hydrogen atoms (more specifically, the single proton in each one's nucleus) in a person's body to make them align and ultimately emit energy in the form of a radio wave. The latter is then detected by the MRI machine and used to create an image. As mentioned in Chap. 4, magnetism can also be used to make small animals like frogs and mice levitate, by acting on water molecules in their bodies [20, 21].

Unfortunately for anyone hoping to make a frog float due solely to the power of his mind (or perhaps "magnetic personality"?) alone, the strength of the magnetic fields used to do this are many orders of magnitude greater than any our brains generate. The powerful magnet used for MRI is often a "superconducting" one, using liquid helium at a temperature of around $-269\,°C$. It produces a magnetic field in the range of about 1.5–7 T—up to about 100,000 times stronger than the Earth's magnetic field of no more than about 65 µT at its surface.

Like magnetism, gravity makes objects attract each other. However, except at the edge of current theory [22] and the realm of science fiction with "antigrav" devices or imaginary materials such as cavorite in H. G. Wells's *The First Men in the Moon* (1901), gravity does not make objects repel each other. While gravity is powerful enough to (generally speaking) keep planets in their place, moving anything by using our minds to alter gravity is even more outlandish than employing magnetism.

As noted in Chap. 4, in classical physics the force of gravity is proportional to the product of the masses of two bodies and inversely proportional to the square of the distance between them. In Einstein's theory of general relativity, expressed in the simplest terms gravity represents a curvature of spacetime associated with mass and radiation (see Chap. 6).

Unfortunately gravity is an extremely weak force. The amount of mass needed to produce a noticeable gravitational effect (e.g. keeping us from floating off to space) is enormous. No matter how much someone eats, the mass of the human body is miniscule compared to Earth's (slightly less than 6×10^{24} kg). Yet despite the latter's huge mass our bodies can temporarily defy it by simply raising an arm.

Moving from established to speculative physics, an alternative mechanism for telekinesis might be that electromagnetic impulses produced by our brains could be amplified by some hitherto unrecognized energy field, e.g. the Force or a hypothetical "psionic" field. This might be compared to how a tiny voltage or current in a radio or other electronic device can be effectively increased in strength by passing through an amplifier circuit and interacting with a far greater power source there. Obvious problems with this analogy include the fact that no such pervasive energy field has been identified, and even if it did exist there may be no way to manipulate it with our minds.[6]

Returning to classical physics, moving an object with our minds or, more conventionally, our hands means that we would have to apply a force to it—but what exactly is a "force"? Newton's second law of motion describes force in terms of what is acted upon—an object with a certain mass—and what is done to it—the object accelerates, with acceleration being its change in velocity over time. Quantitatively this is expressed as:

$$F = ma$$

Here F is force in newtons, m is mass in kilograms, and a is acceleration in meters/second^2. In simple terms, force can be thought of as what pushes or pulls an object, without describing exactly "what" it is that does the pushing or pulling. Expanding this concept, "work" is the energy transferred to or from an object via a force, expressed mathematically as:

$$W = Fd\cos\phi$$

In this equation W is the work in joules, d is displacement (the change in the object's position, including the distance it moves) in meters, and $\cos\phi$ is the cosine of the angle between the force F applied to the object and its displacement.

Moving an object also involves kinetic energy and momentum. Kinetic energy is expressed as:

$$E_k = \tfrac{1}{2}mv^2$$

Here E_k is kinetic energy in joules, m is mass in kilograms, and v is velocity in meters/second. Momentum (P), or more specifically linear momentum, is expressed in kilograms-meters/second and involves a different combination of an object's mass and velocity:

$$P = mv$$

---

[6] "Dark energy" is a hypothetical energy field that may permeate the universe and be responsible for the latter's increasing rate of expansion. At least in its effect (though not in its method or details) of opposing the collapse of the universe due to gravitational attraction of the matter within it, dark energy can *very* loosely be thought of as a form of "negative gravity." However, if it exists dark energy appears to work only at the truly "cosmic" level and not be amenable to manipulation by even the greatest Jedi Masters.

Force, momentum, and velocity are all "vector" quantities—that is, they have both direction and magnitude (how "strong" they are). Thus, an object moved by telekinesis in a particular direction would need to be acted on by a single force acting in such a way as to "pull" it in that direction, or by a net force doing it that would be the vector sum of two or more individual forces. Likewise, if a superhero were to telekinetically deflect a bullet shot at her its momentum would have to be reduced by drastically decreasing its velocity, changing its direction, or both before it struck. Doing both simultaneously would require that one or a combination of forces be used at the same time on the bullet.

Moreover, both the total momentum and energy in a system of objects must be "conserved"—that is, the total sum of each must be the same before and after a force or forces act on them. Momentum can be transferred between or among objects, but the total momentum after that redistribution must be the same. Thus, when a moving billiard ball strikes another the first slows down or stops as part or all of its original momentum is transferred to the moving second ball. Likewise, barring conversion of energy into matter as described by Einstein's famous formulation $E = mc^2$ (see Chap. 1), the total amount of energy in a closed system remains the same after any interactions. One type of energy can be converted into another, such as the "potential" energy of a coiled spring being converted into kinetic energy if it is released. However, that original energy must come from somewhere—and the question with telekinesis remains what exactly that form of energy is and where it comes from.

Another important issue regarding the "force" involved in telekinesis is whether it, like more conventional types, decreases with increasing distance from its source. Gravitational attraction, attracting or repelling electrostatic forces, sound, and electromagnetic waves (such as light) follow an "inverse square law." This means that each one's intensity is inversely proportional to the square of the distance from its source. For example, doubling the distance between yourself and a glowing light bulb would reduce the intensity of its light at your new location to one-fourth of its original value. A magnet's magnetic field shows an even more rapid fall with distance, obeying an "inverse cube law." Thus, doubling the distance from a magnet reduces the intensity of its field to only one-eighth of what it was originally.

Let's apply these ideas to a "practical" situation. Early in *The Empire Strikes Back* (1980) Luke Skywalker has been captured and literally hung up to dry by a hungry Yeti-like wampa. Before that creature can finish munching on a bloody appetizer and get to Luke as the main course, the young Jedi-to-be uses telekinetic energy via the Force to draw his nearby lightsaber to his hand. He moves it just in time to free and defend himself before (literally) disarming the wampa.

By one estimate the lightsaber has a mass of about one kilogram [23]. The distance between Luke's outstretched right hand and the handle of the light-saber appears to be approximately 0.70 m (a little over two feet), and the direction the latter moves is upward at what appears to be roughly a 45° angle with the ground. Whatever force (small "f") Luke exerts on the lightsaber to make it move upward at that angle must counterbalance the downward force exerted by the ice planet Hoth's gravity. Based on how the characters in the film move, the latter appears roughly similar to that of Earth, or 1 g (about 9.81 m/s²). Thus, the gravitational force Luke needs to overcome is strictly downward and approximately (by $F = ma$) 1 kg x 9.81 m/s², or 9.81 N.

Ignoring any friction forces due to the lightsaber being partially imbedded in a pile of snow, Luke would have to exert a net force whose vertical component is greater than the 9.81 newtons of the planet's gravity and whose horizontal component is in his direction. Based on a careful review of the scene, a reasonable estimate is that the lightsaber moves about 0.3 m upward and, if so, by applying the Pythagorean theorem[7] with a hypotenuse of 0.70 m, about 0.60 m horizontally. Let us further assume that the acceleration he imparts to the lightsaber is about 3 m/s² for both the horizontal and vertical components. Thus, the force exerted in the horizontal direction is about 3 N (1 kg mass of the lightsaber multiplied by that acceleration) and 12.81 N in the vertical direction (3 N plus the 9.81 N to overcome Hoth's gravity). The net force applied in the 45° direction is thus, using its horizontal and vertical components as the two sides of a right triangle and that net force as its hypotenuse, about 13.15 N.

The amount of work required to move the lightsaber can then be calculated using the previously cited equation, $W = Fd\cos\phi$, using $F = 13.15$ N, $d = 0.70$ m, and cos 45°. The amount of work done found by using that equation is 6.68 J. This is a surprisingly small amount of energy—only a little more than a tenth of that needed to keep a standard 60 W light bulb glowing at its rated brilliance for a second. Luke must also apply the appropriate forces no longer than needed for the lightsaber to reach his hand and then "turn them off," perhaps even using a slightly smaller total force by letting his weapon "coast" the last few centimeters to his fingers. Fortunately his Jedi training apparently included doing mental calculations on vector sums and the proper application of these forces, work, and energy, since to the wampa's dismay (and subsequent need to call 911) Luke succeeds in his demonstration of classical physics.

---

[7] For a right triangle, $a^2 + b^2 = c^2$, where c is the hypotenuse of the triangle, and a and b are the lengths of the other two sides.

Unfortunately, however, the problem remains—where exactly did the energy to move his lightsaber come from, how does he utilize it solely with his mind, and how does that energy relate to more conventional forms such as thermal energy, acoustic energy, kinetic energy, etc.? Perhaps some things in science fiction are indeed best left unexplained, lest a work like *The Empire Strike Back* lose a major character in its opening scenes.

8.3 Precognition

Precognition—the ability to successfully predict the future—has been attributed to historical figures such as Michel de Nostredame (aka Nostradamus) and mythical ones such as the Trojan princess Cassandra. Their predictions/prophesies are typically couched in ambiguous terms or, even when specific and true as in Cassandra's case, may prove singularly unhelpful to them or others. Phillip K. Dick's short story "A World of Talent" (1954) uses the term "precogs" for individuals with this ability (as well as perhaps other "psionic" powers). His novel *The World Jones Made* (1956) further addresses the ambivalent advantages of precognition, with the title character seeing his own future death. Other examples include those who receive their visions of future events via dreams, such as the aptly named Dream Girl of the 30th century's Legion of Super-Heroes and the main character of Roger Zelazny's 1965 novel *This Immortal*.

More technically, "precognition" can be described as the "conscious cognitive awareness" of a future event, while a "premonition" is the "affective apprehension" (a "feeling" about rather than actual knowledge) of something that has not yet but will happen [24]. However, the question of what particular biochemical, neurological, and other mechanisms in a human brain could be responsible for precognition or the like is difficult to answer, even if only hypothetically.

Postulating such an ability to know what will happen in the future, even if incompletely, also raises questions asked by both philosophy and physics about the nature of time itself, including what the "future" actually is as well and whether it is "set in stone" or malleable. One interpretation of precognition is fully deterministic—there is really only one future. Although we in the present have not reached it yet, it is out there waiting just below our "time horizon" for us to meet (or perhaps crash into?) it. In this "simple" vision (so to speak) someone with precognition is merely taking a "sneak peak" at things to come.

On the other hand, quantum mechanics emphasizes the probabilistic nature of existence at the subatomic level, such as the likelihood of where a single electron is located, or that the mere act of making a measurement determines the outcome of a quantum state or how a photon of light "behaves." For example, one of a number of thought experiments proposed by physicist John Wheeler regarding this point involves the "double-slit" experiment in which a photon acts as either a particle or a wave [1, 25, 26]. One such experiment involves sending a stream of photons one at a time at a barrier that blocks them except for two narrow slits in it. If a photon passes through one slit it acts as a particle and can be recorded by a photodetector. However, if it passes through both slits it will act as a wave and become part of an interference fringe pattern (alternating light and dark regions caused by different light waves interacting with each other).

The question Wheeler proposed was whether the "fate" of a photon to act as a particle or wave could be altered retroactively after it passed that barrier, when (presumably) it would already have been "determined" which way it would behave. He posited doing this by having an "observer" decide whether the photon would either strike a photographic plate (thus demonstrating its wave property and producing part of an interference pattern) or a photodetector, in which case it would act like a particle. This could be done by having the observer decide *after* the photon has passed the slits to either keep the photographic plate in place or swing it aside after it passed the slits but *before* it reached the photodetector. In one sense, the observer can be thought to be altering the photon's past state (wave or particle, depending on whether it originally passed through both or only one slit in the barrier) from its "future." Without going into details, a variation of this finding involves using quantum entangled photons and a beamsplitter to determine their paths in what has been called a "delayed choice quantum eraser" experiment [27].

Extending these results to our macroscopic world (a great extrapolation indeed!), precognition might be described as the ability to "know" beforehand what has happened in the "future" to create what will ultimately become our "present." However, other explanations have also been suggested to explain these results without invoking such "spooky action at a temporal distance" actions, including that the photons involved are in a state of "superposition"— the information within them to act either as a particle or wave is present before they are "measured" [26]. If so, there would be no need to invoke a "future affecting the past" explanation.

Formal experiments have also been done to see if test subjects might demonstrate precognitive abilities. In a recent study [24] a series of different experiments was done involving 9 different experiments and over 1000 participants to assess how well they could predict a particular outcome before it happened. For example, in one experiment each subject sat in front of a computer screen showing pictures of two curtains side by side. All subjects were told that one of the curtains had a picture behind it (which might have "erotic" content) and the other did not. Each subject was instructed to pick the curtain (right or left one) with the picture behind it, over a total set of 36 trials with a pair of curtains. The individuals studied received immediate feedback regarding their "right" or "wrong" choice after making their selection. A computer program randomly selected the types and orders of the curtain/picture pairs for each individual's series of trials, with that specific information not being determined until after the person had made his or her guess. Overall, 8 of the 9 experiments done in this study showed statistically significant results regarding subjects making "correct" choices at a level greater than chance.

However, attempts to reproduce these results using similar methods were unsuccessful [28, 29]. Whether this difference in results reflected subtle changes in methodology, "confounding variables" (unidentified influences such as participant motivation or preconceptions), problems with statistical analysis, etc., this lack of reproducibility leaves the original question of whether precognition exists (or at least can be adequately measured by these particular testing methods) still open and does not establish its existence. It is also a far greater step to postulate (as some science fiction does) that an individual could in a reasonably reliable and predictable manner "know" about complex, specific events happening in the future versus what picture would be in back of one of two curtains on a computer screen (not in itself the most useful superpower). For now, we may need to be satisfied Master Yoda's words about predicting future events—"Difficult to see. Always in motion is the future...."

8.4 Extrasensory Perception

The term "extrasensory perception" (ESP) can be defined in several ways, ranging from the narrow way I will use it here to a broader one including telepathy, precognition, and other paranormal abilities. "Sensory" perception involves use of our five basic senses—sight, hearing, smell, taste, and touch—to gather information about our external environment. Humans also possess senses not localized to single body parts or tissues, including our perception of temperature, injurious stimuli that cause pain, and proprioception (our ability to recognize the relative positions and movements of body parts such as arms and legs).

ESP involves perceiving the external world in ways beyond what can be done by these unaided senses. "Classic" experiments done in the mid-20th century to evaluate this ability in test subjects have included using special Zener ("ESP") cards[8] each marked with one of five different symbols—a circle,

---

[8] These cards were developed by Karl Zener, a psychologist, in the 1930s.

square, cross, three vertical wavy lines, or a star [30] A standard deck of Zener cards consists of 25 of these cards with 5 of each kind. The person being tested would try to sense the type and order of these cards in a hidden pack. This was typically done by having another individual look at a series of cards concealed from the test subject while the latter tried to sense what they were. Giving more correct answers than would be expected by chance alone might then be explained by the use of telepathy (the test subject read the thoughts of the person looking at the cards) or clairvoyance (the ability to obtain information about a person, thing, event, or location by means other than the five basic senses, particularly sight and sound).

Various animals have senses more acute and covering wider ranges of sight and sound than ours. As described in Chap. 6 our eyes detect only that tiny range of the electromagnetic spectrum we call light. Conversely animals such as certain types of snakes, vampire bats, and some insects used special organs and body structures to "see" (or at least sense) infrared light, while ones such as birds and bees can see into the ultraviolet. Our hearing range of up to about 20 to 20,000 Hz is put to shame by bats, which can hear frequencies up to about 212,000 Hz, and even more so by one variety of moth that can detect sounds as high as 300,000 Hz.[31] Dogs have much more sensitive senses of smell than we do.

Some animals have other senses that we lack. The electroreception described previously is one example. Bees, birds, sharks, and some other creatures can detect magnetic fields (magnetoreception). This may help them use Earth's magnetic field to determine their location in what can be likened (in effect though not in details) to a built-in GPS. Bats use echolocation to sense their surroundings, sending and receiving ultrasonic waves as part of their natural sonar system. Some fish have organs that detect changes in pressure while others can sense the direction of water currents.

The ways our senses and those of other animals work are based on known principles of biology and physics. One aspect of ESP, clairvoyance, assumes the existence of an ability to mentally "visualize" or "hear" objects that cannot be seen or heard. It may involve sensing or even having a "vision" of events that are happening or have happened at some distance location. Or it might manifest as a direct "knowing" of factual information about another person without having learned it by conventional sensory means.

Some key difficulties for establishing that such abilities exist include that there is no known, definite physical mechanism for them; the inherent limitations of statistical variability (see below); and that their results are not clearly reproducible at an individual level. It is not absolutely necessary to know how something works in order to use it. For example, navigators on ships used simple compasses as long as a thousand years ago, long before the underlying theory of how magnetism works was developed. However those instruments actually did work, they did so in a predictable fashion, and we now know the fundamental reasons why and how they operate.

As in other types of research, formally testing whether a person has ESP uses standard statistical techniques. A criterion typically used to see if results are significantly related to each other (with one possibly causing the other) is if test results show that there is a less than 1 out of 20 chance that they are not related (the "null hypothesis"). More technically this is expressed as "$p < 0.05$", where this "p-value" refers to probability.

However, evaluating the validity of statistical findings involves many factors. For example, a test must include an adequate number of data points. Successfully guessing a single Zener card from a deck only one time could be due to chance alone. Conversely, if someone being tested for ESP were to successfully identify a large enough percentage of cards as part of testing involving a great enough number of cards the results could be potentially "statistically significant."

But even such "positive" results would still only represent a probability rather than a certainty they were due to a person really having "ESP." Even at $p < 0.05$ there is still that less than 1 out of 20 chance the null hypothesis really is true. Also, if a large enough number of individuals are tested it could mean that slightly less 5 % of them could, on average, have "statistically significant" positive results by chance alone. Such results would also have to be further confirmed by additional testing, demonstrating that they were reproducible.

Moreover, scientific testing involves trying to (as much as possible) eliminate or control all "variables" except those being tested. Phrased less formally, it means that to evaluate whether doing one thing actually causes another to happen or is at least correlated with it, you first have to exclude other things that could also cause that result.

Thus, if a person were to have statistically significant positive ESP results with Zener cards it would be necessary to make sure that those results were not due to some reason other than ESP. For example, could the test subject have been deliberately cheating? If that individual were shown only the back of the cards, could the pattern imprinted on them still be slightly visible? Could the person have, if only subliminally, noted cues as to what card was being used based on the facial expressions of the tester holding or pointing to it? Or was the test subject really a disguised, unscrupulous Kryptonian criminal recently escaped from the Phantom Zone who used his x-ray vision instead of ESP to correctly identify the cards?[9]

---

[9] If Clark Kent were tested he could do that too—but, of course, he would not do anything dishonest.

Finally, here too any mechanisms underlying ESP would have to involve previously unrecognized facets of known biological and physical principles; new ones that do not contradict but expand upon established scientific laws; and/or show that our current theories need to be modified. If the existence of ESP and other paranormal abilities were already "confirmed" via the scientific method (e.g. using strict criteria such as statistically significant positive results during tests, reproducibility, rigorous exclusion of other explanations etc.) then it would be reasonable to formulate theories about how it works and test them further. In the absence of such confirmation, to be useful any theory about how ESP might work would have to be testable—able to be confirmed or refuted to an appropriate level of statistical significance by controlled experiments or, at least, other types of confirming observations.

Anecdotal reports of ESP or "psychic" abilities do not meet these criteria. Experiments to assess the presence of such abilities (e.g. with Zener cards) have not yielded consistent results, and any purported "positive" ones could have been due to methodological errors, the statistical issues described previously, or other factors. This is all compatible with the null hypothesis—that such abilities really don't exist—being true.

Still, it could also mean that we lack enough information to determine or have not yet developed an effective way to demonstrate that clairvoyance and similar things really exist. As powerful as empirical techniques and inductive reasoning are for formulating general laws about matter and energy from many individual observations, there is at least a theoretical chance (however improbable that may be based on established knowledge) that we might be missing something. The scientific method is open to reevaluating concepts of the physical nature of reality if new data are obtained or better theories are formulated, as evidenced by how once "cutting-edge" concepts such as the ether or phlogiston have fallen out of favor.

In the words of the physician–philosopher William James (1842–1910), "If you wish to upset the law that all crows are black, you mustn't seek to show that no crows are; it is enough if you prove one single crow to be white." Nonetheless, given the present lack of reproducible, confirmative empirical evidence or any plausible mechanism for ESP and similar abilities it appears the probability of that particular "white crow" really being out there is at best vanishingly low. Barring new information, it seems to approach a limit of zero and can (at least for now) be rounded off from a practical (although not absolute) standpoint to the null hypothesis that "It's not out there." And even if ESP is "real" but weak and unpredictable, the question of why it is that way rather than strong and reliable would also need to be addressed.

8.5 Teleportation

Teleportation—in simple terms, the ability to instantly transfer an object or person from one place to another—would be another very useful ability to have. Someone possessing it as an innate talent could avoid traffic jams during the morning and evening commute to work.[10] Weary individuals relaxing in their homes watching television could eliminate the short walk from couch to refrigerator to get a snack. As mentioned earlier, Alfred Bester's 1956 work *The Stars My Destination* posits personal teleportation, or "jaunting," being as common an ability in a futuristic society as walking, though limited in range (or so it seems) to Earth. Works with "lone teleporters" include Gilbert Gosseyn in A. E. van Vogt's novel *The World of Null-A* (1948) and David Rice in Steven Gould's novel *Jumper* (1992).

Present-day science has experimentally confirmed the existence of "quantum teleportation." [32–36] Unfortunately this does not mean that an atom or molecule can be directly transferred from one location to another. Instead it involves creating an effect at a distance. More specifically, quantum teleportation deals with interactions between photons or atoms that are "entangled" [37]. As noted previously, this means that quantum states of a pair of (for example) photons or particles such as protons or electrons are interconnected in ways that are dependent on how they are measured but independent of the distance between them. For example, until its quantum state is measured each of two entangled particles (e.g. an electron and a positron produced by decay of a subatomic particle called a "pion") has an indeterminate quantum state, such as whether its "spin"[11] is "up" or "down" [38–40]. Only when that quantum state is measured is it determined to be either up or down. However, with the entangled electron and positron, if one of them is measured as having an up spin the other—no matter how far away it is—will be measured as having a down spin. Likewise, if one is measured as having a down spin the other will be found to have an up spin. The net effect is that, if you know the quantum state of one of those two entangled particles, you also know—without even having to check it—the quantum state of the second, no matter how far away it is from the other.

Quantum teleportation using this property of entanglement is being investigated as a means to send information in the form of "qubits." Unlike the standard "bit" by which digital information is transferred, which has a

---

[10] Of course, an unscrupulous individual with this ability could teleport into bank vaults and out again after collecting "souvenirs" there, thus eliminating the need to hold an honest job.

[11] Put in greatly simplified terms, the "spin" of a particle such as an electron is its "intrinsic angular momentum," analogous to (but not exactly the same) as the Earth's daily rotation (spin) around its north-south axis of rotation.

value of either 0 or 1, a qubit can have either of those values or "both," in a "superimposed" state in which it might potentially take on either value. Using qubits rather than bits in a computer system would greatly increase its computational abilities in many ways. In one basic method, two entangled photons/qubits are created and separated, so that two observers separated by a certain distance has one of the entangled photons.

A third photon/qubit in an unknown state is created by one of the observers. That individual then makes a joint measurement between the photon in the unknown state, which contains the information to be teleported, and the first entangled photon. The results of this measurement are then sent to the second observer via a "classical" (e.g. at a speed no greater than light, such as by radio waves) means. The latter uses that result to manipulate the second entangled photon, which can then reproduce the same unknown state of the third photon at that location. Thus, for example, photons too have a quantum state called "spin" (similar to but not exactly the same as that associated with particles such as electrons and associated with its "polarization"—see Chap. 1) with values of either +1 or –1 that can be manipulated in this way [41, 42].

The net effect is that the third, non-entangled qubit has been effectively "teleported" between the two locations. This is made possible by the fact that measuring one of the entangled photons "automatically" creates a complementary change of a quantum state in the second photon. A recent report described using this technique to transmit qubits a distance of over 143 km. Further research would seem likely to extend this range [35]. Unfortunately in this context, human beings consist of considerably more than single photons or atoms. Thus, while quantum teleportation may be promising for information transfer, it will not serve as a means of sending people from one place to another.

Also, whether teleportation is based on biology or technology (the latter technique more properly deemed "matter transference," notable examples of such devices being the transporter in *Star Trek* or the transmat employed in various episodes of *Doctor Who*), its use by or on people or anything else macroscopic raises serious questions about how exactly it is accomplished. Using concepts borrowed from known physics, perhaps the atoms and molecules in a person's body could be converted to energy, which then travels at the speed of light to a distant location where the process is reversed. Alternatively the arrangement of all the matter in that body could be scanned and stored, the original one destroyed, and a new one recreated from matter at a distant site using the information previously obtained on how it is "constructed." Or a person might teleport by "folding" spacetime, moving through a wormhole that links two separate parts in space, or traveling through a fourth spatial dimension.

Unfortunately all these "explanations" have the same issue—they artificially graft valid (or, as in the case of a fourth spatial dimension, speculative) scientific concepts onto a context where they do not apply. Our bodies do indeed consist of complex arrangements of matter, all of which could in theory be converted to energy. The "simplest" way to do this might be to have all its atoms simultaneously exposed to their antiparticles—in other words, antimatter. When a subatomic particle (e.g. an electron) collides with its antiparticle (in this example a positron), both are annihilated and converted completely to energy in the form of two gamma rays that move in opposite directions.

Setting aside the already challenging problem of creating the antimatter equivalent of a human body, there is a major issue regarding how much energy is created by doing this. Einstein's famous equation relating energy and matter, $E = mc^2$, once again proves helpful. The value of c is slightly less than 3×10^8 m/s, and we will assume that a science fiction victim/character has a mass of 80 kg and ignore that of any clothes worn at the time. Multiplying that mass by c^2 yields a total energy of 7.2×10^{18} J.

Putting that number in perspective, this is roughly 2000 times the amount of energy produced by a one-megaton hydrogen bomb and about one-tenth of the entire annual energy production of the United States [43]. Any advanced technology that could create/store this much energy at one time, transmit it elsewhere, then reverse the process to reconstruct a person would seem to fall under Sir Arthur C. Clarke's definition of "magic," as described in his "Third Law"—"Any sufficiently advanced technology is indistinguishable from magic."

Even if that much energy could be controlled—perhaps focused into a tight beam—and directed elsewhere, it would not be healthy for anyone to stand in the way of that beam, which would essentially be an extremely powerful death-ray. And unless it were used in space, this transmitted energy would interact at the very least with air molecules along its path. This would have dramatic effects on them such as creating a high-temperature plasma and perhaps distorting the beam itself to produce unfortunate errors when the person is recreated.

Such a beam would presumably be made "coherent" and follow a straight line with little "spreading out" over long distances, similar to a laser. However, transmitting such a beam on Earth would also have to take the curvature of our planet into account. When teleporting someone to a location much farther than walking distance (in which case it would be a bit slower but much simpler to just walk there!), the beam might travel high enough above the horizon that, unless a ground-based receiver captured it first, it would wind up going off into space.

For what it's worth, this particular issue has potential "solutions" such as teleporting to and from a spacecraft in geosynchronous orbit, or sending the beam to far locations on Earth by first transmitting it to a relay satellite that then resends it to a terrestrial target. Unfortunately the problems creating, processing, and handling all that energy still remain. Worst of all, if something went wrong and that enormous amount of energy were released into the surrounding area instead of directed to the target location, it would not only be fatal to the attempted teleportee but also create a deadly environment for anyone unlucky to be within a very wide radius of that mishap.

There are also serious issues involved with the method of scanning a person's body to create a "file" of information containing that individual's exact makeup and constituent parts. Sending that information by electromagnetic or other conventional means (e.g. analogous to sending an email over the Internet?) and receiving it elsewhere to create a duplicate of the original from "raw material" would at least obviate the problems of dealing with the huge amount of energy released by merely annihilating a person's body.

However, this technique has problems of its own. At the microscopic level down to the subatomic the human body is far from a static solitary "thing." As mentioned in Chap. 1 the adult human body contains at least an estimated 37 trillion or so cells [44] and many, many more atoms and molecules. The myriad chemicals formed from the latter are in constant motion in the extra-cellular space as well as within cells, and some of the latter (e.g. red and white blood cells) are also continuously moving. These represent "moving targets" to what already seems the insurmountable task of recording not only what a person is made of, but also how all that matter is organized.

Moreover, as noted in Chap. 5, our bodies contain more bacteria than cells. This raises the question of whether they would or should "come along for the ride" during the teleportation process with the body proper. Many of them are symbiotic with humans, and presumably should be teleported if they could be identified as separate entities as part of a hypothetical scanning process. Bacteria too are moving within our bodies, further complicating that scan.

Likewise, should the scanning/teleportation process also involve less savory body substances such as urine, mucus, and feces? The accuracy of the scanning and reconstruction after teleportation would also have to be done within very narrow limits. Misplacing a few atoms in a post-teleportation person's fingernail could be within acceptable health limits. Scrambling too many cells in the brain or heart would not. Similarly, if the teleportation process had a glitch in which it transmitted all the elements in the human body except for oxygen and hydrogen, the end result would be a very dry husk of a corpse at the destination. Even a 100% successful scan would require an enormous capacity to store/send data obtained regarding the exact composition and arrangement of at least all essential matter in a human body.

Also, assuming the scanning process included everything at the skin level of a person's body, how far away from that surface would it extend? Would the hair on our head and other body areas, the clothes a person is wearing, or even the surrounding air molecules be included? And when a person is reconstructed at a distant site what happens to the air molecules there when the body reforms? Are they somehow pushed away—or would they be incorporated into the person's body with potentially unhealthy results?[12]

The human body contains about 26 elements essential for life. Roughly 99 % of it is made of only 6 elements. As noted in Chap. 1 they are, in order of greatest to least abundance by mass, oxygen, carbon, hydrogen, nitrogen, calcium, and phosphorus. These and other needed elements are incorporated into myriad types of molecules, many (such as DNA and proteins) of considerable chemical and structural complexity.

The standard biological method of creating a complete adult from initial "raw materials" takes years. Teleportation via the scanning method would presumably require doing that in seconds. It also raises the issue of what forms the materials needed to recreate a complete human being would be in to do that. While some chemicals in our body, such as sodium and potassium ions or even hormones such as insulin, are "generic," our DNA definitely is not. The reconstruction process would presumably have to make the latter from nucleotide units, assuring at the same time that it was reproduced perfectly in virtually all body cells.

This method of teleportation raises many questions too. If a person's body could be scanned and the data to duplicate it stored on a futuristic version of a hard drive, by eliminating the "transmission" part the technique would theoretically be able to create as many clones of the original as desired. This issue was addressed as a single duplication in the *Star Trek: The Next Generation* episode "Second Chances" (1993), but its more general applications were not explored.

Also, if the original body is destroyed as part of this process, is the duplicate "really" the same as the person who stepped into the teleporter, or is the latter actually dead? Does the scanning process itself destroy the person's body at the initial location or, if it does not, by not optionally destroying it as part of the teleportation process would this be equivalent to "cloning at a distance?"[13]

---

[12] Of course, this problem with air molecules could be addressed by teleporting into a vacuum chamber or space. However, unless the teleportee were sent wearing a spacesuit, as described in Chap. 3 this solution would have safety issues of its own.

[13] If nothing else it would seem prudent to not destroy the "original" prematurely before the "duplicate" is created at the remote location, so the process could be attempted again if something went wrong at the

This method of teleportation assumes the presence of one device to scan/ transmit an individual and a "receiver" somewhere else to reconstruct the individual. Many depictions of teleportation, like the *Star Trek* transporter, do not employ any such receiver, and the scanning/transmission process itself can be done at great distances. While the end result of this method of teleportation and the type just described are the same—a person "disappears" from one location and "appears" at another—they must employ significantly different ways to do this. How exactly a "transporter beam"—presumably consisting of some form of energy, but if so what type?—can do this is uncertain.

Of course, as with any fictional (and real life) technology things can go either accidentally or intentionally wrong. In *Star Trek: The Motion Picture* (1979) the effect on two Starfleet officers beaming onto the refurbished *Enterprise* when its transporter malfunctions is fatal, as is that of a character whose energy pattern is deliberately "scrambled" during transport in the *Deep Space 9* episode "The Darkness and the Light." *The Stars My Destination* describes "blue jauntes" in which a self-teleporting individual accidentally rematerializes inside a solid object.

A device not sophisticated enough to distinguish different types of living tissue could also lead to unfortunate results, as experienced most graphically by the main character in the original movie version of *The Fly* (1958). Note, however, that such macroscopic mixing of *H. sapiens* and *Musca domestica* would be what is known in medical parlance as a "non-survivable injury," thus making that movie a short subject instead of a full-length feature. The 1986 remake of the movie posits a subtler blend of human and fly initially only at the genetic level. However, given the markedly different genetic profiles, physiologies, metabolism, etc. of those two species, from a strictly biological standpoint the main character's subsequent mutation is equally unlikely.

Rather than converting a person's body into energy or literally destroying/ recreating it, other techniques could—at least conceptually—be used to effectively move someone from one location to another. In a manner similar to that experienced by A Square, the two-dimensional protagonist of Edward A. Abbott's 1884 novel *Flatland* who encounters a spherical being, if a person's three-dimensional body could be somehow transferred entirely to a fourth spatial dimension it would seem to simply disappear. By then moving within that four-dimensional space for a distance and then returning to our three-dimensional world, the individual would reappear at that new location. Science fiction works such as Nelson S. Bond's story "The Monster from Nowhere"

destination. However, even after an apparently successful duplication the original might decide at that point that he or she would prefer *not* to be destroyed, either "ever" or at least until receiving irrefutable confirmation that the duplicate was indeed "perfect." Moreover, any imperfections in the latter might not be immediately apparent or detectable, further complicating the situation.

(1939) present a 4D being that can do that to humans, although not with the latter's consent or returning them to our 3D world.

Einstein's theories describe the universe in terms of four dimensions. However, only three of them are the spatial ones familiar to us, while the fourth is time. Superstring theory posits the existence of (at least) 10 spatial dimensions, with the "extra" seven not observed possibly because they are either "compacted" and too tiny to observe, or exist as part of a larger Universe of which our familiar one is only a part [45]. Unfortunately the fourth spatial dimension described in these ideas is not accessible to us mere 3D creatures and thus does not appear to represent a practical means for "teleportation."

A wormhole is a theoretical way in which two otherwise separate locations in spacetime could be directly connected. In theory it could even be used for what would effectively be faster-than-light travel between the ends of the wormhole, though material objects could still only travel at sub-light speeds within the wormhole itself. Moreover, with appropriate manipulation it might even be used as a time machine for travel into the past [46].

Unfortunately, assuming that wormholes can actually be created (or exist "naturally") and possess these characteristics (all currently confined to the realm of theory), the practical details of making one are formidable [46–48]. As a means of getting to the local grocery store in record time, it would be a clear case of overkill and vastly more resource-intensive than the slow-but-sure method of walking or driving there.

The discussion of teleportation so far has neglected to mention any way in which an individual, either by learning that skill, having extra brain tissue with that capability (e.g. Gilbert Gosseyn), or innate talent for it (perhaps employing a hypothetical "teleportation organ") could perform such a feat. The reason for that neglect is that, based on real-world science or any reasonable extrapolation of it, there simply seems to be no biological way a person could actually do that. It would also involve violation of multiple fundamental laws of physics, such as conservation of energy and momentum. Even using advanced technology there are severe problems with doing it, as has been just discussed. While the idea of "jaunting" certainly results in an exciting and imaginative science fiction novel, its scientific plausibility is, to say the least, highly suspect.

8.6 The Bottom Line

Paranormal abilities such as those described previously can be effective story elements in science fiction. However, while they have a clear place in science fantasy, based on current evidence their role in "hard," science-based fiction is, even looked at optimistically, minimal at best. So far such abilities have not been found to be clearly reproducible or to have any rationale based on known physical or biological principles to explain how they work. Assuming hypothetically that they do exist, the magnitudes of their effects appear too low to distinguish them from chance events, eliminate all confounding variables or alternative explanations, etc.

Please note that the above statement does not say that paranormal abilities do not exist. If subtler varieties of paranormal abilities such as telepathy or precognition are rare (e.g. present in only isolated individuals), weak, and unpredictable, then their presence could be intrinsically irreproducible and below the threshold for unequivocal detection. However, to *establish* that such abilities exist requires that they meet enough standard scientific criteria to do that.

Not all phenomena that science recognizes as "true" are, in fact, reproducible. For example, based on our current knowledge it is not possible (or prudent) to recreate the "Big Bang" that created our Universe in a laboratory. However, that theory is consistent with established scientific laws and data (e.g. the presence of the cosmic background radiation). Paranormal abilities appear to not only contradict those laws (albeit with perhaps a slight fuzziness around the edges based on quantum and other theories) but have no clearly, unequivocally supportive data to back up their existence. Thus, for example, if someone actually could demonstrate the mental ability to consistently and reliably lift an X-Wing fighter out a swamp, with no evidence of electromagnets, wires, or other conventional physical means, then that individual's claim to have telekinesis could reasonably be taken seriously.

Likewise if a person could "blink" away from one part of a room to another instantaneously in front of human witnesses, undoctored video recordings, etc., then teleportation might be considered to be a real possibility. Even being able to immediately repeat word-by-word over and over again what another person or group of individuals were thinking would be impressive as a feat of potential telepathy. However, in the absence of such demonstrations those abilities cannot be confirmed.

Science fiction characters with paranormal abilities are far too useful for story purposes to give up solely in the name of scientific realism. However, if they are used, it would be prudent to touch as lightly as possible (if at all) on the "science" behind how they were born with or gained those abilities and the physics and biology underlying how they operate. Any such "explanation" is likely to quickly crash head on into any of a number of (presumptively) unyielding scientific principles and further reduce the "believability" of those paranormal powers.

An alternative is to acknowledge that such capabilities do, in fact, contradict known science. However, given the already highly speculative nature of those abilities from a "realistic" standpoint, this approach too might not be particularly helpful at establishing their plausibility. In my own stories, when I have described aliens with telepathy their ability was simply mentioned in passing, without further explanation. That approach introduces such a "paranormal" element into the plot without otherwise calling attention to itself.

Anyone planning to write science fiction might consider these suggestions after reading other peoples' minds to find what they would like to see in a story before teleporting back home to enter those ideas into a desktop computer by telekinetically moving the keys of its keyboard. But then, you already knew what I was going to write in that last sentence....

References

1. Lerner E. Alternate abilities: the paranormal. Analog Sci Fict Fact. 2014(June). pp. 19–28.
2. Herculano-Houzel S. The human brain in numbers: a linearly scaled-up primate brain. Front Hum Neurosci. 2009;3:31.
3. Roth G, Dicke U. Evolution of the brain and intelligence in primates. Prog Brain Res. 2012;195:413–30.
4. Kandel E. Nerve cells and behavior. In: Kandel E, Schwartz J, Jessell T, editors. Principles of neural science. 3rd ed. New York: Appleton & Lange; 1991. pp. 18–32.
5. Smith SL, Smith IT, Branco T, Hausser M. Dendritic spikes enhance stimulus selectivity in cortical neurons in vivo. Nature. 2013;503(7474):115–20.
6. Kennett R. Modern electroencephalography. J Neurol. 2012;259(4):783–9.
7. Hari R, Salmelin R. Magnetoencephalography: from SQUIDs to neuroscience. Neuroimage 20th anniversary special edition. Neuroimage. 2012;61(2):386–96.
8. Pei X, Hill J, Schalk G. Silent communication: toward using brain signals. IEEE Pulse. 2012;3:43–6.
9. Ayaz H, Onaral B, Izzetoglu K, Shewokis PA, McKendrick R, Parasuraman R. Continuous monitoring of brain dynamics with functional near infrared spectroscopy as a tool for neuroergonomic research: empirical examples and a technological development. Front Hum Neurosci. 2013;7:871.
10. Bunce S, Izzetoglu M, Izzetoglu K, Onaral B, Pourrezaei K. Functional near-infrared spectroscopy. Eng Med Biol Mag IEEE. 2007;26(4):38–46.
11. Kovac L. The 20 W sleep-walkers. EMBO Rep. 2010;11(1):2.
12. EMF. Electric and magnetic fields associated with the use of electric power. 2002. http://www.niehs.nih.gov/health/materials/electric_and_magnetic_fields_associated_with_the_use_of_electric_power_questions_and_answers_english_508.pdf. Accessed 15 April 2015.

13. Hameroff S. Consciousness, the brain, and spacetime geometry. Ann NY Acad Sci. 2001;929:74–104.
14. Pfaff W, Hensen B, Bernien H, van Dam SB, Blok MS, Taminiau TH, et al. Unconditional quantum teleportation between distant solid-state quantum bits. Science. 2014;345:532–5. (1253512 Published online 29 May 2014)
15. Lee KC, Sprague MR, Sussman BJ, Nunn J, Langford NK, Jin XM, et al. Entangling macroscopic diamonds at room temperature. Science. 2011;334(6060):1253–6.
16. Electroreception in fish, amphibians and monotremes. http://www.mapoflife. org/topics/topic_41_Electroreception-in-fish-amphibians-and-monotremes/. Accessed 6 May 2015.
17. Albert JS, Crampton W. Chapter 12. Electroreception and electrogenesis. In: Lutz P, editor. The physiology of fishes. Boca Raton: CRC Press; 2006. pp. 429–70.
18. Pais-Vieira M, Lebedev M, Kunicki C, Wang J, Nicolelis MA. A brain-to-brain interface for real-time sharing of sensorimotor information. Sci Rep. 2013;3:1319.
19. Berger A. Magnetic resonance imaging. Br Med J. 2002;324:35.
20. Berry M, Geim A. Of flying frogs and levitrons. Eur J Phys. 1997;18:307–13.
21. Liu Y, Zhu D, Strayer D, Israelsson U. Magnetic levitation of large water droplets and mice. Adv Sp Res. 2010;45(1):208–13.
22. Villata M. CPT symmetry and antimatter gravity in general relativity. EPL (Europhys Lett). 2011;94(2):20001.
23. Lightsaber. http://starwars.wikia.com/wiki/Lightsaber. Accessed 15 April 2015.
24. Bem DJ. Feeling the future: experimental evidence for anomalous retroactive influences on cognition and affect. J Pers Soc Psychol. 2011;100(3):407–25.
25. Wheeler J. The "past" and the "delayed-choice" double-slit experiment. In: Marlow A, ed. Mathematical foundations of quantum theory. Massachusetts: Academic Press, Inc.; 1978.
26. Ma X, Kofler J, Qarry A, Tetik N, Scheidl T, Ursin R, et al. Quantum erasure with causally disconnected choice. Proc Natl Acad Sci U S A. 2013;110:1221–6.
27. Kim Y-H, Yu R, Kulik SP, Shih Y, Scully MO. Delayed "choice" quantum eraser. Phys Rev Lett. 2000;84(1):1–5.
28. Galak J, Leboeuf RA, Nelson LD, Simmons JP. Correcting the past: failures to replicate psi. J Pers Soc Psychol. 2012;103(6):933–48.
29. Ritchie SJ, Wiseman R, French CC. Failing the future: three unsuccessful attempts to replicate Bem's 'retroactive facilitation of recall' effect. PLoS ONE. 2012;7(3):e33423.
30. Joyce N, Baker D. ESPecially intriguing. Monit Psychol. 2008;39(4):20.
31. Moir H, Jackson J, Windmill J. Extremely high frequency sensitivity in a 'simple' ear. Biol Lett. 2013;9:20130241.
32. Takeda S, Mizuta T, Fuwa M, van Loock P, Furusawa A. Deterministic quantum teleportation of photonic quantum bits by a hybrid technique. Nature. 2013;500(7462):315–8.

33. Steffen L, Salathe Y, Oppliger M, Kurpiers P, Baur M, Lang C, et al. Deterministic quantum teleportation with feed-forward in a solid state system. Nature. 2013;500(7462):319–22.
34. Ralph T. Quantum communication: reliable teleportation. Nature. 2013;500:282–3.
35. Ma XS, Herbst T, Scheidl T, Wang D, Kropatschek S, Naylor W, et al. Quantum teleportation over 143 kilometres using active feed-forward. Nature. 2012;489(7415):269–73.
36. Pfaff W, Hensen B, Bernien H, van Dam S, Blok M, Taminiau T, et al. Unconditional quantum teleportation between distant solid-state quantum bits. Science. 2014;345(6196):532–5.
37. Penrose R. The entangled quantum world. The road to reality. New York: Vintage Books; 2004. pp. 578–608.
38. Griffiths D. Quantum mechanics in three dimensions. Introduction to quantum mechanics. New Jersey: Pearson Prentice Hall; 2005. pp. 143–212.
39. Griffiths D. Afterword. An introduction to quantum mechanics. New Jersey: Pearson Prentice Hall; 2005. pp. 432–46.
40. Penrose R. Quantum algebra, geometry, and spin. The road to reality. New York: Vintage Books; 2004:527–77.
41. Griffiths D. Identical particles. Introduction to quantum mechanics. New Jersey: Pearson; 2005. pp. 213–60.
42. Furusawa A, Sorensen R, Braunstein S, Fuchs C, Kimble H, Polzik E. Unconditional quantum teleportation. Science. 1998;282:706–9.
43. UCLA's ePhysics. Energy scales. http://ephysics.physics.ucla.edu/energy-scales-table. Accessed 6 May 2015.
44. Bianconi E, Piovesan A, Facchin F, Beraudi A, Casadei R, Frabetti F, et al. An estimation of the number of cells in the human body. Ann Hum Biol. 2013;40(6):463–71.
45. Tipler PA, Llewellyn RA. Particle physics. Modern physics. 5th ed. Texas: W. H. Freeman and Company; 2008. pp. 561–618.
46. Thorne K. Black holes & time warps. New York: W. W. Norton & Company; 1995.
47. Halpern P. Cosmic wormholes. The search for interstellar shortcuts. London: Penguin Books; 1993.
48. Gilster P. Centauri dreams. Imagining and planning interstellar exploration. New York: Copernicus Books; 2004.

9
The Biology of Immortality

Do not go gentle into that good night.
Rage, rage against the dying of the light.

Dylan Thomas
"Do Not Go Gentle into That Good Night"

Grow old along with me!
The best is yet to be,
The last of life, for which the first was made…

Robert Browning
"Rabbi Ben Ezra"

The desire for immortality or to at least live a very long time figures prominently in many legends and writings. The ancient *Epic of Gilgamesh* depicts that Sumerian king's fruitless quest to live forever. Early chapters of the Bible describe many individuals living for centuries, up to Methuselah's 969 years. In Greek mythology humans such as Tithonus could be granted immortality by Zeus, but they might live to regret that gift.[1] Jonathan Swift's *Gulliver's Travels* (1726) describes a similar fate for undying "struldbrugs" and the way their country of Luggnagg deals with them.

Science fiction contains many examples of long-lived individuals, groups, and aliens. Some of Robert A. Heinlein's works, such as the novel *Methuselah's Children* (1958), include Lazarus Long and other members of the Howard Families, who can live hundreds or even thousands of years. Unlike Tithonus, Edgar Rice Burroughs's hero John Carter apparently stopped aging at the peak of his physical prowess.[2] Episodes of the original *Twilight Zone* television series such as "Escape Clause" (1959) and "Long Live Walter Jameson" (1960) depict both the advantages and downsides of immortality, and the *Highlander*

---

[1] Eos, goddess of the dawn, asked Zeus to make her human lover, Tithonus, immortal. Unfortunately she neglected to request eternal youth for him too. Tithonus continued to age, eventually becoming senile and frail—in one version of the myth, even turning into a cicada—but could not die.
[2] He also seemed to possess a tremendous amount of luck at avoiding dismemberment and fatal wounds despite many sword fights against Barsoomian foes.

movie series follow the conflicts among a small group of "immortals." The DC comic book villain Vandal Savage was reputedly born over 50,000 years ago, while characters from the worlds of fantasy and horror such as vampires can also live indefinitely.

Long-lived alien races are also common. In *Star Trek* Vulcans typically live up to about two centuries, while members of the Q Continuum are immortal unless they choose to end their existence. All things considered, Yoda is quite spry considering his being 900 years old in *The Empire Strikes Back* (1980). The Arisians of E. E. "Doc" Smith's "Lensman" series also possess lifespans longer than humans, and the race of Guardians depicted in earlier issues of the Silver Age *Green Lantern* comic books have lived for around ten billion years.

But in our real world, even more than taxes the one certainty in life is death. Any time between conception and senescence, the complex biochemical processes separating the quick and the dead can be irreparably disrupted by accident, illness, or aging. Deprive the human body of oxygen and nutrients for sufficient time, traumatize it with enough force, or merely wait for the ravages of time to take their toll, and its cells and complex organic chemicals revert back to their simpler lifeless origins.

We die.

How long that takes to happen differs from person to person but is constrained by natural limits. If disease or accidents do not get us first, aging alone will. The latter can be described in medical terms as "the gradual and overall loss of various physiological functions, leading to the end of lifespan," [1] or "progressive, generalized impairment of function, resulting in an increased vulnerability to environmental challenge and a growing risk of disease and death." [2]

The longest any human has been documented to live is 122 years and 164 days. That milestone was achieved by Jeanne Louise Calment, a woman born in France on February 21, 1875 and dying there on August 4, 1997 [3]. While it is possible that other individuals might have lived longer than this without that fact being documented, based on our current knowledge of aging the "true" maximum for human longevity, even assisted by present-day medical science, seems to be no more than about 125 years [4]. A more realistic upper limit that a significant number of people might achieve has been estimated to be in the range of 110–115 years [5].

But could future advances greatly extend our lifetimes far beyond that limit by keeping us from physically aging after we reach adulthood, or even rejuvenating those of us well past our youth? Great strides have been made recently in identifying factors that can significantly extend lifespan in nonhuman animals. Likewise we have a growing understanding of the molecular, genetic, cellular, and other changes that underlie the human aging process.

However, the ultimate question is how this knowledge might be used to prevent, repair, compensate for, or even reverse the effects of aging. The next several chapters will describe a variety of techniques that might be used individually or in combination, including stem cells, nanotechnology, genetic engineering, and bionics. This chapter will focus on the mechanisms and principles involved with aging, theories of why it occurs, and the many challenges involved with keeping us biologically young, vigorous, and healthy as the calendar changes.

9.1 The Dying of the Light

Until we get sick, injured, or notice our first gray hairs it is easy to take our bodies for granted. Just as we turn the key in a car's ignition and drive with no thought of how its engine and other parts work together to get us to our destination, we expect every molecule, cell, and organ within us to do its "assigned" task. Until a fan belt breaks, or your colon informs you that it has developed constipation, you ignore them.

Chapter 1 showed how complex the human body is at both the chemical and cellular level, with an adult having an average of at least 37.2 trillion cells [6]. These include over two hundred types of cells organized into complex organs (heart, liver, etc.) and tissues, including the many bones, muscles, blood vessels, and nerves that support us both physically and physiologically. Those cells continuously process nutrients, remove wastes, and produce and repair the proteins, nucleic acids, and other substances needed to keep us alive.

And yet the bioengineering marvel that is the human body is never more than seconds from disaster. To keep living we must be constantly supplied with proper amounts of oxygen, food, and water. Subjecting part or all of our bodies to temperatures, pressures, toxins, electric currents, radiation, or acceleration and deceleration forces outside their narrow "design tolerances" results in damage, illness, or death. The ability of skin and mucous membranes to protect internal tissues and organs from trauma and infection is so limited that even the edge of a piece of paper can slice through them. A rhinovirus-laden sneeze in our direction easily overwhelms them to make us catch a cold. And although *H. sapiens* has helped curb or eliminate many natural predators like saber-tooth "tigers," despite science and civilization we all remain vulnerable to violence from the deadliest predators of all—our fellow humans.

Even if we avoid external dangers, our bodies ultimately fail from within. Medical textbooks describe in excruciating detail the many ways each organ system can break down. In some individuals the genetic cards are stacked heavily against them from the very beginning. Serious chromosomal abnormalities such as with Down syndrome (trisomy 21), or inheriting a single defective gene (e.g. Huntington's disease and hemophilia A), can seriously impair health and lead to premature death. Subtler genetic variants can render an otherwise "normal" person more vulnerable to hypertension(high blood pressure), coronary artery disease, infections, and autoimmune diseases(e.g. rheumatoid arthritis) in which an individual's own immune system attacks normal tissues. And besides factors we cannot control, our own poor decisions about diet and exercise, using tobacco and other harmful substances, and engaging in risky behaviors (e.g. those that help spread sexually transmitted diseases) further impair physical wellbeing and may lop years off our lives.

Yet even if we have "good" genes, lead healthy lives, and dodge every figurative (or literal) bullet shot our way—we still die. If disease or accidents don't get us first, aging does. Normally, after birth our size, strength, and ability to learn new skills increase until they peak at various times during childhood, adolescence, and young adulthood. By our mid-thirties, however, our physical capabilities have essentially crested. After that, although the rate at which it occurs in different organ systems varies from person to person, and we might be able to temporarily compensate for some declines in function, the overall trajectory is downhill.

Table 9.1 shows examples of how different organ systems change with age. A few changes are essentially cosmetic, like skin wrinkling or loss and graying of scalp hair. Others, like diminished sexual function, reduce quality of life but not necessarily its length. Still more, such as diminished immune system function and decreased ability of skin and mucosa to act as barriers to microorganisms, significantly increase vulnerability to infection, cancer, and other diseases [7, 8]. Aging is also the leading risk factor for many chronic diseases such as osteoporosis, kidney failure, cardiovascular disease, blindness, dementia, etc. that not only impair but often shorten and ultimately end life [9, 10].

Aging itself is not a disease but a natural process, albeit one whose effects we would like to live without. However, aging is associated with an increased incidence of serious diseases that reflect a lifetime of environmental stresses (e.g. smoking or high cholesterol), decreased ability to ward off new stresses (like infections), and reduced ability to repair or replace damaged and dysfunctional cells [4]. "Healthy," disease-free aging does not necessarily produce major cardiovascular damage or dementia. Unfortunately, serious diseases like atherosclerosis and Alzheimer's disease are currently so common in the elderly they seem "normal." Global changes in our bodies such as disturbances in circadian rhythms (see Chap. 3) can also lead to sleep disturbances and a variety of adverse physiological changes [11].

Table 9.1 Some age-related changes in major organ systems

Organ System	Changes Seen
Skin	Thinning, ↓wound healing, ↓protection against ultraviolet light, ↑blistering, ↓thermoregulation, ↓response to injury, ↓sensation, ↑white hairs
Eyes	Cataracts, glaucoma, macular degeneration, presbyopia ("far-sightedness")
Ears	Presbycusis (↓hearing, especially of higher frequencies), tinnitus ("ringing in the ears"), dizziness
Cardiovascular	Arteries: Dilation, ↑wall thickness and stiffness (predisposing to hypertension) Heart: ↑mass, ↑left ventricular wall thickness, deposition of amyloid protein, calcification, mildly decreased resting heart rate, decreased peak heart rate with exercise
Pulmonary	↓lung recoil, ↓maximal expiratory flows, stiffer chest wall
Gastrointestinal	Difficulty swallowing, ↑risk of choking and aspiration, ↓protection of the gastric mucosa, constipation, diverticulosis, ↓liver size, ↑risk of gallstones (cholelithiasis), ↓mucosal immunity
Genitourinary	↓size, mass and function of the kidney, benign prostatic hyperplasia Male sexual function: ↓testosterone production, delayed and diminished maximal erection, shorter orgasm, ↓ejaculatory force, ↑refractory period, ↓fertility
Musculoskeletal	↓mass and strength of bones and muscles, degeneration of cartilage and intervertebral disks, ↓height
Neurologic	↓brain weight, ↓number and connections between neurons, ↓short-term memory,
Endocrine	↓melatonin, ↓growth hormone, ↓cortisol, ↓aldosterone, ↓DHEA
Hematologic	↓red blood cell production
Immune	↓antibody response, ↑autoimmune response, ↓ability to respond to viral infection and suppress tumor growth and spread
Gynecologic	↓estrogen production, menopause, atrophy of the breasts, vulva, vagina, and urethra

On the bright side, with one exception (the bowhead whale, which can live up to 200 years) [12, 13] humans already have the longest known maximum lifespan—the greatest number of years we potentially can live—of any mammal or bird [12, 14]. With a few exceptions, vertebrate species with the lowest metabolic rates and the highest ratios between brain and body weight have the longest maximal lifespan [15]. Humans rank at or near the top of both categories, combining a low metabolic rate with high brain-body weight ratio. Different species also have various reproductive strategies that may impact average and maximal lifespan. Thus, mice live a short time and rapidly produce many offspring, each one of which has a low probability of survival to adulthood. Humans and other primates, dolphins, and whales have long delays before puberty and produce few offspring but extensively nurture them. These different "paces" of individual development and reproduction are associated with significant differences in life expectancy and lifespan [16].

However, some non-mammalian animals such as certain fish, sharks, clams, alligators, and Galapagos turtles do not seem to "age" at all after they attain maturity. After they reach a certain size adults of those species can maintain their bodies without the deteriorating physiological functions and increasing vulnerability to disease that our species experiences. Unlike humans, who show an overall low mortality rate at young ages that increases significantly as age increases, some types of tortoises, reptiles, amphibians, and fish have high mortality rates when young that fall dramatically with age. For those animals, if they manage to survive the perils of youth the odds of living to a ripe old age are good [17].

Very small (about 1 cm long) fresh-water animals of the genus *Hydra* seem to have the capacity to continuously renew their cells, rendering them eternally "young." [18, 19] Planarians, a type of tiny flatworm, can regenerate a complete body if large parts of it are removed and even "clone" themselves into two new "whole" worms if the original is split in two [20]. They also constantly replace aged differentiated cells. These abilities involve a population of cells called "neoblasts" that, though capable of continuously renewing the simple organs and tissues of planarians, unfortunately has no analog in humans.

However, despite these biological abilities none of those animals live forever. As with us, lack of food, disease, predators, infections, and accidents still eventually kill them. Even clams, which can live for hundreds of years, can fall victim to "involuntary clam-slaughter."[3] And lest we become jealous of the Galapagos turtle's ability to live up to about 170 years, the downside is they have to live all those "extra" years *as* a Galapagos turtle, without the creature comforts and intellectual stimulation we humans might enjoy.

Maximum human lifespan does not appear to have changed significantly since *H. sapiens* developed, or to even be much different from our primate ancestors of several million years ago [4]. "Life expectancy," the number of years on average that a person at a particular age can expect to live, is also clearly less than 120-plus years. Nonetheless, it has increased dramatically in developed countries within the last century [21]. In 1900, life expectancy at birth for men in the United States was 49.6 years and 49.1 years for women

---

[3] See "Scientists accidentally kill world's oldest animal at age 507," http://www.usatoday.com/story/tech/2013/11/15/newser-worlds-oldest-animal/3574863/.

[4]. Estimated figures for 2012 are 76 and 81 years, respectively. Estimated life expectancy at birth is even greater in some other countries (e.g. 81 and 85 years in Switzerland and 80 and 87 years in Japan for men and women, respectively) [22].

Some reasons proposed for the present gap in longevity between men and women include a dramatic fall in mortality associated with pregnancy and childbirth since 1900; that women are generally less likely to engage in destructive behavior, like excessive tobacco use; and that women are intrinsically "hardier" than men. However, although women are more likely than men to live to an advanced age, compared to men who do manage to reach extreme old age those women are, on average, more likely to have greater morbidity and disability. Interestingly, if a person is able to achieve extreme old age (by one definition, between 110 to 114 years old), the annual mortality rate does not increase but remains constant over that age range at about 50 % [23].

Historically, age distribution in human populations has resembled a pyramid, with infants and children being most numerous and the number of adults within a particular age range decreasing with increasing age, with only a tiny fraction at the most advanced ages. Since about 1950, however, the shape of this distribution in developed societies has become closer to a rectangle due to proportionally fewer children being born and more people living to greater ages [21].

Most of the improvement in overall life expectancy since the early twentieth century reflects increased survival rates during infancy and childhood, due to factors such as improved prevention and treatment of infectious diseases from better sanitation and nutrition, immunizations, and discovery of antibiotics. In recent decades a smaller improvement due to healthier lifestyles (e.g. reduction in tobacco use) and better medical care for diseases affecting primarily middle-aged and older adults, like cardiovascular disease and cancer, were the major contributors to improved longevity [4].

But despite all these advances, the underlying question remains: Why do we age at all after we reach adulthood? Why don't we just get older, maintaining indefinitely the same appearance, health, and vigor that we achieve in, say, our early twenties? A number of theories provide possible answers.

9.2 Longevity and Genes

How long animals (including us) live has a genetic component. This is suggested by the fact that natural maximal lifespans are reasonably uniform in all members of a particular species. Current estimates are that inherited genetic factors account for roughly up to 25–30 % of what our age at death will be

[5, 19, 23–26]. Thus, individuals whose parents are long-lived are also likely to live to an advanced age, potentially due to factors such as inheriting "good" genetic variants that delay aging, confer increased resistance to at least some diseases, etc [26]. This concept provides part of the rationale for the long-living members of Heinlein's Howard Families.

But these inherited benefits may contribute little to human longevity before age 60 [5, 25, 26]. Prior to that age environmental (e.g. accidents, infections, nutrition, etc.) and lifestyle factors are predominant. However, from age 60 years old onward any genetic variants that delay such events as cellular dysfunction, protect against major killers such as cardiovascular disease or cancer, etc. play increasingly vital roles in healthy aging.

Specific genetic mutations in a number of species have been identified that significantly improve their longevity. These mutations generally are associated with metabolic effects that help the animal resist starvation and other stresses [27, 28]. For example, in the roundworm *Caenorhabditis elegans* inactivation of the gene *age-1* extends lifespan by about 65 % while inactivating *daf-2* roughly doubles it [15, 24, 29–31]. These two genes and similar ones are associated with the insulin/IGF-1 (insulin-like growth factor) pathway involving metabolic processes that allow those worms to survive food scarcity [32, 33]. Favorable mutations in *daf-2* also keep members of *C. elegans* "looking" young for their age as well as living longer. Eliminating the actions of the gene that codes for PI(3)K, an enzyme involved with cell growth that interacts with the insulin/IGF-1pathway, can even make roundworms enter their "dauer" stage, a period of temporary "suspended animation" that can last for months, during their development as larvae [27]. When those mutated worms exit the dauer stage they can go on to become adults living about 10 times longer than normal [27, 34].

So far several hundred genes that affect the lifespan of *C. elegans* for better or worse have been identified [31]. Mutations involving more than one gene can result in even greater increases in lifespan than those produced by single genes. For example, simultaneous mutations in the *daf-2* and *daf-12* genes can increase lifespan as much as 500 % [35]. The effects of genetic mutations on lifespan may also be organ-specific. For example, increased activity of the *daf-16* gene in the roundworm's neuronal tissue increased longevity by 10 %, while doing it in intestinal tissue resulted in a 50–60 % improvement. The latter may be due to the gut's important roles of taking up energy-providing nutrients as well as protecting against toxins and pathogens [30].

Other species commonly studied in laboratory settings also have genes linked to longevity. Partially reduced activity of the gene *mth* (short for "methuselah") extends the average lifespan of *Drosophila melanogaster* (the fruit fly) by about 35 %, [36] and mutations in its *daf-2* gene also increases longevity [30]. Inhibition of the TOR (target of rapamycin) metabolic pathway decelerates premature senescence in cells whose cycle is blocked and can extend lifespan in *C. elegans, Drosophila*, and mice [9, 37].

Rapamycin is an antifungal antibiotic approved for use in humans to suppress the immune system and treat cancer [38]. It inhibits the activity of the enzyme TOR kinase, which is part of a network that senses cellular nutritional conditions [9, 27, 33, 38–40]. The TOR pathway is involved in such functions as protein synthesis, growth, and the body's responses to a calorie-restricted diet [41]. Blocking the TOR pathway with rapamycin increases lifespan in yeast, *C. elegans, Drosophila*, and mice [42–48]. However, the beneficial effects of that drug on human longevity are uncertain, and it can cause significant side effects such as increased glucose levels and kidney failure [42, 47].

Metformin, a medication used to treat type 2 diabetes, also inhibits the TOR pathway, slowing aging in mice and extending how long members of *C. elegans* stay healthy. However, complete knockout of TOR or related genes is lethal in mice and yeast. It has been suggested that, although the TOR pathway is needed for an organism's development and growth, once those have been achieved it can on balance be harmful instead [43].

Increased levels of another nutrient sensor system, AMPK ("AMP-activated protein kinase"), can increase lifespan in roundworms and mice [33]. Deletion of the gene *S6K1* in mice makes them resistant to obesity and diabetes, thus helping them live longer. Silencing a mouse's *p66^{shc}* gene and thus eliminating the protein it produces increases its lifespan [3, 49]. Mice having mutations associated with decreased levels of IGF-1 or whose cell receptors for growth hormone were deleted have a small size (as low as a third of average), delayed puberty, and reduced fertility. However, these dwarf mice age slowly, have a significantly reduced risk of developing diabetes and cancer, and have a lifespan roughly 15 to 40% greater than normal [27, 39, 41, 43]. Smaller members of some other species (e.g. various breeds of dogs and horses) also tend to have greater maximal lifespans than larger ones [27].

Those latter findings suggest that, at least within a species, size does matter regarding aging, with a smaller size being associated with greater longevity.[4] However, while some forms of dwarfism in humans has been associated with decreased risk of diabetes and cancer, overall average longevity has not been found to be increased. On the other hand, some studies have suggested a negative correlation between body size and longevity in humans [41]. Further study of this association and the mechanisms that may underlie it is still needed.

---

[4] This might also help explain why reached his 900th birthday Yoda did.

The greatest increases in longevity due to genetic effects have been identified in species whose average lifespans are far less than us humans. *C. elegans* normally lives up to about 3 weeks, *Drosophila* roughly 3 months, and mice about 3 years [38, 40]. Thus, a mutation that lets a roundworm live 5 times more than normal means it still only lives 15 weeks—an impressive accomplishment from its viewpoint but not a desirable total lifespan for a human. This raises the question of whether there is an inverse relationship between a species' average lifespan and how much can be done through genetic changes to extend it. If so, we humans might have considerably less intrinsic "wiggle room" to do this than *C. elegans* does.

Compared to the many genes associated with longevity in *C. elegans* and other species, although over a hundred potential candidate genes affecting human lifespan have been looked at, so far only a very few genetic variants have been identified with reasonable consistency as potential contributors to increasing our longevity [5, 18, 19, 23–26, 42, 50]. These include variants of the gene *APOE* and, to a lesser extent, *FOXO3A* genes. For example, *APOE2* has been found to be present more frequently in elderly than in younger individuals, while the *APOE4* variant was found less frequently in individuals living to an advanced age. *APOE4* has also been associated with a higher risk of developing Alzheimer's disease [26]. The *FOXO3A* gene is the human analog to the *daf-16* gene in *C. elegans* that can have mutations prolonging life. Some variants in *FOXO3A* have also been associated with increased longevity [5, 38]. *FOXO* genes have also been associated with increased longevity in *C. elegans*.

Other genetic variants have been associated with increased human longevity in some but not all studies and populations. For example, a group of 7 "sirtuin" genes help protect against cellular stress [50–54]. Overexpression of the *Sir2* gene in yeast and *C. elegans* has been reported to increase their lifespans in some [3, 27, 53] but not all studies [55].

In mammals one sirtuin gene, *SIRT1*, helps control glucose metabolism, how insulin acts, and fat storage [42]. Resveratrol is a chemical found in the skin of red grapes and wine made from them [53]. It may activate *SIRT1* and has been reported to increase the lifespan of rodents [27, 53]. However, its effects on human health and longevity are still uncertain [27, 53, 56]. Age-related reduced activity of the anti-tumor gene *Gadd45* might also contribute to development of cancer and Alzheimer's disease, while its overexpression could potentially extend lifespan [28].

However, unlike studies conducted under tightly regulated laboratory conditions in animals, research to identify specific genetic factors that increase life expectancy in humans cannot control for environmental and other external influences [57]. Mice do not drive cars at high speeds, fight in wars, smoke, or engage in other behaviors that can kill them despite having protective genes. Thus, although statistical methods can be useful for finding associations between specific genetic variants and longevity, establishing a causal relationship would be more difficult. It would also hopefully involve the gene being associated with metabolic or other effects that would be expected to increase longevity.

Studies comparing identical ("monozygotic") twins, who originate from a single fertilized egg and thus start out the same genetically,[5] show that they have lifespans (barring accidents) within about 3 years of each other. Such studies indicate that some diseases have a high degree of heritability (how much genetic factors alone contribute to its development). For example, the heritability of developing Alzheimer's disease after age 65 has been reported to be 79%. However, the heritability of other age-related diseases is significantly lower (e.g. 0 to 6% for Parkinson's disease), suggesting environmental or other factors are also involved [58].

Recent studies also indicate that centenarians do not carry a smaller number of "risky" alleles (e.g. those predisposing to cancer, cardiovascular disease, or type 2 diabetes) than are present in the average population. While this could be accounted for by the presence of other genetic factors protecting ("buffering") against development of those diseases, here too lifestyle effects (e.g. eating a healthy diet, exercise, refraining from smoking, etc.) may be the dominant "protector" instead [5]. It also suggests that the presence of "good" genes may be more important than the absence of "bad" genes for living a long and healthy life. Genetic studies of centenarians have shown increased incidences of certain genes that appear to be protective, e.g. by reducing risk of developing atherosclerosis.

Multiple genes with different metabolic and other effects (e.g. preventing damage to or repairing DNA) are thought to be associated with the aging process. However, the total number and types of genes involved with human aging is unknown. Moreover, defects in only single genes can cause "progeroid" syndromes —genetic disorders that mimic at least some features of premature "aging."

For example, individuals with one such disorder, Werner's syndrome, show early graying and thinning of hair, develop cataracts in their early thirties, and commonly die by their late forties—usually from a myocardial infarction (heart attack) or cancer [24, 59, 60]. A mutation in a single gene (*WRN*) coding for a type of helicase (an enzyme involved in DNA replication and repair)

---

[5] By comparison "dizygotic"("fraternal" or "nonidentical") twins occur when a woman releases two separate eggs during a menstrual cycle that are each fertilized by different sperm cells. Dizygotic twins have the same degree of genetic similarity as siblings produced individually from the same parents.

has been identified as the cause for Werner's syndrome. Hutchinson-Gilford Progeria syndrome involves a different point mutation in a gene that impacts multiple systems including ones involving DNA replication and repair. Individuals with that syndrome demonstrate severe growth retardation and aging-like effects within the first years of life, typically dying of myocardial infarctions and strokes in their teenage years [59].

What, if anything, these particular mutations have to do with "normal" aging is uncertain. However, they demonstrate how impairments in critical systems (e.g. DNA repair) can hasten effects that at least simulate senescence.

9.3 Mechanisms and Theories of Aging

Some age-related changes involve simple "wear-and-tear." Our bodies are limited in how well they can repair damage caused by a lifetime of mechanical stresses to the joints and back. Cartilage and other tissues in them can wear down over time. In some cases various tissues themselves change for the worse, such as the age-related "stiffening" of the walls of arteries that can contribute to high blood pressure. Other tissues, such as those making up skin, can repair and renew themselves to a much greater degree. However, their ability to do this too declines with age, leading to such effects as wrinkling and sagging of skin, graying and loss of hair, etc.

One set of theories postulates that aging is also related to accumulated injuries to cells and their biochemical components, including the lipids, proteins, and nucleic acids that compose them and that they, in turn, produce. This may involve damage and mutations to DNA, causing synthesis of aberrant proteins and other substances with altered or no biological function. These in turn can impair healthy functions of cells, kill them outright, or transform them into malignant (cancerous) ones. Such injuries may result from environmental hazards (e.g. radiation and viruses), or the destructive effects of normal chemicals that cells produce. For example, one potentially toxic element we are exposed to every day is implicated in producing harmful molecules. Using it may impair our ability to stay physiologically young, but we cannot live without it.

That element is oxygen.

Biochemical processes using oxygen (aerobic metabolism) to generate ATP (adenosine triphosphate), the primary chemical energy source for life-sustaining reactions, are much more efficient than those not using it (anaerobic metabolism).[6] The large number of ATP molecules generated by aerobic

---

[6] A single glucose molecule may produce up to a net number of 36 ATP molecules via aerobic metabolism, while anaerobic metabolism in humans has a net yield of only 2 ATP molecules.

metabolism of glucose and other substrates, mainly in mitochondria (see Chap. 1), are an essential "fuel" needed to keep our complex cells functioning properly.

But there is a price for using oxygen to provide our energy needs. Oxidative phosphorylation, electron transport, and other oxygen-using processes produce the "free radicals"—molecules with an unpaired electron—mentioned in Chap. 6, and a related group of chemicals, reactive oxygen species (ROS). These include the superoxide anion O_2^-, hydrogen peroxide (H_2O_2), and especially the hydroxyl radical (OH*) [15, 50, 61, 62]. All of these damage protein, lipids, and DNA, potentially contributing to the aging process [2, 3, 27]. Damage due to ROS has also been linked to a variety of specific diseases such as atherosclerosis, diabetes, and Parkinson's disease [35].

With aging, mitochondrial function declines and free ROS production increases [33, 35]. Some recent reports suggest that low levels of ROS may be needed to keep cells healthy, while higher ones are detrimental [27, 62]. This may be an example of "hormesis," the concept (expressed in simple terms) that a little of a substance is good for health but a lot is harmful [63].

Antioxidants (e.g. Vitamins C and E) and enzymes like superoxide dimutase and catalase help neutralize these destructive substances [61]. Even hydrogen sulfide, the compound responsible for the smell of rotten eggs and poisonous in low concentrations as a gas, is produced by our cells and may be beneficial by inhibiting free radical reactions [64]. Cells also have efficient mechanisms for repairing damage to DNA caused by oxygen free radicals.

While its nucleus contains the bulk of a cell's DNA, mitochondria also have a small amount. Unfortunately, compared to the nuclear variety mitochondrial DNA is more vulnerable to injury and mutation and less able to repair itself [62]. Over time oxygen free radicals may damage or destroy enough mitochondria to seriously impair cellular energy generation and potentially kill the cell [2, 42, 61]. Elderly individuals show evidence of increased oxidative damage to cells and tissues as well as decreased levels of natural antioxidants.

Other naturally occurring substances in the body also damage cells. For example, nitric oxide (NO) is generated within the brain. It helps control neuronal activity in areas like the cerebellum and regulates release of hormones from the pituitary and pineal glands. However, NO is also a free radical—and production of toxic levels (as occurs with infections) can damage the brain.

Thus, over time enough unrepaired injuries occur in cells to account for some of the changes seen with aging. At least partly, how long we live depends on how well our cells resist and repair the damage they inevitably suffer—ultimately, a losing battle.

Another theory of aging involves the concept of "hyperfunction." It proposes that processes functioning at a certain level contribute positively to an organism's growth and reproductive systems early in life. However, maintaining that same level of function later in life can, on balance, be harmful [38, 52]. The "overactive" systems involved are thought to include various "signal transduction pathways" (i.e., those involved with activation of a receptor on or in a cell by a molecule outside it) such as TOR, IGF-1, and growth hormone. Those latter systems are all associated with nutrient sensing and regulation. As noted previously, inhibiting these systems can extend lifespan in *C. elegans*, mice, and other organisms.

Evolutionary forces may also shape how and why we age. From the standpoint of natural selection, the primary function of each member of a species is to pass its genetic material on to the next generation. Traits that increase the odds of an individual surviving to reproductive age and producing more offspring are favored. However, after an individual lives long enough to contribute its germ cells (eggs and spermatozoa) or, in species like us whose young require a long nurturing period after birth, have lived long enough to raise sufficient offspring, in Nature's eyes that adult might be considered expendable.

One theory based on these evolutionary principles—that there are "death genes" that actively kill us after we are old enough to produce and raise the next generation, presumably to "make room" for our children—does not (fortunately) seem to be the case in humans. Likewise there do not appear to be genes that specifically promote aging [65].

On the other hand, "antagonistic pleiotropy,"[7] the concept that genes with good early effects would be favored even if they have bad effects at a later, post-reproductive age, is considered plausible [10, 43, 61]. For example, producing levels of hormones like testosterone and estrogen that enhance fertility in our youth might have the delayed effect of increasing the risks of developing prostate or breast cancer in our "golden" years. The TOR metabolic pathway described previously, with its beneficial effects during youth and potentially deleterious one in adults, might be another instance of this.

Similarly, the "disposable soma" theory postulates that limited metabolic resources are allocated between ensuring reproductive success versus preserving the soma (all cells in the body besides germ cells). For animals like field mice that have a high mortality rate in the wild from predators, disease, starvation etc., survival of the species requires that a high priority be placed on early and successful reproduction. There is no evolutionary "motivation" for them to develop complex, energy-requiring homeostatic processes to prevent the degenerative effects of aging or late development of cancer if they are probably going to die young anyway [4, 27, 38, 61, 65, 66].

---

[7] Working this very technical term into a casual conversation will likely impress your listeners, but not necessarily in a positive way.

The principle also applies to longer-lived species. Truly old animals of any type are rare in their native environment. If they manage to avoid death from natural causes before they age significantly, decreasing strength and swiftness render them less able to compete for food with more vigorous younger members of their kind, more vulnerable to predators, and (as in predators themselves) more likely to die from disease or accidents.

A theory based on these considerations, "mutation accumulation," postulates that, because animals typically do not grow old in the wild, the force of natural selection decreases with increasing age. Thus, alleles of genes that produce late-acting deleterious effects can accumulate and be propagated to future generations because animals die too young for those effects to manifest [9, 33, 65, 66].

But even if animals are kept safe and meticulously cared for in zoos, they still die from the effects of aging. Humans are no different. Reducing mortality from tuberculosis and other historically common causes of death among the young over the past century has allowed many more people to live long enough to "unmask" previously uncommon, delayed causes. Now it is the gradual loss of function in various organ systems, and diseases like the top three killers of the elderly—cardiovascular disease, cancer, and stroke—that do us in.

.There are several basic ways to try to improve how our bodies deal with aging. As noted previously, both individual and average life expectancy can and have improved dramatically due to altering the "healthiness" of our environment via better sanitation, nutrition, advances in medical care, etc. Increased average life expectancy is not guaranteed, however. The increased prevalence of obesity, large number of individuals following unhealthy lifestyle choices (e.g. voluntary lack of exercise, substance abuse, etc.), increased vulnerability to infections (such as from antibiotic-resistant pathogens or a pandemic caused by a deadly virus) and so on could potentially result in a future *decrease* in average life expectancy [67, 68]. For example, in the United States there was a 50 % increase in obesity prevalence per decade among adults throughout the 1980s and 1990s. Obesity rates in 65- to 74-year-old men jumped from 10.4 % in the period 1960 to 1962 to 41.5 % in 2007 to 2010, with women in that age range showing rates increasing from 23.2 to 40.3 % during that time [69].

While it has adverse health consequences at earlier ages, obesity is an especially serious problem after the age of 60 largely due to its association with elevated risk of cancer, cardiovascular disease, type 2 diabetes mellitus, and other diseases that increase both morbidity and mortality. It may also hasten onset of age-related problems with higher brain functions such as information reception and processing, intelligence, etc [69]. Thus, if such life-shortening effects of obesity in individuals and whole populations increase faster than medical science can deal with them, the net effect might even be an overall decrease in average life expectancy during the twenty-first century [67].

On a more optimistic note, changes in diet, widespread performance of at least moderate levels of aerobic exercise, and other lifestyle changes could potentially ameliorate some of these effects. Still more optimistically it has been estimated that, if progress in reducing mortality were to continue at the same rate as over the past two centuries (something that is *not*, however, guaranteed), most children born in the year 2000 will become centenarians [23]. However, for that to happen some combination of many favorable factors (lifestyle modifications, medical advances, avoiding pandemics and catastrophic global disasters, etc.) will need to occur. Even now, if a person endeavors to live a healthy lifestyle the current odds of reaching the age of 100 are still only about a few percent [23].

But as Tithonus discovered, merely living longer can be a curse if it is not accompanied by a good or at least acceptable quality of life. Both "normal" declines in organ system functions and chronic illness contribute to functional impairments and worsened quality of life as we age. Arthritis and other musculoskeletal problems like fractures due to brittle bones, deteriorating sight and hearing, dementia due to Alzheimer's disease and other causes, the long-term ravages of common diseases like diabetes and atherosclerosis, etc. can severely compromise an older individual's ability to lead a pain-free life that is meaningful to him or her.

Not surprisingly, individuals aged 65 years and older use hospitals and nursing home resources far more than younger ones. In 2010 the mean length of stay in a hospital for those at least 65 years old in the United States was 5.5 days—and many (like those with severe chronic problems like congestive heart failure) require multiple admissions each year [70]. Based on a survey conducted in 2004, at that time about 3.63 % (1.3 million) American aged 65 or older resided in nursing homes [71]. However, many of those living outside nursing homes could require strong support from family or friends to help them perform basic activities of daily living—or even merely survive. Even the healthier, more independent elderly require on average great amounts of expensive medical care for physician visits, medicines, diagnostic tests, etc.

Thus, even in the absence of an increase in maximum human lifespan it would still be highly desirable to increase what has been called "healthspan"— how long we stay free of major diseases and debility as we age. This too is an area where appropriate changes in lifestyle and advances in medical care can contribute significantly. Interestingly, individuals who live exceptionally long also tend to stay reasonably healthy well into advanced age, presumably due to some combination of environmental effects, lifestyle, and genetics [5].

9.4 The Quest for Eternal Youth

Science fiction often deals with what is currently the most challenging and least certain way to improve aging—greatly reducing the rate of, halting, or even reversing the aging process. How science fiction does this is typically technological, with writers postulating dramatic advances in organ transplantation, nanotechnology, cloning, genetic engineering, and even more speculative areas such as downloading a person's "mind" into a new body.

However, while some of these methods can indeed increase an individual's life expectancy (e.g. via a heart transplant or gene therapy), so far no single one or combination of them have been able to increase the *maximum* human lifespan. There are many reasons why the latter is so challenging. Later chapters will focus on the current status of specific methods that might be employed to do this, such as nanotechnology. Here we will concentrate on the limited means currently available that might, at least modestly, prolong "youth" and health.

Hormone replacement therapy and dietary supplements have been used to try to prevent or reduce some age-related problems and possibly increase life expectancy. One concept underlying such use is that, because levels of certain hormones and other chemicals fall in the body with increasing age, restoring them to more "youthful" levels will retard or perhaps even reverse some deleterious effects of aging. For example, taking supplemental calcium can be used in both older men and women with significant age-related decline in bone mass (osteoporosis) and those at risk of developing it.

In post-menopausal women estrogen replacement therapy may help prevent osteoporosis. Estrogen can also be used with another hormone, progesterone, to reduce symptoms associated with menopause. However, estrogen replacement can also be associated with significant adverse effects, including an increased risk of heart disease, formation of blood clots, stroke, and breast cancer. The decidedly mixed effects that can be produced by artificially restoring more "youthful" levels of estrogen illustrate the principle that as our bodies change at the molecular, cellular, and tissue level with age, our responses to hormones and other chemicals may also change—perhaps for the worse. Unless we can make more fundamental changes in how our bodies work as we age, introducing higher levels of a hormone into that unaltered age-related physiological environment may cause more harm than good.

Other types of replacement therapy can also have uncertain, mixed, or predominantly harmful effects. Growth hormone (GH) is a hormone secreted by the pituitary gland, located at the base of the brain. GH stimulates production of insulin-like growth factors and interacts with other hormones. Deficiency of GH causes increased body fat, reduced protein synthesis, and decreased lean body and bone mass. There is a trend for serum levels and normal pulsa-tile secretion of GH to fall with increasing age.

In several short-term clinical studies of elderly individuals, subcutaneous injections of either growth hormone or insulin-like growth factors were as-sociated with mild increases in lean body mass and skin thickness, particu-larly in those with GH levels below average for their age. However, these changes were not associated with any functional improvement, and they were frequently accompanied by significant side effects (e.g. fluid retention, breast enlargement in males, joint pains, and hyperglycemia). At present, using GH as "therapy" for age-related declines in this hormone is not recommended [42].

Several other hormones whose levels decrease with age are commonly avail-able as oral "dietary supplements." General issues with these preparations in-clude that in the United States they are not regulated by the Food and Drug Administration (FDA) regarding their contents and the actual dosage they contain. They may also interact with prescription medicines and potentially cause adverse effects.

Dehydroepiandrosterone (DHEA) and its sulfate (DHEA-S) are the most abundant steroids in the human body and available as dietary supplements. In humans and several other primates they are secreted in large amounts by the adrenal glands and converted into potent androgens (e.g. testosterone) and estrogens in peripheral tissues. While a few small animal and clinical trials have shown improvements with use of DHEA (e.g. improved memory in mice and increase bone mineral density in the lumbar spine in women), overall results from multiple studies have not shown a consistent, sustained benefit in muscle strength, ability to think, or other parameters [72].

Melatonin, a hormone secreted by the pineal gland (located within the brain), is a powerful antioxidant and helps induce sleep. Blood levels fall with aging. Some research suggests that taking melatonin as a supplement can help treat insomnia in the elderly [72, 73]. However, melatonin levels can also be boosted by a variety of simple, non-pharmacological methods such as expo-sure to daylight, avoiding bright lights at night, and not smoking or using ethanol.

Overall, the current general assessment is that replacing hormone levels that fall "naturally" with age has little or no benefit and that using such hor-mones should be limited only to individuals who have a true hormone defi-ciency independent of aging [72].

Other substances have also been used to try to delay the effects of aging. Epidemiological studies have shown that individuals with diets rich in an-tioxidant nutrients—vitamins and other substances that help neutralize the damaging free radicals and ROS described previously—have a significantly reduced risk of many diseases associated with aging. The latter include cardiovascular disease, gastrointestinal and lung cancers, Parkinson's disease, Alzheimer's disease, and cataracts.

Perhaps on the questionable theory that if a little is good a lot is better, antioxidant vitamins like retinol, ascorbic acid, tocopherols (vitamins A, C, and E, respectively), and beta-carotene are available as dietary supplements in doses much higher than are found in fruits and vegetables. However, clinical studies using such doses have not always shown them to be beneficial, and in some cases they were found to be harmful [62]. For example, in both the beta-Carotene and Retinol Efficacy Trial (CARET) and the Alpha-Tocopherol, Beta-Carotene (ATBC) studies of persons at risk for developing lung cancer, the doses of beta-carotene given were associated with an increased incidence of lung cancer and both cardiovascular and all-cause mortality [74, 75]. In CHAOS (Cambridge Heart Antioxidant Study), individuals with known coronary artery disease given 400 IU (international units) per day of Vitamin E had a reduced incidence of subsequent heart attacks compared to those receiving placebo, but no improvement in their risk of dying from a cardiac cause. A third group given 800 IU/day of Vitamin E actually had a greater incidence of both of these undesirable events. Overall results from clinical trials regarding this use of Vitamin E have shown at most a small benefit [76].

Taking even large doses of these vitamins may not be as effective as getting them from fruits and vegetables because those foods also contain many other beneficial antioxidants. Moreover, the protective effects of supplemental vitamins may be neutralized if they are taken as part of diets high in processed foods containing xenobiotics (e.g. chemical additives, flavorings, and colorings), which increase production of free radicals. Also, under certain circumstances (e.g. intracellular acidosis), Vitamin C can actually act as a source of free radicals. Finally, it is uncertain if taking these substances when they are not part of natural foodstuffs leads to adequate absorption and delivery in useful concentrations to cells.

Overall, as with replacing particular hormones whose levels decline with age, there is currently no convincing evidence that ingesting large doses of vitamins will significantly delay or reverse any of the effects of aging. However, this is an active area of research, and future studies might change this overall assessment.

In the meantime, two methods have been reasonably well established to significantly extend lifespan in animal models. Reducing the number of calories in the diet an organism ingests has been found to extend lifespan in a number of species, including yeast, *C. elegans*, *Drosophila*, mice, rats, dogs, and rhesus monkeys. Some strains of mice and rats have been reported to live about 40% longer than average (a range of about 30 to 50%) if they are fed a diet containing all essential nutrients, but only 60–70 % of the calories they would ingest if allowed to eat as much as they wished [39]. Such diets may involve reducing not only the total number of calories provided (a "calorie restriction" diet) but also decreasing the amounts of particular dietary constituents such as amino acids (a "dietary restriction" diet) [38].

A recent analysis of many such studies showed somewhat weaker effects (14–45 % increase in median lifespan in rats and 4–27 % in mice), but overall calorie restriction remained beneficial. Maintaining food intake at just above malnutrition levels also seems to reduce the rodents' incidence of major diseases. These effects may be due to multiple potential causes, including reduced levels of ROS, changes in the TOR and other pathways, etc [77]. However, the benefits of such diets may come at the potential expense of slowing growth and development as well as reducing fertility [15, 42, 43, 61, 78–82].

The effects of calorie-restricted diets on primates are also being studied. Rhesus monkeys have an average lifespan in captivity of about 27 years, with a maximal lifespan of some 40 years. Two recent studies on rhesus monkeys fed calorie-restricted diets reported differing results regarding longevity and development of age-related diseases [39, 78, 79]. One study using adult rhesus monkeys reported that those fed a 30 % calorie-restricted diet (one with 30 % fewer daily calories than the monkey would have eaten on its own) and followed over 20 years had delayed disease onset and mortality compared to controls [78]. However, a similar study using a group of young, middle-aged, and older monkeys followed for a similar period showed no statistically significant differences in these factors compared to the control group regarding longevity, cardiovascular benefits, or incidence of diabetes [39, 79]. The latter study did, however, indicate the calorie-restricted monkeys had a lower incidence of cancer and delayed onset of age-related diseases.

The effects of long-term use of a calorie-restricted diet on maximal lifespan in humans is unknown [3]. Based on our much longer average and maximum lifespans than rhesus monkeys, it would require many more decades than the latter to confirm a beneficial effect in humans. Moreover, it will take extremely motivated volunteers to maintain diets poised at a near-starvation level for year after endless year. It is also uncertain how many moments of weakness at a fast-food restaurant it would take to at least partly negate all those years of dietary sacrifice [38].

Much shorter term studies have and are being done to at least assess the effects of a calorie-restricted diet on "biomarkers" of aging such as fasting insulin concentrations (e.g. higher ones can correlate with type 2 diabetes). Initial results from human volunteers randomized to eat a 20 to 30 % calorie-restricted diet over a 6-month period as part of the CALERIE (Comprehensive Assessment of Long-Term Effects of Reducing Intake of Energy) study did show improvement in some biomarkers. The full results from a 2-year follow-up study of 218 individuals randomized to a 25 % calorie-restricted diet are pending as of this writing [80, 83].

However, many human studies have shown that not ingesting excess calories over daily needs, as well as eating particular kinds of foods that help promote healthier biomarkers (e.g. those that lower cholesterol) and decreasing intake of those that worsen them, can both reduce morbidity from cardiovascular and other diseases and increase life expectancy [81]. Thus, even if calorie-restricted or other diets do not significantly increase maximal lifespan, they could reduce morbidity and increase healthspan. Also, outside of controlled studies, individuals who follow a calorie-restricted diet (typically as low as 1400 calories/day for men and 1120 for women) in middle age for 8 years are anecdotally reported to have better than age-average blood pressures and metabolic profiles. This at least suggests the possibility (but unfortunately does not prove) that calorie-restricted diets can potentially delay biological aging and improvement longevity. A calorie-restricted diet may also reduce the rate of aging in the brain [84].

The type of diets elderly individuals eat and even the varieties and distributions of microbes in their intestines may also have an impact on how well they age. In one study the frailest older people had diets high in fat and lacking in fiber, as well as having particular gut bacteria different from younger individuals. Whether the latter in particular is a contributor to or effect of frailty is uncertain, however [85].

Finally, in studies involving mice, calorie-restricted diets are typically started at an early age, e.g. just after they are weaned. However, determining whether any beneficial effects on human lifespan and health would be maximized if calorie restriction were started during childhood instead of adulthood would be very difficult to study. It would also raise the question of where to draw the line between restricting enough calories to improve future longevity but not so much as to impair a child's growth and development—or whether an acceptable balance can even be made between these two health considerations.

A second, less well-established potential method for life extension would have distinct downsides for humans. Castration (aka "gonadectomy") of some animals prior to puberty prolongs lifespan [32]. However, any controlled study applying this method to human children for assessment of its effects on longevity would entail obvious ethical and other issues. Hypogonadism (decreased function of the testes in men and ovaries in woman), which has some hormonal effects similar to those produced by castration, is often associated with obesity and diabetes—factors that in themselves would not extend lifespan [32]. However, some studies involving reviews of historical records that compared age of death between castrated men and "intact" men of similar socio-economic status found that castration was associated with a significantly longer lifespan [32, 86]. At this time, the potential beneficial effects of castration on extending lifespan remain uncertain, and the detrimental results of that procedure are all too clear-cut.

9.5 The Hayflick and Other Limits

Normal human cells do not live forever. This is self-evident when the cells are part of a person's body. However, through most of the twentieth century it was thought that if those cells were grown under laboratory conditions in tissue cultures, with all the nutrients and other conditions required for life, they would live and divide indefinitely.[8]

However, a series of experiments reported over 50 years ago using fibroblasts disproved that idea. Fibroblasts are cells commonly found in "connective tissues." The latter are composed primarily of intercellular material and help maintain the body's structure. They include bones, cartilage, and the subcutaneous tissue beneath the skin, where fibroblasts are mainly found. The latter are usually the most numerous cells present in "general" connective tissue, and synthesize most of the material (such as collagen and elastin fibers) found within it. Fibroblasts divide and proliferate actively during wound repair after the skin is cut.

In 1961 Leonard Hayflick and Paul Moorhead described how normal fibroblasts from human embryos grown in tissue cultures had a limited ability to divide [29, 61, 87]. Those individual cells were allowed to multiply in culture bottles until they covered the floors of these vessels. Half of the cells were placed into a new culture bottle, where they were allowed to grow, divide, and double their population in that second vessel. This process was repeated until, after an average of 50 population doublings, the fibroblasts stopped dividing, lived a few months in a stable non-dividing ("senescent") state, then died.

Subsequent experiments showed that human fibroblasts derived from middle-aged individuals, which had already undergone a number of divisions naturally during the donor's lifetime, could divide only about 20–30 times before becoming senescent. Similar cells from very aged people like centenarians can divide only perhaps 10 more times before "giving out."

These results suggest that at least some of the effects of aging may be due to the human body's cells reaching the "end of the line" in their ability to proliferate—the "Hayflick limit." [3, 33] One mechanism for why our cells

---

[8] This idea is used in a 1927 story, "The Tissue-Culture King," by Julian Huxley, a biologist and older brother of Aldous Huxley, the author of *Brave New World* (see Chap. 10).

possess a finite ability to divide was proposed in the early 1970s. As noted in Chap. 1, cell replication requires duplicating the DNA located in chromosomes within its nucleus. The two strands forming the well-known double helix separate, and a ribonucleic acid (RNA) "primer" consisting of a short sequence of nucleotides attaches to one end of the strand to be duplicated. This RNA primer is then used by a group of enzymes, DNA polymerases, as a starting point to begin synthesizing a new strand of DNA, sequentially adding nucleotides of the proper type and in the correct order to it. As described in Chap. 1, these nucleotides consist of one of four bases—adenine, guanine, cytosine, and thymine—linked to a sugar and phosphate group, and are abbreviated A, G, C, and T, respectively.

For the duplicated DNA to "code" for production of the right proteins in the new cell, it should be an exact reproduction of the original strand, with the same sequence, types, and order of nucleotides. DNA polymerases do an excellent but not perfect job in this regard, "proofreading" the new strand of DNA for errors while it is being synthesized and repairing any mistakes. However, those enzymes cannot duplicate the tiny section at one end of the original DNA strand where the RNA primer attaches. Therefore, the duplicated DNA strand (and the chromosome it resides in) is slightly shorter than its "parent."

The ends or "caps" of chromosomes are called "telomeres." [88–90] They are needed for normal cell function, such as preventing the tips of chromosomes from "sticking" to each other and breaking down, and perhaps anchoring chromosomes to the inner membrane of the nucleus [14]. Telomeres consist of several thousand repeated sequences of the same nucleotides. In humans and other mammals this repeat sequence is TTAGGG.[9] For the reasons described above, with each successive generation of new cells some of these repeat sequences are lost, and telomeres become progressively shorter [88, 91, 92].

Although other factors such as DNA damage can contribute to this process, an important reason why cells might reach the Hayflick limit and become senescent is that their telomeres eventually shorten to a critical length [3, 9, 33, 88]. Some human studies have shown that long-lived individuals have longer or at least less variable telomere lengths compared to controls, suggesting they might have better than average genetic or other mechanisms for maintaining their telomeres [5]. In that case, if there were some way to repair and lengthen the telomeres again, perhaps "old" cells might regain their "youthful" ability to divide.

---

[9] The initials refer to bases within DNA—"T" for thymine, "A" for adenine, and "G" for guanine.

In fact, our bodies do produce an enzyme, telomerase, that can make telomeres longer again [29, 93]. Normally telomerase activity is seen in germ cells (eggs and sperms), but not or to a much smaller degree in "somatic" cells—all the other types that make up our bodies. However, studies using cultures of normal human fibroblasts exposed to telomerase showed that their telomeres do indeed lengthen again. Those treated fibroblasts could then go on to have several hundred population doublings—far beyond the Hayflick limit.

So is telomerase the elixir of youth? Will taking a swig of the delicious new beverage Telomerade (patent *not* pending) rejuvenate our cells, and let us watch in a mirror as the wrinkles on our faces flatten out again to the smoothness of a baby's posterior?

Unfortunately it is not that simple. Some manifestations of aging do appear to be related to repeatedly dividing cells (like fibroblasts) losing enough of their telomeres to end their replicative life and becoming senescent. As we will see in Chap. 13, this is also an issue with the "stem cells" present in many tissues that can, when needed, differentiate into new mature cells of a particular type. However, this is not the case with many other important types of cells. The highly differentiated cells of many body organs either have a limited ability to replicate (e.g., hepatocytes in the liver), or do not appear to replicate at all after they have differentiated into their final form (for example, myocardiocytes in the heart or neurons in the brain). Their telomeres can remain long in the aged—and telomerase would seem to have little place for maintaining or restoring their "youthfulness." In addition, telomere dysfunction may be triggered by other stresses besides shortening alone [94].

Also, there may be a good reason why normal human somatic cells not only express little if any telomerase activity, but many cells actively suppress it. On balance, it might be "healthier" for the body as a whole for these aging cells to let their telomeres shorten and to become senescent rather than expose them to telomerase.

Cells with shortened telomeres that have stopped dividing at the Hayflick limit (also known as Mortality stage 1, or M1) and become senescent can be induced to divide again under certain conditions. Senescent cells may contain increased levels of proteins like p53 and the pRb that are essential for establishing and maintaining senescence [10, 14]. When genes coding for these proteins are suppressed (e.g. from exposure to certain viruses), these cells can divide again (with more shortening of their telomeres) another 10–20 times until stopping at another stage—Mortality stage 2, M2. As its name implies, nearly all cells reaching that stage die, due to development of genetic mutations that render the cell nonviable.

However, a tiny fraction of cells develop mutations that allow them to keep dividing beyond M2, by activating telomerase or via other mechanisms to restore telomere length and structure. These mutated cells thus become "immortal," potentially capable of replicating indefinitely. Their growth and function are no longer controlled by the body's normal regulatory systems, and they may even spread to distant sites within a person. Their numbers can increase to the point that they interfere with the activities of or even kill nearby "normal" cells. Given enough time those "immortal" cells can even bring about their own demise by causing the death of the body they live inside [88, 89, 92, 95].

The medical term for what those mutated cells do is "cancer."

Overall, limiting the number of times cells like fibroblasts can divide may be a compromise between maintaining "youthful" characteristics in a tissue or organ as long as possible and the risk of developing cancer. Every time a cell divides there is a chance that one of its "daughters" will have a single mistake or multiple critical errors in its duplicated DNA that will let it mutate into a cancer-causing (malignant) cell. The odds of this happening increase in proportion to how many times a particular cell and its progeny have divided—especially when they approach or (as described above) bypass M1—as well as the total number of dividing cells in a particular tissue.

Thus, even if telomeres are restored to a cell many generations removed from the "original," that genetically "old" cell or another produced a few more divisions down the road may have already accumulated or possibly soon will accumulate enough mutations to become cancerous. Making telomeres long again does nothing in itself to repair any genetic damage already within the cell's DNA. Therefore, the short-term beneficial effects of having normal somatic cells produce telomerase might be more than counterbalanced by its potential to both create a cancerous cell and facilitate its subsequent growth and spread of the resulting cancer.

Cells may become senescent for reasons other than telomere shortening, including detecting damage and mutations in DNA, inadequate nutrients or growth factors, and increased levels of damaging free radicals [10, 14, 87]. While doing this may be protective against cancer, it has its own bad effects. Besides no longer performing their usual functions, senescent cells may adversely alter the microenvironment in their vicinity, reducing the function of nearby cells that are not senescent and inducing a harmful state of low-level chronic inflammation [7, 14, 33, 38]. Finding a way to remove such senescent cells might delay some effects of aging. However, this possibility remains speculative along with any current way to actually do it, although methods such as developing drugs that selectively destroy senescent cells while leaving normal ones unharmed have been suggested [10].

Another approach might be to delay development of senescent cells by interventions such as a healthy diet and regular exercise, thus (to some degree) reducing the effects of aging. Previous studies have indicated that telomeres shorter than normal were associated with increased risk for cardiovascular disease, and they can also be shorter than normal in tobacco users [88, 96, 97]. Shorter telomeres have also been suggested as a risk factor for obesity, and any dietary or other interventions that reduce the odds of developing it might be especially important in individuals with that finding [98, 99].

The *p53* gene is also an important influence on cell aging. It is a powerful tumor suppressor, inhibiting cells with potentially damaged DNA from dividing. The effects of *p53* on cellular senescence may be less clear. By causing arrest of the cell cycle when it detects DNA damage or other issues within a cell, *p53* may either indirectly or directly contribute to the cell becoming senescent [43, 87, 92]. The *p53* gene can in some cases repair cellular DNA and in others help induce senescence, apoptosis (induction of cell death if the cell is beyond repair), or quiescence (a temporary state of reduced activity and cell division arrest short of permanent senescence) in a damaged cell [37, 52, 87, 100–103].

In short, Nature may have "decided" that letting our cells "age" was the lesser of two evils compared to the greater risk of developing cancer, via the actions of our cells' *p53* tumor suppressor system and others [29, 82, 101, 104]. Although we still eventually die, on average that may take longer to happen if our cells become senescent rather than being allowed to divide more and accumulate enough mutations to end our lives sooner as wrinkle-free but cancer-ridden corpses.

Instead, at least for the near future, far from using it to keep us young or even rejuvenate us the major role for telomerase for increasing lifespan may be via drugs that inhibit it. About 85 to 95 % of common human cancers (including those of the prostate, breast, colon, and lung) show telomerase activity. Higher levels may correlate with worse prognosis. Thus, theoretically telomerase inhibitors would act predominantly on malignant cells—removing their ability to continuously repair their telomeres and making them "mortal" again. As with normal cells, after a limited number of additional divisions the telomeres of the cancerous ones would shorten to the point where they would become senescent and no longer multiply.

Unfortunately, there are also potential problems with using telomerase inhibitors for cancer therapy. The few (but important!) normal cells that in particular express telomerase—egg cells, spermatozoa, and somatic cells like epithelial cells, stem cells, and lymphocytes—could also be adversely affected by these drugs. Also, by the time a cancer is detected it may be too advanced for a telomerase inhibitor to successfully halt its growth. For example, if the drug could stop cancer cells from dividing after 20 more divisions, but it takes only 10 more divisions for the tumor to grow large enough to kill a person, it would be "too little, too late."

It is possible that, as we gain more knowledge of the potential benefits and risks of telomerase for "rejuvenating" cells, we might be able to maximize its beneficial effects and minimize its risks. Perhaps combining a telomerase "treatment" with techniques for repairing DNA damage in "old" cells might safely restore them to functional "youth" without causing cancer. In one study, mice that were genetically modified for increased cancer resistance and to express telomerase had a 43% increase in lifespan compared to controls. In another study genetically induced re-expression of telomerase in mice was associated with mild extensions of lifespan, better fitness, and no significantly increased risk of cancer [105–107]. However, we are far from knowing whether such methods could be used in a practical, effective, and safe way in humans.

Finally, at birth telomere lengths in white blood cells are the same regardless of gender. However, at later ages males tend to have shorter telomeres than age-matched females, suggesting that it may be a causal factor or at least a marker for other processes that contribute to the shorter average lifespan of males versus females [108].

Aging is associated with other effects beyond wear-and-term, DNA damage, telomere shortening, etc. It has also been suggested that changes in the hypothalamus, a part of our brain that (among other things) helps control release of important pituitary gland hormones such as GH, might have a significant role in aging and be a potential target for measures to increase lifespan [1].

Epigenetic factors—those that regulate how genes act, without changing the DNA sequence of the gene itself—may also play important roles [109]. One such factor, DNA methylation, involves attachment of a methyl group (one carbon atom with three of hydrogen) to cytosine in DNA [5, 110]. This usually causes a gene to be "silenced"—that is, there is reduced or no transcription of the gene by RNA to produce a protein or other product [36]. The decreased activity ("expression") by particular genes that is produced by DNA methylation correlates with and may contribute to age-related diseases such as osteoarthritis, cancer, and ones involving degeneration of the nervous system. Changes in DNA methylation are associated with age, and its presence has been suggested as a "biological clock" to estimate the age of tissues [111]. However, there is said to be no reported evidence so far that human lifespan can be extended by actually altering patterns of DNA methylation [33].

Recent studies also suggest that genetic contributions to longevity can arise not only from one's parents but from previous generations of ancestors too—what has been called "transgenerational inheritance." A report on *C. elegans* found that roundworms with a particular mutation in the *H3K4me3* complex that increases their lifespans produced progeny that were similarly long-lived despite not having that mutation [112, 113]. How this trait is actually passed on is unclear but may involve epigenetic changes. Some human studies also suggest transgenerational inheritance—for example, the observation that nutritional and smoking habits in paternal grandparents could influence a person's own lifespan [113].

9.6 The Future of Aging

Modern medicine can significantly improve life expectancy. In both young and old, improving techniques for diagnosing and treating potentially fatal health problems can prolong health as well as lengthen survival in many illnesses. However, it is estimated that, even if the three major causes of death in the elderly—cardiovascular disease, cancer, and stroke—were eliminated, average life expectancy would increase by only about 15 years [4]. That is a good improvement, but well short of "immortality."

Based on our current knowledge, it is possible there may be intrinsic limitations to human biology that, despite our best efforts, renders living well past the known maximum (so far) of about 122 years in a reasonably healthy state inordinately expensive, difficult, or even (for all practical purposes) impossible. While the fact that (as far as we know) no person has succeeded in living 150 years does not prove it cannot be done, it does suggest that living that long is not an inherent capability of our bodies. Perhaps, given enough time, too much damage will occur in enough cells that even the best DNA repair proteins and other treatments will not be able to handle them all. How long it would take for this to happen on average for an individual or in a large population cannot be realistically predicted now. It may be possible that, despite great cost and effort, we may be able to add only a few years to that 122-plus maximum.

Aging is also not a uniform process in all types of our cells and organs. Failure in only one of a number of critical systems can kill us or drastically reduce our quality of life. It would be of little value if, by using medicines, surgery, or more advanced methods we could routinely maintain healthy hearts or livers for 200 years, but keep our brains and nervous system from deteriorating for only about 100.

Unfortunately cognitive decline associated with aging likely begins in the late 40s, particularly involving the ability to recall facts and experiences [114, 115]. Although neurons are lost with aging, particularly in specific regions such as the hippocampus and neocortical areas, much age-related cognitive decline may involve loss of synaptic connections instead [114]. By one estimate, even in the absence of concomitant diseases, by an age of around 120 the number of neocortical synapses could fall to the level seen in Alzheimer's disease, representing a 40 % loss of synaptic connections [84].

However, the human brain does have a degree of "neuroplasticity"—the ability to reorganize itself and its functions after suffering injury, neuronal losses, or other insults. The term "cognitive reserve" has been used to describe factors such as intelligence, education, and mental stimulations that help the brain adapt to such changes and maintain its function [84, 114].

Nonetheless, although it can be delayed to some degree, deterioration of brain function is part of the spectrum of "normal" aging. Permanent damage to the brain can also occur suddenly and catastrophically due to a severe head injury as well as a cerebrovascular event ("stroke") or anything else that deprives irreplaceable cerebral tissue of the nutrients, oxygen, and other substances it needs to function. Parts of the brain, such as those involved with strictly motor and sensory functions, could potentially be repaired by sufficiently advanced techniques using stem cells, nanotechnology, etc. However, if enough neurons are lost in the parts of our brain that "code" for our most individual traits—our unique personalities, memories, intellectual abilities, and so on—even if those neurons are repaired or replaced part or all of what is needed to make each of us an "I" would be irretrievably lost.

Bruce Sterling's novel *Holy Fire* (1996) presents a late-twenty-first century society beginning to go beyond advanced life-extension techniques to actually rejuvenate an entire aged body. Its very detailed description of that latter process gives a good flavor of how complex and thorough that process would have to be, as well as its psychological effects. The character who undergoes that cutting-edge procedure at a still reasonably healthy age of 94 must submit to many weeks of treatment while unconscious. The latter includes sterilizing her body both externally and internally, removing and re-growing all her skin, finding and repairing all the accumulated DNA damage in her cells, restoring telomeres in appropriate cells, etc. The relevant passages do an excellent job of showing how much and how long a person might have to "drink" from the "Fountain of Youth."

As detailed as the description of the rejuvenation process is, however, far more than it portrays would actually need to be done to truly make someone anatomically and physiologically "young" again. For example, the effects of aging in many parts of the body include replacement of one type of tissue for another, e.g. "stiffer" fibrous tissue for more elastic types in blood vessels, heart valves, the heart and other muscles, etc. As noted previously cell functions also change with age independently of changes in DNA alone, involving other alterations within cells and organelles (e.g. the mitochondria). As will be discussed in Chap. 13, pools of stem cells in many tissues that are necessary for repair functions inside the body would also have to be regenerated. And as Chaps. 11 and 12 will describe, methods to identify abnormal cells inside a person's body—much less repair them—are still in their infancy, particularly at the vast scale of an adult's 37 trillion-plus cells with all their myriad types.

However, to "work" in a science fiction story, a description like the one in that novel does not have to be comprehensive nor take every potential difficulty into account. It is enough to give the reader a general taste based on salient biological details about what would be involved with "rejuvenation"—and the author does this well.

9.7 The Bottom Line

Science fiction has the luxury of being able to ignore the biological difficulties involved with markedly extending healthy human lifespan. It can simply assume the success of what are now still strictly hypothetical and highly uncertain technological or other means for making us live much longer. By glossing over (or perhaps even contradicting) the science involved, science fiction can skip right to the "fiction" and explore the ramifications of what "immortality" or "eternal youth" might mean to individuals and societies.

And those ramifications may be good, bad, or ambiguous—well worth exploring as a "thought experiment" within a fictional work. For example, "The Trade-Ins" episode (1962) of *The Twilight Zone* presents the issues an elderly couple must confront when deciding whether or not to trade their old bodies in for a pair of new, young ones.

My story "The Best is Yet to Be" explores reasons why another married couple in their mid-70s might—or might not—decide to accept a rejuvenation treatment that would restore their bodies to what they biologically were in their 20s.[10] Knowing the difficulties described previously involved with actually doing that, I decided to keep my description of that process very general, skip the technical details, and concentrate on the human drama.[11] The wife in my story is suffering from depression—an all too common problem in the elderly. She ruminates about a family tragedy that occurred many years ago—the death of a child. The husband feels guilty because he had wanted to have that child when his wife was uncertain she was. The story depicts the couple's difficulties dealing with a shared past, their uncertainties about whether it would be worth resetting their biological clocks to face a much longer future, and how their decisions play out.

Such stories illustrate a common conundrum regarding extending human lifespan for centuries or more. The obvious benefit of doing that is, you don't die. The equally obvious problem with it is—*you don't die*. Whether expressed

---

[10] *Analog Science Fiction and Fact* December 1996.
[11] Of course, if I ever really developed a treatment for restoring biological youth I would look much younger and spend much of my time admiring the medal I acquired in Stockholm.

poetically as the "seven ages of man" in Shakespeare's play *As You Like It* or by other means, humans go through a limited gamut of developmental stages and activities during their finite lives. Infancy, childhood, puberty, active adulthood, reproduction, age-related physical decline, and death are, in the most general terms, what we humans can do within lifetimes of a century plus or (far more likely) minus some decades. Increase that span to many centuries, and the basic problem is, how would one spend all that extra time?

Science fiction might answer it by presenting characters that use their longevity to explore the Galaxy or acquire power for good or evil purposes. Those individuals might be able to formulate long-term plans that could take many normal lifetimes to come to fruition. On the other hand, in the "real world" some individuals might have no wish to do any of those things. "Immortality" might mean merely doing the same mundane things like eating, sleeping, enjoying entertainment of various types, etc. over … and over … and over again. Instead of adults facing typical challenges such as raising children or working at jobs for a matter of decades and then "winding down" to at least some degree in one's later years, such activities along with both the enjoyment and stresses associated with them might continue for centuries.

Eventually an individual might reach the psychological point of asking what is the point of doing the same things (with minor variations) again and again. Likewise maintaining interpersonal relationships with family members, friends, spouses, or other strong personal ties would be difficult if not impossible, particularly for an "immortal" in a world where those other individuals were either mortal or died due to accidents and other causes. Immortality could prove to be a very lonely "gift," with new relationships never completely erasing the loss of old ones.

And during all those additional years the world at large will not be standing still. In some ways living far into the future might prove something more to be feared than welcomed. Overpopulation, global warming, depletion of natural resources—these are insidious problems. Their impact is not felt in a day or week, but gradually, cumulatively, over years, with the occasional crisis to force our attention temporarily to them. They tend to be ignored or addressed sluggishly because today and tomorrow will be pretty much the same whether society does anything about them or not. Perhaps the world will run out of petroleum by 2100. But as long as one can fill his or her vehicle's gas tank today, it may not seem that important. It might be easy to take the attitude, even unconsciously, that because such issues probably will not become critical until one has died of old age, they can be left to later generations to solve. Except—if a person is "immortal", that individual *will* live long enough to be affected by such issues. And if they are not addressed adequately by society, he or she will have to live with the resulting economic and social turmoil.

Living for centuries will also mean experiencing more history. In some times and places, such as ancient Egypt or China before the twentieth century, societies and cultures might stay reasonably stable for hundreds of years. But far more commonly, periods of relative social calm are measured in decades —a fraction of an average lifetime. The greatest political power of classical Athens lasted barely 76 years, from the battle of Salamis in 480 BC to crushing defeat by the Spartans in 404 BC. The Roman Empire prospered under its longest series of "good" emperors, from Nerva to Marcus Aurelius, for only 84 years before starting to fragment under the pressure of internal and external conflicts. Such events as the Mongol conquests in China, the Spanish conquistadors' encounters with the native population in the "New World," the French Revolution and subsequent Napoleonic wars, the English and American Civil Wars, etc. all produced revolutionary changes in cultures, untold social upheavals, and tremendous human misery. Looked at retrospectively, some good also resulted from such monumental events. But the people who lived through them in "real time" had no assurance of any such good— and millions did not survive the experience at all.

Will "immortal" individuals be able to adapt as the world of their original youth changes either subtly or violently? For example, take a baby boy born into a prosperous middle-class family in Berlin c. 1898. He grows up happy, carefree, and well fed in a stable society at peace. Move ahead to late 1918—and that young man is now a battle-hardened soldier huddling in the final trenches of a disastrous global war. Another 5 years and he is back in Berlin, penniless and hungry in a chaotic nation with an economy shattered by hyperinflation. Ten more and his nation has a new political system that within less than another decade would launch a second war and bring most of Europe under its sway. Suddenly it is June 1945 and a now middle-aged man comes back to a Berlin in ruins, with many more decades of simmering conflict ahead for that city.

Such swings of fortune and misfortune, tragedy and triumph can occur within the bounds of a single "standard" lifetime. Now imagine living through centuries of such events—and immortality may appear less appealing. Assuming one survives such experiences, the old curse "May you live in interesting times" will include many more such times than a single person would ever have had to endure before.

Science fiction stories may include the idea that only one individual or a small group of them can be immortal, such as in the *Star Trek* episode "Requiem for Methuselah" (1969) or the Howard Families. In others extended longevity is possible for everyone but may have prohibitive costs or other limitations that restrict its use, such as the limited supply of human organs for transplantation described in some of Larry Niven's "Known Space" works. Or whole societies may consist of "immortals," as in Damon Knight's "World Without Children" (1970).

Having billions of people become immortal also leads to questions such as how to deal with overpopulation and risk taking. In *An Essay on the Principle of Population* (1798) Thomas Malthus described potential problems with a large enough population outstripping its food supply and economic issues such as too large a labor force for the work available. In science fiction, with members of whole societies living much longer than in the past, "solutions" include prohibiting them from having children (except to make up for any losses due to deaths such as from accidents) or rendering them (either voluntarily or involuntarily) sterile. The absence or rarity of children in such a society would indeed be expected to have profound impacts on what its members do for good or ill.

Overpopulation issues could also be addressed by having long-lived humans travel to the stars. Besides issues of how many people could realistically leave Earth, this intrinsically risky action raises the further question of how many humans would be willing to do that. Losing one's life while doing something dangerous like exploring alien worlds is different if one has only an additional 4 decades or so of potential life to lose as opposed to centuries. In some individuals "immortality" might increase the drive for survival over having new but potentially deadly experiences. It could potentially increase creativity and progress as individuals acquired more experience over a longer lifetime. Or it might introduce an element of stagnation with the loss of new, young generations with novel perspectives on issues, fresh ways of looking at the arts and sciences, etc.

In addition, the natural "turnover" of jobs as well as positions of power and authority with the passing of each generation would either no longer apply or have to be "artificially" enforced. Otherwise there could be conflict between "younger immortals" and those older but still youthful ones entrenched in business, government, etc. for decades or even centuries with no personal need to step down due to age or simply through death. Likewise, having an Einstein or Mozart live in good health for a century or two would benefit the world. On the other hand, history could have been far worse if Hitler or Stalin had lived that long.

Finally, no matter how long humans might eventually live, death is inevitable. A few more centuries or millennia of active healthy living, while certainly an improvement on current limits, is less than an instant compared to the age of the Universe. Even such difficult-if-not-impossible science fiction concepts as copying one's mind for transfer to new bodies or keeping it in an artificial medium would only delay the inevitable, even if it meant stretching "existence" for thousands or millions of years. In science fiction and in real life, a principle written by the first-century AD Stoic philosopher Lucius Annaeus Seneca is still worthy of consideration: "Men do not care how nobly they live, but only how long, although it is within the reach of every man to live nobly, but within no man's power to live long." If he were alive today due to an immortality treatment or brought here via a time machine, Seneca would likely add that this idea still applies no matter how "long" long is.

References

1. Zhang G, Li J, Purkayastha S, Tang Y, Zhang H, Yin Y, et al. Hypothalamic programming of systemic ageing involving IKK-beta, NF-kappaB and GnRH. Nature. 2013;497(7448):211–6.
2. Kazachkova N, Ramos A, Santos C, Lima M. Mitochondrial DNA damage patterns and aging: revising the evidences for humans and mice. Aging Dis. 2013;4(6):337–50.
3. Nemoto S, Finkel T. Ageing and the mystery at Arles. Nature. 2004;429:149–52.
4. Hayflick L. The future of ageing. Nature. 2000;408:267–9.
5. Brooks-Wilson AR. Genetics of healthy aging and longevity. Hum Genet. 2013;132(12):1323–38.
6. Bianconi E, Piovesan A, Facchin F, Beraudi A, Casadei R, Frabetti F, et al. An estimation of the number of cells in the human body. Ann Hum Biol. 2013;40(6):463–71.
7. Boraschi D, Aguado M, Dutel C, Goronzy J, Louis J, Grubeck-Loebenstein B, et al. The gracefully aging immune system. Sci Transl Med. 2013;5(185):185ps8.
8. Montecino-Rodriguez E, Berent-Maoz B, Dorshkind K. Causes, consequences, and reversal of immune system aging. J Clin Invest. 2013;123(3):958–65.
9. Niccoli T, Partridge L. Ageing as a risk factor for disease. Curr Biol. 2012; 22(17): R741–52.
10. Tchkonia T, Zhu Y, van Deursen J, Campisi J, Kirkland JL. Cellular senescence and the senescent secretory phenotype: therapeutic opportunities. J Clin Invest. 2013;123(3):966–72.
11. Costa I, Carvalho H, Fernandes L. Aging, circardian rhythms and depressive disorders: a review. Am J Neurodegner Dis. 2013;2(4):228–46.
12. Deweerdt S. Looking for a master switch. Nature. 2012;492:S10–S1.
13. Rare Whales Can Live to Nearly 200, Eye Tissue Reveals. 2006. http://news.nationalgeographic.com/news/2006/07/060713-whale-eyes.html. Accessed 15 April 2015.
14. Ohtani N, Hara E. Roles and mechanisms of cellular senescence in regulation of tissue homeostasis. Cancer Sci. 2013;104(5):525–30.

15. Finkel T, Holbrook N. Oxidants, oxidative stress and the biology of ageing. Nature. 2000;408:239–47.
16. Jones JH. Primates and the evolution of long, slow life histories. Curr Biol. 2011;21(18):R708–17.
17. Baudisch A, Vaupel JW. Evolution. Getting to the root of aging. Science. 2012;338(6107):618–9.
18. Boehm A, Khalturina K, Anton-Erxlebena F, Hemmricha G, Klostermeierb U, Lopez-Quinteroa J, et al. FoxO is a critical regulator of stem cell maintenance in immortal Hydra. Proc Natl Acad Sci. 2012;109(48):19697–702.
19. Nebel A, Bosch T. Evolution of human longevity: lessons from Hydra. Aging. 2012;4(11):730–1.
20. Reddien PW. Specialized progenitors and regeneration. Development. 2013; 140(5):951–7.
21. Scully T. To the limit. Nature. 2012;492:S2–S3.
22. Life Expectancy Data by Country. http://apps.who.int/gho/data/node.main. 688?lang=En-US. Accessed 15 April 2015.
23. Vaupel JW. Biodemography of human ageing. Nature. 2010;464(7288):536–42.
24. Chung W, Dao R, Chen L, Hung S. The role of genetic variants in human longevity. Ageing Res Rev. 2010;9:S67–S78.
25. Eisenstein M. Great expectations. Nature. 2012;492:S6–S8.
26. Murabito JM, Yuan R, Lunetta KL. The search for longevity and healthy aging genes: insights from epidemiological studies and samples of long-lived individuals. J Gerontol A Biol Sci Med Sci. 2012;67(5):470–9.
27. Kenyon CJ. The genetics of ageing. Nature. 2010;464(7288):504–12.
28. Moskalev AA, Smit-McBride Z, Shaposhnikov MV, Plyusnina EN, Zhavoronkov A, Budovsky A, et al. Gadd45 proteins: relevance to aging, longevity and age-related pathologies. Ageing Res Rev. 2012;11(1):51–66.
29. Finkel T, Serrano M, Blasco MA. The common biology of cancer and ageing. Nature. 2007;448(7155):767–74.
30. Rera M, Azizi MJ, Walker DW. Organ-specific mediation of lifespan extension: more than a gut feeling? Ageing Res Rev. 2013;12(1):436–44.
31. Tissenbaum HA. Genetics, lifespan, health span, and the aging process in Caenorhabditis elegans. J Gerontol A Biol Sci Med Sci. 2012;67(5):503–10.
32. Hansen M, Flatt T, Aguilaniu H. Reproduction, fat metabolism, and lifespan: what is the connection? Cell Metab. 2013;17(1):10–9.
33. Lopez-Otin C, Blasco MA, Partridge L, Serrano M, Kroemer G. The hallmarks of aging. Cell. 2013;153(6):1194–217.
34. Ayyadevara S, Tazearslan Ç, Bharill P, Alla R, Siegel E, Shmookler Reis R. Caenorhabditis elegans PI3K mutants reveal novel genes underlying exceptional stress resistance and lifespan. Aging Cell. 2009;8(6):706–25.

35. Hekimi S, Guarente L. Genetics and the specificity of the aging process. Science. 2003;299(5611):1351–4.

36. Guarente L, Kenyon C. Genetic pathways that regulate ageing in model organisms. Nature. 2000;408:255–62.

37. McCubrey J, Demidenko Z. Recent discoveries in the cycling, growing and aging of the p53 field. Aging. 2012;4(12):887–93.

38. Gems D, Partridge L. Genetics of longevity in model organisms: debates and paradigm shifts. Annu Rev Physiol. 2013;75:621–44.

39. Bourzac K. Live long and prosper. Nature. 2012;492:S18–S20.

40. Lapierre LR, Hansen M. Lessons from C. elegans: signaling pathways for longevity. Trends Endocrinol Metab. 2012;23(12):637–44.

41. Bartke A. Healthy aging: is smaller better?—a mini-review. Gerontology. 2012;58(4):337–43.

42. Barzilai N, Huffman DM, Muzumdar RH, Bartke A. The critical role of metabolic pathways in aging. Diabetes. 2012;61(6):1315–22.

43. Blagosklonny MV. Revisiting the antagonistic pleiotropy theory of aging: TOR-driven program and quasi-program. Cell Cycle. 2010;9(16):3151–6.

44. Harrison DE, Strong R, Sharp ZD, Nelson JF, Astle CM, Flurkey K, et al. Rapamycin fed late in life extends lifespan in genetically heterogeneous mice. Nature. 2009;460(7253):392–5.

45. Hughes KJ, Kennedy BK. Cell biology. Rapamycin paradox resolved. Science. 2012;335(6076):1578–9.

46. Lamming DW, Ye L, Katajisto P, Goncalves MD, Saitoh M, Stevens DM, et al. Rapamycin-induced insulin resistance is mediated by mTORC2 loss and uncoupled from longevity. Science. 2012;335(6076):1638–43.

47. Lamming DW, Ye L, Sabatini DM, Baur JA. Rapalogs and mTOR inhibitors as anti-aging therapeutics. J Clin Invest. 2013;123(3):980–9.

48. Smith K. A pill for longer life? 2009. http://www.nature.com/news/2009/090708/full/news.2009.648.html. Accessed 15 April 2015.

49. Kaeberlein M, Kapahi P. Cell signaling. Aging is RSKy business. Science. 2009;326(5949):55–6.

50. Merksamer P, Liu Y, He W, Hirschey M, Chen D, Verdin E. The sirtuins, oxidative stress and aging: an emerging link. Aging. 2013;5(3):144–50.

51. Abdellatif M. Sirtuins and pyridine nucleotides. Circ Res. 2012;111(5):642–56.

52. Berman A, Leontieva O, Natarajan V, McCubrey J, Demidenko Z, Nikiforov M. Recent progress in genetics of aging, senescence and longevity: focusing on cancer-related genes. Oncotarget. 2012;3(12):1522–32.

53. Poulsen MM, Jorgensen JO, Jessen N, Richelsen B, Pedersen SB. Resveratrol in metabolic health: an overview of the current evidence and perspectives. Ann N Y Acad Sci. 2013;1290:74–82.

54. Sebastian C, Satterstrom FK, Haigis MC, Mostoslavsky R. From sirtuin biology to human diseases: an update. J Biol Chem. 2012;287(51):42444–52.

55. Burnett C, Valentini S, Cabreiro F, Goss M, Somogyvari M, Piper MD, et al. Absence of effects of Sir2 overexpression on lifespan in C. elegans and Drosophila. Nature. 2011;477(7365):482–5.

56. Marchal J, Pifferi F, Aujard F. Resveratrol in mammals: effects on aging biomarkers, age-related diseases, and lifespan. Ann N Y Acad Sci. 2013;1290:67–73.

57. Partridge L, Gems D. Benchmarks for ageing studies. Nature. 2007;450:165–7.

58. Steves CJ, Spector TD, Jackson SH. Ageing, genes, environment and epigenetics: what twin studies tell us now, and in the future. Age Ageing. 2012;41(5):581–6.

59. Coppede F. The epidemiology of premature aging and associated comorbidities. Clin Interv Aging. 2013;8:1023–32.

60. Martin G, Oshima J. Lessons from human progeroid syndromes. Nature. 2000;408:263–6.

61. Bonsall MB. Longevity and ageing: appraising the evolutionary consequences of growing old. Philos Trans R Soc Lond B Biol Sci. 2006;361(1465):119–35.

62. Bratic A, Larsson NG. The role of mitochondria in aging. J Clin Invest. 2013;123(3):951–7.

63. Calabrese E, Iavicoli I, Calabrese V. Hormesis: why it is important to biogerontologists. Biogerontology. 2012;13(3):215–35.

64. Zhang Y, Tang ZH, Ren Z, Qu SL, Liu MH, Liu LS, et al. Hydrogen sulfide, the next potent preventive and therapeutic agent in aging and age-associated diseases. Mol Cell Biol. 2013;33(6):1104–13.

65. Kirkwood T, Austad S. Why do we age? Nature. 2000;408:233–8.

66. Rando TA. Stem cells, ageing and the quest for immortality. Nature. 2006;441(7097):1080–6.

67. Olshansky S, Passaro D, Hershow R, Layden J, Carnes B, Brody J, et al. A potential decline in life expectancy in the United States in the 21st century. N Engl J Med. 2005;352(11):1138–45.

68. Gaziano J. Global burden of cardiovascular disease. In: Libby P, Bonow R, Mann D, Zipes D, Braunwald E, eds. Braunwald's heart disease. 8th ed. United States: Saunders Elsevier; 2008. pp. 1–22.

69. Chan JS, Yan JH, Payne VG. The impact of obesity and exercise on cognitive aging. Front Aging Neurosci. 2013;5:97.

70. Older Persons' Health: Health Care Utilization. 2014. http://www.cdc.gov/nchs/fastats/older-american-health.htm. Accessed 15 April 2015.

71. The National Nursing Home Survey: 2004 Overview. 2004. http://www.cdc.gov/nchs/data/series/sr_13/sr13_167.pdf. Accessed 15 April 2015.

72. Morley J. Scientific overview of hormone treatment used for rejuvenation. Fertil Steril. 2013;99(7):1807–13.

73. Wade A, Ford I, Crawford G, McConnachie A, Nir T, Laudon M, et al. Nightly treatment of primary insomnia with prolonged release melatonin for 6 months: a randomized placebo controlled trial on age and endogenous melatonin as predictors of efficacy and safety. BMC Med. 2010;8:51.

74. Martano G, Bojaxhi E, Forsteniehner C, Huber N, Bresgen P, Eckl H. Validation and application of sub-2 micrometer core-shell UHPLC-UV-ESI-Orbitrap MS for identification and quantification of beta-carotene and selected cleavage products with preceding solid-phase extraction. Anal Bioanal Chem. 2014;406(12):2909–4.

75. Goodman G, Thornquist M, Balmes J, Cullen M, Meyskens F, Omenn G, et al. The beta-carotene and retinol efficacy trial: incidence of lung cancer and cardiovascular disease mortality during 6-year follow-up after stopping beta-carotene and retinol supplements. J Natl Cancer Inst. 2004;96(23):1743–50.

76. Pruthi S, Allison T, Hensrud D. Vitamin E supplementation in the prevention of coronary artery disease. Mayo Clin Proc. 2001;76:1131–6.

77. Swindell WR. Dietary restriction in rats and mice: a meta-analysis and review of the evidence for genotype-dependent effects on lifespan. Ageing Res Rev. 2012;11(2):254–70.

78. Colman RJ, Anderson RM, Johnson SC, Kastman EK, Kosmatka KJ, Beasley TM, et al. Caloric restriction delays disease onset and mortality in rhesus monkeys. Science. 2009;325(5937):201–4.

79. Mattison JA, Roth GS, Beasley TM, Tilmont EM, Handy AM, Herbert RL, et al. Impact of caloric restriction on health and survival in rhesus monkeys from the NIA study. Nature. 2012;489(7415):318–21.

80. Redman L, Ravussin E. Caloric restriction in humans: impact on physiological, psychological, and behavioral outcomes. Antioxid Redox Signal. 2011;14:275–87.

81. Trepanowski JF, Canale RE, Marshall KE, Kabir MM, Bloomer RJ. Impact of caloric and dietary restriction regimens on markers of health and longevity in humans and animals: a summary of available findings. Nutr J. 2011;10:107.

82. Vijg J, Campisi J. Puzzles, promises and a cure for ageing. Nature. 2008;454 (7208):1065–71.

83. Stewart TM, Bhapkar M, Das S, Galan K, Martin CK, McAdams L, et al. Comprehensive assessment of long-term effects of reducing intake of energy phase 2 (CALERIE Phase 2) screening and recruitment: methods and results. Contemp Clin Trials. 2013;34(1):10–20.

84. Mora F. Successful brain aging: plasticitiy, environmental enrichment, and lifestyle. Dialogues Clin Neurosci. 2013;15(1):45–52.

85. Hughes V. Cultural differences. Nature. 2012;492:S14–S5.

86. Min K, Lee C, Park H. The lifespan of Korean eunuchs. Curr Biol. 2012;18:792–3.

87. Qian Y, Chen X. Senescence regulation by the p53 protein family. Methods Mol Biol. 2013;965:37–61.

88. Calado R, Young N. Telomere diseases. N Engl J Med. 2009; 361:2353–65.

89. DePinho R. The age of cancer. Nature. 2000;408:248–54.

90. Pollard T, Earnshaw W, Lippincott-Schwartz J. Chromosome organization. In: Pollard T, Earnshaw W, Lippincott-Schwartz J, editors. Cell Biology. 2nd ed. Philadelphia: Saunders Elsevier; 2008. Pp. 193–208.

91. Kong CM, Lee XW, Wang X. Telomere shortening in human diseases. FEBS J. 2013;280(14):3180–93.
92. Tumpel S, Rudolph KL. The role of telomere shortening in somatic stem cells and tissue aging: lessons from telomerase model systems. Ann N Y Acad Sci. 2012;1266:28–39.
93. Nandakumar J, Cech TR. Finding the end: recruitment of telomerase to telomeres. Nat Rev Mol Cell Biol. 2013;14(2):69–82.
94. Herbig U, Ferreira M, Condel L, Sedivy J. Cellular senescence in aging primates. Science. 2006;311:1257.
95. Aubert G, Lansdorp PM. Telomeres and aging. Physiol Rev. 2008;88(2):557–79.
96. Babizhayev M, Savel'yeva E, Moskvina S, Yegorov Y. Telomere length is a biomarker of cumulative oxidative stress, biologic age, and an independent predictor of survival and therapeutic treatment requirement associated with smoking behavior. Am J Ther. 2011;18:209–26.
97. Fyhrquist F, Saijonmaa O, Strandberg T. The roles of senescence and telomere shortening in cardiovascular disease. Nat Rev Cardiol. 2013;10(5):274–83.
98. Njajou OT, Cawthon RM, Blackburn EH, Harris TB, Li R, Sanders JL, et al. Shorter telomeres are associated with obesity and weight gain in the elderly. Int J Obes (Lond). 2012;36(9):1176–9.
99. Tzanetakou I, Katsilambros N, Benetos A, Mikhailidis D, Perrea D. "Is obesity linked to aging?": adipose tissue and the role of telomeres. Ageing Res Rev. 2012; 11:220–9.
100. Armanios M. Telomeres and age-related disease: how telomere biology informs clinical paradigms. J Clin Invest. 2013;123(3):996–1002.
101. Hasty P, Christy BA. p53 as an intervention target for cancer and aging. Pathobiol Aging Age Relat Dis. 2013;3. http://www.ncbi.nlm.nih.gov/pmc/articles/PMC3794078/. Accessed 15 April 2015.
102. Li W, Vijg J. Measuring genome instability in aging—a mini-review. Gerontology. 2012;58(2):129–38.
103. Rufini A, Tucci P, Celardo I, Melino G. Senescence and aging: the critical roles of p53. Oncogene. 2013;32:5129–43.
104. Strauss E. Cancer-stalling system accelerates aging. Science. 2002;295:28–9.
105. Boccardi V, Herbig U. Telomerase gene therapy: a novel approach to combat aging. EMBO Mol Med. 2012;4:685–7.
106. de Jesus BB, Blasco MA. Potential of telomerase activation in extending health span and longevity. Curr Opin Cell Biol. 2012;24(6):739–43.
107. de Jesus BB, Vera E, Schneeberger K, Tejera A, Ayuso E, Bosch F, et al. Telomerase gene therapy in adult and old mice delays aging and increases longevity without increasing cancer. EMBO Mol Med. 2012;4:691–704.
108. Barrett E, Richardson D. Sex differences in telomeres and lifespan. Aging Cell. 2011;10:913–21.
109. Ben-Avraham D, Muzumdar RH, Atzmon G. Epigenetic genome-wide association methylation in aging and longevity. Epigenomics. 2012;4(5):503–9.

110. Johnson AA, Akman K, Calimport SR, Wuttke D, Stolzing A, de Magalhaes JP. The role of DNA methylation in aging, rejuvenation, and age-related disease. Rejuvenation Res. 2012;15(5):483–94.

111. Horvath S. DNA methylation age of human tissues and cell types. Genome Biol. 2013;14:R115.

112. Greer EL, Maures TJ, Ucar D, Hauswirth AG, Mancini E, Lim JP, et al. Transgenerational epigenetic inheritance of longevity in Caenorhabditis elegans. Nature. 2011;479(7373):365–71.

113. Mango S. Generations of longevity. Nature. 2011;479:302–3.

114. Jellinger K, Attems J. Neuropathological approaches to cerebral aging and neuroplasticity. Dialogues in Clin Neurosci. 2013;15(1):29–43.

115. Sweatt JD. Neuroscience. Epigenetics and cognitive aging. Science. 2010;328 (5979):701–2.

10
Sex in Science Fiction

Love conquers all things; let us too surrender to Love.

Virgil
Eclogues

Captain, it has been my job to study this planet, and it has been most difficult, for it is unique. It is so unique that I can barely comprehend its facets. For instance, almost all life on the planet consists in species of two forms. There are no words to describe it, no concepts even. I can only speak of them as first form and second form. If I may use their sounds, the little one is called 'female,' and the big one, here, 'male,' so the creatures themselves are aware of the difference... And, Captain, in order to bring forth young, the two forms must cooperate.

Isaac Asimov
"What Is This Thing Called Love?" (1961)

This is the only chapter in the book that deals with a single human organ system's relationship to science fiction. Readers who have experienced the phase of development called "puberty" will presumably be able to think of reasons for this focus on the human reproductive system and its functions.

Many science fiction works involve sexual behavior as a primary or at least important plot element. H. G. Wells's novel *In the Days of the Comet* (1906) describes changes in sexual attitudes resulting from humanity's exposure to the gases from a passing green comet. Aldous Huxley's *Brave New World* (1932) is an early example of a future society in which sexual activity is officially divorced from reproduction. Many of Robert A. Heinlein's later works such as *Stranger in a Strange Land* (1961) and *I Will Fear No Evil* (1970) deal extensively with sex-related issues. Ursula K. Le Guin's *The Left Hand of Darkness* (1969) depicts a space-based human civilization whose members are, similarly to some fish (e.g. clownfish and angelfish) and other animals, "ambisexual" and able at times to assume either male or female reproductive attributes.

Organisms on Earth adopt various strategies for reproduction (sexual or otherwise), some of which are employed in science fiction by "modified" humans (e.g. as in *The Left Hand of Darkness*) or aliens. Other works describe reproductive strategies without exact terrestrial analogs. For example, Isaac Asimov's *The Gods Themselves* (1972) describes a race with three genders having distinct roles in reproduction. The alien Puppeteers in Larry Niven's and coauthors' "Known Space" works also have three genders, although the "female" one is not sentient and acts as a host body for offspring produced by the other two.

Science fiction works with aliens having reproductive cycles different from ours can also be used to comment on human sexual practices and attitudes. Asimov's short story "Playboy and the Slime God" (1961), later retitled as "What is This Thing Called Love?", describes a Galaxy filled with extraterrestrials who reproduce only by asexual budding, and an "experiment" conducted by such visiting aliens to try to figure out the mysterious alternate means by which a human male and female produce offspring. My own stories "Naked Came the Earthling" and "To Save Man" also satirize humanity's interest in sex from an extraterrestrial's point of view.[1]

Both in the "real" world and science fiction, gender and sexual behavior have myriad complex ramifications that go far beyond biology, with aspects involving culture, interpersonal relationships, art, politics, economics, etc. However, the focus here will be on sexual reproduction in its most fundamental biological aspects and its core function as the means humans and other organisms use to continue their species and evolve. In recent decades new reproductive technologies have expanded how offspring can be produced beyond the "classic" method. Science fiction can posit development of additional methods (e.g. artificial wombs) as well as sexual activity in extraterrestrial environments, human-alien couplings, etc. However, before getting into these more speculative topics it is best to start with basic concepts about sex and its mechanics.

10.1 Why Does Sex Exist?

The key biological function of sex in humans and other animals is reproduction. Individual members of any species inevitably die due to disease, accidents, predators, or old age. For a species to continue it must produce an

---

[1] These stories appeared in the July/August 2001 and July/August 2012 issues, respectively, of *Analog Science Fiction and Fact*.

adequate number of new, young individuals to eventually replace older ones lost through attrition [1].

Some simple animals such as sea anemones, sponges, and flatworms can reproduce either without sex (asexually) or by sexual reproduction [2]. Ones like earthworms, leeches, or certain types of fish (e.g. sea bass and clownfish) can either change during their life cycle from male to female or vice versa, or be both simultaneously (hermaphrodites).[2] However, mammals and nearly all other types of vertebrates (animals with a backbone) must permanently develop into one or the other gender for successful reproduction.

Unicellular organisms such as bacteria and other "prokaryotes" (see Chap. 1), all of which lack a membrane-bound nucleus, use only the asexual method. They replicate by "binary fission," with the original cell duplicating itself and splitting into two new identical ones. Bacteria can vary their genetic makeup— a critical effect in offspring produced by sexual reproduction—by incorporating foreign DNA into their own, spontaneous mutations, or by transferring genetic material between two individual bacteria via asexual processes. Members of another major group of prokaryotes, the Archaea, can also transfer DNA in both directions between two cells.

"Eukaryotes" are unicellular and multicellular organisms that have a membrane-bound nucleus in their cells. They include fungi, single-celled protozoans, and all types of multicellular invertebrate and vertebrate animals, including mammals like us. Some eukaryotes can reproduce asexually. Cut off the arm of certain varieties of starfish or split off part of specific kinds of flatworms (e.g. planarians) and a complete new individual can grow from the severed portion. This new animal is a true "clone," genetically identical to its parent. Asexual reproduction also works well for animals such as sea anemones that have limited mobility. Those creatures frequently attach themselves to a rock and stay there for a long time, reducing their chances of finding a mate. This problem is avoided by budding off a new individual. Asexual reproduction can also be used to produce many offspring in a short time.

However, asexual reproduction has a major disadvantage if environmental conditions change. Because offspring are genetically identical to the parent, a new disease or a stressful change in living conditions means they will all be vulnerable to it. One the other hand, sexual reproduction results in greatly increased "genetic diversity." Each offspring shares some, but not all, characteristics of its parents. It is a new combination of different genetic material obtained from each parent. In that case, if a new disease develops or the

---

[2] Although earthworms and other normally hermaphroditic animals possess the equivalent of both ovaries and testes, they typically do not self-fertilize but can assume either male or female functions in sexual reproduction with another member of their species.

environment changes significantly, based on their different, unique genetic makeup some of the species' offspring will be more vulnerable or likely to die than their parents would. Others will have about the same risk. But some progeny will, by chance, have genetic characteristics that make them more likely to survive. These latter individuals will be most likely to reproduce and pass on those same characteristics to their offspring. This process of "natural selection" increases the odds that a species as a whole will survive changes in its environment.

This major advantage of sexual reproduction compared to the asexual variety may help explain why nearly all multicellular eukaryotes employ sexual reproduction. Some eukaryotes, ranging in complexity from insects up to sharks and komodo dragons, can also either routinely or rarely reproduce by "parthenogenesis," in which an egg cell develops into a new organism without fertilization.

However, sexual reproduction is the only natural way that mammals reproduce. They consist of many more highly specialized cells than simpler animals like a sea anemone. The cells in muscle, nerves, and other parts of the body cannot on their own change to a form that could grow into a new mammal, nor do the egg cells of females spontaneously start to develop into new offspring without the aid of a male's sperm cells. As will be discussed later, some mammals can be cloned—a form of asexual reproduction. However, this requires artificial techniques that do not occur in the wild. Otherwise, with uncommon exceptions such as identical twins, every baby mammal is genetically unique.

Human beings are, of course, mammals. We too are subject to natural selection and share the same basic biological reason for having sex as other animals do. Like aardvarks and anteaters, bears and bison, *H. sapiens* must produce enough offspring for our species to survive. And also like other mammals, our brains and bodies are "programmed" with an instinct and drive to have sex.

With one major exception, every organ system in our bodies—e.g. the cardiovascular system's heart and blood vessels, the respiratory system's lungs, the gastrointestinal system's stomach and intestines—performs functions needed to keep us alive. That exception is the reproductive system. There is an absolute biological need for a person to breathe, eat, and eliminate waste to maintain life. In contrast, although humans typically consider it highly desirable, from a strictly biological standpoint sexual activity is optional for any given individual. Instead sex is essential only at the level of a species as a whole, requiring only that enough members successfully reproduce to ensure the species' survival—but not necessarily that any particular member participate in that activity.

In its natural form, reproduction in mammals requires performance of sexual intercourse ("coitus") by a male and female who are both fertile. In humans it is the final stage of a process that begins at conception when the chromosomal path for an individual's gender is set. A female's egg cell (an "oocyte" if it is immature and an "ovum" if mature) normally has a single X chromosome, while males produce sperm cells with either X or Y chromosomes. The union of an ovum and an X chromosome-containing sperm cell will thus produce a "zygote" with the sex chromosome combination XX. In the usual course of events the resulting offspring will develop female internal reproductive organs (e.g. two ovaries each connected by a fallopian tube to a uterus, aka a "womb"), as well as female genitalia and external sexual characteristics, such as breast enlargement and a characteristic distribution of pubic hair at puberty [1].

Conversely the fusion of an ovum with a Y chromosome-bearing sperm cell will result in an XY combination. In that case the expected results in offspring are formation of male reproductive organs, including the testes (where sperm are produced), penis, and various other associated structures such as the prostate gland, epididymis, vas deferens, and other parts associated with storage, transport, and production of semen, the final fluid containing sperm cells.

Development of female and male anatomic characteristics involves the effects of different hormones—primarily estrogen and progesterone in females, and testosterone in males. These act both before birth to produce the organs needed for later reproduction and in a more complex way at puberty, when organisms develop "mature" sexual characteristics and become fertile. Some parts of the male reproductive system can be thought of as analogs to ones in the female reproductive system. Thus, under the influence of testosterone, whose production in males is primarily associated with the Y chromosome, tissue that develops into ovaries and the clitoris in a female become testes and the penis, respectively, in a male.

A wide variety of genetic, chromosomal, hormonal, and other intrinsic as well as environmental factors can lead to an individual not developing along the path to fertile puberty. For example, an individual with Turner syndrome has only a single, X chromosome (XO) and a female phenotype (external appearance). However, this syndrome is also associated with lack of menstrual periods as well as infertility and sterility due to non-functional ovaries. Various other atypical chromosomal patterns that have been reported include XXX, XXXX, XXY (Klinefelter syndrome), and XXYY. Individuals with only X chromosomes have a female phenotype, while those with at least one Y chromosome have a male one. Depending on the particular chromosomal combination fertility may or may not be significantly impaired [3–7].

Androgen insensitivity syndrome is associated with physiological and other changes resulting in cells being partially or completely unable to respond to androgens such as testosterone. In individuals who are XY, increasing degrees of androgen insensitivity can have effects ranging from a normal male phenotype with impaired sperm production and overall male secondary sexual characteristics in mild cases, ambiguous external genitalia associated with partial insensitivity to androgens, to a female phenotype with internal testes and no uterus with complete insensitivity to them.

In an adult female, fertility requires a functioning menstrual cycle. In human females the first menstrual period (menarche) typically occurs between the ages of around 9 and 15 (average about age 12). While females have had oocytes in their ovaries since birth, at least one now matures each month into an ovum and is capable of being fertilized. The normal menstrual cycle occurs over a period of about 28 days, plus or minus about 7 days.

At the beginning of each menstrual cycle, rising blood levels of "follicle-stimulating hormone" stimulate up to seven oocytes to start developing into ova. Each ovum is protected inside a tiny fluid-filled capsule called a follicle. An ovum is the largest cell by diameter in the human body, measuring about 0.1 mm (1/250 of an inch) across. Only one follicle and the ovum inside it usually mature during a menstrual cycle, while the others are destroyed. A mature follicle produces estrogens that change the lining of the uterus (the endometrium) such that, if a fertilized egg reaches it at the right time, it can attach itself and grow there. The follicle also triggers the pituitary gland to secrete "luteinizing hormone" for several days. This causes the follicle to release the ovum into the fallopian tube—a process called "ovulation." The ovum then migrates through the fallopian tube to the uterus.

What happens next depends on whether the ovum encounters a sperm cell during the latter's journey toward the uterus and fallopian tubes during about a 24-h period after the ovum's release from the ovary. If the ovum does not, it will ultimately be destroyed. But if it does and becomes fertilized (thus initiating a pregnancy), the resulting "zygote" can then continue its movement toward the uterus for implantation about 6–12 days after it was created. There it can eventually grow through later stages of development as an embryo (the first 8 weeks after fertilization) and fetus (8 weeks post-fertilization to birth) before finally being born as a baby.

Whether or not fertilization ultimately occurs, after the ovum is released from it the follicle changes into the "corpus luteum." The latter secretes both estrogen and progesterone for about the next 2 weeks, during the second half of the menstrual cycle. Those hormones prepare the uterus to accept a fertilized egg. If fertilization and implantation occur, the corpus luteum secretes progesterone to sustain the pregnancy. However, if fertilization does not occur it stops producing progesterone and degenerates into an area of scarring within the ovary, the "corpus albicans." This also results in the tissue along the inner wall of the uterus that was waiting for what would have been a "blastocyst" (formed after the original zygote has divided into several hundred cells) sloughing off. In most female animals this tissue is broken down and absorbed internally. However, in humans and some other primates a mixture of this tissue and blood flows out of the female's body through the vagina—an event called "menstruation."

Females develop their maximum number of oocytes (about 7 million in humans) before birth and continuously lose them thereafter. At birth their number has fallen to about 1–2 million and to around 400,000 at puberty. At "menopause" the remaining oocytes are lost and menstrual cycles cease.

Fertilization of an ovum that a female adult releases typically requires participation by a fertile male during sexual intercourse. Sexual arousal in the female preceding this event involves activation of cerebral and peripheral nerve pathways, primarily the pudendal nerve and sacral plexus, leading to and from the spinal cord. This process can involve mental (e.g. thinking erotic thoughts) and physical stimulation of sensitive body parts. Parasympathetic nerves dilate the arteries supplying blood to erectile tissues (e.g. the clitoris) and stimulate secretion of mucus into the introitus, the opening of the vagina.

The male's contribution to coitus is centered on achieving and maintaining a penile erection sufficient to enter the female's vagina. That erection results from actions involving the vascular and nervous systems of the penis. Mental and physical stimulation cause parasympathetic impulses to pass from the sacral spinal cord through the pelvic nerves to the penis. These cause arteries in the penis and smooth muscle in its erectile tissue to relax. The latter consists of large enclosed spaces ("cavernous sinusoids") that are normally nearly empty of blood. Those sinusoids dilate tremendously when arterial blood flows into them under pressure and venous outflow is partially occluded. Since the erectile tissue is surrounded by strong fibrous tissue, the net result is hardening and elongation of the penis.[3]

Orgasm (aka "climax") in both females and males requires activation of the sympathetic nervous system. During coitus, in males it is typically accompanied by forceful expulsion ("ejaculation") of semen into the female's vagina. The amount of semen produced can be up to 5 to 10 ml but may be considerably less. A sperm cell in semen consists of a cellular "head" and a tail that propels it. The head part represents the smallest cell in the human body, containing all of the sperm cell's genetic material.

---

[3] Needless to say describing these prerequisites for human coitus in dry medical terms tends to drain them of erotic content.

Human males produce millions of sperm cells each day, with about 500 million released during ejaculation. After coitus these myriad cells can live for up to several days within the vagina and, if they make it that far by propelling themselves with their tails, the uterus and fallopian tubes.

Humans and some other primates (e.g. gorillas, chimpanzees, and orangutans) also differ from nearly all other mammals regarding when coitus occurs. In non-primate mammals sex is seasonal, occurring only at specific times. Females of those other species have an "estrous cycle" instead of a menstrual cycle. Sexual intercourse generally occurs only during one phase of the estrous cycle—the "estrus" phase, when the female is "in heat." This is when her eggs are either ready to be fertilized or, as in cats and some other animals, production of eggs that can be fertilized is caused by sexual intercourse itself. Thus, in these other animals sex is intermittent, with long periods of time when little or no sexual activity occurs.

Roughly the opposite state of affairs occurs in humans. Adult members of *H. sapiens* can decide to be sexually active at any time they are physically capable of doing it—and they often do, under a wide variety of circumstances. While the essential biological raison d'être for coitus is reproduction, humans clearly choose to do it for other reasons too.

10.2 Technology and Sex

Over the past several decades various methods have been developed to assess health issues regarding human reproduction and to "artificially" assist with it. These include prenatal screening of prospective parents for genetic issues (e.g. being a carrier for a gene associated with a known defect), treating impaired fertility in both males and females, and various techniques to produce fertilization outside a woman's body. One example of the latter is "in vitro" fertilization, one type of "assisted reproductive technology" [8, 9]. This technique involves stimulating a woman to produce a greater than normal number of ova, removing them from her body, then fertilizing at least some of them to produce zygotes. One or more of the latter can then be introduced into the same or another woman's (a "surrogate mother's") uterus for implantation and development. This method was first successfully used in humans in 1977, with the resulting baby being born in 1978. Alternatively, in "intracytoplasmic sperm injection" a sperm cell is injected directly into an ovum prior to implantation.

As mentioned in Chap. 7, sperm cells, oocytes, and embryos can be cryogenically stored for long periods of time for later use. Science fiction stories involving travel to distant stars may involve transporting future colonists in that considerably less massive and resource-using form than as children or adults. Once the starship reaches its destination, in vitro fertilization or perhaps other means can then be used to expand that alien world's human population more rapidly and with greater genetic diversity than would be possible from adult spacefarers using only the "classic" method for reproduction. For example, Edward M. Lerner's recent *Dark Secret*, serialized in *Analog Science Fiction and Fact* in 2013, uses this method to colonize an extrasolar world after humanity is nearly wiped out by a cosmic disaster.

A further expansion of this concept is the "artificial uterus." In *Brave New World* embryos develop in "decanting bottles" before birth, although the novel is vague on the details of that technology beyond the functional role it serves.[4] Similar devices are depicted in *Star Wars: Attack of the Clones* (2002) as the "birthplaces" of future soldiers of the Republic's clone army. "Uterine replicators" are commonly used by extraterrestrial societies in Lois McMaster Bujold's "Miles Vorkosigan" works. But unlike in vitro fertilization followed by implantation into a biological uterus, which allows a zygote to develop in an intrinsically nurturing environment after its creation, simulating that function by technological means—"ectogenesis"—to produce healthy offspring is not currently feasible.

Challenges to developing a functional artificial uterus are not limited to that organ itself. The uterus serves important functions for the developing embryo and fetus prior to birth. A blastocyst must implant in the inner wall of the uterus, the endometrium, whose lining tissue has (as previously described) been specially prepared for that event. To sustain growth, the blastocyst forms specialized cells that will help form the placenta. Both it and the umbilical cord that develops later ultimately connect the mother's blood circulation to the developing embryo and fetus.

The uterus also serves as a "safe haven" within the mother's body for the developing embryo/fetus. As the latter grows the uterus increases dramatically in size. That organ also helps "protect" the mother from the growth of the embryo/fetus within her pelvis and abdomen as well as isolating it from the rest of the mother's body. In later stages of pregnancy the fetus is also "cushioned" by amniotic fluid within the bag-like membranes of the amniotic sac inside the uterus.

---

[4] That novel also describes other reproductive technologies, such as multiple forced division of a fertilized egg to produce a large number of identical embryos—"clones"—that also (at least for now) still remain in the realm of science fiction.

At birth the uterus undergoes a series of programmed contractions whose function is to push the developed fetus through the mother's dilated cervix (the opening at the bottom end of the pear-shaped uterus) and out through the vagina. As part of this process the amniotic sac ruptures, releasing its fluid. If this sequence of events is accomplished successfully the fetus is born, and the attached umbilical cord is cut. The placenta, now no longer needed, is subsequently expelled from the uterus.

The enclosing/protecting functions of the uterus and cushioning ones of amniotic fluid for the embryo/fetus might potentially be mimicked by artificial means. Birth could then be accomplished as in *Brave New World* by "decanting." However, the functions of the placenta/umbilical cord are considerably more complex and represent important limiting factors in creating a true artificial uterus. Some of those functions already can be reproduced to some degree by technological means, but other important ones cannot.

The placenta and umbilical cord provide the developing embryo/fetus with essential nutrients needed for its survival and growth via their continuity with the mother's arterial and venous blood vessels. Before and even after the fetus develops lungs that could function on their own, it does not use them to breathe before birth. Instead it relies on its mother's circulating blood to continuously provide it with oxygen and remove carbon dioxide from its own blood.

Oxygenated arterial blood passes from the mother via the single umbilical vein in the placenta and umbilical cord into the fetus as the latter's own circulatory system develops. After passing through its circulatory system, venous blood in the fetus flows into the right side of its heart. There blood can pass through an opening, the foramen ovale, between that organ's upper two chambers, the right atrium and left atrium, and thus bypassing the lungs. From there blood passes into the fetus's arterial circulation.

Venous blood can also flow into the right ventricle and pulmonary artery. Here too, however, instead of passing through the lungs it can go directly back into the fetus's arterial system via a short tubular connection between the pulmonary artery and the aorta called the ductus arteriosus. After passing through the fetus's arterial circulation this blood, now deoxygenated, returns to the mother's circulation via the umbilical cord's two umbilical arteries. That blood eventually returns to her heart and lungs for gas exchange. From there her oxygenated arterial blood goes back into the fetus to complete that whole circuit.[5]

The placenta also acts as a protective barrier and "gatekeeper." It can prevent some toxins and potentially harmful chemicals (e.g. medicines) from passing into the embryo/fetus. Conversely the placenta allows nutrients and other helpful substances, including critical hormones (e.g. thyroid hormone), needed by the developing embryo/fetus to enter its circulation. Waste products such as uric acid produced by the fetus are also transferred via the umbilical cord and placenta to the maternal circulation for clearance.

The placenta itself produces important hormones, including "human chorionic gonadotropin" (hCG) that stimulates the corpus luteum to continue to secrete progesterone, as well as estrogen. This maintains the ability of the endometrial tissue lining the inner wall of the uterus to sustain pregnancy. hCG, whose presence in blood and urine very early during gestation is used as a standard "pregnancy test," also helps prevent the mother's body from rejecting the placenta as "foreign" tissue. Another hormone produced by the placenta, "human placental lactogen," helps maintain a healthy metabolic state in the fetus and assists its growth.

The placenta and umbilical cord also help protect the developing embryo/fetus from infections. They can do that directly by preventing at least some microbes from reaching the embryo/fetus, and by transferring some of the mother's own antibodies against various pathogens to it. After birth those antibodies provide the baby with "passive immunity" against some infectious diseases for at least several months after birth.

A few but not all placental functions could potentially be provided by artificial means [10–12]. Some wastes produced by a fetus can be removed via techniques equivalent to dialysis. Oxygen-carbon dioxide gas exchange might be performed using another well-established technique, extracorporeal membrane oxygenation (ECMO). Experiments reported in the late 1990s employed such methods to keep fetal goats alive for up to nearly 10 days. However, the fact that all those fetal goats still died (due to circulatory failure

---

[5] Not long after the baby is born and its lungs take over the function of breathing, these two "shortcuts" that blood follows through its heart—the foramen ovale and the ductus arteriosus—normally close off. However, as described in Chap. 3, up to about a third of people have a "patent foramen ovale," in which that opening is not completely closed or is covered by a potentially mobile flap of tissue that could, under the right circumstances (such as increased pressure in the right atrium), unseal it. Likewise the ductus arteriosus may not close in a small percentage of individuals. Because pressures in the systemic circulation (e.g. the aorta) are higher than in the pulmonary circulation after birth, a "patent ductus arteriosus" results in some arterial blood from the aorta passing back into the pulmonary circulation. If the amount of blood "shunted" in this way is great enough it can significantly reduce how much oxygenated blood the rest of the body receives.

and other causes) indicated that significant refinements to the method are still required.

A potential step in growing a zygote and eventually an embryo/fetus outside the mother's body could include developing a scaffolding lined with appropriate endometrial tissue (or a functional equivalent) as a substitute for the uterus. The placenta and umbilical cord develop from cells originating in the blastocyst itself. However, for functioning versions of those tissues to be created some way would have to be found to make them "link" with endometrial tissue in an artificial uterus or otherwise connect the developing embryo/fetus with a vascular supply to provide it with everything it needs for life. Overall, actually using solely technological means (biological or otherwise) to perform all the critical functions the material circulation does naturally in providing essential nutrients, assisting immune function, and maintaining the many complex functions (e.g. regulating hormone levels) necessary to maintain homeostasis[6] and support healthy growth in the very vulnerable embryo/fetus represents a major challenge.

10.3 Cloning

Cloning of animals such as amphibians and some mammals is no longer confined to science fiction [13–16]. Dolly the sheep, whose birth was reported in 1996, lived for nearly 7 years after her birth. She was created using "somatic cell nuclear transfer" (SCNT). This involves removing the nucleus from an ovum and then inserting the nucleus from a somatic cell—any cell except those that produce sperm and ova (germ cells)—from an adult animal into that "enucleated" ovum.[7] In the particular method used to produce Dolly, that new ovum-nucleus combination was induced to divide using an electric shock[8] [14, 17, 18].

Introduction of that new nucleus into the environment of the ovum "reprogrammed" it back to a stage of development when it could form all the types of tissues present in an animal's body, just as an embryo's can. Overall,

---

[6] "Homeostasis" is the ability of a biological system to maintain constant, healthy conditions within it by adjusting to changes in its environment. For example, the human body's adjustments (e.g. shivering) to a cold environment described in Chap. 7 involve homeostasis.

[7] The somatic cell used to create Dolly was a mammary gland cell obtained from a female sheep. Dolly was named by the investigators after the singer-actress Dolly Parton, whose anatomic attributes vis-à-vis mammary glands are readily apparent.

[8] This technique invites comparison with the laboratory-based method used to bestow life by unconventional means depicted in the 1931 movie *Frankenstein*. However, instead of assembling human body parts derived from multiple corpses (most notably and regrettably a criminal's brain), doing research on a literally "sheepish" species whose members are not known for going on murderous rampages might be considered a more prudent choice.

Dolly had three "mothers," with one providing the enucleated ovum, a second the somatic cell, and the third acting as a surrogate by having the ovum-somatic cell implanted into her uterus after dividing to the blastocyst stage [14].

Since then other species of mammals, including pigs, deer, horses, cattle, mice, dogs, and cats, have been successfully "cloned"[9] [15, 16, 19, 20]. However, there are a variety of issues with creating clones by this method. For many species the process is very inefficient in producing offspring that survive to birth and are at least reasonably healthy. Dolly was the sole successful product of 277 attempts (a success rate of about 0.4%) to produce a cloned sheep that survived to adulthood. There were only 4 other live births from those attempts, with these additional clones dying shortly after birth. Dolly herself died in 2003 at an age about half the average one for her particular kind of sheep. Her death was attributed to virus-induced tumors, a common cause of mortality in her breed of sheep. There was no definite indication that her origin as a clone was a significant factor in her demise. However, it would require additional studies involving larger numbers and types of animals to assess the overall risk of premature death in clones [17, 18].

The reported success rates for cloning different types of animals via SCNT vary from no successful results to a rate of live births as high as 20 to 25% in cattle and, more recently, mice [16, 21, 22]. Overall rates of live births in mammals that have been cloned are in the range of 1 to 2% [19, 21]. Unsuccessful attempts typically are lost via spontaneous abortions and are associated with developmental abnormalities. Serial cloning has also been done, in which a somatic cell from a cloned animal has been used to produce another clone. This has been reported to be successful so far for only up to about two generations removed from the original clone in cats and cattle. However, a recent study extended this process to at least 25 generations in mice, with production of over 500 viable offspring from a single somatic cell derived from the original donor mouse [22].

So far no documented case of human cloning resulting in a live birth has been confirmed. Nonetheless, recent reports have described creation of human blastocysts using modified SCNT techniques based on what part of the cell cycle the receiving ovum was in when a somatic cell nucleus was introduced [23–26]. However, no blastocyst created in this way was implanted into a woman's uterus to assess its potential for further growth and development. Only time will tell whether this will be done in the future and what the results will be.

Regardless of the science involved, the current inefficiency of the SCNT method for producing viable offspring in most animals tested raises serious

---

[9] More specifically, this technique is termed "reproductive cloning."

ethical questions about its use in humans. This technique would require harvesting a large number of ova from women and implanting one or a few in many potential "mothers." Based on the experience with other species, there is a significant likelihood that the vast majority of any resulting pregnancies would result in spontaneous abortion or live birth of developmentally impaired babies. The resources needed to potentially produce a "normal" human clone—assuming that is possible using SCNT or similar techniques, which is still not a certainty—would be great and the anticipated overall results poor. To reliably produce a clone army similar to that in *Attack of the Clones* would require as yet unknown modifications to the SCNT method or perhaps an entirely new one.[10]

Moreover, a "clone" of any animal or person produced by SCNT would be genetically similar to but not an exact biological duplicate of the adult providing the somatic cell nucleus. Although the vast majority of a mammalian somatic cell's DNA is in its nucleus, as noted in Chap. 9 a very small but important amount is located in its energy-producing mitochondria. Any human or other cloned mammal's mitochondrial DNA will be derived from the recipient's ovum rather than the donor's somatic cell.

Also, although the nuclear DNA's sequences in the donor's cell would serve as the template for those of the prospective clone, the degree to which the donor's cell is "reprogrammed" back to a state allowing it to develop a clone may vary. More importantly, how the genes in the transferred DNA function can also be different due to epigenetic effects (see Chap. 9) and environmental factors.

As we will see in Chap. 12 regarding true clones produced by Nature—"monozygotic" siblings, produced by a zygote splitting at least once early in development—the influence of genetic factors regarding development of different traits and diseases varies from mild to strong. With them too epigenetic and external environmental factors, such as nutrition, upbringing, education, etc. can contribute significantly to their overall development, personalities, etc. that go beyond genetics alone.

---

[10] Conversely, as noted previously hundreds of mice have been produced by serial cloning from a single (Jango Fett-like?) donor. This suggests that it might indeed be possible to create a clone army consisting of mice. However, issues regarding their size, paw dexterity, general lack of a militaristic attitude, etc. would need to be addressed before they would be ready to conquer the Galaxy.

10.4 Sex in Space

Humans venturing away from Earth into an extraterrestrial environment—space itself or alien worlds—will continue to perform the same biological functions they do in less exotic locales. Besides eating, breathing, etc., at some point they will also engage in other commonplace (though not necessarily always prudent) actions, including sexual activity [27, 28].

In science fiction, sex in space may take place within an otherworldly setting where conditions of temperature, atmosphere, gravity, etc. are so close to typical ones at Earth's surface that they present no significant impediment to performance. On the other hand, barring some highly imaginative technology, sexual relations during extravehicular activity or on a world with little or a poisonous atmosphere where the participants must wear spacesuits to survive would not be feasible. Likewise, an alien world that subjected human visitors to uncomfortable extremes of temperature, making them too hot or too cold despite countermeasures such as less or more clothing, could make sexual activity very unpleasant if not physically impossible.

Sex in space presents both difficulties and risks both now and in the near future. These include physical ones, such as the effects of microgravity and high radiation levels, as well as potentially significant psychological issues. Science fiction involving sexual activity away from Earth often avoids issues with microgravity by setting the story on a spaceship with artificial gravity or a world whose natural gravity is not too different from ours. Issues with the high radiation levels in space may be ignored altogether. The drama and conflict produced by sexual relationships, love triangles, jealousies, frustrations, etc. among characters in space usually occur in contexts where they present far less physical danger than they would in present-day scenarios such as a trip to Mars.

Current and near-term capabilities for spaceflight involve space stations of various sizes and small craft that reach LEO or potentially land on the Moon or Mars. In a modular structure as large as the International Space Station (ISS), achieving privacy for sexual activity might be difficult but not impossible.[11] This would be more of an issue for a multi-person crew heading to Mars in a much smaller vehicle. Common psychological responses to being in space include overexcitement, anxiety, and other stressors that might also inhibit sexual performance.

Reports involving mixed-gender crews in isolated terrestrial settings, such as in Antarctica, indicate that sexual relationships, including extramarital ones,

---

[11] See my story "Hearts in Darkness" (*Analog Science Fiction and Fact* March 2002) for an example of this.

can develop and may lead to conflict among crewmembers. While such activities might make for exciting reading, on proposed real-life space missions they could prove disastrous for crew morale and efficiency, further endangering an already dangerous endeavor [29, 30]. Conversely, other reports indicate that having mixed-gender crews could also encourage more harmonious relationships than might occur in those with only one gender [30]. Variations in attitudes and behaviors among different individuals in a particular crew could lead to widely varying outcomes. Factors in interpersonal relationships that might cause problems would ideally be identified by observing how individuals interact together prior to flight [31].

As described in Chap. 4, microgravity has multiple effects that could dampen both desire and ability to perform sexual activity. The headaches, dizziness, nasal congestion, nausea, vomiting, and other discomforts associated with space motion sickness (SMS) could render an astronaut at least temporarily too sick to contemplate sex. Hormonal factors could also play a role. Testosterone levels fell to less than 20 % of control values in male rats flown on a Soviet biosatellite, COSMOS 2044, for 14 days in 1989 [32]. Although the effects of prolonged microgravity on testosterone levels in human males are not established, a similar fall might reduce libido, performance, and possibly fertility [33]. The mild anemia and relative hypovolemia (reduced blood volume) that develop in microgravity might also have mild adverse effects on penile erection, which requires an adequate blood volume. Nonetheless, despite these issues male astronauts reportedly are capable of having erections in microgravity [34].

As noted previously, successful coitus involves stimulation of both parasympathetic and sympathetic nerves. Exposure to microgravity has relatively little effect on parasympathetic activity, and may actually increase it. Conversely sympathetic activity may be reduced, with decrease in vascular tone. How much these responses may affect erection or climax is unclear, but given that the former requires appropriate changes in tone of both arteries and veins this is a potential concern.

The mild muscle atrophy and deconditioning that occur in space despite vigorous exercise routines can increase both oxygen consumption and the amount of work performed during sexual activity, especially by the more active partner. On Earth coitus produces overall moderate and gradual increases in heart rate and blood pressure, with peak levels occurring at climax—usually in the range of 130 beats/min and 170–180 mmHg systolic blood pressure. Energy requirements are about 2–3 METs[12] during foreplay and up to around

---

[12] MET = metabolic unit, defined as a "normal" oxygen consumption at rest of 3.5 ml of oxygen per minute per kilogram of body weight.

4–5 METs during orgasm. This represents a modest level of work—on average, roughly equivalent to walking at a rate of about 3–4 miles/h.

However, these hemodynamic responses and peak workload can increase significantly when sexual activity is performed with an unfamiliar partner in a new environment. In that case both heart rate and systolic blood pressure may increase well above the ranges just cited, and maximum energy requirements could rise to at least 6 METs. Coitus itself carries a very slightly increased risk of myocardial infarction (incidence of about 1–2 in a million) and sudden death, primarily in males, associated with the increased workload needed to perform it. Since astronauts are physically fit and screened for coronary artery disease and other risk factors prior to flight, this danger seems minimal but is not zero.

As noted in Chap. 4 microgravity is also associated with mild decreases in immune function, and the high radiation levels in space described in Chap. 6 could result in microorganisms becoming more resistant to antibiotics. Astronauts are routinely screened prior to flight for sexually transmitted diseases (STDs). If present, STDs caused by bacteria such as *Chlamydia trachomatis*, *Neisseria gonorrhoeae* (responsible for gonorrhea), *Treponema pallidum* (the cause of syphilis), or the protozoan parasite *Trichomonas vaginalis* can be treated with antibiotics before going to space.

However, curative treatment is not an option for astronauts infected with STDs caused by viruses. These include genital herpes, produced by the herpes simplex 2 virus, and infection with human papillomavirus (HPV). There is now an effective vaccine for the latter, but no cure for either. Both of these STDs have a high prevalence. According to the United States Centers for Disease Control, about 1 out of every 6 people aged 14–49 years have genital herpes. An estimated 79 million Americans are currently infected with HPV, and about 14 million people become newly infected each year [35].

Unfortunately this means that some astronauts might already have these latter STDs. Besides abstinence, which is (obviously) 100 % effective for preventing coitus-based infection, a basic method to reduce risk of transmission to a partner is to use a barrier device (e.g. a condom). Also, coitus alone can cause urethritis and cystitis (inflammation and/or infection of the urethra and bladder, respectively), particularly in women due to the anatomy of their external genitalia and shorter urethra compared to men. Preventive measures can include drinking liberal amounts of fluid and urinating after coitus. However, if such an infection does occur treatment typically requires use of oral antibiotics. They, like other medications, will be in short supply in space, and the bacteria causing a urinary tract infection will likely be resistant to at least some available antibiotics. Worse, cystitis can lead to pyelonephritis, a more serious and potentially life-threatening infection involving a kidney that may require intravenous antibiotics.

While NASA and other space agencies screen for and deal with STDs prior to flight, it is uncertain whether those same standards will be applied to future "space tourists" going to privately run "orbiting hotels" (e.g. proposed inflatable habitats). Hopefully such health measures will be taken, along with informing individuals going to them about the physical difficulties and other risks associated with sexual activity in microgravity to be described soon.

Another issue involves liquids such as saliva, perspiration, vaginal secretions, and semen released during coitus and non-coital sexual activity. As described in Chap. 4, fluids in microgravity continue to "float" until they strike a surface. They could potentially be inhaled, indirectly enter the mouth of an astronaut, or settle on equipment as well as the clothing and skin of crewmates not involved in the sexual activity when they enter the area afterwards. Use of a condom during coitus would help contain semen, but not other fluids. Likewise pubic and other hairs could also be released during intercourse, although because they are not intrinsically moist they may not adhere to any surface and thus remain suspended.

Body heat dissipation may also be decreased within the confines of a spaceship or space station due to reduced circulation of air. Thus, prolonged close bodily contact could make astronauts engaged in foreplay and coitus uncomfortably hot in the purely temperature sense. Ambient noise levels on the ISS due to fans and other equipment are relatively high. This might prove a distraction during coitus, but it could also help mask the sounds produced by participants from uninvolved crewmates. Unpleasant odors produced by partners (e.g. various body secretions, flatus, etc.) may also be somewhat more noticeable due to the relatively small internal volume of a spacecraft or a module of the ISS.

Besides its adverse hormonal, hemodynamic, and other biological effects, microgravity represents a physical challenge to performing coitus. While some might find the idea of a coupling man and woman floating free inside a spaceship an interesting fantasy, in reality it could prove difficult and dangerous due to Newton's Third Law of Motion (see Chap. 5). In practice this means that both partners would have to carefully coordinate their movements and keep themselves secured to each other to reduce the chances of decoupling, or of one or both rocketing into the side of their spacecraft during their most vigorous actions. Hitting one's head or other sensitive area unexpectedly while attention is focused entirely on sexual activity might not only break the mood of the moment but could also injure the participants.

A misguided movement might also cause lateral or backward buckling and fracture of the erect penis. This injury involves tearing of the latter's fibrous tissue and would render further coitus impossible. A penile fracture produces a sharp snapping, cracking, or popping sound, followed by excruciating pain, swelling, bleeding, and deformity of the penis. Initial treatment involves cold compresses, pressure dressings, splinting, and analgesics—all of which, as with any medical supplies, will be in limited supply in space.

Definitive management of a penile fracture ideally involves surgical repair. However, the latter would probably not be an immediate option on the ISS or any orbiting craft even in the unlikely event a urologist-astronaut were on board, due to lack of appropriate facilities and supplies for performing the operation. The injured crewmate would have to be evacuated to Earth for such emergent treatment—something that would obviously not be a viable option for astronauts on their way to or on Mars. If surgery could not be performed, complications might include development of an abscess and long-term ones such as erectile dysfunction and permanent curvature of the penis.

A less adventuresome but safer approach would be to ensure both partners are safely "secured" and confined to a relatively small area. One method would be a fixed, cage-like structure with padded sides and an interior volume only slightly greater than the two partners. Alternatively one partner could be secured (e.g. with straps or other restraining devices) against a stationary structure such as a wall of the craft or space station while the other is more loosely restrained. The latter, in a "superior" position, could grip appropriately placed handles or other fixed objects above or below that individual to prevent unwanted motion away from the other partner. Both partners might also be tucked within a secured sleeping bag-type enclosure or mesh secured to a wall. Side-to-side positions may require less extensive restraining systems.

The technical issues that dolphins and other aquatic mammals face when mating in water are in some ways similar to those humans would encounter in microgravity. A third dolphin may assist a mating pair by giving a well-timed push or restraining the backwards motion of one of its colleagues while the latter and another engage in intercourse. While employing such a helper is not required during human coitus in a terrestrial environment, having a crewmate "lend a helping hand" in this way would be a potential option in microgravity.

Nonetheless, while the psychological, physiological, and physical effects of the space environment could impede performance of intercourse, there is no theoretical reason why, with proper planning and care, it could not be performed successfully by suitably motivated and prepared humans. Whether coitus actually has happened yet is uncertain. The first time a male-female crew was in space within reach of each other occurred in 1982 aboard the Soviet Salyut 7 space station. A three-person crew consisting of the second woman in space,[13] Svetlana Savitskaya, and two male crewmates stayed aboard Salyut 7 for several days. She later returned to that space station with two different male crewmates on a second mission in 1984, where she became the first woman to perform a spacewalk. Mixed gender crews were routine on Shuttle flights and some missions on the Soviet space station Mir, and remain so on the ISS. STS-47, a Space Shuttle mission in 1992, included a husband-wife pair of astronauts.

However, despite these at least physically possible opportunities, there is no official confirmation that human coitus has occurred in the final frontier. NASA appears to have the equivalent of a "Don't ask, don't tell" policy. However, given the potential risks associated with coitus and the negative impact they might have on a mission, it is reasonable that, for purely practical reasons, coitus would be discouraged on "official" spaceflights. Practicing abstinence for long periods of time, while certainly within human capability, might generate psychological stress on long-duration flights. Sexual stimulation performed alone (e.g. via manual means), paired or group activity not involving coitus (such as oral-genital contact), or other sexual actions involving non-coital penetration (with the exception of anal intercourse, which would presumably have at least the same level of physical risks in microgravity as coitus) might either relieve stress or, conversely, perhaps increase psychological frustration.

These are real-life issues that, particularly for very long-term missions such as those to Mars, will need to be at least considered. One proposal for a flyby of Mars involves sending a two-person, husband-wife crew. However, while their marital status might be of some help in risk reduction (e.g. known sexual compatibility/prior experience with each other), it would still not reduce the dangers of urinary tract infections, physical injury during coitus within microgravity, etc. Eventually, if it has not already happened, humans will perform coitus for either reproductive (e.g. colonizing an alien world) or, along with non-coital sexual activity, for non-reproductive reasons. Science fiction stories typically ignore the practical risks and consequences involved with such issues. However, until they can be addressed by technological means such as artificial gravity, or humans either actually colonize suitable worlds or

---

[13] The first woman in space, Valentina Tereshkova, was launched into orbit in 1963 aboard the Soviet Vostok 6 space capsule. A male cosmonaut, Valery Bykovsky, was in orbit at the same time. However he was in his own separate one-person capsule, Vostok 5, and no physical contact was possible between them.

live in large space habitats/spacecraft with a sufficient medical infrastructure, they will remain significant.

On the other hand, some nonhuman animals already have had sex in space. Various invertebrates such as fruit flies (*Drosophila melanogaster*), worms (e.g. *Caenorhabditis elgans*, see Chap. 9), and snails (*Biomphalaria glabrata*) have successfully mated in microgravity[14] [36, 37]. A group of 5 geckos was launched on the Russian Foton-M4 satellite in 2014 on a 44-day orbital mission to see if they could mate in space. Unfortunately, when the capsule returned to Earth all of the geckos were dead, and because there were no live feeds from the satellite it was uncertain whether they engaged in sexual activity while in orbit.

On the other hand, small group of rats launched for an 18.5 day flight on a Soviet biosatellite, Cosmos 1129, in 1979 did copulate [37, 38]. Five female and two male rats were sent into orbit within a container that initially kept the two genders apart. On the second day of the flight the pair of doors separating the two groups were opened and all seven were allowed to mingle. Nature subsequently took its course, and this intrepid crew of space explorers bravely went where no rat had gone before.

Findings of this study on rodent reproduction included that both ovulation and fertilization occurred in the female rats. The embryos produced apparently underwent early development. However, they were then absorbed long before birth and lost. The reason for their failure to survive is unknown [37]. About 5 days after landing the male rats on that flight were mated with females who had stayed on Earth. The latter had larger litters than controls, but offspring had a higher rate of developmental abnormalities, including growth retardation, bleeding, bladder enlargement, and abnormal location of their kidneys. However, those male rats produced normal offspring after mating again about 2.5–3 months post-flight [33]. The latter results might have been due to factors in the space environment (e.g. radiation) that affected existing, mature spermatozoa but not those that matured after return to Earth. The implications of these findings in humans, particularly on bringing an embryo/fetus to term, are uncertain.

However, medaka fish (average length about 2–4 cm) aboard the STS-65 Space Shuttle flight in 1994 not only successfully mated in microgravity but produced viable offspring [37]. Being already in an aqueous environment that mimics some effects of microgravity may have facilitated their task. However, these studies involving rats and fish provide little help assessing either the feasibility of humans performing coitus in space or producing offspring.

---

[14] Presumably, however, the snails took their time doing it.

10.5 Pregnancy in Space

On Earth coitus between a fertile man and an ovulating woman can result in pregnancy. However, there has been no reported research on how being in space would affect the probability of a woman becoming pregnant after successful coitus. Given the absence of hard (so to speak) data on this subject, its likelihood can only be speculated on by considering factors that could have an impact on it. In microgravity ejaculated spermatozoa should "stay put" within the vagina before moving into the uterus and fallopian tubes. Transport of spermatozoa in the latter is accomplished by the muscle contractions and ciliary action ("waving" movement of hair-like structures) within the tubes and the spermatozoa's own tail-propelled motility. These should be little affected by microgravity.

If pregnancy is not a desired goal following coitus in space—and, for multiple reasons to be discussed soon, it is strongly contraindicated at the present time—contraception must be considered. Surgical sterilization (e.g. vasectomy in men and such measures as tubal ligation in women) is an option for astronauts who do not wish to have children in the future. Some astronauts may also be physically incapable of causing pregnancy or becoming pregnant for other reasons. These include male astronauts with azospermia (no detectable spermatozoa in semen) or female astronauts who are post-menopausal, have had a hysterectomy (surgical removal of the uterus), or bilateral oophorectomy (excision of both ovaries).

Oral contraceptives (aka "birth control pills") containing the female hormones estrogen and progesterone, either alone or in combination, are highly effective if used properly on Earth. However, their effectiveness in microgravity has not been established. As noted in Chap. 4, microgravity might significantly affect absorption, duration of action, and peak blood levels of medications. These factors could alter the timing of administration and dose of birth control pills needed for efficacy. Extended-release medroxyprogesterone acetate (Depo-Provera) or levonorgestrel implants (Norplant) might be a more dependable choice. Both oral contraceptives and a urine pregnancy test are included in the medical supplies stocked on the ISS [39]. It should be noted, however, that oral contraceptives can also be used for other purposes such as regulating a woman's menstrual cycle [29, 40].

Barrier devices (condoms and diaphragms) with or without spermicidal agents, as well as intrauterine devices (IUDs), would be expected to have similar efficacy in space as they do on Earth. "Coitus interruptus" and the "rhythm method" are significantly less effective contraception techniques than these others, and their successful performance in microgravity would be even more problematic. The former could also result in globules of semen floating within the spacecraft that would need to be confined and eliminated. Increased menstrual cycle unpredictability in space (see below) would make the rhythm method particularly impractical for preventing pregnancy.

The space environment itself may have contraceptive effects. Formal studies on the effects of microgravity on menstruation in human females are lacking. However, because it has some physiological effects similar to microgravity, prolonged bed rest has been used to indirectly evaluate how being in space might impact menstrual and related functions. As described in Chap. 4, both prolonged bed rest and entry into microgravity normally shift up to 1.5–2.0 l of fluid from the lower to the upper body. This soon triggers a small degree of urinary loss of what the body considers "excess" fluid as well as moving that fluid from blood vessels into the extravascular space. Bed rest studies have shown that initiating such rest just before the stage of a woman's menstrual cycle when ovulation occurs produces either an increase or only a small decrease in plasma volume for the first 4 days. Thus, by analogy, a female astronaut's entry into space at that time in her cycle might result in a delay in the expected fall in blood volume.

Other bed rest studies have shown no significant change in menstrual cycle length. Eventually studies of these and other effects on menstruation and ovulation (including serial measurement of serum gonadotropins, progesterone, and estrogen levels) will need to be conducted in microgravity itself over at least one complete menstrual cycle (up to about 30 days). A prolonged stay on the ISS would make this practical. At least anecdotally, however, woman astronauts have reported no significant new issues during menstruation in microgravity compared to its occurrence on Earth [29, 40].

Still, menstrual dysfunction might occur in individual female astronauts from several factors. Alterations in normal circadian rhythms, the intensive exercise required both before and during flight to perform mission objectives and reduce muscle atrophy, overall stress during the mission itself, and possibly direct actions of microgravity on the complex system of hormonal secretion that regulates the menstrual cycle could have various effects on it. These might include menstrual irregularity, oligomenorrhea or amenorrhea (infrequent or no menstrual periods, respectively), lack of ovulation during a period, and menorrhagia (excessive menstrual bleeding).

It has been theorized that microgravity might induce retrograde menstruation (regurgitation of uterine blood and tissue backwards through the fallopian tubes and onto pelvic and abdominal organs, e.g. the ovaries) and endometriosis (presence of the type of tissue lining the uterus in abnormal locations, such as within the pelvis), potentially causing long-term infertility after leaving space. However, there is no clear evidence yet that these latter effects occur or at least are significant [40]. Menstrual blood flow can be managed in microgravity as on Earth via use of pads and tampons. Oral and other forms of pharmacological contraception can suppress menstrual bleeding.

Effects of microgravity on male fertility are also not well established. Endocrine disturbances (e.g. the fall in serum testosterone in rats described previously) and changes in sperm motility might decrease fertility. Toxins used in life-support systems, propellants, and so on in spacecraft may reduce sperm counts. Conversely, microgravity-induced fluid shifts may decrease varicoceles (abnormally dilated veins of the spermatic cord), resulting in increased sperm counts.

Exposure to increased radiation levels in space (see Chap. 6) could temporarily reduce sperm counts in men or contribute to premature ovarian failure in women. Spermatozoa-forming cells and spermatozoa themselves are among the most radiosensitive cells in men. Acute doses of as little as 10 rad, or cumulative exposure to lower doses (e.g. during prolonged stays in space) might impair male fertility for a prolonged period. However, to produce azospermia or permanent sterility requires much higher doses of radiation—perhaps high enough to prove fatal to a man. Mature spermatozoa are also sensitive to potential radiation-induced genetic damage that could produce defects in offspring. However, based on studies in men exposed to radiation from the atomic bombs used on Japan, this risk appears to be small. The risk could be further reduced if a man exposed acutely to high-dose radiation refrains from trying to cause pregnancy until at least after the normal approximately 74-day period for a full cycle of spermatogenesis. Libido itself does not seem to necessarily be affected by even high acute radiation doses of a few hundred rad.

A woman's ovaries lie about 5–7 cm below the skin, with abdominal tissues (e.g. fat) attenuating the radiation dosage the ovaries receive. Testes have considerably less "cushion" provided by the scrotum, although a scrotal (lead) shield could be worn to protect them. Oocytes are also somewhat more resistant to radiation-induced genetic defects because, unlike spermatozoa-producing cells, they are not actively dividing and have good enzymatic repair systems.

Conversely, while males can create new spermatozoa, eventually recovering normal counts, females only have a limited supply of oocytes—if they are destroyed, they cannot be replaced. A single dose of 300–400 rad usually results in complete elimination of oocytes and estrogen production from ovaries. In those exposed to smaller, repeated doses, effects on fertility are cumulative. Radiation could also cause endometriosis, impairing fertility long after exposure.

Astronauts wishing to have children after prolonged stays in space could cryopreserve spermatozoa, oocytes, or fertilized embryos beforehand. These would then be available for inducing pregnancy via the in vitro fertilization and implantation techniques described previously. Potentially, and certainly for science fiction purposes, cloning might also become an option in the future.

NASA's current medical standards specifically disqualify any pregnant woman from going into space [40]. Microgravity, radiation, and other hazards represent significant risks both to the woman and her embryo/fetus. Pregnancy-induced nausea and vomiting ("morning sickness") is common early in pregnancy, usually resolving by about the 16th week of gestation. Among its other effects SMS can also cause nausea and vomiting ranging from mild to serious. The combination of "normal" morning sickness and SMS could significantly worsen those symptoms. At worst it might cause "hyperemesis gravidarum," with a woman experiencing severe vomiting that has the potential for causing marked dehydration and even liver damage.

Heart rate and blood volume increase during pregnancy on Earth. Fluid retention usually is greater than the increase in red blood cell production, resulting in a relative anemia. Microgravity itself results in cardiovascular deconditioning (loss of vascular tone), the previously mentioned reduction of intravascular fluid, and an approximately 15 % reduction in red blood cells during the first 60 days in weightlessness. The combination of these normal changes during pregnancy with those caused by microgravity could seriously impair blood flow and oxygen delivery to the fetus, particularly if the woman were to return to Earth later during pregnancy. At best fetal growth might be impaired, with low birth weight. At worst it could cause neurological impairment or death of the fetus.

On the positive side, microgravity-induced fluid shifts would decrease the incidence of varicose veins and peripheral edema (leg swelling) often seen in the later stages of pregnancy. Back pain and difficulty moving in the final months of pregnancy would also be reduced. Conversely, pregnancy-related complications that are routinely manageable on Earth, such as urinary tract infections and gestational diabetes, would be more difficult to deal with in space due to limited supply of medications. Also, during pregnancy basal metabolic rate increases with associated increases in oxygen consumption and generation of body heat, thus putting greater demands on life-support systems in spacecraft.

The typical levels of reduced muscle and bone mass caused by microgravity (see Chap. 4) present special problems in pregnant women. Reduced mass and strength of pelvic musculature could impair vaginal delivery of a baby. To support bone growth in the fetus a pregnant woman should take supplemental calcium, otherwise her body will obtain that mineral from her own skeleton. This "normal" tendency toward bone demineralization and possible osteoporosis would be worsened by additional calcium losses due to microgravity.

Besides its adverse effects on the pregnant woman herself, the typical radiation levels described in Chap. 6 that are present in space, the Moon, or Mars—much less the far higher ones produced by coronal mass ejections or solar particle events—might have significant teratogenic (birth defect-inducing) effects on the embryo/fetus or be fatal to it. High acute radiation doses within the first 2 weeks of gestation will most likely cause spontaneous abortion rather than an eventual live birth with birth defects. Although it can have serious effects at any stage, radiation is most likely to impair embryonic/fetal development between 2–25 weeks after conception, with the greatest risk between 2–15 weeks. Cataracts, abnormal genitalia, microphthalmia (small eyes), decreased lifespan, microcephaly (small head), and mental retardation may occur, as well as a possible mildly increased risk of leukemia developing during childhood [41–43].

The threshold level of radiation exposure during pregnancy below which fetal risk is minimal has not been well established but may be around 5 rad. The total radiation dose (at least 10 rad) a woman could receive over the roughly 9 months of pregnancy during long-duration orbital or deep space missions would thus represent a significant risk to embryonic/fetal development. For comparison, the maximum cumulative radiation exposure limit to the fetus for all 9 months of pregnancy in terrestrial female radiation workers is 0.5 rem (roughly equivalent to 0.5 rad)[15] [44].

Studies of children born to pregnant women exposed to radiation from the atomic bombs dropped on Hiroshima and Nagasaki showed increased rates of mental retardation. The highest risk for this was associated with radiation exposure at 8–15 weeks gestation, with even small doses of radiation increasing that risk. The incidence of severe mental retardation ranged from 4 % from acute exposure to 10 rad to 60 % for up to 150 rad [43]. Even in children without mental retardation there was a radiation dose-dependent reduction in intelligence quotient (IQ), with the greatest risk again being associated with exposure between 8 and 15 weeks gestation. However, these effects appeared to be minimal after 25 weeks gestation.

The toxins within spacecraft described in Chap. 3 represent significant health risks for both the woman and, even more so, the fetus. If the mother inhales excessive levels of chemicals like carbon monoxide, nitrogen tetroxide, or hydrazine both she and the fetus could experience hypoxia or anoxia (partial or complete reduction of oxygenated blood, respectively, to part or all of the body) and respiratory distress. The developing fetus is particularly sensitive to anoxic encephalopathy (brain damage caused by lack of oxygen) from these or other causes.

---

[15] Rem = roentgen equivalent man. See Chap. 6 for more details.

As described in Chap. 3, decompression sickness could result from sudden loss of cabin pressure in a spacecraft. Nitrogen bubbles come out of solution in blood and other body fluids, causing symptoms such as difficulty breathing, pains in the limbs, and convulsions. Some but not all reports have suggested that women may be more susceptible to decompression sickness than men [29]. The fetus is particularly vulnerable to the effects of nitrogen bubbles in both the maternal circulation and its own. Bubbles that form in its veins will move to the arterial system via the foramen ovale and ductus arteriosus. Those arterial bubbles could then impair the blood supply to critical organs like the brain.

Studies involving pregnant women who developed decompression sickness during scuba diving reported that they had an increased incidence of cardiac arrhythmias and congenital malformations involving the fetus. Treatment of decompression sickness primarily involves use of hyperbaric oxygen, which unfortunately is teratogenic in the first trimester. This is currently a moot point, however, since current spacecraft and even the ISS do not have hyperbaric chambers to treat an affected woman astronaut or her fetus.

The lack of standard medical facilities in space would also limit the care available for women with higher-risk pregnancies and deliveries. All the complications that can threaten the life and health of both woman and fetus on Earth could also occur in space, with potentially disastrous results.

Blood pressure during pregnancy is typically around 90/60 mmHg or a little higher—less than is "normal" in women who are not pregnant. In microgravity blood pressure may also be lower than preflight due to the fluid shifts, reduced sympathetic tone of blood vessels, etc. described previously. Here too, as with nausea and vomiting produced by both SMS and morning sickness, the combination of hemodynamic effects produced by microgravity and pregnancy could be additive to the point of impairing the health of both mother and embryo/fetus. Also, in microgravity there would be no additional beneficial hemodynamic effects to instituting bed rest for increasing blood supply to the uterus and fetus due to any potentially helpful fluid shifts having already occurred.

A pregnant woman might instead develop hypertension (high blood pressure). On Earth it can be managed with specific blood pressure medications known to be safe for the embryo/fetus (not all such medications are) and bed rest. The latter's beneficial effects include helping reduce blood pressure and increasing blood perfusion to the uterus and fetus. Once again, however, such measures in a spacecraft or on the ISS would be limited by the availability of medications and some effects of microgravity.

"Preeclampsia" develops in up to about 6–8 % of pregnancies, usually during the second and third trimesters. It manifests as hypertension, protein in the urine (proteinuria), headache, and visual disturbances in the mother. Preeclampsia can injure the fetus by reducing the amount of oxygen it receives. Though its effects are typically not life-threatening, in rare instances preeclampsia can cause serious complications for the mother such as cerebral hemorrhage (bleeding in the brain), stroke, or temporary blindness. The greatest danger is that it might worsen into "eclampsia," with development of seizures and potentially coma and death. Treatment for preeclampsia includes antihypertensive medications. Magnesium sulfate given as an initial intravenous bolus dose followed by a continuous infusion can be used for both severe preeclampia and for eclampsia. Definitive treatment for either condition is delivery of the fetus, with development of eclampsia requiring that it be done as soon as possible.

Significant bleeding due to partial separation of the placenta from the uterine wall ("abruptio placentae") or other causes would also be major problems, especially with limited blood supplies in space and the relative anemia and hypovolemia caused by microgravity. As noted in Chap. 5, the incidence of ectopic pregnancy (implantation of a blastocyst outside the uterus, such as within a fallopian tube or elsewhere in the abdomen) might be increased by microgravity. Sometimes it can be treated with a medication, methotrexate, given either orally or by injection. However, that latter medication is not part of the standard medical supplies on the ISS, and an ectopic pregnancy can become a surgical emergency. The effects of microgravity or acceleration/deceleration (e.g. during launch or reentry) on premature or term rupture of the amniotic sac—the membrane containing the fetus and amniotic fluid within the uterus—are not known.

Delivering a baby in microgravity would, from a practical standpoint, require that the woman herself, medical personnel, and instruments be secured and not floating free. Fetal monitoring equipment would be needed. Maintaining sterility and controlling the normal release of blood and other bodily fluids (e.g. amniotic fluid) at birth in microgravity would also be challenging. A modified "birthing chair," perhaps with an airflow system providing negative pressure (similar to the toilet on the ISS) to trap fluids, might be used. As described in Chap. 5, intravenous or local (e.g. for regional blocks) anesthetic agents could be given and would be expected to be effective. Conversely, epidural and spinal anesthesia would have significant issues regarding safety/efficacy in microgravity.

These difficulties would be compounded if a cesarean section became necessary due to fetal distress or other causes. This would require not only an expert assessment that the procedure was needed, but also that someone trained to perform it be present. Issues regarding maintaining a sterile field for the operation would be critical in that situation, and the woman's recovery time after birth could be longer than on Earth.

10.6 Space Kids

Some science fiction stories involve children and teenagers having exciting (and perhaps dangerous) adventures in space and worlds with less than Earth's gravity. Robert A. Heinlein in particular wrote a number of fine examples of this subgenre, including stories such as "The Menace from Earth" (1957) and novels like *Farmer in the Sky* (1950), *The Rolling Stones* (1952), and *Podkayne of Mars* (1963). Such works can reflect a "sense of wonder" regarding space travel as seen from the unique perspective of young people.

Unfortunately, in real life the same hazards (especially microgravity and radiation) that threaten the health of adults carry even greater short- and long-term risks to babies, young children, and adolescents. Unlike adults, the long bones and muscles of children are still growing and developing. Long-term exposure to micro- or low gravity (e.g. on the Moon or Mars) could result in growth retardation due to lack of the stimulating effect of Earth's gravity on increasing muscle mass and both the mineralization and growth of bones, particularly the long bones of the thighs and legs. The overall potential effects of reduced gravity on the skeletal system could include delayed closure of the fontanels of a baby's skull (the "soft spots" between the bony plates of the skull before those bones fuse), short stature in adulthood due to decreased lengthening of long bones associated with premature closure of epiphyseal (growth) plates, and abnormal development of vertebrae, possibly leading to nerve compression syndromes.

Experiments with young, growing rodents in microgravity have shown the expected decrease in bone mass and reduction in at least radial growth of bones [36, 37, 45]. How much these findings and the theoretical considerations described above actually apply to human children in space remains uncertain. This is based on both the practical reason that no such studies are contemplated in the foreseeable future, and because the risks to growth and development cited above, while not certain, have a strong enough basis in established mechanisms that they can be reasonably considered too significant to evaluate directly.

The fall in muscle and bone mass that occurs in microgravity and is thought to occur to lesser degrees in lunar and Martian gravity can be ameliorated, though not necessarily eliminated, by countermeasures such as vigorous exercise programs. However, babies and toddlers obviously could not perform such countermeasures (e.g. walking on a treadmill or doing resistive exercises) in that environment, and thus would be unprotected from the expected high loss (or lack of initial development) of muscle and bone mass.

Infants raised in either microgravity or, most likely, lunar gravity would not learn how to walk or develop "normal" neurovestibular reflexes. They could well have great difficulty adapting should they later go to Earth. Gradual adaptation and learning while still in space using an artificial gravity system generating a significant fraction of 1 g might at least partially reduce this problem. The question of how prolonged exposure to micro- or low gravity would affect overall neurological (particularly brain) development of children is also unknown. Rat embryos exposed to microgravity while their vestibular system was becoming functional showed delayed development, and retinal cells of newly born rats also showed damage [36, 46]. How much such adverse effects might relate to development in humans is, again, uncertain, but these findings are worrisome.

Other potentially serious physiological effects of reduced gravity include cardiovascular deconditioning and endocrine disturbances. The former is not by itself likely to be significant in microgravity, but it could make adaptation to Earth's gravity more difficult in children than in adults. Likewise the effects of micro- or reduced gravity on the hormonal changes that induce puberty are also unknown and might impair its development.

Babies and children raised in space, the Moon, or Mars would be even more vulnerable to injury from the high levels of radiation there than adults are. Even at their lowest those levels are great enough to potentially cause dysfunctional development of different organ systems and impair growth. Radiation might also significantly increase both the short- and long-term risk of developing cancer (particularly given the child's potentially long lifespan during which malignancies can develop), as well as cause mutations in germ cells (e.g. a girl's oocytes) that could be passed on to progeny when the children reach adulthood. At early ages a child's immune system is also less well developed. Radiation could potentially impair it enough to further weaken it and leave the child more susceptible to infections.

Another intrinsic hazard of having children in space is their emotional and mental immaturity relative to adults. Adolescents and younger children in science fiction (e.g. many of Heinlein's "juvenile" works) may demonstrate remarkable intelligence, responsibility, and resourcefulness. However, in our real world this is not necessarily the case. Every child is unique in the way and rate he or she learns, develops life skills, grows psychologically, etc. How and when these happen can also be uncertain and, in a statement that will come as no shock to those of us who are parents, a child may not act predictably or follow even the most important rules designed to ensure his or her safety. This is an extremely important (and potentially fatal) consideration in outer space and the potentially habitable real estate (e.g. Mars) within our Solar System, where a hostile environment would make disaster never more than seconds away for a child and others.

Overall, it is difficult to reasonably predict from current knowledge of growth and development what the full spectrum and severity of potential problems—anatomic, physiological, psychological, etc.—for children raised in gravity significantly lower than Earth's would be. So far no species of vertebrate animal, much less a human being, has been raised in microgravity from fertilization to development of sexual maturity [36]. Thus, for many years to come the actual effects of doing this will remain speculative, although not beyond the range of reasonable extrapolation. Based on all the considerations described here and barring new information to the contrary, it will require measures such as a great increase in space-based infrastructure (e.g. large space habitats) or methods still far in the future (for example, advanced genetic engineering—see Chap. 12) to deal with them and safely create and raise children away from Earth.

10.7 Sex and Aliens

Unless there is some validity to anecdotal reports of human-extraterrestrial couplings as part of alien abductions, intercourse of the sexual variety between denizens of different worlds is still solely within the sphere of science fiction. The latter may make its job "easier" by having beings from another planet look entirely human, thus helping to explain the attraction between Kal-El of Krypton and Lois Lane of Earth.[16] Likewise the relationship that develops into love and marriage between Edgar Rice Burroughs's John Carter and Dejah Thoris might be enhanced by both the overwhelming similarities and subtly exotic differences between their two species.

Individual humans might be sexually stimulated by interactions with one or more different genders, partners spanning the full spectrum of ages, animals, and inanimate objects (including those that might once have been alive.) Based on these considerations, it is not unreasonable to assume that if alien beings (and not necessarily sentient ones) were encountered, one or more humans might initiate a sexual interaction. Complementary anatomy

---

[16] Larry Niven's essay "Man of Steel, Woman of Kleenex" (1969) enumerates many of the biological challenges involved with a Kryptonian male empowered by Earth's yellow sun and a considerably more fragile human female sharing the ultimate sexual intimacy. The dangers faced by the more vulnerable partner would be similar if the genders of alien and human were reversed. The point is that, just as one must make many suspensions of disbelief regarding the laws of physics and chemistry to enjoy a story involving Superman at his most powerful, one must do the same regarding obvious facts of biology.

for penetration-reception might prove helpful, but other methods might also be employed.

However, while sexual activity between humans and aliens, including actions simulating coitus, is not implausible, science fiction may choose to leave unmentioned certain biological and psychological aspects of such interactions in ways that rival Kryptonian-human ones. Although particular humans might still decide to ignore one or more of these points, there are distinct reasons why such opposites might not attract. These include a particular individual's sexual orientation(s) and how the potential alien partner meets (or, far more likely) fails to meet them; physical, cultural, social, psychological, etc. concepts of sexual attractiveness and "beauty" between beings from two different planets; and basic biological compatibility.

For example, does the alien have the mass of an Apatosaurus and, if nothing else, present a risk of crushing a human partner? Does it require an atmospheric pressure, environmental temperature, or other physical parameters incompatible with human life? Are human sweat or other bodily fluids toxic to it, or vice versa (e.g. an alien secreting sulfuric acid from its skin when sexually aroused)? Does the extraterrestrial, whatever its gender (assuming it has one) practice post-mating behavior similar to a terrestrial female black widow spider or praying mantis?

In the same vein, such subtle factors as secreting certain chemicals called pheromones can, via smell (along with other sensory input such as sight, touch, hearing, or even taste), act as sexual attractants. If an alien species had pheromones based on hydrogen sulfide (both noxious in smell and poisonous to humans), or an extraterrestrial were repulsed by the odor of human perspiration, such factors would seem to reduce the odds of interspecies sexual activity occurring.

It is reasonable to assume both in science fiction and "reality" that alien life (intelligent or otherwise) could employ a broader range of reproductive activity than humans or other species on Earth do. For example, natural selection via genetic diversity in offspring might, in an otherworldly species comparable to humans in size that was asexual and reproduced by fission, be accomplished by environmental factors specific to that world rather than the "mixing" that occurs with sexual reproduction. Chemical mutagens, great enough levels of ionizing radiation from the planet's star, etc. might be sufficient to create a sufficiently high rate of mutations in the single "parent's" germ cells that "cloned" children would have significant differences in their overall makeup.

As mentioned previously, terrestrial bacteria can exchange genetic material without sex, and their sheer numbers greatly accelerate the pace of their evolution (e.g. in developing resistance to particular antibiotics). Variants in DNA compared to the terrestrial variety, life based on other chemistries (e.g. solvents other than water), and other biochemical differences could well be reflected in aliens' sexual practices (or lack thereof). Likewise, although humans have two basic (and dimorphic, with two forms) genders, with one of each having appropriately functioning anatomy required for successful reproduction, as suggested by the examples from Asimov and Niven cited at the beginning of this chapter this might not necessarily apply to alien species. Having three or more genders raises issues regarding how such a more complex arrangement (relative to the minimum number of two required to create adequate genetic diversity in offspring) might have evolved, but certainly does not exclude its possibility.

On the other hand, with apologies to Mr. Spock and many later characters in the *Star Trek* universe as well as other science fiction, the chance of a human-nonhuman mating producing offspring is extremely unlikely. That would require a degree of overall biological compatibility far beyond the level of anatomy, cells, and biochemistry, to that of the very elements in the bodies of human and alien sexual partners. While there is now DNA evidence that *H. sapiens* and *H. neanderthalensis* could and did interbreed, they were closely related members of the same genus, with fundamentally minimal differences in physiognomy and physiology—essentially very minor variations on a single biological theme.

However, any natural mating activity between a human and even the most closely related nonhuman primate is intrinsically sterile due to sufficient differences in their germ cells, DNA, the particular proteins and other substances their bodies produce, and many other factors. No matter how humanoid one may look, beneath the surface such differences between an alien and a human would be tremendously greater and far beyond superficial traits such as the pointed ears of Vulcans and Romulans. As we will see in Chap. 12, a change in only a single enzyme the human body produces can cause death and disability. Even if they serve similar physiological functions, the underlying incompatibilities between the actual amounts and types of trace elements, chromosomes, cells, etc. between two species evolved on different worlds would make it, from a practical standpoint, impossible for them to produce viable offspring. Put another way, a theoretical child produced by such a union would at the very least represent an entirely new species *sui generis*—drastically different in certain ways from either of its progenitors. Unfortunately those differences in biochemistry, physiology, etc. would be simply too great to create a living organism.

In futuristic societies like that depicted in *Star Trek*, one might posit "assisted" reproduction via genetic engineering or "hidden" genetic compatibilities, such as those suggested in the *Star Trek: The Next Generation* episode "The Chase" (1993), which reveals that ancient aliens "seeded" many worlds with similar genetic material. However, as described in previous chapters, DNA basically acts as a template for eventual production of a wide range of different proteins from specific amino acids. These in turn, acting in such capacities as individual enzymes and hormones, assist in life-enabling metabolic activities, synthesis of carbohydrates and lipids, etc. whose combination of overall details are specific for a given species. Assuming they make proteins, the particular amino acids alien life may use, whether their chirality ("handedness") is right or left (see Chap. 1), and the proteins thus created need not be identical to that in humans or any other form of terrestrial life. Put another way, no matter how advanced the genetic engineering techniques used, blending the DNA of an amoeba and an apple will not produce an aardvark—and any alien DNA would be expected to be even more incompatible with a human's in the "coding" it contains than that.

Science fiction may try to get around such issues in other ways. In the movie *Starman* (1984) the titular alien uses the hair (or, presumably, the DNA derived from its root) [7] of a woman's recently deceased husband to transform itself into a copy of him as he was just before he died. At the risk of giving away a plot spoiler, the alien-now-in-a-human-body and the woman have an idyll of intimacy. Very soon afterwards the alien tells her that she has conceived their child who, because the former is a biological copy of her dead spouse, will (at least genetically) be the latter's.

Such plot elements combine the at least superficially factual with the wildly implausible. If the alien's body were indeed the genetic twin of the female character's husband, there might be a slight biological validity to its statement.[17] Even looked at most generously, at best the situation might be considered somewhat similar to one in which her husband had left cryogenically preserved spermatozoa prior to his demise that she used for in vitro fertilization or artificial insemination. However, the major biological "miracle" involved is how the alien actually, within the space of a few minutes, creates an entire human body from a tiny amount of DNA. As noted in previous chapters, our bodies contain many trillions of cells that are created, differentiate, and develop over (as one example) 40 years to produce a 40-year-old human being. The obvious questions of how even the most "advanced" alien (via technology or its own intrinsic biology) can do this so quickly, where it gets the raw material to do this (never mind transforming it into highly specific

---

[17] But only slight. Even "identical" twins, who start out genetically identical at the earliest stage of embryonic development, become increasing less so over time due to factors such as epigenetic changes, environmental influences, random inactivation of particular genes within their cells, etc. Also, there is more to the concept of being an individual person than genetics. Thus, if one of a pair of male monozygotic twins (see Chap. 9) fathers a child, the other one could not reasonably be considered the child's father too.

types and amounts of proteins, carbohydrates, lipid, cellular organelles, etc.), and so on are left unanswered.[18] The reason for showing but not explaining how the alien does it appears to be that while this metamorphosis is necessary for the plot, it cannot be explained by any "conceivable" scientific means. As with other science fiction works that contain such "miracles," one can either choose to stop reading or watching at that point, or continue and see how well the story proceeds after deciding on a willing (but very great!) suspension of disbelief.

In short, as the title of a 1967 movie suggests, Mars may need women. However, for the reasons described previously and others, barring dubious plot twists such as the Martians actually being survivors of an advanced Atlantean or other human civilization who colonized the Red Planet—an "explanation" actually used in the movie *Fire Maidens of Outer Space* (1956)—they should not expect any progeny resulting from their quest. The reason why the bug-eyed monsters ("BEMs") of early twentieth century pulp science fiction lusted after human females may well have had a biological rationale, but one based on the reproductive interests of its human male readers rather than those of the peculiarly xenophilic aliens themselves. Even in the 1958 film *I Married a Monster from Outer Space* (an excellent example of an unambiguously plot-summarizing title) the union between alien and human proves fruitless.

And yet—despite the biological implausibility of his origin, the original *Star Trek* series would have been poorer without Mr. Spock's efforts to fit into both human and Vulcan societies. Here again, even if a science fiction plot contains only a smidgen of "real" science but a generous serving of drama and thought-provoking action, the final result can still be very effective.

10.8 The Bottom Line

All the reproductive issues described in this chapter represent significant challenges in the real world of medical science, with great uncertainty as to how, when, or even if they can be met. Science fiction has it much easier by virtue of its ability to immediately "cut to the chase" and assume that the artificial uterus, routine human cloning, etc. have become realities and then use them as plot devices.

Problems associated with microgravity and radiation can already (at least potentially) be dealt with reasonably well using present-day technology and known scientific principles. Difficulties in dealing with them are instead due largely to economic, political, and other non-scientific factors. For example,

---

[18] These issues also overlap with the problems with "shapeshifting" discussed in Chap. 2.

as noted in Chap. 4, given enough money and other resources it would be possible to build a space habitat large enough to house thousands of individuals and protect them from radiation. It could also be rotated at a velocity high enough to provide sufficient artificial gravity and have enough medical resources on board to effectively minimize the concerns about coitus, pregnancy, child-rearing, etc. in space described previously. In this particular example, "merely" increasing the scale of size and resources above current capabilities in space stations "solves" these problems.

On the other hand, future colonists on the Moon and Mars will still have to work within certain fixed parameters (e.g. those worlds' ambient gravity). They will also have to implement countermeasures to reduce risks from or work around other dangers (such as from radiation) that, short of very advanced and complex means such as the Martian terraforming depicted in Kim Stanley Robinson's "Mars" trilogy of novels, will only partially ameliorate them. Nonetheless, works like Heinlein's "juvenile" novels or his other writings such as *The Moon is a Harsh Mistress* (1966) that depict children growing up and biologically thriving in extraterrestrial settings within the Solar System can still be enjoyed as stories in spite of those biological issues.

References

1. Stratmann HG, Stratmann M. Sex and your heart health. Springfield: Starship Press, LLC; 2007.
2. Clément G, Slenzka K. Animals and plants in space. In: Clément G, Slenzka K, editors. Fundamentals of space biology. Sacramento: Microcosm Press and Springer; 2006. pp. 51–80.
3. Davenport ML. Approach to the patient with Turner syndrome. J Clin Endocrinol Metab. 2010;95(4):1487–95.
4. Groth KA, Skakkebaek A, Host C, Gravholt CH, Bojesen A. Clinical review: Klinefelter syndrome–a clinical update. J Clin Endocrinol Metab. 2013;98(1):20–30.
5. Kim I, Khadilkar A, Ko E, Sabanegh E. 47, XYY syndrome and male infertility. Rev Urol. 2013;15:188–96.
6. Otter M, Schrander-Stumpel CT, Curfs LM. Triple X syndrome: a review of the literature. Eur J Hum Genet. 2010;18(3):265–71.
7. Tartaglia NR, Howell S, Sutherland A, Wilson R, Wilson L. A review of trisomy X (47,XXX). Orphanet J Rare Dis. 2010;5:8.
8. Pandey S, Shetty A, Hamilton M, Bhattacharya S, Maheshwari A. Obstetric and perinatal outcomes in singleton pregnancies resulting from IVF/ICSI: a systematic review and meta-analysis. Hum Reprod Update. 2012;18(5):485–503.
9. Assisted Reproductive Technologies. A guide for patients. 2014. http://www.reproductivefacts.org/uploadedFiles/ASRM_Content/Resources/Patient_Resources/Fact_Sheets_and_Info_Booklets/ART.pdf. Accessed 15 April 2015.

10. Bulletti C, Palagiano A, Pace C, Cerni A, Borini A, de Ziegler D. The artificial womb. Ann NY Acad Sci. 2011;1221:124–8.

11. Pak S, Song C, So G, Jang C, Lee K, Kim J. Extrauterine incubation of fetal goats applying the extracorporeal membrane oyxgenation via umbilical artery and vein. J Korean Med Sci. 2002;17:663–8.

12. Sakata M, Hisano K, Okada M, Yasufuku M. A new artificial placenta with a centrifugal pump: long-term total extrauterine support of goat fetuses. J Thorac Cardiovasc Surg. 1998;115:1023–31.

13. Briggs R, King TJ. Transplantation of living nuclei from blastula cells into enucleated frogs' eggs. Proc Natl Acad Sci U S A. 1952;38:455–63.

14. Wilmut I, Schnieke A, McWhir J, Kind A, Campbell K. Viable offspring derived from fetal and adult mammalian cells. Nature. 1997;385:810–3. (February 27)

15. Institute NHGR. Cloning fact sheet. 2014. http://www.genome.gov/25020028. Accessed 15 April 2015.

16. Niemann H, Lucas-Hahn A. Somatic cell nuclear transfer cloning: practical applications and current legislation. Reprod Domest Anim. 2012;47(Suppl 5):2–10.

17. Whitfield J. Obituary: Dolly the sheep. 2003. http://www.nature.com/news/2003/030218/full/news030217-6.html. Accessed 15 April 2015.

18. Wilmut I, Beaujean N, de Sousa P, Dinnyes A, King TJ, Paterson L, et al. Somatic cell nuclear transfer. Nature. 2002;419:583–6.

19. Thuan N, Kishigami S, Wakayama T. How to improve the success rate of mouse cloning technology. J Reprod Dev. 2010;56:20–30.

20. Ogura A, Inoue K, Wakayama T. Recent advancements in cloning by somatic cell nuclear transfer. Philos Trans R Soc Lond B Biol Sci. 2013;368(1609):20110329.

21. Kues W, Rath D, Niemann H. Reproductive biotechnology goes genomic. CAB Rev. 2008;3:1–18.

22. Wakayama S, Kohda T, Obokata H, Tokoro M, Li C, Terashita Y, et al. Successful serial recloning in the mouse over multiple generations. Cell Stem Cell. 2013;12(3):293–7.

23. Chung Young G, Eum Jin H, Lee Jeoung E, Shim Sung H, Sepilian V, Hong Seung W, et al. Human somatic cell nuclear transfer using adult cells. Cell Stem Cell. 2014;14(6):777–80.

24. Tachibana M, Amato P, Sparman M, Gutierrez Nuria M, Tippner-Hedges R, Ma H, et al. Human embryonic stem cells derived by somatic cell nuclear transfer. Cell. 153(6):1228–38.

25. Yamada M, Johannesson B, Sagi I, Burnett LC, Kort DH, Prosser RW, et al. Human oocytes reprogram adult somatic nuclei of a type 1 diabetic to diploid pluripotent stem cells. Nature. 2014;510:533–6.

26. Baker M. Stem cells made by cloning adult humans. 2014. http://www.nature.com/news/stem-cells-made-by-cloning-adult-humans-1.15107. Accessed 15 April 2015.

27. Stratmann HG. Sex in space: The fantasy and the reality. Analog Sci Fict Fact. 1998;118(2):45–59.

28. Woodmansee L. Sex in space. Burlington: CG Publishing, Inc.; 2006.

29. Buckey JC. Space physiology. Oxford: Oxford University Press; 2006.

30. Kanas N, Manzey D. Space psychology and psychiatry. 2nd ed. Berlin: Springer; 2010.

31. Longnecker D, Molins R. A risk reduction strategy for human exploration of space. Washington, D.C.: The National Academies Press; 2006. http://www.nap.edu/catalog/11467/a-risk-reduction-strategy-for-human-exploration-of-space-a. Accessed 6 May 2015.

32. Amann R, Deaver D, Zirkin B, Grills G, Sapp W, Veeramachaneni D, et al. Effects of microgravity or simulated launch on testicular function in rats. J Appl Physiol. 1992;73(2):S174–85.

33. Tou J. Models to study gravitational biology of mammalian reproduction. Biol Reprod. 2002;67(6):1681–7.

34. Planel H. Space and life. An introduction to space biology and medicine. Boca Raton: CRC Press; 2004.

35. Genital Herpes–CDC Fact Sheet. http://www.cdc.gov/std/herpes/STDFact-Herpes.htm. Accessed 6 May 2015.

36. Clément G. Fundamentals of space medicine. 2nd ed. Berlin: Springer; 2011.

37. Horn E. Animal development in microgravity. In: Clément G, Slenzka K, editors. Fundamentals of space biology. Sacramento: Microcosm Press and Springer; 2006. pp. 171–226.

38. Wolgemuth D, Murashov A. Models and molecular approaches to assessing the effects of the microgravity environment on vertebrate development. ASGSB Bull. 1995;8(2):63–71.

39. Taddeo T, Armstrong C. Spaceflight medical systems. In: Barratt M, Pool SL, editors. Principles of clinical medicine for space flight. Berlin: Springer; 2008. pp. 69–100.

40. Jennings R, Baker E. Gynecologic and reproductive concerns. In: Barratt M, Pool SL, editors. Principles of clinical medicine for space flight. Berlin: Springer; 2008. pp. 381–90.

41. Williams P, Fletcher S. Health effects of prenatal radiation exposure. Am Fam Physician. 2010;82:488–93.

42. Mole R. Consequences of pre-natal radiation exposure for post-natal development. A review. Int J Radiat Biol. 1982;42:1–12.

43. Otake M, Schull W. In utero exposure to A-bomb radiation and mental retardation; a reassessment. Br J Radiol. 1984;57:409–14.

44. Standards for Protection Against Radiation. http://www.gpo.gov/fdsys/pkg/CFR-2011-title10-vol1/xml/CFR-2011-title10-vol1-part20.xml#seqnum20.1201. Accessed 6 May 2015.

45. Morey-Holton E, Hill E, Souza K. Animals and spaceflight: from survival to understanding. J Musculoskelet Neuronal Interact. 2007;7:17–25.

46. Ronca AE, Fritzsch B, Bruce LL, Alberts JR. Orbital spaceflight during pregnancy shapes function of mammalian vestibular system. Behav Neurosci. 2008;122(1):224–32.

47. Hughes C. Challenges in DNA testing and forensic analysis of hair samples. 2013. http://www.forensicmag.com/articles/2013/04/challenges-dna-testing-and-forensic-analysis-hair-samples. Accessed 15 April 2015.

11

The Promises and Perils of Medical Nanotechnology

To see a world in a grain of sand…

William Blake
"Auguries of Innocence"

Nanotechnology involves manipulating matter at the scale of atoms and molecules. More specifically it deals with materials and devices with sizes in the range of 1–100 nm (nm, 1 nanometer = 1×10^{-9} m), as well as processes operating at that level [1–3]. Many molecules within the human body fall within that range of sizes. For example, proteins have dimensions between 1 and 20 nm, the width of a DNA helix is about 2.5 nm, and even ribosomes, the protein-constructing organelles within cells, have diameters of about 2–4 nm. Indeed, our bodies are living models of "natural" nanotechnology in action.

However, the concept of nanotechnology, including its possible medical applications, is a relatively recent one in both the worlds of science and of science fiction. Some of its concepts can be traced back to a talk and subsequent paper by the renowned physicist Richard Feynman over 50 years ago [4]. In it Feynman described potential applications regarding miniaturization and working at essentially the nanoscale, including the possibility of "swallowing" a tiny mechanical surgeon that could evaluate and operate on a diseased heart valve. The term "nanotechnology" is said to have originated in a paper by Norio Taniguchi in 1974. Beginning in the 1980s K. Eric Drexler described and popularized many principles and potential uses for nanotechnology [5–7].

For example, the 1966 movie *Fantastic Voyage*[1] depicts a fanciful version of Feynman's internally operating surgeon. In it the *Proteus* (a high-tech submarine that in some ways makes the *Seaview* of the 1964 television series *Voyage to the Bottom of the Sea* look like an original Model T car by comparison) and its human crew are shrunk to, if not quite the nanometer level, at least small

---

[1] The novelization of this film was created by an author with some prior experience writing science fiction, one "Isaac Asimov."

enough to be safely injected into the bloodstream of a normal-sized patient. The mission of the medical personnel and others onboard is to navigate their craft to that unconscious individual's brain and blast a blood clot blocking one of his arteries with a rifle-like laser, thus restoring blood flow and (hopefully) "cure" him.[2]

Somewhat more realistically, the future of nanotechnology has been projected to include development of self-replicating "nanobots" imbued with artificial intelligence, other types of submicroscopic machines, new materials for uses as "mundane" as clothing and as exotic as space elevators, and medical applications ranging from treating diseases to reversing the effects of aging [6, 8, 9]. This chapter will focus on how nanotechnology intersects with human biology—a subject that science fiction has also explored in recent decades. For example, Neal Stephenson's *The Diamond Age* (1995) depicts the realization of what are currently still unfilled promises of both medical and nonmedical nanotechnology, such as ubiquitous "matter compilers" that create food, drugs, and many other materials.

On the other hand, the creations of medical nanotechnology in science fiction can also develop dangerous "minds" of their own. Greg Bear's novel *Blood Music* (1985) posits the creation of "noocytes," lymphocytes (a type of white blood cell) genetically altered at the nanoscale that develop an aggregate intelligence. Working inside a person's body, the noocytes start with physically improving it but ultimately transform people into something far more than human. Michael Crichton's *Prey* (2002) and Edward M. Lerner's *Small Miracles* (2010) depict nano-size devices within a person's body interacting with the brain in ways they were not originally designed to do. In the episode "The New Breed" (1995) of the relaunched version of *The Outer Limits* nanobots become overenthusiastic about "improving" an ill individual's body. The Borg in the later *Star Trek* series are notorious for practicing medicine via nanoprobes on the inhabitants of whole worlds with little attention to having their "patients" sign consent forms beforehand. In my story "To Save Man"[3] the human race is at the receiving end of an unrequested "treatment" using similar technology, with ambivalent results.[4]

---

[2] Although they look nothing like the one shown in the movie, modern medicine does sometimes employ lasers as part of a catheter system to help destroy plaque and blood clots in blocked arteries. Effective "thrombolytic" medications are also available for this purpose. Unfortunately miniaturized submarines are not yet part of standard medical practice.

[3] *Analog Science Fiction and Fact* July/August 2012.

[4] However, unlike the Borg I was not practicing medicine without a license when I did it.

11.1 Nanotechnology and the Human Body

All things considered, the biological nanotechnology we are born with works amazingly well. As described in Chap. 9, at its best it can potentially keep us alive and reasonably healthy for a little over 120 years. Our RNA, DNA, proteins, hormones, and other organic molecules act as endogenous nanobots in complex systems that maintain all the complex "homeostatic" processes needed to sustain life. There are many feedback loops within our bodies "engineered" to produce just the right amount of hormones, initiate blood clotting when we bleed, retain or eliminate water and electrolytes as needed, and fight infections.

Unfortunately, for all the good they do those same molecules and systems can sometimes be "fooled" into hurting instead of helping, or even just standing by when danger threatens. For example, as noted in Chap. 2 there are many types of "autoimmune" diseases, which involve a person's own immune system mistakenly destroying healthy tissue such as the thyroid gland in Graves' disease, or joints, kidneys, skin, etc. in systemic lupus erythematosus. The "coagulation system" makes blood clot appropriately to stop bleeding after a cut. That same system is activated when a partially obstructing atherosclerotic "plaque" ruptures in a coronary artery. However, in the latter case the clot produced can block off that artery and the blood it supplies to the heart enough to cause a potentially fatal myocardial infarction.

Likewise, the "renin-angiotensin-aldosterone system" involves a complex set of actions by those three substances and others to help maintain blood pressure when it falls too low. Among other effects, arteries "clamp down" and the kidneys are stimulated to retain fluid and maintain blood volume. Those kinds of responses are helpful for certain kinds of problems, such as an injury that causes mild to moderate bleeding. However, when the problem is due to other causes like the heart's left ventricle not pumping as "strongly" as it should, too much fluid may be retained and leak out into the lungs, making the person's overall situation worse rather than better. Our immune systems can fail to recognize and destroy cancer cells. And, as seen in Chap. 9, our bodies' abilities to stave off and repair the effects of aging have very definite limits.

Medical nanotechnology holds great potential for improving on these and other shortcomings of our built-in nanotech apparatus. This type of nanotech would make use of our extensive and growing knowledge of human biochemistry, anatomy, physiology, cell function, etc. to either assist our own body systems or perhaps even supplant them with "new and improved" versions. Researchers in the field have suggested that a sufficiently advanced medical technology could cure cancer, make wholesale "improvements" in how our bodies look and function, and provide healthy lives lasting centuries. The science fiction works cited previously and others explore the implications and sometimes dangers of those and other possible results of medical nanotechnology. However, there remains a huge gap between what those methods can currently do and what they can be imagined to do in the future.

11.2 Medical Nanotechnology: The Dream

K. Eric Drexler's book *Engines of Creation*, first published in 1986, is frequently cited as laying out many principles and promises of nanotechnology.[5] Drawing in some cases from analogies involving biological processes that create proteins and DNA, Drexler's book postulated the development of different types of "nanomachines" such as "assemblers" and "replicators." As described in Chap. 1, our cells contain a system for synthesizing proteins that includes collecting individual molecules of amino acids, transporting those "parts" and the information on how to put them together to "factories" (ribosomes), and then "assembling" them into a specific protein. Likewise DNA has the ability to "replicate" itself and direct the creation of whole new cells.

Enzymes are biological molecules, primarily proteins, that are present within and outside cells. They act as "assemblers" and "disassemblers" of various substances that our bodies need, greatly accelerating chemical reactions that help to either construct or break down molecules (the "substrate") that are specific for particular enzymes [10]. Drexler describes designing enzymes and "protein machines" to assemble new biological and other materials that go well beyond those involved in life processes. He hypothesized that "molecular machines" could also be controlled by electronic or mechanical computers operating at the nanoscale level that could direct their actions to produce a material or perform some other action.

Possible medical applications of nanotechnology were described as including cell repair machines that can enter a cell, detect abnormalities in DNA or other molecular components within it, and correct them. This concept included intracellular repair devices that could perform only a single type of repair, or more complex ones capable of multiple repair functions. The latter could be guided by a nanocomputer programmed to assess what needs to be done and issue "directions" to do it.

Such nanodevices could also potentially act as monitors to continuously assess the health of the body's cells, work with or independently of the built-in

---

[5] Drexler has given an updated assessment of the potentials of nanotechnology in a more recent book, *Radical Abundance. How a Revolution in Nanotechnology Will Change Civilization* (2013).

immune system to battle infections, reverse errors in DNA and malfunctions in metabolic processes to eliminate a variety of diseases and cancer, and either prevent or repair the effects of aging. Some of these potential beneficial effects could also be performed outside of cells, such as by nanobots removing atherosclerotic plaques from the inner linings of arteries.

Delivering on some of these possibilities for nanomedicine could include creating machinery at the molecular level to provide motion and other mechanical actions, as well as nanobots with computational capabilities. Such devices might involve production and use of rods, shafts, cylinders, bearings, motors, etc. analogous to macroscopic machines but created and functioning at the molecular level [7, 8, 11, 12]. Both biological and non-biological molecules and machinery might be combined to perform those functions and even improve on those provided by Nature.

For example, a "respirocyte" is a proposed artificial red blood cell that could store and release oxygen and carbon dioxide better than the natural kind [8, 13]. It would have the equivalent of pressure tanks, pumps, and filters within it in a size (about one cubic micron in volume, a micron being 1×10^{-6} m) able to pass through the tiniest blood vessels, the capillaries. The respirocyte would be powered by glucose, a fuel ubiquitous in the bloodstream. If enough respirocytes were in the blood they could provide a greater "reserve" of oxygen than conventional red blood cells do, allowing a person to hold his or her breath for much longer periods of time (e.g. while swimming underwater).

A similar theoretical nanodevice that improves on natural cells is the "microbivore," a microscopic machine that physically destroys pathogenic bacteria, viruses, and other infectious agents [8]. These would have binding sites on their surfaces specific for a particular pathogen. After physically grabbing the latter the microbivore would pull it into an internal chamber where it would be "digested" (think of what a blender does to food) into component parts that would in turn be released back into the bloodstream for use and clearance by the body.

For more complex medical nanobots to work they may need the equivalent of sensors to acquire information about their environment, including the presence and structures of specific molecules. Techniques for doing this can be based on biochemical models such as that used by enzymes and ribosomes, or the "lock and key" binding between antigens (e.g. "foreign" substances within the body) and antibodies (proteins a person's immune system produces to "connect" to specific antigens and interact with them). Sensors based on physical phenomena such as displacement, motion, force, pressure, temperature, electrical and magnetic fields, etc. might be incorporated into "medical nanorobots"[13]. These could involve springs and gyroscopes machined at the nanoscale to measure velocity and acceleration, nanopendulums to assess force, or materials used to measure temperature based on the latter's effects on their expansion or contraction. All of these could involve totally non-biological materials in their design and manufacture.

Artificial nanobots could require an energy source. This might be provided by mechanical means (e.g. pendulums and flywheels) or by chemical, thermal, acoustic, electrical, magnetic, or even nuclear energy. One form of energy might also be converted into another (such as electrical into mechanical) for the nanobot to do its work. It might also be helpful for nanobots to "communicate" with each other by chemical methods or even by having built-in radio frequency transceivers. For example, if one nanobot were to detect a cancer cell or a harmful bacterium it could then "call for reinforcements" to help it assess the extent of the problem and deal with it.

Nanobots would require the ability to physically navigate throughout the body to go where they are "needed." Our blood vessels and the body's parallel network of lymphatic system vessels offer the obvious "superhighway" for them to do this. However, as noted previously, to circulate throughout the body nanobots must be small enough to pass through the tiniest blood vessels. They must also be "smart" enough to know when to "get off the road" and into an appropriate body part, even down to the intracellular level. Meanwhile they must not be neutralized or attacked by the cells and organs of the immune system, stopped by "roadblocks" and accumulate or be trapped within organs other than their targets, avoid reacting with the biological molecules, cells, and substances they are traveling with, and not be prematurely cleared from the body by the kidneys, gastrointestinal tract, etc.

It would also be helpful for nanobots to be able to move independently rather than being just passively swept through the bloodstream or extracellular space. This might be done by mechanical means that mimic those found in Nature, like the tail of a sperm cell. Once they arrive at their target, nanobots could also benefit from having "manipulators"—limb-like structures or other types of "tools" to physically work with a cell's organelles, damaged DNA, etc. These too might be modeled after those found in living organisms, such as the tiny hair-like cilia of some bacteria or nanoscale versions of tentacles. Pneumatic (pressure-based) manipulators using changes in volumes of gases or liquids, or telescoping ones that extend and retract via mechanical means are other possibilities.

One variation on the concept of self-constructing and replicating nanomachines is that of a "nanofactory" working within a person's body [14]. It entails creation of artificial "pseudo-cells" composed of shell-like particles introduced into an individual that are capable of synthesizing substances needed by the body using the latter's own biomolecules. Such a nanofactory would be designed to detect the presence of infections, cancer cells, or other abnormalities. It would then create a drug or other material from the body's own biochemical "raw materials" that would in turn be delivered to where it could provide treatment.

While doing its work the nanofactory would, like more "conventional" nanobots, need to avoid destruction by the immune system or premature clearance from the body. It would also either have to sense when it needed to stop production of a therapeutic substance or have the ability to be "turned off" from outside the body when its work was done.

A similar concept involves using DNA to assemble very small (less than 6 nm), individual nanoparticles made from inorganic materials into larger "superstructures" within the body. After interacting with target tissues (e.g. delivering a drug to a cancer cell) these superstructures could be designed to then break down into smaller components that would be more easily recycled or eliminated by the body [15].

Another possibility is an autonomous DNA robot that can be "folded" into a container holding a chemotherapeutic agent as its payload. By coupling it with a type of molecule designed to attach specifically to a cancer cell the DNAbot could penetrate it and release its lethal contents into that cell [16].

11.3 Medical Nanotechnology: The Reality

However, before or even if such advanced capabilities using nanotechnology might become reality, a large number of practical issues need to be addressed. These include how difficult nanobots might be to make, their durability, the possibility they could develop problems with their programming, and their potential for (like conventional medications) causing unexpected "side effects."

Individual nanobots or other nanodevices would also have to "know" what cells to target to fulfill their functions and be able to reach them. The latter might be especially difficult in "compacted" tissues with little extracellular space, such as the liver or brain, whose cells roughly border each other. As noted previously, nanoscale devices would need to be small enough to pass throughout the circulatory system and cross cell membranes while avoiding destructive interactions with and premature clearance by the body. Other issues include whether a "pool" of them could or should be "stored" somewhere in the body for future activation and, if so, determining where that would be and how to let them "know" that they were needed.

Another important question is, if nanodevices had the ability to self-replicate using the body's own biological materials,[6] how would that be regulated? For example, the human body can increase the rate of release and production of certain types of white blood cells (WBCs) from bone marrow when its immune system detects the presence of an infection. Many of those WBCs will be destroyed as part of their infection-fighting activities, and the number created will decrease to normal levels after the infection resolves. If enough of these WBCs "escape" this control mechanism and replicate unchecked, the result is leukemia, a type of malignancy.

Nanodevices too would need effective controls to ensure that neither too many nor too few were available to do their work inside the body, and that they did only what they were designed to do. Drexler used the term "gray goo"[6] to describe the results of self-replicating nanodevices processing materials in their environment in an uncontrolled fashion, "outcompeting" other systems (including biological ones, like the molecules and cells inside our bodies) in using those materials and growing. Inside the human body, the gray goo concept could be considered analogous to a cancer that (especially if the uncontrolled nanodevices replicate faster than human cells) would be much more virulent than the worst naturally occurring variety.

Even assuming this "worst case scenario" did not occur and nanodevices inside us acted as designed, there might also be practical limits to how well they might work. For example, if they were programmed to make repairs inside the body following severe trauma, they would have to reach the damaged areas (presumably at least partly via the bloodstream) and treat injuries rapidly enough to keep the individual alive. However, if the trauma involved significant internal or external blood loss along with low blood pressure, with arteries within body parts and organs constricting and reducing flow to them as a response, it might be difficult to impossible for nanobots or similar devices to reach damaged areas. The "window of opportunity" for the nanobots to reduce and hopefully halt ongoing damage to vital organs before injuries became fatal could be very narrow—perhaps too short for even the best-designed nanodevices.

Even with less severe damage it would still take nanobots a finite amount of time to make repairs. Moreover, they would need to be "smart" enough to both identify what specific problems potentially many millions of cells were having and also have the proper "tools" to deal with them. The latter would have to include an extremely broad range of individualized repair plans based on the sheer complexity of the human body, with its many different kinds of

---

[6] This capability would make them, at least in one sense of the term, what have been called "von Neumann machines."

cells having their own particular characteristics and functions, responses to hormones and other substances, varying vulnerabilities to injury, etc. Such internal repair systems might require both "general practitioner" nanobots for broader levels of monitoring/repairing the body that are common to many types of cells and organs, as well as "specialist" nanobots designed to deal with problems specific to particular ones.

In short, the sheer scale of the task even the most advanced nanobots would be expected to do—"policing" the 37 trillion-plus cells of the human body, with its over 200 different types of cells in myriad configurations, to ensure they all work well and fixing them if they are not—is staggering. As noted previously, this also assumes that the nanobots themselves can be injected/ mobilized quickly enough to do their work within the intrinsic limits of the human body to maintain its own homeostasis (including the finite amount of time needed for cell division, something not all its cells—like the terminally differentiated ones of the heart and brain—are capable of doing) and that the nanodevices will not cause problems of their own. What might work in principle at the nanoscale may be far more difficult to implement in practice at the complex, macroscopic scale of an entire human body.

Medical nanotechnology in our present-day world has a drastically different meaning than the (so far) purely conceptual uses described previously. Current medical nanotechnology focuses on using nanoscale materials rather than devices. According to one report, as of 2011 there were over 1300 commercially available products containing nanomaterials. These include cosmetics, food containers, sunscreens, hair sprays, and similar items without futuristic or science fiction connotations [17]. No nanobots or miniature medical submarines are on that list.

Present-day "nanomedicine"[3] focuses on using particles of various types with at least one dimension at the nanoscale level. The small size alone of nanoparticles gives them the advantage of being able to enter cells, tissues, and other parts of the body better than larger particles. This can enable them to deliver medications and other materials to otherwise inaccessible areas to diagnose and treat a variety of conditions, especially cancer [1, 18].

A key advantage to medical nanomaterials is that they have a very large ratio of surface area to volume [3, 19]. For example, a solid cube with dimensions of 1 cm on a side could be divided into 1×10^{21} smaller ones that are 1 nm on a side. The total volume of that single cube and all those smaller ones would be the same, but the total surface area of the smaller cubes would be 10 million times more than the original one. The much greater overall surface area of nanoparticles compared to larger ones allows many more biologically active molecules such as medications to be coated onto or otherwise attached to their surfaces than would be possible with the smaller surface area of larger particles [2].

Moreover, the physical and chemical properties of a particular material's particles may be significantly different at the smallest scales [19]. For example, iron oxide nanoparticles 20 nm or less in size generate stronger magnetic fields than larger ones due to differences in the electron spins associated with their respective size ranges [2, 20]. This difference has practical effects in how well various sizes of iron oxide nanoparticles can be used for diagnostic medical purposes, such as improving the quality of images obtained with magnetic resonance imaging (MRI).

Another example is the titanium dioxide and zinc oxide used in sunscreen preparations to protect against ultraviolet light from the Sun. Conventional preparations of these compounds with larger particles have a white color when applied to skin. In contrast, nanoparticles of these compounds are transparent and thus more acceptable aesthetically when used in sunscreen lotions, creams, etc [19].

Many kinds of nanoparticles with different compositions, shapes, and sizes are being evaluated for diagnosing and treating illnesses, including delivering drugs and even genes to specific cells in the body [21–23]. Metallic ones are often made of gold due to that element's particularly useful properties in biological systems. Nanoparticles using gold can also be created in a variety of diameters and shapes, including rods and shells.

Gold nanoparticles have relatively low toxicity inside the body and can be used to delivery drugs to cells within it [24]. They might also be employed for "photothermal" therapy. This involves using an external laser or other source of light energy to stimulate gold nanoparticles to vibrate and heat up—a physical property that larger gold particles do not possess [19]. The heat produced in those nanoparticles can be great enough to destroy a tumor cell they are either in contact with or are inside. The light source can send its energy at power levels sufficient to "activate" gold nanoparticles but too low to damage healthy tissue. That light can be in the near-infrared range (wavelengths of about 750–900 nm), which penetrates skin and cells beneath it (e.g. a tumor) deeper than visible light. For example, near-infrared light can penetrate through skin about 3 cm into muscle and brain, and as much as 10 cm into breast tissue. This technique thus may be a potential means for noninvasively treating relatively superficial tumors involving those areas [25].

Gold nanoparticles may also be used for diagnostic purposes [24]. This includes employing them as contrast agents for computed tomography (CT) systems or treating them in ways that let them detect metabolic events associated with tumor growth. Sensor arrays using gold nanoparticles have been used to analyze the exhaled breath of patients to detect organic compounds associated with lung cancer [26]. This method could potentially serve as a simple, rapid, and noninvasive way to screen for the presence of lung cancer in appropriate at-risk patient populations.[7]

Silver nanoparticles could be applied externally to the skin to treat burns and open wounds. They can kill bacteria, reduce inflammation, and increase the rate of healing better than silver preparations using larger particles and may also have fewer side effects than them. Silver nanoparticles are used in commercially available products such as burn wound and surgical dressings [3, 19]. Due to their reported greater efficacy for promoting healing and protecting against infection they may also require less frequent dressing changes, thus also reducing the cost of care.

Magnetic nanoparticles, particularly those made with iron oxide, may also be used as carriers to deliver drugs within the body [24]. As mentioned previously, they can also serve as a component of contrast agents to improve the quality of images obtained by MRI [3, 27]. Such iron oxide nanoparticles may be "superparamagnetic"—that is, they become magnetized when subjected to a strong magnetic field but are not magnetic when the field is stopped. This property can enhance their value as contrast agents.

Another type of nanoparticle called a "quantum dot" uses semiconductor materials made of compounds of zinc, cadmium, or other metals [24, 28, 29]. Quantum dots are used in biological research as imaging agents. They emit light whose wavelength depends on their size (typically in the range of 2–10 nm) rather than their composition. Cells, DNA, and other biological material can be "marked" with quantum dots and tracked via the light the latter emit.

"Mesoporous silica," with molecules consisting of silicon, carbon, hydrogen, and oxygen, can be made into nano-sized spheres and rods with patterns of pores on them. These silica nanoparticles can be taken up by cells (e.g. in tumors) and hold either drugs or fluorescent dyes used for imaging to identify the cells [24, 29].

Nonmetallic nanoparticles may be based on different kinds of "polymers" —molecules comprised of repeating subunits of smaller molecules [2, 24]. Several medications approved for clinical use by the United States Food and Drug Administration for treatment of breast and other types of cancer include chemotherapeutic agents encapsulated within nanoparticles made of albumin (a protein), lipids, or other substances [21, 29, 30].

---

[7] Another "health benefit" of gold nanoparticles might also be as an effective defense against attacks by Cybermen, as suggested by the *Doctor Who* serials "Revenge of the Cybermen" (1975) and "Earthshock" (1982).

One type of nanoparticle composed of biological material, a liposome, is typically created from two layers of lipid molecules, similar to the arrangement in a cell membrane [1, 31]. The "lipid bilayer" of a liposome is roughly shaped in the form of a sphere. It can contain drugs within it that are "hydrophilic," that is, soluble in water. Water-soluble drugs may have difficulty crossing a cell membrane and entering a cell, whereas lipid-containing material may be able to cross it much more easily.

The importance of a liposome's dual-nature, with a lipid surface enclosing a chemotherapeutic or other drug, is that the latter can "come along for the ride" and readily enter the cell when the liposome itself does. The liposome would then break down and release the drug within the aqueous environment inside the cell to do its work [1, 30]. The net effect is that much higher concentrations of a chemotherapeutic agent bound to liposomes can reach cancer cells than if the agent were simply given on its own. Preparations of liposomes with a chemotherapeutic agent (e.g. doxorubicin) are clinically available in the United States for treatment of breast, ovarian, and other types of cancer [29].

Another type of biological nanoparticle is a "dendrimer," which has tree-like branches. It is composed of large molecules such as sugars, nucleotides (part of the structure of DNA and RNA), or amino acids. Dendrimers are currently used only as research tools, with no approved clinical applications as yet [32].

An important kind of non-biological nanoparticle is the carbon nanotube. It is made from "graphene," a one-atom thick "sheet" of carbon atoms bound together in a fixed hexagonal pattern [33, 34]. The graphene is rolled up into a cylindrical shape with a typical diameter of about 1 nm (range 0.2–2 nm) but a length that can be many million of times longer. Carbon nanotubes may be single-walled, or multi-walled with more than one layer in their walls. Multilayered carbon nanotubes may have an inner diameter of 1–3 nm but an outer one ranging from 2 to 100 nm [28, 35].

Because they are insoluble in water and other solvents present in living organisms, the surface of a carbon nanotube must be coated with molecules (e.g. lipids or polymers) that make them able to be used within animal models and, ultimately, as part of human clinical trials. Other molecules such as a chemotherapeutic agent can be bound to a carbon nanotube, which could then potentially target, penetrate, and kill a cancer cell by delivering that drug inside it [36]. Carbon nanotubes could also act as carriers for other molecules such as vaccines, radiotracers for imaging, etc [3, 37]. Like gold nanoparticles, carbon nanotubes can absorb near-infrared radiation and may do it more efficiently than gold. For example, one animal study using a low injected dose of carbon nanotubes followed by exposure to near-infrared light from a laser safely increased the local temperature in a tumor to 60 °C and damaged it [38]. Thus, carbon nanotubes too might one day be used for photothermal therapy of cancer cells [24, 29].

Carbon nanotubes are a subset of a more general group of carbon nanoparticles, "fullerenes," which comprise any molecule consisting of carbon atoms arranged in various shapes. A "buckyball" is a hollow, roughly spherical arrangement of carbon atoms that resembles the type of geodesic dome popularized by Buckminster Fuller, whose first name appears in that term and last name in "fullerene." A spherical fullerene with 60 carbon atoms ("C_{60}") and attached hydroxyl groups (combinations of one oxygen and one hydrogen atom) is soluble in water and thus could potentially be used in biological systems. Like nanotubes it might serve as a carrier for chemotherapeutic agents targeting cancer cells and protect against some of those agents' side effects on healthy tissue [39].

Nanoparticles used for either diagnostic or therapeutic medical purposes can be introduced into the body by various routes. Oral administration has to take into account the low pH (highly acidic environment) in the stomach and how well nanoparticles and any drugs or other substances carried by them can be absorbed through the intestines. Fortunately "nanocarriers" can be modified in such a way to help protect their "cargo" and perhaps even enhance their uptake in the gastrointestinal tract [40].

Nanoparticles larger than about 100 nm may not be able to pass through the skin, limiting transdermal administration [40]. Smaller nanoparticles might be able to get through the skin barrier, but how effective this method is for getting them to the desired area of the body is uncertain. A subcutaneous (just beneath the skin) injection may also have limited value for getting at least some types of nanoparticles into the bloodstream. This can be better accomplished by injecting them into the abdominal cavity or, of course, by direct administration into a vein. Inhaling nanoparticles into the lungs may also be an effective route for both local effects and eventual uptake into the blood, if those particles are small enough and include materials that help them evade macrophages and other cells that normally remove foreign materials inside the lungs.

Once inside the body, nanoparticles can enter cells for both diagnostic and therapeutic purposes via different mechanisms. For example, they can be actively engulfed by a cell ("endocytosis"), physically penetrate the cell's membrane, or first attach to that membrane and then be absorbed through it [28]. Once within the cell the nanoparticle can perform the types of actions described previously, such as "marking" the cell or, in the case of a malignant one, destroying it.

Nanoparticles within the human body might be recognized as "foreign" and activate its immune system. Larger nanoparticles (e.g. those with a diameter of 100 nm or more, such as some multi-walled carbon nanotubes) in the bloodstream are more likely to be detected and "attacked." White blood cells such as macrophages could then engulf them or otherwise prevent nanoparticles from getting to the part of the body (e.g. a tumor) they were designed to reach. Still larger nanoparticles (more than 200 nm) can accumulate in the spleen before reaching their target site in the body [20, 41]. Nanoparticles composed of a wide range of materials can be treated with special substances such as glycolipids and polyethylene glycol that make them less likely to be detected and stopped by the immune system before they reach their target [27, 29]. Most nanoparticles are also small enough to do a good job of evading the immune system based on their size alone.

Current research using nanoparticles focuses strongly on treating various types of cancer. Those particles can stay in the bloodstream longer than conventionally administered chemotherapeutic and other drugs, thus increasing their ability to reach tumors and deliver those medications to malignant cells and destroy them [21]. Nanoparticles can also potentially penetrate deeper into solid tumors and deliver higher doses of chemotherapeutic agents inside them than conventional administration of those drugs can. Compared to healthy tissues, tumors may have increased leakage of fluids from the new blood vessels they create to supply themselves with nutrients and oxygen, potentially allowing chemotherapeutic drugs to be better able to reach the tumor's cells. However, other parts of the tumor might have a poor blood supply, and increased pressure in the fluid surrounding the blood vessels may limit movement of drugs out of them to the cancer cells. Those cells can also be tightly packed together or bound by materials such as collagen that reduce the ability of drugs to penetrate them [20].

Nanoparticles can be modified to overcome those barriers better than chemotherapeutic drugs alone can do. Those particles can be made of materials that the cancer cells are "fooled" into actively taking inside them. Moreover, nanoparticles can be coupled with antibodies, proteins, and other substances that bind specifically with particular "biomarkers" associated with tumor cells but not healthy ones [27]. Alternatively a tumor-specific material may be injected first that then attaches to and "labels" tumor cells. Nanoparticles coupled with a chemotherapeutic agent and another substance that combines with the one originally given can then be administered [35]. Either method enables the nanoparticles to distinguish between normal and abnormal cells and "lock on" to the latter. Cancer cells can also be "tagged" by an imaging agent for identification or targeted for destruction by release of a chemotherapeutic agent (e.g. after being absorbed by the cancer cell) carried by a nanoparticle.

Conventional chemotherapeutic agents can produce significant amounts of "collateral damage" to some healthy tissues, leading to side effects and other problems. This is because the same methods they use to inhibit and destroy cancer cells (e.g. interfering with their division or metabolism) can also affect healthy ones, particularly those that are rapidly dividing. The latter include hair follicles (leading to hair loss), blood cells (resulting in anemia and reduced immune function), and the cells lining the stomach and intestines (producing nausea, vomiting, diarrhea, etc.) Nanoparticles can be much more selective for cancer cells than these "shotgun" effects of standard chemotherapeutic agents, and thus less likely to injure or kill normal cells. Much larger than normal doses of chemotherapeutic drugs can also be delivered via nanocarriers due to the latter's selectivity for cancer cells, thus allowing administration of a lower total dose to reduce adverse effects on healthy ones [2].

Conventional formulations of chemotherapeutic agents may contain materials required in their manufacture that can reduce how effectively they are distributed in the body, and those materials can also have their own side effects. For example, a clinically available nanoparticle formulation of the cancer-treating drug paclitaxel can be made without potentially toxic substances needed to manufacture a non-nanoparticle based preparation of that medication [20]. The nanoparticle version will thus not produce some of the side effects the other type can, enabling higher doses of it to be safely given and to be infused over a much shorter period of time (30 min versus 3–24 h for the conventional formulation). It may also be more effective for treating breast and other types of cancer due to its being better able to reach malignant cells and staying active in the body for a longer time than a non-nanoparticle formulation [29].

Because nanoparticles have such a large total surface area relative to their overall diameter, they can have many therapeutic (e.g. drug) molecules, imaging agents, etc. attached to them in an overall very compact volume. Drugs and other molecules can also be dissolved or dispersed within the nanoparticle itself [29]. Potentially two or more drugs could be coupled with individual nanoparticles for combination therapy of cancer and other diseases [27]. Alternatively a nanoparticle could carry a "prodrug," a substance that by itself is inert but that becomes active under the right circumstances. For example, the prodrug could be "activated" after it and its nanoparticle carrier enter a cancer cell. Certain prodrugs can themselves become part of a "nanoassembly," a type of nanoparticle whose structure includes the prodrug combined with other substances without a separate carrier nanoparticle (e.g. a carbon nanotube) [41].

Nanoparticles can also have other important properties for potential medical use. Besides using light and magnetic forces to "activate" gold nanoparticles and destroy cancer cells, some nanoparticles can also respond to external physical stimuli such as heat and sound as well as the chemical characteristics of their environment, including its pH. For example, in one "proof-of-principle" laboratory study rod-like "nanomotors" made of gold attached to and then penetrated a preparation of cancer cells [42, 43]. An external ultrasound beam was then used to make the nanomotors move around and spin within the cells, and magnetic forces were even used to steer them. Doing this vigorously enough within the cancer cell could kill it by physically damaging its organelles and other internal structures or puncturing its cell membrane.

Iron oxide nanoparticles suspended in water can be injected directly into tumors. They generate heat when subjected to a sufficiently strong external oscillating magnetic field [1]. Human research trials are now being conducted to determine how effective this method is for destroying cancer cells by increasing their ambient temperature from a normal value of about 37 °C to the minimum lethal value of around 45 °C [25, 29].

Besides carrying chemotherapeutic drugs into cancer cells, nanoparticles can introduce chemicals called "photosensitizers" into them [29]. Photosensitizers absorb light at a particular wavelength and use that energy to interact with oxygen to ultimately produce superoxide, hydrogen peroxide, and other substances within the cell that are toxic to it—a form of treatment called "photodynamic therapy."[25]. The light energy to do this may, however, need to be provided internally by means of instruments with fiber-optic cables such as an endoscope. Besides acting as passive carriers for photosensitizing chemicals, some types of nanoparticles themselves, such as quantum dots, can also act directly as photosensitizers too.

Nanoparticles might also be used to enhance the effects of radiation treatments for cancer. For example, delivery of chemotherapeutic agents by nanocarriers to cancer cells can make cells not directly destroyed by the agent more sensitive to destruction by radiation therapy. Gold nanoparticles by themselves may "amplify" the effects of ionizing radiation on cancer cells through electron scattering and other physical means [24, 32].

Current research includes using materials created with nanotechnology to help grow new tissues. In one study neurons were able to grow on a framework composed of carbon nanotubes, and those cells even formed tight contacts with their membranes similar to those in normal neuronal tissue [44]. This technique might potentially help in repairing a person's damaged nerve or even in developing nerves that are at least partially synthetic. Nanomaterials may also be useful for stopping bleeding quickly and healing chronic wounds

[45–47]. An animal study showed that injecting single-walled carbon nanotubes could potentially reduce brain damage caused by a stroke [48].

Nanoscale materials might even be used outside the body to help create part or all of complex organs like the lungs and heart. This could involve isolating healthy cells from a patient, cultivating them, then seeding the cells onto a scaffold structure that includes nanomaterials such as wires or particles. The scaffold would then serve as a mechanical support as the cells, assisted by added growth factors and other molecules, grow into connected, functional forms for eventual transplantation into the patient to replace damaged tissue [49–51]. Such transplants could potentially range from "patches" to the ultimate goal of creating a whole new heart or other organ.

Other tissues such as bone might also be repaired within a person's body using scaffolding created from nanomaterials. In theory, placing an appropriate scaffolding within damaged bone could attract stem cells (see Chap. 13) within the body that would then divide and change into "osteoblasts," the type of cell that lays new bone tissue [3].

Still another use for nanoparticles is as part of a gene delivery system [29, 30, 52]. Gene therapy is being evaluated as a method for treating cancer.[8] It involves introducing new sequences of genetic material into malignant cells that alter their function and make them less likely to divide and spread. Combining these bits of DNA, RNA, or other genetic material with nanoparticles can increase the efficiency with which those injected complexes enter cancer cells, and thus may improve their overall effectiveness. This same method is also being considered for treatment of infections with the human immunodeficiency virus (HIV), the cause of AIDS (acquired immunodeficiency syndrome) [41]. In an animal model specially formulated magnetic nanoparticles were "steered" to tumor cells using an external magnet [53, 54]. They were directed well enough to deliver RNA genetic material to those cells that inhibited their function.

Nanoparticles could also be used for "theranostics," a portmanteau word that combines "therapeutics" and "diagnostics"—that is, the ability to both diagnose and treat a medical problem [1, 24]. For example, as mentioned previously gold nanoparticles can identify cancer cells via imaging methods and potentially destroy them via either photothermal therapy or mechanically using external ultrasound and magnetic fields. Gold nanoparticles also heat up when exposed to radiofrequency ("radio") waves, which pass through skin deeper than near-infrared light can and could possibly be used to treat more deep-seated cancers. Similarly, carbon nanotubes could be coupled with quantum dots, with the latter being used to help locate cancer cells and the

---

[8] See Chap. 12 for more details about gene therapy.

former killing them by releasing a chemotherapeutic agent or via photothermal therapy [28]. Nanoparticles could also be used as "biosensors" to detect cells that require repair or, as in the case of cancer cells, destruction [55].

But as potentially important as all these areas of research are, the nanoparticles used in current medical research and clinical practice fall well short of the nanoscale "computers" and autonomous devices envisioned by Drexler and others. They are still literally just particles, modified to blindly do only a single job or tiny range of tasks. Far from being long-lasting "sentinels" patrolling our bloodstream, tissues, etc. looking for abnormalities in cells and other problems they can then efficiently correct, these nanoparticles are present and useful within the body for only a very limited time. For example, a nanoparticle carrying a chemotherapeutic agent into a cancer cell is either destroyed or loses its therapeutic value when the drug is released and does its job. As we will see in the next section, non-biodegradable nanoparticles such as carbon nanotubes would then need to be cleared by the body in a timely and safe manner lest they accumulate and cause problems of their own.

A possible next step in making nanoparticles "smarter" is to modify them in ways that assist and perhaps augment the body's own repair and defense mechanisms. For example, nanoparticles might potentially be treated with substances that help our blood coagulation system in stopping a degree of internal bleeding that would overwhelm our "built-in" capabilities. As part of theranostics, nanoparticles might carry chemical "sensors" that can detect biochemical conditions associated with tissue damage and inflammation. They could then release therapeutic substances to help treat the problem. For more serious tissue injury nanoparticles with sufficient "intelligence" might diagnose the problem and release chemicals designed to stimulate the body's own cells to differentiate or divide and create new, healthy ones [50].

In short, there are now many known techniques for getting nanoparticles into cells. While this capability is now used primarily to find and destroy cancer cells, it also means that sufficiently "smart" nanoparticles might be able to help to heal damaged or dysfunctional "normal" cells. This might be done short of creating "nanobots" by coupling nanoparticles with bits of RNA, DNA, or other biological materials capable of diagnosing a problem within the cell, and repairing or replacing the latter either directly or by stimulating the cell itself to do it. Nanoparticles below about 40 nm in diameter could even enter the cell nucleus itself to possibly "tweak" its DNA and other internal components [56].

The use of nanoparticles for gene therapy mentioned previously is one model for this, but Nature provides another. Viruses are essentially nanoscale machines consisting of a short length of genetic material (DNA or sometimes RNA) protected by a protein-containing shell [57]. They are parasites that inject their genetic material into a cell and, in the Borg sense of the term, "assimilate" it. An infecting virus replaces the cell's DNA-based control function with its own, taking over the intracellular machinery for protein synthesis and other resources to make the nucleic acid and other parts for new copies of the virus. The latter can even permanently incorporate its own DNA into that of the host cell.

But while pathogenic viruses act only to the detriment of the cell they invade, it might be possible to create a human-made, nanoscale "virus" designed to help it. As noted in Chap. 9, part of the aging process includes changes to and dysfunction of DNA and other intracellular systems. At least theoretically, the genetic code within the cell might be "upgraded" by introducing new instructions via a virus-like nanocarrier of genetic material or other biochemical means. Rather than doing a "hostile takeover" as a "natural" virus does, this "therapeutic" version could help the cell by repairing its DNA, reverse changes that increase the risk of developing cancer, and possibly help "senescent" cells to return to normal function and replicate again. This method might largely bypass the need for computer-like nanobots and use better-established biochemical and genetic means.

However, actually putting these principles into practice will take a great deal more research. Our knowledge of the intracellular changes associated with disease and aging, while growing, is still very incomplete. Without knowing more about the processes involved and their effects it is uncertain whether the net results of a particular genetic change would be beneficial or harmful. As will be discussed in Chap. 12, current therapy involving genetic "repairs" within a cell is limited to correction of isolated "mutations" that can be as little as a substitution of a single critical molecule for another in the DNA of particular cells. Making what are by comparison wholesale changes in DNA, intracellular repair systems, etc. will, even considering only the scale involved (much less its complexity and uncertainties), be much more challenging. And, as with any new method, there is the risk that making an "improvement" in one function of the cell might have unforeseen negative consequences elsewhere.

A less speculative use for nanotechnology is to help create better "microarrays" used for DNA sequencing. A DNA microarray typically consists of tiny areas of gene sequences placed in parallel rows on a slide. These sequences interact with certain RNA molecules obtained from a cell to see how well they "match" with the gene sequences on the microarray. This in turn indicates how well they correspond to active or inactive gene sequences from the original cell and provides valuable information about its genetic composition. The genetic information obtained can then be used for clinical purposes such as evaluating the potential response to a particular drug based on the presence or absence of specific genes involved in its metabolism or effects, or to identify biomarkers in tumor cells to guide therapy. Nanoscale manufacturing can enable more DNA sequences to be placed on a slide to allow faster evaluation of the presence and status of a cell's individual genes [27, 58].

Other methods using nanomaterials are used to test for and measure the amounts of molecules in biological samples [2]. Such "in vitro" (outside the body) laboratory tests include the use of gold nanoparticles to detect human chorionic gonadotropin as part of a urine pregnancy test. Similar techniques are employed to evaluate HIV and malaria. Gold nanoparticles are also used for certain types of genetic screening tests.

Nanomaterial and nanotechnology can also be applied to creation of "microdevices"—miniature devices that can be integrated with cell and tissue cultures, or used to create nanomaterials as well as evaluate their biological efficacy and toxicity [59]. For example, by using appropriate cells they can act as an "organ-on-a-chip" to evaluate the potential beneficial and adverse effects of new drugs, nanomaterials, etc. outside the body before actually testing them in real, whole humans. Such devices can also be used with modified nanoparticles (e.g. bound with an antibody or fluorescent dye) for early diagnosis of cancer cells in blood samples taken from a patient.

11.4 Risks of Medical Nanotechnology

Nanoparticles may have health and environmental hazards [17, 60]. As noted previously the physical and chemical properties of a particular substance's nanoparticles may change appreciably depending on whether they are separate from each other or "stuck" together by water or other solvents into larger agglomerations. This change could alter their potential toxicity. For example, nanoparticles in sunscreens released into bodies of water could prove harmful to fish and other aquatic life in them. Aerosolized nanoparticles in hair spray and similar products could be inhaled and cause problems with the lungs. Potentially toxic nanoparticles could be ingested directly or indirectly (such as by eating fish that have been exposed to contaminated waters), or absorbed through the skin [19, 61]. None of these risks are as dramatic as the "gray goo" described by Drexler caused by unrestricted replicators or the "nanobots gone wild" threats posed in works such as Michael Crichton's *Prey* (2002), but they are important considerations in the real world.

While some types of nanoparticles are made of materials that are biodegradable, others are not. Inorganic nanoparticles can be cleared from the body by the kidneys through excretion in urine. They could also be eliminated by the liver and gallbladder in bile released into the duodenum, with ultimate passage to the outside via the large intestine [20, 25]. Smaller nanoparticles (e.g. less than 10 nm in diameter) are more likely to accumulate in and be removed by the kidneys, which is generally a more rapid route of elimination than the liver and may result in a lower risk of toxicity. Larger nanoparticles are more likely to accumulate in other tissues and organs such as the lungs, liver, and spleen. Having too many of these nanoparticles in them could cause damage [62].

For example, while their overall toxicity is low, under the right circumstances enough gold nanoparticles might accumulate within the body to produce toxic effects. Injections of gold salts used to treat rheumatoid arthritis have been associated with kidney damage, reduction in blood cell counts, and other problems, suggesting that having a sufficiently high number of gold nanoparticles in the body might cause similar side effects [27].

As noted previously, nanoparticle complexes of liposomes and chemotherapeutic drugs have been approved for clinical use in the United States. However, the liposomes themselves can be associated with side effects, such as being detected by the body and evoking an immune response with production of antibodies against them [1, 20].

Nanoparticles above about 200 nm in width might even cause damage inside the body by physically blocking off tiny blood vessels (capillaries). Even smaller nanoparticles might potentially activate a person's own blood coagulation system, resulting in formation of blood clots large enough to obstruct larger blood vessels [20].

Carbon nanotubes might also have adverse effects [2, 37, 63]. For example, they might accumulate in the liver and other normal tissue and injure them. Moreover, carbon nanotubes are structurally similar to asbestos fibers, and those inhaled from the environment might present similar risks of lung damage and cancer [20, 27, 64]. Animal studies suggest that inhaling multi-walled carbon nanotubes at a sufficiently high concentration may reduce the function of T cells, an important part of the immune system [64]. However, the overall safety profile of carbon nanotubes based on their size, site of administration (e.g. intravenously), routes of clearance from the body (such as by the liver or kidneys, etc. is still being studied.

Another important issue with nanoparticles is if or should they cross the "blood-brain" barrier [20]. This is an anatomical and physiological system by which the human body protects the brain against "foreign" material, including many medications, large molecules, infectious bacteria, etc. by preventing them from entering it. Molecules that can pass through the blood-brain barrier have an atomic mass no greater than about 500 atomic mass units (by comparison that of water is 18). They are also primarily "lipophilic"—soluble in lipids (fats) rather than water, although water itself and other very small molecules like oxygen can pass through that barrier.

However, it may be desirable to "breach" that barrier with nanoparticles designed to provide treatment for particular diseases [65]. For example, Alzheimer's disease affects primarily older adults and can produce dementia, behavior problems, and confusion. It is associated with development of anatomic abnormalities such as plaques along with degeneration and death of neurons in the brain. Parkinson's disease involves selective loss of neurons in the "substantia nigra," a part of the brain involved in coordinating movements. It can cause tremors, difficulties with balance, and limb rigidity. In multiple sclerosis there is loss of myelin, the protective sheath around the extensions of neurons called axons. The type and severity of symptoms depend on what and how many neurons are involved, and they may include weakness, visual problems, and difficulties with coordination and balance.

New drugs being developed to treat these and other neurological conditions may not be able to cross the blood-brain barrier on their own. By coupling them with nanoparticles small enough (e.g. smaller than 35 nm)[56] and modified to do this, the drugs can reach the brain and hopefully provide treatment. For example, a type of antibody specific for crossing the blood-brain barrier can be attached to the surface of nanoparticles, or that surface can be coated with a substance designed to pass through into the brain. Thus, while the nanoparticles themselves may not have a direct therapeutic benefit, they can serve as the "key" to unlock the "door" into the brain and carry medications to treat diseases there.

In contrast to some other types, iron oxide nanoparticles are thought to be reasonably safe. A person's body can actually use them to make hemoglobin for red blood cells [27].

The way a particular kind and size of nanoparticle interacts with a person's immune system may also be difficult to evaluate using standard testing techniques [66]. During their manufacture nanoparticles may be contaminated with "endotoxins"—harmful materials associated with the cell walls of some bacteria. Assays done to check for the presence of endotoxins and other substances (e.g. "pyrogens"—ones that cause fever) may fail to detect them due to interference from the nanoparticles themselves. Specific changes in standard tests and testing protocols might need to be made to adjust for these effects.

As described previously, nanoparticles can be modified to be selectively taken up by cancer and other target cells rather than by other, healthy ones. Nonetheless some may still be absorbed or otherwise interact in undesired ways with various types of normal cells throughout the body. For example, in one study normal human fibroblasts (see Chap. 9) showed damage to their DNA and chromosomes after exposure to nanoparticles made of a cobalt-chromium alloy and about 30 nm in diameter [67]. The nanoparticles did not penetrate the fibroblasts but instead appeared to injure them due to indirect effects that even worked through a "barrier" of other cells between the nanoparticles and fibroblasts. The clinical concern is that this cobalt-chromium alloy is used as part of some artificial joint replacements in patients. Normal "wear" of the metal in those replacements could release cobalt-chromium nanoparticles that in turn might damage normal cells.

11.5 The Bottom Line

Medical nanotechnology is yet another area in which there is a great gap between concepts of what might be possible in the future and what can be done now. Using nanotechnology to produce cosmetically pleasing sunscreens with nanoscale titanium dioxide particles is a far cry from having ever-vigilant nanobots patrolling our bloodstreams and continuously repairing and rejuvenating our cells to keep us perpetually "young." Many, many things would have to go "right" on a vast scale for nanotechnology to produce such beneficial final results, with many more ways they might possibly not work at all or go wrong. Nonetheless, despite the enormous practical difficulties involved, because nothing absolutely forbids them (as far as we know yet) from a biological/physical standpoint, the concepts proposed by Drexler and others are "fair game" for the thought experiments and drama of science fiction.

Given its at least proposed capabilities, "nanotechnology" or the devices associated with it—"nanites," "nanobots," and the like—can be used in science fiction as "buzzwords" to "scientifically explain" quasi-magical biological and other effects. For example, how the Borg's injection of such devices can create such dramatic and seemingly near-instantaneous changes in beings they "assimilate" is never actually explained but only shown. On the other hand, a "harder" science fiction approach might at least lay some of the theoretical and factual background behind what nanotechnology applied to human beings can do.

Two works by Arlan Andrews, Sr. demonstrate effective ways to do the latter. In "Hail, Columbia!" (1993) homeless people in a future Washington, D.C. are given money coated with therapeutic nanodevices. Once absorbed through the skin those devices cure the recipients of any psychiatric or major physical ailments they may have, as well as improve their mental health by eliminating the emotional content of painful past memories. "Other Heads" (1993) posits nanobots in the bloodstream that continuously maintain health, as well as a neural interface in the form of a ring that, when placed on a finger, "connects" with the user's brain.[9] Thus, when a character is described as wearing such a device on a ring (fourth) finger, presumably it uses the peripheral nerves innervating that digit (the ulnar and median nerves) as electrical "conduits" back to the brain.

Both stories provide just enough scientific background information to establish the plausibility of these technologies, while concentrating instead on describing how they work and especially the far-reaching effects they have on individuals and society. A paper submitted to a science journal or a patent would need to provide much more and "real" detail about how to make such devices, including what was done to meet the many challenges to creating them described previously in this chapter. However, as long as no actual scientific errors (or at least ones that are not immediately recognizable as standard story conventions) are included, adding too many such details could not only bog down a story but render any scientifically "shaky" ones vulnerable to specific criticisms. Instead these two works appropriately add only the amount of realistic science needed to make the plot possible—and then proceed to simply tell the story.

Another good example of this "less-is-more" principle is Edward M. Lerner's novel *Small Miracles*. It begins by introducing a character wearing a jumpsuit with advanced capabilities for protecting him using nanotechnology. When he suddenly falls victim to a horrific explosion that kills hundreds in the surrounding area, nanites in the jumpsuit's fabric sense the blast and stiffen it enough that he experiences serious injuries but not the fatal ones he would have received without the suit. That suit also saves his life by injecting him with painkillers and "nanobots" that rapidly find and deal with sites of internal bleeding, allowing him to survive long enough to receive definitive medical attention for his broken bones and other injuries.

The nanobots this character received were at least partially composed of carbon nanotubes with enough innate overall computer processing power built in to act in a "smart" fashion to treat him. They were designed to be destroyed or cleared by the body within a day or so after doing their job and to not cross the blood-brain barrier into the brain. The nanobots work exactly as designed and are quickly gone from the character's body after saving him. He goes on to live a happy, healthy, uneventful life—

No, of course he doesn't. The nanobots that saved his life proceed to act in ways their designers never anticipated, passing from his bloodstream into his brain by breaching the blood-brain barrier where they…well, I suggest reading the novel to find out what imaginative, dangerous things happen after those

---

[9] Chapter 14 will describe the current status of such interfaces.

nanobots literally develop a mind of their own. The point is that, by anchoring what ultimately becomes a speculative danger of nanomedicine in terms of actual scientific/biological principles (using carbon nanotubes to construct nanobots, the desirability of having some way of clearing them from the body after they have done their work, etc.), the events in the story become more "realistic" than if only the idea of "nanotechnology" were invoked without laying at least some "real world" basis for it. If that is not done, nanotechnology and nanomedicine run the risk of being used as all-purpose "magic dust" that can do whatever the plot demands with minimal or no concrete grounding in realistic science. Too casual use of nanotechnology could make it seem like the latest replacement for how "radiation" was used to "explain" highly dubious changes in the human body, animals, etc. in many mid-twentieth century science fiction works.

Such use of those two concepts is, however, not entirely the same. Exposure to radiation will not, barring wholesale changes in well-established concepts of physics and biology, lead to creation of giant carnivorous ants (see Chap. 6). Conversely, it is possible that nanotechnology might eventually fulfill at least some of the expansive claims made for it. But because giant ants require complete suspension of disbelief while futuristic nanomedicine does not necessarily need it (even if it does require great "optimism" about future advances), including at least some validly realistic scientific elements in science fiction works dealing with nanomedicine can strengthen them. The trick is to strike an appropriate balance, as *Small Miracles* and the two stories cited previously in this section do, between the realistic and the speculative in a way that furthers and deepens the plot—not a small feat.

References

1. Ranganathan R, Madanmohan S, Kesavan A, Baskar G, Krishnamoorthy YR, Santosham R, et al. Nanomedicine: towards development of patient-friendly drug-delivery systems for oncological applications. Int J Nanomed. 2012;7:1043–60.
2. Kim B, Rutka J, Chan WC. Nanomedicine. N Engl J Med. 2010;363(25):2434–43.
3. Wong KK, Liu XL. Nanomedicine: a primer for surgeons. Pediatr Surg Int. 2012; 28(10):943–51.
4. Feynman R. There's plenty of room at the bottom. Eng Sci. 1960;23:22–36.
5. Drexler KE. Molecular engineering: an approach to the development of general capabilities for molecular manipulation. Proc Natl Acad Sci. 1981;78(9):5275–8.
6. Drexler KE. Engines of creation. The coming era of nanotechnology. New York: Anchor Books; 1986.
7. Drexler KE. Nanosystems. Molecular machinery, manufacturing, and computation. New York: Wiley; 1992.

8. Hall JS. Nanofuture. What's next for nanotechnology. Amherst: Prometheus Books; 2005.

9. Jain K. The handbook of nanomedicine. New York: Humana Press; 2008.

10. Kleinsmith L, Kish V. Energy and enzymes. Principles of cell and molecular biology, (Chapter 2). New York: HarperCollins; 1995.

11. Nanotechnology. Research and perspectives. Cambridge: The MIT Press; 1992.

12. Prospects in Nanotechnology. Toward Molecular Manufacturing. New York: Wiley; 1995.

13. Freitas R. Nanomedicine. Basic capabilities. Austin: Landes Bioscience; 1999.

14. Leduc P, Wong M, Ferreira P, Groff R, Haslinger K, Koonce M, et al. Towards an in vivo biologically inspired nanofactory. Nat Nanotechnol. 2007;2(January):3–7.

15. Chou LY, Zagorovsky K, Chan WC. DNA assembly of nanoparticle super-structures for controlled biological delivery and elimination. Nat Nanotechnol. 2014;9(2):148–55.

16. Toumey C. Nanobots today. Nat Nanotechnol. 2013;8(7):475–6.

17. McCall MJ. Environmental, health and safety issues: nanoparticles in the real world. Nat Nanotechnol. 2011;6(10):613–4.

18. Soppimath K, Betageri G. Nanostructures for cancer diagnostics and therapy. In: Gonsalves K, Halberstadt C, Laurencin C, Nair L, editors. Biomedical nanostructures. Hoboken: Wiley; 2008. pp. 409–37.

19. Labouta HI, Schneider M. Interaction of inorganic nanoparticles with the skin barrier: current status and critical review. Nanomedicine. 2013;9(1):39–54.

20. Desai N. Challenges in development of nanoparticle-based therapeutics. AAPS J. 2012;14(2):282–95.

21. Kim TH, Lee S, Chen X. Nanotheranostics for personalized medicine. Expert Rev Mol Diagn. 2013;13(3):257–69.

22. Yu X, Valmikinathan C, Rogers A, Wang J. Nanotechnology and drug delivery. In: Gonsalves K, Halberstadt C, Laurencin C, Nair L, editors. Biomedical nanostructures. Hoboken: Wiley; 2008. pp. 93–113.

23. Guan J, He H, Yu B, Lee L. Polymeric nanoparticles and nanopore membranes for controlled drug and gene delivery.In: Gonsalves K, Halberstadt C, Laurencin C, Nair L, editors. Biomedical nanostructures. Hoboken: Wiley; 2008. pp. 115–37.

24. Wang LS, Chuang MC, Ho JA. Nanotheranostics–a review of recent publications. Int J Nanomedicine. 2012;7:4679–95.

25. Tong R, Kohane D. Shedding light on nanomedicine. Wiley Interdiscip Rev Nanomed Nanobiotechnol. 2012;4(6):638–62.

26. Peng G, Tisch U, Adams O, Hakim M, Shehada N, Broza YY, et al. Diagnosing lung cancer in exhaled breath using gold nanoparticles. Nat Nanotechnol. 2009;4(10):669–73.

27. Zhang XQ, Xu X, Bertrand N, Pridgen E, Swami A, Farokhzad OC. Interactions of nanomaterials and biological systems: Implications to personalized nanomedicine. Adv Drug Deliv Rev. 2012;64(13):1363–84.

28. Madani SY, Shabani F, Dwek MV, Seifalian AM. Conjugation of quantum dots on carbon nanotubes for medical diagnosis and treatment. Int J Nanomedicine. 2013;8:941–50.

29. Thakor A, Gambhir S. Nanooncology: The future of cancer diagnosis and therapy. CA Cancer J Clin. 2013;63:395–418.
30. Waite C, Roth C. Nanoscale drug delivery systems for enhanced drug penetration into solid tumors: Current progress and opportunities. Crit Rev Biomed Eng. 2012;40(1):21–41.
31. Akbarzadeh A, Rezaei-Sadabady R, Davaran S, Joo S, Zarghami N, Hanifehpour Y, et al. Liposome: classification, preparation, and applications. Nanoscale Research Letters. 2013;8(102):1–9.
32. Miller SM, Wang AZ. Nanomedicine in chemoradiation. Ther Deliv. 2013; 4(2): 239–50.
33. Novoselov KS, Fal'ko VI, Colombo L, Gellert PR, Schwab MG, Kim K. A roadmap for graphene. Nature. 2012;490(7419):192–200.
34. Peplow M. The quest for supercarbon. Nature. 2013;503:327–9.
35. Mulvey JJ, Villa CH, McDevitt MR, Escorcia FE, Casey E, Scheinberg DA. Self-assembly of carbon nanotubes and antibodies on tumours for targeted amplified delivery. Nat Nanotechnol. 2013;8(10):763–71.
36. Liu Y, Wang H. Nanotechnology tackles tumours. Nat Nanotechnol. 2007;2(January):20–1.
37. Kostarelos K, Bianco A, Prato M. Promises, facts and challenges for carbon nanotubes in imaging and therapeutics. Nat Nanotechnol. 2009;4(10):627–33.
38. Moon HK, Lee SH, Choi HC. In vivo near-infrared mediated tumor destruction by photothermal effect of carbon nanotubes. ACS Nano. 2009;3(11):3707–13.
39. Grebowski J, Kazmierska P, Krokosz A. Fullerenols as a new therapeutic approach in nanomedicine. Biomed Res Int. 2013;2013:751913.
40. Da Silva A, Santos R, Xisto D, Alonso S, Morales M, Rocco P. Nanoparticle-based therapy for respiratory diseases. Ann Braz Acad Sci. 2013;85:137–46.
41. Parboosing R, Maguire GE, Govender P, Kruger HG. Nanotechnology and the treatment of HIV infection. Viruses. 2012;4(4):488–520.
42. Johnson D. Nanomotors could churn inside of cancer cells to mush. 2014. http://spectrum.ieee.org/nanoclast/biomedical/devices/nanomotors-could-churn-inside-of-cancer-cells-to-mush. Accessed 15 April 2015.
43. Wang W, Li S, Mair L, Ahmed S, Huang TJ, Mallouk TE. Acoustic propulsion of nanorod motors inside living cells. Angew Chem Int Ed. 2014;53:3201-4.
44. Silva G. Shorting neurons with nanotubes. Nat Nanotechnol. 2009;4(February):82–3.
45. Hartgerink J. New material stops bleeding in a hurry. Nat Nanotechnol. 2006;1(December):166–7.
46. Koria P, Yagi H, Kitagawa Y, Megeed Z, Nahmias Y, Sheridan R, et al. Self-assembling elastin-like peptides growth factor chimeric nanoparticles for the treatment of chronic wounds. Proc Natl Acad Sci U S A. 2011;108(3):1034–9.
47. Ruan L, Zhang H, Luo H, Liu J, Tang F, Shi YK, et al. Designed amphiphilic peptide forms stable nanoweb, slowly releases encapsulated hydrophobic drug, and accelerates animal hemostasis. Proc Natl Acad Sci U S A. 2009;106(13):5105–10.

48. Higgins P, Dawson J, Walters M. Nanomedicine: nanotubes reduce stroke damage. Nat Nanotechnol. 2011;6(2):83–4.

49. Dvir T, Timko BP, Brigham MD, Naik SR, Karajanagi SS, Levy O, et al. Nanowired three-dimensional cardiac patches. Nat Nanotechnol. 2011;6(11):720–5.

50. Dvir T, Timko BP, Kohane DS, Langer R. Nanotechnological strategies for engineering complex tissues. Nat Nanotechnol. 2011;6(1):13–22.

51. Jaconi ME. Nanomedicine: gold nanowires to mend a heart. Nat Nanotechnol. 2011;6(11):692–3.

52. Hosseinkhani H, He W-J, Chiang C-H, Hong P-D, Yu D-S, Domb AJ, et al. Biodegradable nanoparticles for gene therapy technology. J Nanopart Res. 2013; 15(7):1794.

53. Namiki Y, Namiki T, Yoshida H, Ishii Y, Tsubota A, Koido S, et al. A novel magnetic crystal-lipid nanostructure for magnetically guided in vivo gene delivery. Nat Nanotechnol. 2009;4(9):598–606.

54. Plank C. Nanomedicine: silence the target. Nat Nanotechnol. 2009;4(9):544–5.

55. Morrison D, Dokmeci M, Demirci U, Khademhosseini A. Clinical applications of micro- and nanoscale biosensors. In: Gonsalves K, Halberstadt C, Laurencin C, Nair L, editors. Biomedical nanostructures. Wiley; 2008:439–60.

56. Dawson K, Salvati A, Lynch I. Nanoparticles reconstruct lipids. Nat Nanotechnol. 2009;4(February):84–5.

57. Kleinsmith L, Kish V. Chapter 1. Prologue: cells and their molecules. Principles of cell and molecular biology, 2nd edn. New York: HarperCollins; 1995.

58. DNA Microarray Technology. 2011. http://www.genome.gov/pfv.cfm?pageID= 10000533. Accessed 15 April 2015.

59. Hashimoto M, Tong R, Kohane DS. Microdevices for nanomedicine. Mol Pharm. 2013;10(6):2127–44.

60. Behra R, Krug H. Nanoparticles at large. Nat Nanotechnol. 2008;3(May):253–4.

61. Lee Y, Cho M. Application of nanotechnology into life science: benefit or risk. In: Gonsalves K, Halberstadt C, Laurencin C, Nair L, editors. Biomedical nanostructures. Hoboken: Wiley; 2008. pp. 491–501.

62. Minchin R. Sizing up targets with nanoparticles. Nat Nanotechnol. 2008;3(January):12–3.

63. Zhao Y, Xing G, Chai Z. Are carbon nanotubes safe? Nat Nanotechnol. 2008; 3(April):191–2.

64. Elder A. How do nanotubes suppress T cells? Nat Nanotechnol. 2009;4(July):409–10.

65. Kanwar JR, Sriramoju B, Kanwar RK. Neurological disorders and therapeutics targeted to surmount the blood-brain barrier. Int J Nanomedicine. 2012;7:3259–78.

66. Dobrovolskaia MA, Germolec DR, Weaver JL. Evaluation of nanoparticle immunotoxicity. Nat Nanotechnol. 2009;4(7):411–4.

67. Bhabra G, Sood A, Fisher B, Cartwright L, Saunders M, Evans WH, et al. Nanoparticles can cause DNA damage across a cellular barrier. Nat Nanotechnol. 2009; 4(12):876–83.

12

Genetic Engineering: Tinkering with the Human Body

If anything is sacred the human body is sacred.

Walt Whitman
"I Sing the Body Electric"

Science fiction sometimes includes a human character who, by chance or design, is biologically superior in at least one distinct way to a conventional example of *H. sapiens*. These "improvements" may be due to natural processes, such as the slow changes produced by evolution. For example, modern humans have brains over twice as large as members of their ancestors, *H. habilis*, over 1 million years ago. While the latter had only enough intelligence to create simple stone tools, ours has increased enough to create a more complex and at times deadly technology.

With that example of our own development from ape-like creatures, stories like Edmond Hamilton's "The Man Who Evolved" (1931) and "The Sixth Finger" (1963) episode of *The Outer Limits* depicted humans of the far future having larger heads and brains, thus giving them far greater intelligence than our species.[1] Those works and others such as Olaf Stapledon's *The Last and First Men* (1930) were less consistent regarding how our bodies would otherwise change, with a range that included becoming physically weak to having greatly improved strength and manual dexterity (e.g. from an additional, functional digit on each hand).

A more immediate potential way to "improve" humans is via selective breeding, or what has been termed "eugenics." This makes use of the idea that having two parents with a "desirable" characteristic may increase the chance of offspring having it too, and perhaps to a higher degree than either parent. Such breeding programs in science fiction may be designed to improve

---

[1] Such depictions may include the "evolved" human expressing arrogant contempt for the modern-day variety and losing the ability to grow hair on much or all of an enlarged cranium. While the former trait might be understandable though not laudable, the evolutionary value of baldness is less clear.

a particular trait, such as longevity in Robert A. Heinlein's Howard Families (see Chap. 9), military ability as in Gordon R. Dickson's "Dorsai" and Orson Scott Card's "Ender's Game" series, or even "luck" in some of Larry Niven's "Ringworld" novels. The Bene Gesserit, a powerful organization of women in Frank Herbert's "Dune" series, also uses a breeding program for its own ends, as do the Arisians in E. E. "Doc" Smith's "Lensman" series. Heinlein's *Beyond This Horizon* (1948) posits that this method could become so widespread and literally "reproducible" as to create a whole society of "superior" humans. As we will see, however, in the "real world" such methods are far less efficient and predictable than these depictions.

A third method of creating *H. superior* is the "spontaneous" mutation, with offspring possessing traits and abilities that neither parent had, due to a random rearrangement of genetic material. Isaac Asimov's "Mule" character in the "Foundation" series arises as such an unforeseeable factor to disrupt the predictions of psychohistory, as do comic book characters such as Adam Blake (aka "Captain Comet," introduced in *Strange Adventures* #9 in 1951). However, due to the one-of-a-kind nature of this process, the odds of such characters passing on their unique talents to a significant number of descendants is low or, in the case of the Mule, zero.

Genetic engineering in science fiction takes a more direct approach to "improving" humans. Rather than trusting solely in Nature to do the job, an individual's genetic makeup might be directly altered before or after birth in "desired" ways. Examples include Khan Noonien Singh and Julian Bashir from the *Star Trek* universe; Hugo Danner, with his amazing physical abilities similar to and predating those of the original conception of Superman, in Philip Wylie's *Gladiator* (1930); and the hyperintelligent VJ Frank in Robin Cook's *Mutation* (1990). A. E. van Vogt's novel *Slan* (1946) describes genetic manipulation being performed to create an entire new race of physically and mentally superior beings, and the movie *Gattaca* (1997) describes a future permeated by genetic profiling and alterations.

Besides changing humans, genetic engineering can also be employed on nonhumans as in H. G. Wells's *The Island of Dr. Moreau* (at least in its effects if not in its exact methods), or a variety of different plants and animals as in George R. R. Martin's Haviland Tuf stories, whose protagonist travels through space on a "seedship" with advanced ecological engineering capabilities that include genetically modifying plants and animals. However, this chapter will focus on the current status and future prospects of genetic engineering in humans.

12.1 Genes, Chromosomes, and Nature

The DNA in our cells' nuclei can be compared to the operating system and software of a computer. The "programs" it runs set in motion processes such as cell division, creation of proteins and other essential substances, and ultimately regulation of all the "hardware" that makes up our bodies. As mentioned in Chap. 1, the information to do this is contained in discrete sequences in DNA that contain linked chains of "nucleotides." In humans each nucleotide consists of one of four "bases"—adenine, guanine, cytosine, and thymine—coupled to a sugar molecule, "2-deoxyribose," and a third molecule, a "phosphate." Specific sequences of nucleotides are organized as "genes." A gene is the basic unit in DNA that codes for the amino acids used to create the proteins needed for life. In human cells they are located overwhelmingly, though not exclusively, in chromosomes within a cell's nucleus. Within a gene, a series of three bases called a "codon" codes for a specific amino acid. Genes are separated by other sequences of nucleotides that do not code directly for protein creation but, as we shall see, may exert a significant influence on sequences that do. The total genetic content present in an organism's cell is called its "genome" [1, 2].

A gene coding for a particular protein may have alternative forms, differing from each other by the substitution of at least one nucleotide at a particular location. These variant genes are called "alleles." For example, the gene coding for production of a part of hemoglobin (the oxygen-carrying protein in our red blood cells) called the "beta chain" usually has adenine at one location of its DNA sequence. This produces the most common type of hemoglobin, called Hemoglobin A. However, if that gene has thymine instead of adenine at that same location a different variant, Hemoglobin S, is produced instead.

That single change in a base results in production of beta chains with a different amino acid at one point, glutamic acid in Hemoglobin A and valine in Hemoglobin S. Red blood cells with Hemoglobin S are more fragile and likely to rupture than those containing Hemoglobin A. This can lead to anemia depending on how much Hemoglobin S is present in the body. Many other variants in the gene coding for beta chains also exist, resulting in creation of other forms of hemoglobin such as Hemoglobin C or E or in reduced production of beta chains.

While such genetic variants can occur spontaneously during earliest development before birth, they are most commonly inherited from one's parents. The basic principles of how genetic information, including different alleles, is passed on to offspring were first described by a monk, Gregor Mendel, in a paper published in 1865 [3, 4]. He described experiments on how pea plants inherited various measurable traits such as the shape of the peas and their pods, the color of their flowers and pea pods, etc. by cross-pollinating plants having various discrete characteristics. His observations were based on noting the peas' "phenotype," defined as the visible or otherwise identifiable characteristics of a living organism (including humans) produced by one or more genetic factors.

The underlying "laws" Mendel discovered involved the "genotype," the set of genetic factors that create the phenotype. These include various types of alleles that result in different appearances in the peas he studied as well as how other individual traits are inherited in plants and animals. Mendel's "law of segregation" states that these alleles are inherited as discrete units that do not affect each other and are passed on separately into gametes (e.g. sperm cells and oocytes in humans). Each of the 22 different "autosomal" chromosomes (all except the two sex-related X and Y chromosomes—see Chap. 10) in our cells includes one allele for a particular gene located on it. Since there are normally two of each kind of autosomal chromosome, our cells contain two alleles for a particular gene. These alleles may be the same or different [5, 6].

When gametes are produced each receives only one of a particular kind of chromosome, and thus only a single allele. Which of an individual's two alleles goes into a particular gamete occurs randomly, with each having a 50–50 chance of becoming part of a gamete. When two gametes combine, the resulting offspring receives an allele from one parent and a second from the other. This reflects Mendel's "law of independent assortment" that different alleles (and thus the traits and other characteristics each possess) are transmitted independently of each other from the parents to offspring.

While the way alleles for particular genes are transmitted is straightforward, how they express themselves and ultimately affect the phenotype of an individual is more complex. Mendel's experiments dealt with single, clearly visible traits such as shapes and colors involving pea plants. In humans and other organisms, some variant alleles may not cause any obvious change in appearance or affect health, while others (e.g. Hemoglobin A versus Hemoglobin S) cause significant changes. An allele may be "dominant"—that is, its effects will be expressed regardless of whether the other allele in a cell is the same or different. If the other allele is " recessive" it will not be expressed if a dominant allele is present and will only do so if cells have two identical copies of the recessive allele. An individual with two copies of the same allele, either dominant or recessive, is called " homozygous," while someone having two different alleles is "heterozygous"[2, 7].

Thus, when two parents are homozygous for a particular allele all offspring will also have two copies of the same allele and be homozygous too. Expressed another way, all offspring will "breed true" with the same phenotype as their parents. Mendel took pea plants that, when bred together, always produced yellow seeds and another type that always had green ones. In each case the plants were homozygous for a particular seed color, with two identical alleles in each plant. However, when a pea plant producing yellow seeds was cross-pollinated with one that produced green ones, the first generation of offspring all had yellow seeds. Each had an allele for producing yellow seeds from one "parent" and one for making green seeds from the other. They were thus heterozygous for that trait. However, the allele for yellow seeds was dominant while that for green seeds was recessive. The net effect was that those offspring's phenotype was only for yellow seeds, showing Mendel's "law of dominance."

However, when a second generation of pea plants was bred by cross-pollinating members of that first, only heterozygous generation, not all had yellow seeds. Due to random assortment of which allele went into a particular gamete from a parent, 25 % of that second generation were homozygous for the allele producing yellow seeds due to receiving the same yellow seed-producing allele from each parent. Likewise 25 % were homozygous for having green seeds for the same reason—receiving a green seed-producing allele from each parent. The remaining 50 % were heterozygous, with 25 % receiving a yellow seed-producing allele from Parent No. 1 and a green seed-producing allele from Parent No. 2, and the remaining 25 % reversing which parent provided a yellow or green seed-producing allele. Thus, the phenotype of 75 % of this second generation was to produce yellow seeds—25 % from the homozygous offspring and 50 % from those heterozygous for this dominant trait. Only the 25 % of offspring with two alleles for the green seed-producing recessive trait had the phenotype for green seeds. The net ratio of dominant to recessive phenotypes in this second generation was therefore 3:1 [2].

While modern genetics starts with these core laws of Mendelian inheritance, it does not end there. The traits studied by Mendel reflected "complete" dominance, with the dominant allele contributing all and the recessive one none to a heterozygous offspring's phenotype. In human genetics, many conditions with "autosomal dominant" inheritance (the gene associated with the condition is on an autosomal chromosome) have "variable expressivity." For example, Marfan syndrome is due to a variation in a gene that codes for fibrillin-1, a protein needed for making elastic fibers in connective tissue [8, 9]. Some individuals with this genetic variant will have only mild manifestations, such as being only tall and slender with fingers and toes that are longer and some joints more flexible than usual. Others with that same variant can also develop more severe physical problems such as visual difficulties, abnormal curving of the spine, and even life-threatening ones such as widening and tearing of the first part of the aorta that can cause death at an early age.

A related concept is that of "penetrance." This refers to the percentage of individuals having a particular genetic variant that predisposes them to a certain trait or disease who actually go on to demonstrate that trait or develop that disease. While expressivity relates to the degree to which an allele affects an individual's phenotype (including a disease associated with it), penetrance describes the probability of any changes (e.g. a disease) occurring at all. The degree of penetrance may range, for example, from an individual having a low probability of getting a disease to virtual certainty he or she eventually will. Thus, as noted in Chap. 5, some inherited variants of the *BRCA1* and *BRCA2* genes increase a woman's lifetime risk of developing breast or ovarian cancer, but not all women with those variants actually do develop cancer.

Alleles may also show "incomplete" dominance, in which the phenotype produced by one allele is only partially expressed over the other [2]. For example, crossing a snapdragon that produces red flowers (homozygous for an allele that produces red pigment) with another that makes white ones (homozygous for an allele that does not produce that pigment) will result in all offspring having pink flowers. This is due to them having only one allele that produces the red pigment, compared to red flowers with two that can thus make twice as much of that pigment. Offspring of two such heterozygous pink flowers will produce 25 % red flowers (homozygous, with two red pigment alleles), 25 % white flowers (homozygous, with two alleles that do not produce red pigment), and 50 % heterozygous ones with pink flowers.

Other alleles are "co-dominant," with each contributing equally to an individual's phenotype. A classic example for this is the ABO blood group system in humans, with different alleles producing variations of a "glycoprotein" on red blood cells. The A and B alleles create different modifications involving that glycoprotein, while the O allele does not change it. The latter is "recessive" to either the A or B allele. Thus, someone whose genotype is AO will have type A blood, and BO will have type B. However, a person who is AB produces both types of modified glycoprotein equally, with neither A nor B dominating over the other. Thus, the A and B alleles are co-dominant.[2]

As mentioned previously an allele may undergo mutation (e.g. due to an "error" in the creation of DNA derived from one or both parents) early in development. In that case the genotype of offspring is different than would be expected based on the parents' genotypes. For example, about 25 % of individuals with Marfan syndrome do not have a parent with it, indicating that it was caused by a spontaneous mutation [9].

Different alleles of genes located on the sex-related X and Y chromosomes have their own pattern of inheritance. Females typically have two X chromo-

---

[2] See Footnote 3 in Chap. 13 for more information about the ABO blood group system.

somes while males usually have one X and one Y chromosome.[3] If the allele is dominant and on an X chromosome, a male parent with that allele will contribute it to all daughters, who receive their second X chromosome from their mothers, but not to his sons, since he contributes a Y chromosome to them. Conversely the female parent will contribute that allele on an X chromosome to half her offspring, with the other half getting the X chromosome without that allele. The half that do get that allele can be either male or female, since the male parent's X or Y chromosome will determine the gender of the offspring but not whether they get the dominant allele on the X chromosome from the female parent.

X-linked recessive alleles, however, affect only male and homozygous female offspring. The female parent may be a heterozygous "carrier," with one X chromosome without that allele and another with it. Thus, she would not be expected to show clear changes in phenotype due to the recessive allele. If her X chromosome with that allele combines with an X chromosome provided by the male parent their offspring will be a heterozygous female. But if the male parent supplies a Y chromosome, on average half of male offspring will receive the X chromosome with the recessive allele and thus have the condition associated with it, while the other half will get the X chromosome without it. Red-green color blindness and certain types of hemophilia (a blood clotting disorder) have X-linked recessive inheritance.

While diseases associated with them are rare, variants in alleles on Y chromosomes will be inherited based on the gender of offspring. Only males, with a usual genotype of XY, can have this type of disorder, while females (who have only X chromosomes) will not.

As mentioned in previous chapters, a small amount of DNA is present in each cell's mitochondria and, at least in humans, is inherited only from the female parent. Like those involving nuclear DNA, mutations in mitochondrial DNA can also be associated with genetic diseases [10].

As if these patterns of inheritance were not complicated enough, the genotype and phenotype of offspring do not rely solely on having different alleles of single genes. Gametes produced by either the male or female parent may have chromosomes that did not replicate exactly or "normally" from the parent's original ones. This could mean that chromosomes in the gamete might have areas corresponding to multiple genes that have been deleted, duplicated, inverted, or have other changes compared to the parent's original one.

For example, in Trisomy 21 (Down syndrome) one human parent's gamete has two copies of chromosome 21—one complete and the other partial or complete. After adding a single copy from the other parent, offspring can thus

---

[3] See Chap. 10 for more details about sex chromosomes.

have three of those chromosomes for a total of 47 instead of the usual 46. As noted in Chap. 10, the X and Y chromosomes may also occur in combinations other than XY and XX, including XO (Turner syndrome, with a female phenotype) and Klinefelter syndrome (XXY, with a male phenotype) [11].

Based on the first complete mapping of the human genome early in this century, the total number of genes in our cells is now estimated to be in the range of a little less than 23,000, although the exact number is still uncertain [12]. Genes vary greatly in how long their sequences of DNA are, and there can be very many different alleles for individual genes. As noted preciously alleles may vary by only a single nucleotide ("single-nucleotide polymorphism"), as with Hemoglobin A and Hemoglobin S. However, more extensive variations in alleles can also be present. As noted in Chap. 1, only a little over 1 % of DNA actually codes for proteins, the sections called "exons." Additional sequences of DNA associated with the coding sections, "introns," are transcribed along with the exons by RNA but are later removed during production of proteins coded by exons.

Many non-gene areas of DNA are also involved in regulating the activities of the genes. Based on this regulation genes may increase or decrease the amount of a protein produced and also make subtle alterations in those proteins, thus potentially changing their activity. Some genes too can even modify the effects and actions of more distant genes. Other types of molecules within the cell may also do this, as can environmental and other changes. The term "epigenetics" is used to describe such alterations in gene activity without structural alterations in the gene itself—that is, the DNA sequence in the gene remains unchanged but what the gene does changes [13, 14].

Furthermore, certain genes may become inactivated and no longer function. In females one of the two X chromosomes in a particular cell becomes essentially inactive. Selection of whether this happens to the X chromosome derived from the male or the female parent is random. Thus, women who are heterozygous for an X chromosome-based allele that results in no production of an enzyme may produce roughly half of the normal amount, due to inactivation of about half of the "normal" X chromosomes that do code for it. That reduced amount of the enzyme is typically still enough to prevent any clinically significant effects, however, with the woman grossly showing no sign of having that biochemical abnormality.

Variations in certain single (e.g. dominant) genes alone can result in specific enzymes and other proteins not being produced in adequate amounts or functioning well enough to maintain health. This can result in effects and traits involving multiple areas of the body—"pleiotropy"—as shown by the effects of Marfan syndrome on the lengths of limbs as well as abnormalities involving the eyes and aorta. However, many diseases and traits are "polygenic"—that is, due to many different genes and their interactions. Rather than producing single, discrete changes such as the yellow or green colors of Mendel's peas, they express themselves over a wider, continuous range. For example, an individual's eye color is the result of interactions among as many as 16 genes coding for different pigments and other substances [15]. A person's height, skin and hair color, and genetic contributions to overall "intelligence" and life expectancy are all polygenic. Disorders such as schizophrenia and some other psychiatric illnesses, diabetes, and high blood pressure also involve multiple genes and their effects.

In short, the fusion of a man's spermatozoon and a woman's ovum produces a cell with all the initial genetic information needed to produce a unique human being. This information is a mixture of that derived from both parents as well as any changes introduced by potentially less than 100 % faithful replication of the genetic material from each one. At conception the "program" contained within that cell's DNA begins to "run" immediately, leading to cell division, differentiation, and growth. However, due to any further structural or functional changes in that DNA, parts of that original program can be effectively rewritten.

Moreover, environmental factors also influence how the program runs as well as affecting its "results," i.e. the tissues and other parts of our bodies. Poor nutrition, infections, trauma, psychological stress, toxins, radiation, and other detrimental effects on a person's body may prevent fulfillment of the "best" that individual's genotype might have produced. Monozygotic ("identical") twins derive from a single fertilized ovum that splits into two early during subsequent development, and thus start out essentially genetically identical. However, although their overall phenotypes may remain similar, variations in gene activity and environmental factors subsequently make them different in varying ways and degrees. The latter start with their respective developments in the mother's uterus and continue after birth. The net effect is that monozygotic twins do not have a 100 % correlation in their interests, intelligence, or development of disorders (e.g. schizophrenia or high blood pressure), with some factors having higher and others lower commonality between them.

12.2 Potential Applications of Gene Therapy and Genetic Modification

What all this information means is that, although there are basic principles regarding how genes work and are inherited, there are also great complexities in how a person's genotype affects what he or she ultimately becomes both physically and mentally. To reliably perform genetic engineering for a particular purpose requires first knowing not only what genes are involved but how they function. Although we now know the overall structure of the human genome, there are so many variations in the sequences of DNA that comprise those roughly 23,000 genes—their alleles—and factors involved in what they ultimately do that we have only scratched the surface of learning how all those genes act individually and interact with each other.

There are now established methods for modifying the genetic structure of cells. However, current procedures and capabilities fall far short of the nanobots busily repairing and improving DNA described in Chap. 11 or the wholesale changes in DNA leading to radical changes in human phenotypes and abilities depicted in some science fiction. Instead research in this area is moving ahead but in very focused areas, concentrating on therapeutic uses for genetic disorders rather than "improving" the human body over its basic healthy state and form. Human clinical trials using gene therapy must of necessity move slowly due to our still very incomplete knowledge of how to most effectively provide that therapy and the need to reduce its risks to patients as much as possible. Such research also concentrates on the most serious genetic disorders that might be amenable to treatment now and in the near future.

There are several fundamental rationales for performing genetic interventions. Current research focuses on developing ways to treat genetic issues in somatic cells—that is, all the cells in a person's body except germ cells (sperm and egg cells). One important area is to treat individuals with a genetic variant that has debilitating and even life-threatening results. For example, a clinical trial whose results were first reported in 2000 described successful correction of the underlying genetic defect in children with X-linked severe combined immunodeficiency (SCID) [16, 19]. This is a potentially fatal disease associated with a solitary genetic variation that results in certain lymphocytes (a type of white blood cell that is an important component of the immune system) not developing. Those affected by this disease experience recurrent infections beginning very early in life and typically die within several years after birth.

Other "monogenic" diseases—those due to isolated genetic defects—that have been reported to be partially treated by replacing cells having DNA causing the disease with genetically modified cells without that defect include X-linked adrenoleukodystrophy[4] and immunodeficiency syndromes such as chronic granulomatous disease. The former involves a genetic defect that results in accumulation of particular types of fatty acids within the body, with the most severe effects usually occurring in the myelin sheath of cells in the central nervous system, and other cells in the adrenal gland. In different vari-

---

[4] The title character in the 1992 movie *Lorenzo's Oil* had this genetic disease.

eties of chronic granulomatous disease some cells of the immune system can engulf certain pathogenic bacteria and fungi but cannot generate the chemicals needed to kill them, thus reducing the body's ability to fight infections. Other clinical trials have reported some success with treating Leber congenital amaurosis, a cause of blindness, and beta-thalassemia, which involves problems with producing hemoglobin [17, 20, 22].

Genetic modification is also being used to treat cancer. Genes can be introduced into malignant cells to make them produce substances that stimulate a person's own body to mount an immune response against them, particularly by triggering cytotoxic ("cell-killing") T lymphocytes to destroy them [23, 24].

Besides performing gene therapy on somatic cells after a person is born, another approach would be to do it prior to birth [25]. Some genetic defects can be so severe they might cause some degree of irreversible damage to a developing fetus. These can include diseases that are associated with significant neurological problems such as Tay-Sachs disease (an autosomal recessive disorder involving reduced activity of a single enzyme, resulting in excessive accumulation of substances called "gangliosides" in the brain) and Lesch-Nyhan syndrome (an X-linked genetic variant associated with an abnormality in production of an enzyme that results in overproduction of uric acid in the body). Other cell types with DNA variants known to cause serious diseases, particularly those involving blood cells, might also be targeted.

The major advantages of such "in utero" gene therapies include "curing" a genetic disease before it causes appreciable damage, and the potential for the fetus's body to be able to accept genetically modified cells better than when its own immune system is more "mature" and thus more likely to reject them. A further refinement of this method might include performing such a therapeutic genetic repair on an embryo. Fewer and less differentiated cells would need to be genetically altered that early in gestation, potentially making the procedure more effective [23, 25].

However, there are also potential risks to both the mother and the embryo/fetus from such in utero gene modifications [25]. Direct risks to the former may be minimal, although there still might be a possibility that the virus vector used for gene therapy (see Chap. 11) could infect placental tissue or pass into her bloodstream via the placenta and cause harm. The mother could also be harmed by ongoing toxicity from a fetus with a severe genetic defect such as homozygous alpha-thalassemia, a disease that results in creation of abnormal forms of hemoglobin unable to provide oxygen to tissues. If gene therapy is not adequate to treat such a fetus it may, as part of the course of the disease, either die before birth or be so critically ill that it affects the mother's health before dying itself not long after birth.

For the fetus, the actual injection of a gene-carrying vector into it during the second trimester of pregnancy carries reasonably low risk. However, without adequate study it is uncertain whether the new genes could interfere with normal development of organs and tissues, or lead to other, unwanted changes in DNA resulting in eventual development of cancer or other abnormalities.

Other possible venues for gene therapy are on the very earliest stage of development, a fertilized ovum, and on germ cells. Animal studies have shown that injecting genes directly into a fertilized ovum could not only result in them being incorporated into all of the resulting animal's cells but even be passed on to its offspring [23]. Such a method could help prevent transmitting a genetic defect to offspring. For example, the autosomal dominant disorder Huntington's disease commonly does not cause symptoms until individuals who have that abnormal gene are in their mid-30 s or older, and any offspring have a 50 % chance of inheriting it. Monogenetic defects such as Tay-Sachs disease and Lesch-Nyhan syndrome, and retinoblastoma, a genetically transmitted (autosomal dominant) form of cancer involving the eye, would also be potential targets in individuals who carry or have those disorders. Modifying an individual's germ cells prior to conception to correct the genetic defect they have could potentially prevent the latter from being inherited by any offspring.

The next step beyond "repairing" a genetic defect in somatic or germ cells, thus preventing or treating their deleterious effects on health, is to go beyond therapy to "improving" a person's "healthy" cells. At the "simpler" end of the scale, certain individual and perhaps groups of alleles may confer above average benefits in otherwise healthy people. For example, a genetic variant associated with lower total blood levels of cholesterol and of a particular kind, low-density lipoprotein (LDL) cholesterol, could help protect against development of coronary artery disease (CAD, see Chap. 2) [26]. A person without such a relatively protective variant could potentially have it introduced by genetic engineering and gain this benefit too.

Likewise, some individuals have a genetic variant of the *CCR5* gene that markedly reduces the ability of the human immunodeficiency virus (HIV) to enter and infect T lymphocytes [27]. A person who is homozygous for this variant is resistant to infection with HIV and thus development of acquired immunodeficiency syndrome (AIDS). Heterozygotes who already have HIV infection typically have a longer course before AIDS may develop. A study involving a small number of HIV-infected individuals who received T lymphocytes genetically "edited" to have that protective variant of the *CCR5* gene showed good results with this method [27].

Besides engineering resistance to disease, genetic modification of a "healthy" person might be used to enhance what Nature gave. Thus, changing appropriate genes or introducing new ones might be used to increase muscle mass and efficiency, strengthen bones, or produce similar physical enhancements to augment athletic ability. Vision could perhaps be improved by genetically modifying retinal and other cells to allow us to see into the ultraviolet and infrared range. Adding a fourth type of photoreceptor cell to the three the human eye normally use to see color would increase the number of shades and hues we can distinguish [28]. Our built-in cell repair mechanisms and immune systems might be augmented to make us better able to resist infections, detect and destroy cancer cells, and eliminate the wear-and-tear and other effects of aging to keep us anatomically and physiologically young. An advanced enough understanding of the neuronal and other factors underlying intelligence and artistic talent could also potentially employ genetic modifications to augment them too.

Science fiction may include these kinds of upgrades to the human body but might also go beyond them to envision major modifications to its "standard" form and functions. This might be done for cosmetic, recreational, or ideological reasons, as in Charles Sheffield's "Proteus" series, beginning with *Sight of Proteus* (1978), which depicts a future in which individuals can change themselves in virtually any way they can imagine. G. David Nordley's short story "In HIS Image" (1992) shows how some might react negatively to development of such body-changing techniques, while the Shapers of Bruce Sterling's novel *Schismatrix* (1985) and other stories set in that work's universe are enthusiastic users of such methods. Such methods could also be used to rigidly stratify whole societies, as in Frank Herbert's *The Eyes of Heisenberg* (1966).

Genetic manipulation could also have the "practical" use of helping humans adapt to and live in otherwise inhospitable environments. Clifford D. Simak's 1944 short story "Desertion" describes the radical transformation of humans into "Lopers" able to live on the surface of Jupiter. As mentioned in Chap. 4, Lois McMaster Bujold's novel *Falling Free* (1988) describes the creation of "Quaddies," genetically engineered to work in the microgravity of space by having their legs modified into a second pair of arms. James Blish's "Pantropy" stories, collected as *The Seedling Stars* (1957), also explore this theme in a variety of exotic extraterrestrial settings.

Here too such genetic "improvements" could be made at different stages of life, from shortly after conception to long after birth. By incorporating those changes into an individual's germ cells these techniques could be used to create a whole new "race" of humans with no or at least less genetic manipulation being required for future generations.

12.3 Current Status of Gene Therapy and Genetic Modification

After that brief description of some uses of genetic engineering in science fiction, let's examine what can currently be done with that technique, the real-world risks involved, and the many challenges to increasing our capabilities for performing it.

Most clinical trials involving gene therapy have been to treat cancer [17]. These have included studies on malignancies involving the lung, breast, skin, and blood cells (e.g. leukemia). A commonly used target is to transfer a tumor suppressor gene, particularly the *p53* gene, into cancer cells. As described in Chap. 9, along with other functions dealing with development of cell senescence and apoptosis the *p53* gene helps maintain the structural stability of DNA and prevent mutations in it during cell division. In about half of all cancers, particularly those involving the ovaries, colon, head, and neck, *p53* no longer acts in its normal manner [29]. This may be due to mutations in the *p53* gene itself that can make it fail to suppress development of cancer and potentially even assist its growth. Introducing a normally functioning *p53* gene into cancer cells can cause them to either die or be recognized by the body's own immune system and destroyed.

Gene therapy has also been used to help improve blood supply to areas of the heart that are not receiving enough blood due to CAD. These involve adding genes producing substances such as vascular endothelial growth factor (VGEF) that stimulate growth of more blood vessels (a process called "angiogenesis") to those areas. Gene therapy might also be used to help produce new myocardiocytes (heart muscle cells) after the heart is damaged by a myocardial infarction or other cause [17].

Introducing a new gene into a cell is usually accomplished by either coupling it with or incorporating it into a particular carrier or "vector." The gene and its vector can be injected directly into a patient (an "in vivo" technique), where they then hopefully go to the target cells and make the desired genetic modifications to them. Alternatively cells from the person (e.g. white blood cells) can be obtained from a blood or tissue sample, treated with the vector-gene outside the body ("in vitro"), then the "corrected" cells injected back into the person [30].

A variety of viruses possess the natural ability to inject DNA into cells and are the most common type of vector used [24, 30]. Adenoviruses have been employed in clinical trials for over 20 years [17, 31, 32]. A person infected with an adenovirus in its "natural" form typically develops an upper respiratory infection (e.g. involving the throat). Symptoms are usually mild and may include fever, sore throat, conjunctivitis (inflammation of the clear tissue covering the visible part of the eyes and inner eyelids), and malaise. Over 50 different varieties ("serotypes") of adenovirus may cause illness in humans. Some can also cause pneumonia or involve the gastrointestinal tract, and rarely they can produce life-threatening or even fatal infections.

Prior to being used as vectors for gene therapy, adenoviruses are modified to make them safer by removing genes associated with infection. Those genes include ones that allow adenoviruses to replicate on their own, as well as those that could stimulate a self-harming response by the body's immune system. However, clinical trials in the 1990s using the adenovirus as a vector showed that even in an altered form it could produce serious and even fatal reactions [33]. In one research trial an adenovirus was used to introduce a normal gene into patients with cystic fibrosis. This is an autosomal recessive genetic disorder associated with production of thick secretions that are primarily associated with infections and damage to the lungs, but that can also injure other organs (e.g. the pancreas and liver). In 1993 one volunteer in the trial developed a severe inflammatory reaction shortly after administration of the adenovirus, but survived.

However, an even greater risk of gene therapy was demonstrated in 1999 when an 18-year-old volunteer in another trial using an adenovirus vector became acutely ill and died [33, 34]. The trial was designed to treat an X-linked recessive genetic defect in which a person does not make enough ornithine transcarbamylase (OTC), an enzyme that helps removes ammonia from the blood. Although mild cases such as the male volunteer had can be controlled by a strict diet and medications, OTC deficiency could cause severe problems such as lethargy, liver damage, seizures, and death. It was determined that the volunteer's death was due to the adenovirus causing a severe immune response that fatally damaged his internal organs. This event, in fact, triggered an intensive review of gene therapy trials to improve their safety.

Overall, adenoviruses have been found to possess both significant advantages and disadvantages as a vector. They can be used for gene therapy on either dividing cells, such as fibroblasts or those that make red and white blood cells, or non-dividing ones (e.g. neurons and myocardiocytes). Adenoviruses can also be given in large "doses," with each virus particle carrying the new gene, and are taken up well by cells, allowing the gene each carries to interact with nuclei and other intracellular components. Because so many adenoviruses and the therapeutic genes they carry can enter a large number of target cells, those genes can potentially produce a strong beneficial effect.

However, because adenovirus infections are common, a person may already be immune to some serotypes due to a prior infection with it, and thus need to use one to which he or she has had no prior exposure. Moreover, as described previously there is also a risk of inducing a serious or even fatal immune response. If more than one dose of a particular serotype of adenovirus is given the chances of it being less effective increase, and it is more likely to trigger a harmful immune response in the person. These issues have been addressed by using different serotypes of adenovirus if multiple injections are needed, removing as many viral genes as possible, and by blocking the activity of those that remain.

Adenoviruses and the therapeutic genes they carry are not incorporated into the DNA of the cells they enter. This makes it more likely the new gene will be "silenced" by the host cell and that it will not be present in new cells when the "infected" one divides. In either case the new gene will no longer produce the protein/enzyme for which a person has a deficiency. On the other hand, this inability to join with the host cell's DNA reduces the risk of the adenovirus damaging that DNA, perhaps enough to cause the cell to become cancerous. When adenoviruses are used to introduce a gene designed to kill cancer cells their potential for producing only short-term effects is, however, not an issue.

"Adeno-associated viruses" (AAV) are also used as vectors for gene therapy [24, 31, 32]. AAV can enter a wide variety of different cells in the body, including (like adenoviruses) either dividing or non-dividing cells. Unlike adenoviruses, AAV do not cause disease in humans, can incorporate themselves into the host cell's DNA at a specific site, and induce at most only a mild immune response. In their "natural" state AAV require that the cell they enter must also be infected by a "helper" virus—an adenovirus or herpes simplex virus—before they can replicate. When replicating AAV to produce a sufficient "dose" prior to injection into a patient, only the genes from an adenovirus needed to make AAV reproduce are used. AAV's ability to incorporate itself into an infected cell's DNA can also be removed if desired.

A disadvantage of AAV is that, though they do not cause disease, infections by them are common and thus make it likely a person may already have developed "immunity" to them. Any virus vector can also only carry a limited length of the DNA that constitutes the therapeutic gene to be delivered—that is, there is only so much "room" inside the virus capsule for that new genetic "payload"—but AAV have a particularly small capacity to do this compared to other viruses. Nonetheless, gene therapy using this vector has been successful in improving vision in patients with Leber congenital amaurosis [20, 35] and treating other diseases such as Parkinson's disease [36].

Herpes virus vectors, derived primarily from the herpes simplex virus type-1 (HSV-1) that causes "cold sores" around the mouth (not the HSV-2 variety responsible for genital herpes), also have their own advantages and disadvantages. HSV-1 requires extensive modification to remove its disease-inducing and other nonessential genes, which can then be replaced by a particularly long length of DNA that includes a therapeutic gene [24, 32]. The herpes virus can enter many types of cells, particularly neurons, and be produced in large amounts to provide an effective dose. Like the adenovirus it is not integrated into a host cell's DNA, and thus the therapeutic gene it carries may only be functional for a limited time. Similarly, it can also enter either dividing or non-dividing cells.

However, because more than 70 % of people have active immunity to it due to prior exposure, both the herpes vector and the cells containing it are likely to be targeted and destroyed by a person's immune system, thus reducing the effectiveness of the gene it delivers. It might also interact with a "natural" HSV-1 virus that has been "dormant" within a previously infected cell and activate it, thus potentially causing the person being treated to have a new episode of active infection.

A herpes vector can be made to contain an "oncolytic" gene—one designed to lead to the death of cancer cells—and the virus itself may also kill them. This vector has been successfully used to treat cancers such as glioblastoma (a tumor derived from the glial cells that support neurons in the brain) and melanoma (a very dangerous type of skin cancer) [24].

Another type of viral vector, retroviruses, includes many kinds of similar viruses that include RNA in their normal makeup instead of the DNA contained within the ones described previously. One retrovirus commonly used as a vector for gene therapy is the "murine leukemia virus." It can cause cancer in mice but has not been definitively found to cause disease in humans. Retroviruses have the advantages of being able to enter a wide variety of cells, are unlikely to stimulate a significant immune response, and can be produced in large numbers.

Unlike adenoviruses, AAV, or the herpes virus, some types of retroviruses can infect only dividing cells [31]. They normally integrate themselves permanently into the infected host cell's DNA but, unlike an AAV, they do this at random locations in it. This might produce no significant effects on the DNA, kill the cell if the retrovirus disrupts a critical location (e.g. one coding for a vital protein)—or cause the cell to become cancerous.

The latter apparently did occur in four of nine patients treated for SCID with a retrovirus vector [18, 37]. All of these patients were initially treated with bone marrow cells that were infused back into them after the cells were genetically modified. Although 8 of the 9 individuals had successful correction of their decreased immune function, four of the nine subsequently developed "acute T-cell lymphoblastic" leukemia thought to be due to the retroviruses' effects on the modified bone marrow cells' "native" DNA. One of those four patients died from leukemia, while the other three were successfully treated for it.

These results highlighted both the benefits and risks of gene therapy using this vector, prompting additional research to modify it further to increase its safety [24, 38]. In a later study of ten patients with SCID treated with a similar retrovirus vector, one of them also developed that same variety of leukemia [31]. Safety has increased with use of different techniques during more recent studies [38]. For example, over 100 patients treated with T cells genetically modified outside the body using a retrovirus vector have shown no evidence of subsequent malignant transformations of cells [39].

Lentiviruses are a particular variety of retroviruses that, unlike the murine leukemia type, can enter either dividing or non-dividing cells [24, 40]. They too can integrate permanently into a host cell's DNA, or be modified to not do this. Unlike the retrovirus vectors just described, if a lentivirus does combine with the cell's DNA it has a lower likelihood of doing that in a location that causes the cell to become cancerous. To further reduce that risk lentiviruses can be modified so that they do not "turn on" genes near their insertion site into native DNA that could induce cancer. Lentiviruses also have little risk of inducing an immune response and can be "optimized" to target specific cells.

The lentiviruses used as gene vectors are ones that either do not infect humans or, in one case, is a heavily modified version of one that does—HIV, the virus that causes AIDS. The lentivirus vector derived from HIV has been found to be reliable and efficient for delivering new genes into cells. However, for obvious reasons particular attention is given to make sure the modifications applied to lentiviruses prevent them from replicating on their own.

Lentivirus vectors have been used to treat breast, hematological, and other types of cancer as well as genetic diseases [24]. Recent reports indicate that this vector can be used to provide long-term correction of multiples types of genetic disorders [41, 42].

Yet another type of retroviral vector is the "foamy virus." This type of retrovirus may be especially useful for gene therapy involving "hematopoietic stem cells," precursors of all types of red and white blood cells [43]. Foamy viruses are not thought to cause disease in humans, can be used for gene therapy in a wide range of different cells, and can carry a large amount of genetic material as their "payload." However, they require the presence of dividing cells to do their work, and thus may not be ideal for tissues that do not normally do this, such as "terminally differentiated" muscle cells. This latter point illustrates the principle that specific vectors might be best employed to do specific jobs in different types of cells.

Various viruses of the pox family and others have also been used as viral vectors. These include the vaccinia virus, used in the smallpox vaccine. Poxviruses can also be used to create other types of vaccines and, when used for gene therapy, have characteristics that enhance their ability to destroy cancer cells.

Another method employed for gene therapy is to use "naked" DNA—that is, a gene not "protected" by the proteins of a virus's capsule. It can be introduced in the form of a "plasmid," which typically takes the form of a single circular molecule of DNA. An important limitation of this technique is that it has been found to be less efficient for getting that DNA into target cells than using viral vectors [32]. For example, naked DNA might be rapidly destroyed by a person's own immune system or degraded by certain enzymes within the bloodstream.

One technique studied to "protect" naked plasmid DNA until it reaches and can "infect" a target cell is to enclose the DNA within a microbubble composed of a lipid, albumin, or other material. Microbubbles of such substances are routinely used as "contrast" agents in taking ultrasound images of the heart, or "echocardiography." They show up on images as tiny bubbles that reflect the ultrasound beam to help better see where the exact inner border of the left ventricle is, detect the presence of blood clots inside it, etc.

Thus, ultrasound could be used to produce images showing when microbubbles containing DNA reach their target organ. Mildly increasing the ultrasound energy could then make those microbubbles rupture, releasing a large dose of DNA near the target cells. Potential uses of this method for cardiovascular diseases including treatment of atherosclerotic plaques or improving blood supply to the heart [44].

Somewhat similarly, small artificial human chromosomes can also be produced [45, 46]. These can be created either by altering a pre-existing chromosome or building one "from scratch" using DNA as well as proteins and other substances that make up a "normal" chromosome. While viral vectors can carry only a limited amount of DNA, such artificial chromosomes could contain a much larger quantity of it and be more complex.

Other nonviral methods used for gene delivery include associating them with inorganic particles or those with a specific type of organic structure that does not cause infection. Such particles may have a size on the nanoscale and be injected directly into a target tissue rather than having to reach it indirectly through the bloodstream. These other methods have generally been associated with lower efficiencies for having genes reach the desired cells than by using viral vectors [47]. Nonetheless, many of these substances such as chitosan (a polysaccharide) and polyethylenimine have other attractive characteristics such as low toxicity and good biodegradability that make them suitable for further study [47, 48].

Besides issues with the vectors used, the effectiveness of gene therapy may be reduced by other factors. As mentioned previously the recipient's immune system might identify genetically altered cells as "foreign" and try to destroy them, thus reducing their effectiveness. This could require at least temporarily suppressing the person's immune system, with its own attendant risks of reducing the recipient's ability to ward off infections. Also, some genetically modified cells may have more difficulty than others proliferating and establishing themselves within a patient's body [17].

New genes do not necessarily need to be introduced directly into humans to be helpful. Another approach is to genetically modify bacteria, plants, and animals to produce substances that can be used for medical purposes [49, 50]. Techniques include inserting human genes that produce useful hormones and other proteins such as insulin, human growth hormone, and blood clotting factors into bacteria like *Escherichia coli*, which then act as biological "factories" to produce those substances. "Transgenic" animals have been created by genetically modifying them to secrete useful human proteins such as blood clotting factors in their milk, which can then be processed and used for medical purposes. In some cases these animals can then be cloned to increase the efficiency of human protein production.

A large number of gene therapy human research trials—over 1900 between 1989 and 2013—have been conducted or are currently in progress [17, 51, 55]. As of this writing, however, only one gene therapy treatment has been approved for clinical use in Europe, and the Federal Food and Drug Administration (FDA) in the United States has not yet allowed any except for research purposes. The sole exception approved in Europe uses an AAV vector to treat lipoprotein lipase deficiency, a disease that can be inherited as an autosomal recessive genetic disorder. Individuals with this genetic disorder typically have abdominal pain, fat deposits beneath the skin ("eruptive xanthomas"), pancreatitis, enlargement of the spleen and liver, diabetes, and are at increased risk of developing blockages in the arteries of the heart. The gene therapy agent approved will involve a one-time series of up to 60 injections into the muscles of the leg, along with treatment with medications to suppress the immune system for 3 days before and 12 weeks following administration of that agent [56]. The estimated cost of that course of treatment has been reported to be in the range of $ 1,000,000 [57]. However, this may still be less costly overall than conventional treatments for this disease, as well as being more effective at relieving pain and suffering.

12.4 Genomic Medicine

Increasing knowledge of the human genome and how it differs in details from one person to another has also led to new ways to assess an individual's risk for developing different diseases, predicting response to medications, and guiding treatment for cancer. The term "genomic medicine" is used to describe medical care guided by knowledge of both general and individual human genetics. The more general one "genomics" refers to the study of how DNA is structurally organized into "sequences" (i.e. which nucleotides are present in a length of DNA and the order they are linked together within it), what functions those sequences have, and the techniques used to duplicate, manipulate, or otherwise change them [1, 4, 58].

While many potential applications for genomic medicine remain only within the realm of molecular, animal, and human research, others have reached the stage of everyday medical practice. For example, *BRCA1* and *BRCA2* genes produce tumor suppressor proteins that repair DNA. Certain mutations in these genes can reduce the cells' ability to do this and significantly increase the risk of developing certain forms of cancer [59, 60]. Thus, the chance of a woman who has a harmful mutation in *BRCA1* having breast cancer during her lifetime is estimated to be 55–65 % compared to an overall risk of 12 % in the general population of women. Likewise the risk of ovarian cancer occurring before a woman reaches the age of 70 increases from an average value of 1.4 % to one of 39 % if she has a harmful mutation of the *BRCA1* gene, and to between 11–17 % with a certain mutation to *BRCA2* [61]. Screening for these mutated genes can be used to assess a woman's risk of having these cancers and thus guide management of her medical care.

Genetic screening of prospective parents can also help assess risk to offspring. There are over 1000 rare disorders that are known to be due to recessive genes that follow Mendelian inheritance [58]. Thus, if both parents are heterozygous for a particular recessive gene, on average 25 % of their offspring will be homozygous for it and have the disorder associated with it. This information can help them regarding subsequent family planning. It is estimated that each person has on average 2.8 mutations for known recessive disorders, so everyone carries at least a very low risk of producing offspring that are homozygous for one of them [58].

Likewise, genetic screening of offspring prior to and at birth can also act as "early warning" systems regarding the presence of significant disorders and help to guide management. Prenatal screening can be done when indicated by performing an amniocentesis, in which a small amount of the fluid surrounding the fetus in the uterus is removed with a needle and syringe, or by "chorionic villus sampling" in which a sample of the placental tissue is obtained. These latter procedures are only done selectively, however, due to a very small risk of serious complications involving the fetus, mother, or both. Some degree of genetic screening of embryos prior to in vitro fertilization (See Chap. 10) can also be performed.

Going beyond "simple" genetic disorders and chromosomal abnormalities (e.g. trisomy 21), screening for diseases and traits that are far more complex in their genetic characteristics, due to being associated with many genes and genetic variants as well as epigenetic and environmental factors, is much more difficult. These include the risk of developing cardiovascular disease, diabetes, as well as characteristics that present obvious difficulties in measuring before or at birth such as intelligence [58, 62]. Development of the latter can be impaired by a wide variety of major genetic and environmental issues such as Mendelian disorders that directly or indirectly affect development of the brain, poor nutrition, illness early in life, etc. "Healthy" development of intelligence has been found to have a significant degree of heritability but to also be highly polygenic, with a very wide range of genetic variability. Thus, rather than intelligence correlating with one or a few "supergenes," it is associated with the interplay of many genes, most of which are common rather than rare variants.

This complexity presents formidable challenges for genetically engineering "enhanced" intelligence. Conversely, using gene therapy to correct genetic disorders that cause damage to the brain and reduce intelligence (e.g. Tay-Sachs disease) could help preserve the level of intelligence that the affected individual would otherwise have if the disorder were not present.

The 1991 novella (later novel, 1993) "Beggars in Spain" by Nancy Kress presents an imaginative variation on this theme. In a future society where genetic modification is becoming common a small number of babies are engineered to no longer need sleep or even be capable of sleeping. This is explained as being possible based partly on the discovery that sleep is a vestigial function of the brain that can be safely eliminated. Besides the obvious benefit of having more waking and hence productive time, performing this modification is also found to enhance intelligence, have generally positive effects on personality, and even lead to possible perpetual youth. The novel delves deeply into the individual, political, and social implications of the growing number of the "Sleepless."

However, current medical opinion is that sleep is, in fact, not "optional" but a necessary function to maintain health. This is based on extensive clinical observations regarding the effects of severe insomnia and sleep deprivation, which include increased risks of obesity, diabetes mellitus and glucose intolerance, cardiovascular events such as myocardial infarction, and detrimental effects on mood and behavior [63]. A specific genetic disorder, fatal familial insomnia, in which individuals develop worsening insomnia and can ultimately lose the ability to sleep is associated with symptoms such as panic attacks, hallucinations, dementia, and finally death [64, 65]. Research involving the metabolic and physiological role of sleep indicates that lack of it can cause activation of a person's immune system and produce a harmful inflammatory response [66, 67]. Sleep is also thought to play an important role in consolidating memories and clearing the brain of toxic metabolic byproducts [68, 70].

However, while the central idea that losing the ability to sleep would be beneficial is not supported by current information, an effective science fiction work like "Beggars in Spain" can still explore the ramifications of what would happen *if* that were true in a thought-provoking way. A more recent story by the same author, "Pathways," (2014) does involve fatal familial insomnia and a potential future treatment for it.

Genetic information can also be used to guide therapy for the treatment of cancer and responses to different medications. For example, the HER-2 biomarker found in about 30 % of breast tumors has been associated with worse outcome and, when present, can be selectively targeted by medication. This is also done for mutations in another biomarker, EGFR, in non-small cell lung cancers, whose presence indicates that certain chemotherapeutic agents would be particularly helpful. Conversely, individuals with colorectal cancer that has spread elsewhere within their bodies and have the *KRAS* genetic mutation within that cancer are nearly always resistant to certain chemotherapeutic drugs, and thus alternative drugs would be used instead [58]. The discovery of a particular mutation in the gene *BRAF* associated with melanoma led to development of a drug designed to inhibit it and thus treat the cancer [1].

Such biomarkers can be obtained from tissue samples of the tumor and, in some cases, from blood samples. This technique can be potentially helpful both for early diagnosis of a cancer as well as assessing recurrence after previous treatment for it.

"Pharmacogentics" involves how an individual's genetic makeup affects the way his or her body responds to particular medicines [71]. Single genetic variants can correlate with how a person metabolizes a medication, either rapidly or more slowly, and thus help guide what dose should be used. This is the case with the commonly used blood thinner warfarin, with some individuals having genetic variants that make them more resistant to its effects and thus requiring higher doses. Tables listing appropriate starting doses based on the presence of combinations of two such genes that affect warfarin's effectiveness, *CYP29* and *VKORC1*, are now available [1].

Many other medications have similar recommendations regarding adjusting their doses based on the presence or activity of particular genes. The FDA currently lists over 140 different medications with prescribing information that involves genetic determinants [72]. For example, clopidogrel is a frequently used "antiplatelet" medication that inhibits an early stage (one that includes the "clumping" together of cells called platelets) of the complex process by which blood clots. Clopidogrel is, along with aspirin (which also inhibits platelets by a different mechanism), a mainstay for treating individuals having or threatening to have a myocardial infarction. It is also used to prevent this from happening again after a partially blocked coronary artery is opened by inflating a balloon within it, a procedure called "coronary angioplasty." A metal mesh called a "stent" is sometimes inserted into the artery to help keep it open, and a combination of both aspirin and clopidogrel is used to reduce the chances of clot forming in the stented area.

However, it is now recognized that many people have a variant of the gene *CYP2C19* that inhibits clopidogrel from doing that important task. This gene must function properly to convert clopidogrel to an active product ("metabolite") before it can do its job. Having even just one "defective" copy of that gene (being heterozygous for the *CYP2C19*2* allele) has been associated with an increased risk of blood clots forming in stents and elsewhere [73, 74]. However, because no large clinical trials have confirmed the exact level of this risk, current FDA recommendations advise that alternative therapy be used for individuals homozygous for that particular allele, but they also do not require genetic testing to assess this. Alternative methods for dealing with this issue include increasing the standard dose of clopidogrel in individuals heterozygous for that allele. This is an area in which more research needs to be done to determine the best approach to take in the presence of a particular genetic profile based on its actual clinical effects.

Likewise, other genetic variants are associated with significant adverse effects from particular medicines (e.g. the *HLA-B*5701* genotype with abacavir, a medicine used to treat HIV), which should thus be avoided in those patients. A widely prescribed class of drugs, statins, is used to reduce blood levels of LDL cholesterol and has other beneficial effects in patients with partially blocked arteries. One rare but potentially serious side effect of statins is breakdown of muscle, "rhabdomyolysis." Patients who are homozygous or heterozygous for a variant of the *SLCO1B1* gene were found in one study to have a 16.9- and 4.5-fold increase, respectively, in this side effect with high doses of statin therapy compared to those did not have that allele [73, 74]. Groups of multiple genes associated with specific conditions, such as the risk of developing cardiovascular disease or to assess rejection after a heart transplant, can also be tested, although their overall clinical value is still being assessed [58].

Even both the benign and the disease-causing varieties of bacteria that live inside us can have their genomes sequenced to see how they affect our health. For example, some types of bowel disease may be associated with a person's body mounting an immune response to a particular variety of bacteria. Modifying the types of bacteria that occupy our guts may help in dealing with such diseases. Genomic sequencing could also be used to more rapidly identify which harmful organisms are causing an infection than could be done by growing them in cultures, as well as assessing their resistance to specific antibiotics [58].

Overall, as the complexity and costs of current methods for sequencing DNA decrease, it may soon become a routine practice to map everyone's genome to better identify his or her relative risks of developing various diseases and to guide use of medications. A current limitation of such "personalized" medicine is that the interactions among large groups of genes and their many variants may make it more difficult to quantify risks associated with more complex conditions. Also, maintaining the privacy of such genetic records is an issue, particularly to prevent their use as determinants of insurance coverage, employment, etc [75]. In the United States the Genetic Information Nondiscrimination Act was passed in 2008 and specifically prohibited health insurers from using an individual's genetic information to determine eligibility or premiums for health insurance, or even from requesting or requiring genetic testing. However, this act did not cover any issues regarding genetic testing for life, disability, or long-term care insurance [1]. Further refinements of such regulations may be made in the future.

12.5 Repairing and Enhancing the Human Body

Gene therapy may be useful in regenerating bone to help heal fractures, particularly those "nonunion" ones in which bone fragments never fuse correctly and can cause ongoing pain. Compression fractures involving the bones (vertebrae) of the back are also potential candidates for this type of treatment. Genes used for this purpose can involve production of "recombinant human bone morphogenetic proteins," substances that can be injected directly but may be delivered more efficiently and safely via a vector-associated gene producing them [47]. Genetically modifying "mesenchymal stem cells" obtained from a person's own body to express a *BMP* gene and thus create these bone morphogenetic proteins has also been used to promote healing in bone [76].

However, these methods have so far only been studied in animal models, and their usefulness in humans has yet to be determined [32].

Diseases involving the retina are also potential targets for gene therapy. Over 200 genes can show mutations that produce inherited retinal diseases [22, 77, 78]. Development of the most common form of this problem, retinitis pigmentosa, has been associated with at least 30 genes itself. Gene vectors can be delivered into the eye at many potential sites, including eye drops as well as by injection into the lacrimal ("tear") glands, cornea, vitreous humor (the substance that fills most of the eye's globe), or retinal region [79]. A retina-blood barrier also exists that helps protect against an immune response generated by the rest of the body. AAV and lentiviruses have been used as vectors. However, based on animal studies their effectiveness may depend partly on age, with lentiviruses reported to be more effective in developing rather than adult mice. How such results might translate into therapy in humans remains uncertain.

Gene therapy may also eventually play a role in treating skin damage. Potential methods include using genes expressing substances that could help wounds and burns heal better and more rapidly, as well as reducing formation of scar tissue. Such techniques employing a variety of viral and nonviral vectors have so far been primarily used in animal models, but they could potentially cross over into more human trials. Similarly other animal studies have introduced genes to help restore cartilage, make damaged tendons grow back stronger, and both regenerate and restore function in nerves [32].

Diseases involving muscles, such as the various types of muscular dystrophies, may be particularly challenging to treat due to the need to target the large number and mass of skeletal muscles in the human body. To treat these problems, gene therapy directed at the mutated gene responsible for skeletal muscle degeneration might involve either modifying that defective gene so that it functions properly or replacing it [80].

Gene therapy might also be employed to treat atherosclerosis and conditions producing cholesterol profiles that increase the risk of developing it. The blockages in arteries caused by atherosclerosis are produced by a complex set of interactions that involve both inflammation of the inner lining of arteries and accumulation of lipids there, a process particularly linked with increased levels of LDL cholesterol. Preventive measures to reduce the chance of developing atherosclerosis could include making genetic modifications to change how the liver handles lipids and various substances involved with producing different components of cholesterol.

Once atherosclerotic plaques are present, however, methods of treating them might involve different types of genetic manipulations. Genes overexpressed in these plaques might be targeted. These include genes that regulate further uptake of lipids within the plaque that help it increase in size and others involved in the ongoing inflammatory process associated with it. Besides decreasing the rate at which it increases in size, "stabilizing" the plaque so that it is less likely to rupture or otherwise induce a blood clot forming on it (thus rapidly and markedly increasing the degree of blockage in an artery) might also be a goal for gene therapy.

Important issues for genetically preventing atherosclerosis and treating it when present include determining when such treatments should be started and how long they must be given. Since atherosclerosis can begin as early as the second decade of life, the "ideal" way to prevent it might include making a thorough assessment early in life of a person's genetic risk for developing it. Genes that involve handling of cholesterol, lipids, etc. as well as the inflammatory process involved could then be modified as needed to substantially reduce that risk.

Genetic techniques involving the human brain have to take into account that it is composed of various types of cells organized into a complex and, in its fine details, unique structure. Many cells in the human body are functionally interchangeable, such as fibroblasts or red blood cells. The neurons of the brain have considerably less redundancy. The intricate synaptic and other connections among them that encode our memories as well as the ability to reason and use "higher" functions cannot be reproduced by genetic "tinkering" alone. Those techniques might, however, reduce the chance of injury to the brain and also correct specific physiological abnormalities that produce major psychiatric disorders.

For example, major depressive disorder is a common and potentially life-threatening illness. While its causes are multifactorial, based on many genetic factors as well as environmental stresses and events, one aspect of it involves problems with how particular areas of the brain respond to serotonin, a neurotransmitter. Specifically, depression might be improved by enhancing the effects of serotonin in one brain region, the nucleus accumbens, by increasing the number of neuronal receptors there for it. Introducing a gene coding for the protein p11 in that region using an AAV vector was found to have encouraging effects in a study involving mice [81]. Such a method could potentially result in long-term benefits superior to current medical therapy in humans, but it would require further study to confirm this.

Other major psychiatric conditions such as schizophrenia and bipolar disorder are also thought to have multiple genetic components [82]. The presence of many genes in these disorders interacting with each other in myriad ways and also being influenced by epigenetic and environmental factors render potential options for gene therapy more difficult to evaluate and implement. However, if variants of one or a few genes could be identified as significant contributors to such psychiatric problems even without being completely responsible for them, targeting them for genetic modification might still produce a good clinical response. Indirect methods such as identifying how genetic variants influence both the tolerability and efficacy of medications used to treat these conditions could also be useful.

Epilepsy, a condition in which an area of the brain shows uncontrolled "firing" of neurons, might also be treated by introducing genes coding for proteins that help reduce the chances of that abnormal neuronal activity initiating and sustaining a seizure. Such an approach could require that the brain region responsible for seizures be identified first and then targeted with gene therapy [54].

Development of intellectual disabilities can be associated with many types of anatomic, physiological, biochemical, and other abnormalities involving the brain. Those disabilities can occur with conditions that destroy neurons and other supporting cells in the brain, or impair their normal development and creation of synaptic connections among them. For example, the most common cause of inherited intellectual disabilities is the Fragile X syndrome [83]. It is associated with a single-gene defect on the X chromosome in which there is failure to produce enough of a particular protein needed to properly regulate the growth and interactions of dendrites, which are extensions from the main body of a neuron that help it receive and transmit electrical impulses. Gene therapy targeting the defective gene might possibly correct such an isolated defect. However, other causes associated with multiple genes, such as those on the partial or complete "extra" copy of chromosome 21 in Down syndrome, could represent greater challenges to treat.

As noted previously, intelligence has a significant genetic component. However, it is related to many genes and their complex, poorly understood interactions. Intelligence is also related to many anatomic factors (e.g. numbers and functions of neurons in the brain, their synaptic connections, the "specialized" functions of different parts of the brain, etc.), metabolic details (including the actions of a wide variety of neurotransmitters that act differently in different parts of the brain), and physiological functions whose details are still incompletely understood.

Genetic modifications to the brain might be relatively general (e.g. producing selective increases in the numbers and synaptic connections of neurons in particular areas of the brain) or more targeted, such as modifying specific metabolic pathways to enhance cerebral function. However, our knowledge of the specific details of how our brains process information and "think," including the particular neuronal connections, interactions via synaptic networks and neurotransmitters, and other anatomical and physiological factors underlying "genius" level performance in performing calculations, creating music and art, abstract thought, etc., is still very incomplete. While we are continuously learning more about how different parts of the brain are involved with specific abilities we are still far from having a complete picture of the incredibly complicated interactions occurring at the molecular and cellular that produce them.

Moreover, even if we did have that information our knowledge of how to make genetic modifications to the brain is also not nearly enough to reliably make those enhancements. Worse, an "improvement" in one area of the brain might cause undesirable effects in other parts due to interfering with their activities. Likewise, genetic engineering used to modify cerebral function (and possibly anatomy) could have its own undesirable side effects, such as an increased risk of developing cancer, making neurons more vulnerable to injury, etc. In short, while development of "superintelligence" in humans produced by genetic modification may not be impossible and thus falls within the purview of science fiction, the challenges involved with actually doing it are, based on current knowledge, enormous.

Other potential physical modifications to the human body would present their own unique challenges. Such alterations would invariably have to deal at least initially with the intrinsic limits of the human body's cells and how they work together. Certain tissues and organs, such as skin and the liver, are capable of a significant degree of regeneration. On the other hand, as noted in Chap. 9, humans cannot regrow whole limbs or large body parts as, say, a starfish or planarian can because we lack special cells preprogrammed to do that. Moreover, our limbs and the rest of our bodies are considerably more complex in their components, size, physiology, and 3D organization than in those creatures. While human cells clearly can initially do this as part of normal development prior to birth, trying to get them to correctly grow just enough of the right kinds of tissues in just the right ways after loss of a limb would be extremely complicated.

Despite descriptions of such abilities in some science fiction, it would be even more difficult to make cells do things they originally never had the ability to do, such as grow feathers or wings. Making even the former would require wholesale changes to the variety of cell types that comprise normal skin as well as involve issues of how feathers would affect our ability to sweat and otherwise assist thermoregulation (i.e. keeping internal body temperature controlled). For wings to be functional, even on the Moon or other low-gravity world they would require major modifications in the human musculo-skeletal system. These might include new muscles large and powerful enough to make an effective connection between wings and the back, as well as the ability to extend and retract to allow "flapping" or at least gliding.[5] Compared to our bones, those of flying birds are thinner and denser, with less total mass than in a human adult. They also are more likely to be fused together, limiting mobility compared to their human counterparts. Also, the energy birds need for flight requires a higher rate of metabolism than humans, reflected in faster respiratory and heart rates as well as more energy-efficient physiological processes.[6]

In short, creating functional wings for humans would require radical changes in multiple tissues, organs, and biochemical/physiological systems with potential effects that may represent limitations in other contexts. For example, having an intrinsically higher metabolic rate could significantly increase requirements for food and water and potentially result in more rapid damage to some cells, including those associated with aging, via increased production of free radicals. Such ancillary effects might need to be addressed with still more genetic engineering to maintain overall health. The critical point here is that, while the concept of giving humans wings is simple and has obvious analogs in the animal kingdom, successfully implementing such changes in the basic form the human body takes and the underlying ways it works would be far from simple. This also raises both the technical and ethical questions of how many "failures" would occur before successful techniques were developed. In a way, this and similar marked modifications might come under the rubric of "You can't make a silk purse out of a sow's ear."[7]

Likewise, when looked at closely even a "practical" and, at first glance, "simple" modification such as Lois McMaster Bujold gives her Quaddies is far more technically difficult than it might first appear. As noted in Chap. 4, the idea that replacing our lower extremities with a pair of arms would be a functional improvement in microgravity is a biologically sound one. Our feet, legs, and thighs are optimized for weight bearing in 1 g. Likewise the fact toes are much less dexterous than fingers is fine for locomotion on the Earth's surface, even if that means we can no longer climb trees as easily as some other primates—a reasonable trade-off, particularly if one does not live in a forest

---

[5] As described in Chap. 4, the "standard" human body also has serious issues with maintaining even "normal" muscle mass in a gravity lower than Earth's, further compounding the challenges for genetically modifying muscles enough to allow flight.

[6] Of course, there would also be practical issues about what to do about those wonderful wings when we are merely walking, driving a car, taking a bath, sleeping, etc.

[7] On the other hand, Edmond Hamilton's 1938 story "He That Hath Wings" makes this idea seem plausible by mentioning many of the adaptations such as a higher metabolic rate, more powerful muscles in the upper back, etc. that would be needed. Having made these particular allusions to "real" biology, Hamilton can then ignore all the other factors strongly arguing against the idea of his main character actually developing these adaptations and growing wings (in this case due to a natural mutation rather than genetic engineering), and concentrate on its implications for him.

or jungle. The greatest amount of bone and muscle loss during prolonged exposure to microgravity is, as expected, in the lower extremities, pelvis, and lower back, which literally bear the brunt of our usual upright posture in 1 g. Conversely, our arms undergo much less musculoskeletal loss in microgravity because they are already adapted to maintain themselves in the presence of much lower levels of "normal" gravitational stress.

However, replacing legs with arms would require extensive and far-reaching anatomical changes "beneath the surface." The interface between the genetically engineered extra pair of arms and the pelvis represents a significant challenge. Both the normal shoulder and hip joints are of the ball and socket variety. In each type of joint the rounded end of a long bone—the humerus in the upper arm and the femur in the thigh—inserts into a concave (cuplike) surface, the glenoid fossa of the scapula at the shoulder and the acetabulum of the pelvic bone at the thigh. The bones at each of these joints interface via specific muscles, ligaments, and other tissues such as fluid-filled, sac-like bursae that all must attach in specific ways for the joint to be functional.

Thus, the head of the humerus is neither the right size nor are its typical attachments at the shoulder designed to either physically fit or work well with the acetabulum of the pelvic bone. A shoulder joint is considerably more mobile than a hip joint due to the humerus being much more loosely connected to the scapula and clavicle ("collarbone") than the femur is to the pelvis. Likewise the musculoskeletal and other anatomic architectures of the wrists and fingers are very different from those of the ankle and toes. Even the knee and elbow, while both "hinge" joints, differ in their details (e.g. the patella, or "knee cap," has no analogous bone in the elbow).

There also major differences in the sizes and courses of major blood vessels and nerves in upper versus lower extremities. Even more importantly, the fact that fingers are more sensitive and dexterous than toes is not only due to differences in their muscles, bones, blood vessels and peripheral nerves but also how much of the cerebral cortex is devoted to their sensory and motor functions. Thus, much larger amounts of the brain and many more neurons are devoted to the far more complex movements fingers perform than there are to make toes wiggle slightly. In short, modifying the human body to have a quartet of equally functional arms instead of only a pair would also require major functional and anatomical changes in the human brain too.

A technical question is whether this particular modification would be better done by genetically modifying the cells of a developing lower extremity to turn it into an upper one instead, or by simply inducing the body to grow an upper extremity where a lower one would normally be. Both represent formidable challenges because of the many differences between the two types of limbs, their associated joints, and the other factors just described.

Considering how different in shape and functional abilities ankles, feet, and toes are compared to our wrists, hands, and fingers (e.g. an opposable thumb versus a definitely non-opposable great toe), it might seem better to induce the body to grow an arm and adapt its interface to the pelvis rather than modify cells that would have formed a knee to become an elbow instead, change an ankle into a wrist, or transform toes into fingers. However, even that method would involve wholesale alterations in bones, muscles, supporting tissues, blood vessels, peripheral nerves, and the brain and spinal cord. Here too, while perhaps not theoretically impossible, successfully doing this would require solving many, many problems.

Of course, as has been mentioned in previous chapters, such "futuristic" genetic engineering is by no means the only area in which science fiction works may drastically underestimate or ignore the nuts-and-bolts problems involved with translating such an imaginative and elegant concept as the Quaddies into actual reality. As a crude analogy, such works might be thought as describing a sleek, shiny, beautiful car and telling about how well it runs, all its luxury features, etc. What those descriptions do not include are the many essential elements of what it took to make that car real, such as the miners who had to find the ore for its metals, the refining process needed to isolate them, the factories required to shape them into parts for the vehicle, all the other myriad industrial capabilities needed to make its tires, engine, brake system, electronics, as well as the many scientific discoveries that had to be made to develop those techniques, etc. In short, a large number of processes had to be discovered and many sophisticated procedures followed to produce the end result of a car whose fundamental features might be described in only a few sentences.

Similarly, making advanced genetic engineering techniques a "routine" reality will require not only the knowledge about human genetics we have already gained but far, far more—and there is no guarantee that we will not run into major roadblocks to what we can do based on absolute or simply practical (too time consuming, too expensive, too risky, etc.) limitations.

12.6 Dangers of Genetic Engineering

Some of the potential and real dangers of genetic engineering have been described in passing previously. These include the particular risks of current methods using various viruses, such as producing infections, triggering a life-threatening immune response, or promoting development of a malignancy. In individuals with otherwise severely debilitating and ultimately fatal illnesses, the great and potentially lifesaving benefits of gene therapy make accepting some level of risk reasonable, with every available means used to keep the overall risk as low as possible.

On the other hand, using genetic engineering to try to make an already healthy person "healthier" by using genetic manipulations to modify muscle mass/efficiency to enhance certain kinds of athletic ability, or even improve how that person's body deals with cholesterol to reduce risk of atherosclerotic disease, requires a closer look at the risks involved. Thus, a "good news-bad news" scenario in which a particular method were to reduce risk of CAD but increase that of dying from cancer would not be a good bargain. This would be an example of the medical maxim "It only takes one bad thing to kill you," with an excellent lipid profile at the time of death being no consolation to the corpse. Also, as with any new but more conventional medications and procedures, the overall spectrum and likelihood of adverse effects from a particular genetic manipulation may not be known until well after it is employed in large numbers of people.

One way to reduce the risk of genetically modified cells causing problems is for them to have a built-in "self-destruct" switch. For example, as part of their modification those cells could have a gene included that, under predetermined conditions occurring either within the body or introduced from outside it, could make them self-destruct via apoptosis if they produced adverse effects.

In one study, a particular type of immune response called "graft-versus-host disease" (GVHD) was treated with this strategy in patients who had previously undergone stem cell transplantation from a healthy donor for a relapse of acute leukemia [84]. GVHD can occur after a stem cell or bone marrow transplant when the transplanted cells detect the recipient's own cells as "foreign" and attack them—the reverse of the usual immune response. In this study T lymphocytes given for later treatment of these patients were first genetically modified to have the capacity to self-destruct if a patient were given a particular medication. When some patients developed GVHD from those cells, a dose of that medication rapidly killed the modified T lymphocytes and treated the GVHD.

Some of the much more extensive genetic modifications described in certain science fiction would also hopefully have "fail-safe" systems to reverse the manipulation should something go wrong. As described previously with the Quaddies, however, the more complex the change the more difficult it would be to reverse if there were a problem with it or even if the person who received it changed his or her mind. Even more problematic from both a scientific and ethical standpoint would be genetic manipulations performed prior to birth. As described previously, making modifications in the still undeveloped embryo or fetus might be easier and more effective than after birth, but they would also carry far greater potential risks of things going wrong. If an embryo/fetus has a genetic defect that will kill it a relatively short time after or even before birth and a genetic modification could save its life, the risk-benefit ratio in such a situation is acceptable for attempting treatment. In such cases, although the embryo/fetus obviously has no say in the matter as to what is done to it, the presumption that it would want it done is reasonable.

However, if the proposed genetic manipulation were one that only might improve an otherwise healthy embryo/fetus but carries either uncertain or known significant risks or, worse, not be in its but someone else's best interest, the ethical issues involved become much different. For example, Bujold's Quaddies are described as being created solely for use as specialized biological tools. The company that created them viewed them solely as property. There are many potential "justifications" (though ones that violate standard medical ethics) for this attitude. For example, because the Quaddies owe their very existence to the company and are supported by them, having them work for the company is fair recompense. The company might also take the position that Quaddies, while human to some degree, were designed to be of less intrinsic value than "free" or "normal" humans and are thus not entitled to the same rights. Or it might either choose to not recognize the Quaddies as fundamentally human at all or, even if it did, to simply disagree with one formulation of Immanuel Kant's philosophical "categorical imperative" that human beings should be treated as ends in themselves and not only as means to an end.

Alternatively, echoing Thrasymachus in Plato's *Republic* or the Athenians in Thucydides's "Melian Dialogue" from his *History of the Peloponnesian War*, the company might make a simple "might makes right" argument—the fact that it has the power to enforce its will on the Quaddies renders the need for any justification irrelevant. If Quaddies are only tools, then it is "reasonable" to simply dispose of them when they are not needed or become obsolete. This indeed is a key plot point in Bujold's novel *Falling Free*, when an effective antigravity system is developed that renders the Quaddies' adaptations to work in space within a microgravity environment no longer useful or profitable.

Even if it looked at them only as biological tools, the company could also consider whether it would be feasible and cost-effective enough to genetically re-engineer the Quaddies by changing their additional pair of arms into legs, thus making them more suitable for working outside of microgravity as an alternative to discarding/destroying them. Unfortunately for the Quaddies, genetically modifying them to develop altered limbs as part of their "normal" growth as an embryo and fetus would most likely be much "easier" than changing a huge number of multiple types of cells after they had fully differentiated and grown into arms, with important associated changes in the pelvis, brain, etc. As with many current electronic devices ranging from cell phones to televisions, the cost and complexity of repairing them may be far greater than simply disposing of them and getting a replacement. Not good news for the Quaddies, who in *Falling Free* must use other means to survive.

In summary, assessing the overall benefits and risks of genetic engineering must entail what that method can actually do at a given time (currently very limited, but expanding); the risks and benefits involved; and who makes the judgment of what constitutes a risk or benefit, which hopefully will include the individual involved or a surrogate who places the interests of that individual first.

12.7 The Bottom Line

Rather than the relatively "straightforward," focused, single gene manipulations that are currently possible, science fiction may postulate genetic engineering techniques that are many orders of magnitude more complex; associated with astoundingly high success rates or, alternatively if needed for the plot, incredibly dire and extreme consequences; or are anatomically, physiologically, metabolically, etc. extremely challenging regarding their feasibility if not (at least from a practical standpoint) impossible. On the other hand, describing the individual, social, ethical, and other issues involved if such manipulations *could* be done—the classic science fiction question of "What if…"—is, of course, the core raison d'être of such speculations.

For example, if genetic engineering were to become so successful that genetic defects could be identified/corrected and enhancements made in vitro at a very early stage in embryonic development prior to uterine implantation, would the considerably more random genetic results of conception via the "traditional" method become, by comparison, the less ethical alternative?[8] Going back even before conception, could individual sperm cells and eggs cells be pre-screened too and modified to make them more likely to produce a "better" embryo? One of science fiction's fundamental roles is to examine the consequences of new scientific capabilities, and certainly genetic engineering represents a fertile (so to speak) avenue for this.

A more near-term, real world issue might be "gene doping." If a person's "natural" athletic or, for that matter, eventually intellectual or artistic abilities could be enhanced in some way by genetic manipulations, would it be fair for them to compete with other individuals equipped with only the "original" DNA their parents gave them, or for such "gene-plus" individuals to compete

---

[8] Widespread use of this ability would also to some degree bypass, for good or ill, the process of "natural selection" described in Chap. 10. It would also raise the issue of whether human genetic manipulators and Nature might have different ideas of what constitutes a "beneficial" trait in different environmental and other contexts.

at all? This could mirror to some degree recent controversies regarding use of anabolic steroids and "blood doping." However, while the latter measures required that the individual involved seek them out in some way, such genetic manipulations could be performed prior to birth or during early childhood by other individuals to bulk up muscles, strengthen bones, increase height, improve the peripheral nervous system to speed up reflexes, etc. to create a future "superstar" in some sport.

It is a truism that scientific advances might produce as many if not more problems than they solve. Although it is too early to tell, genetic engineering may also on balance fall into this category. Fortunately these possibilities can be explored extensively in science fiction before those potential benefits and risks become reality, in the hope the former can be maximized and the latter minimized.

References

1. Hudson K. Genomics, health care, and society. N Engl J Med. 2011;365(11): 1033–41.
2. Lewin B. Part 1: DNA as information. Genes VI. Oxford: Oxford University Press; 1997. pp. 49–150.
3. Mendel G. Experiments on plant hybridization. 1865. http://www.mendelweb. org/Mendel.html. Accessed 6 May 2015.
4. van Cleve F. The fabulous fruits of Mendel's garden. Analog Sci Fict Fact. 2013;133(7/8):71–7.
5. Sen S, Kar D. Mendelian inheritance. In: Johri B, ed. Cytology and genetics. Oxford: Alpha Science International Ltd.; 2005. pp. 109–24.
6. Sen S, Kar D. Gene expression and interaction. In: Johri B, ed. Cytology and genetics. Oxford: Alpha Science International Ltd.; 2005. pp. 125–41.
7. Korf B. Genetics. In: Runge M, Patterson C, editors. Principles of molecular medicine. 2nd ed. New York: Humana Press; 2006. pp. 3–8.
8. Bolar N, Van Laer L, Loeys B. Marfan syndrome: from gene to therapy. Curr Opin Pediatr. 2012;24:498–504.
9. von Kodolitsch Y, Robinson PN. Marfan syndrome: an update of genetics, medical and surgical management. Heart. 2007;93(6):755–60.
10. McCandless S, Cassidy S. Nontraditional inheritance. In: M, Patterson C, editors. Principles of molecular medicine. 2nd ed. New York: Humana Press; 2006:9–18.
11. Zinn A. Sex chromosome disorders. In: Runge M, Patterson C, editors. Principles of molecular medicine. 2nd ed. New York: Humana Press; 2006. pp. 446–52.
12. Pertea M, Salzbert S. Between a chicken and a grape: estimating the number of human genes. Genome Biol. 2010;11:206–12.

13. Pollard T, Earnshaw W, Lippincott-Schwartz J. Introduction to cells. Cell Biology. 2nd ed. Philadelphia: Saunders Elsevier; 2008. pp. 3–16.
14. Pollard T, Earnshaw W, Lippincott-Schwartz J. DNA packaging in chromatin and chromosomes. Cell Biology. 2nd ed. Philadelphia: Saunders Elsevier; 2008. pp. 209–30.
15. White D, Rabago-Smith M. Genotype-phenotype associations and human eye color. J Human Genet. 2011;56:5–7.
16. Cavazzana-Calvo M. Gene therapy of Human Severe Combined Immunodeficiency (SCID)-X1 disease. Science. 2000;288(5466):669–72.
17. Ginn SL, Alexander IE, Edelstein ML, Abedi MR, Wixon J. Gene therapy clinical trials worldwide to 2012—an update. J Gene Med. 2013;15(2):65–77.
18. Hacein-Bey-Abina S, Le Deist F, Carlier F, Bouneaud C, Hue C, De Villartay J, et al. Sustained correction of X-linked severe combined immunodeficiency by ex vivo gene therapy. N Engl J Med. 2000;346(16):1185–93.
19. Mukherjee S, Thrasher AJ. Gene therapy for PIDs: progress, pitfalls and prospects. Gene. 2013;525(2):174–81.
20. Cideciyan AV, Jacobson SG, Beltran WA, Sumaroka A, Swider M, Iwabe S, et al. Human retinal gene therapy for Leber congenital amaurosis shows advancing retinal degeneration despite enduring visual improvement. Proc Natl Acad Sci U S A. 2013;110(6):E517–25.
21. Dong A, Rivella S, Breda L. Gene therapy for hemoglobinopathies: progress and challenges. Transl Res. 2013;161(4):293–306.
22. McClements ME, MacLaren RE. Gene therapy for retinal disease. Transl Res. 2013;161(4):241–54.
23. Wivel N, Walters L. Germ-line gene modification and disease prevention: some medical and ethical perspectives. Science. 1993;262:533–8.
24. Vannucci L, Lai M, Chiuppesi F, Ceccherini-Nelli L, Pistello M. Viral vectors: a look back and ahead on gene transfer technology. New Microbiologica 2013;36:1–22.
25. Zanjani ED. Prospects for in utero human gene therapy. Science. 1999;285 (5436):2084–8.
26. Guella I, Asselta R, Ardissino D, Merlini PA, Peyvandi F, Kathiresan S, et al. Effects of PCSK9 genetic variants on plasma LDL cholesterol levels and risk of premature myocardial infarction in the Italian population. J Lipid Res. 2010;51(11):3342–9.
27. Tebas P, Stein D, Tang WW, Frank I, Wang SQ, Lee G, et al. Gene editing of CCR5 in autologous CD4 T cells of persons infected with HIV. N Engl J Med. 2014;370(10):901–10.
28. Bennett J. Gene therapy for color blindness. N Engl J Med. 2009;361(25): 2483–4.
29. Suzuki K, Matsubara H. Recent advances in p53 research and cancer treatment. J Biomed Biotechnol. 2011;2011:978312.
30. Kalburgi S. Recent gene therapy advancements for neurological diseases. Discov Med. 2013;15(81):111–9.

31. Nienhuis AW. Development of gene therapy for blood disorders: an update. Blood. 2013;122(9):1556–64.

32. Giatsidis G, Dalla Venezia E, Bassetto F. The role of gene therapy in regenerative surgery: updated insights. Plast Reconstr Surg. 2013;131(6):1425–35.

33. Marshall E. Gene therapy on trial. Science. 2000;288(5468):951–7.

34. Raper SE, Chirmule N, Lee FS, Wivel NA, Bagg A, Gao G-p, et al. Fatal systemic inflammatory response syndrome in a ornithine transcarbamylase deficient patient following adenoviral gene transfer. Mol Genet Metab. 2003;80(1–2):148–58.

35. Maguire A, Simonelli F, Pierce E, Pugh E, Mingozzi F, Bennicelli J, et al. Safety and efficacy of gene transfer for Leber's congenital amaurosis. N Engl J Med. 2008;358(21):2240–8.

36. LeWitt PA, Rezai AR, Leehey MA, Ojemann SG, Flaherty AW, Eskandar EN, et al. AAV2-GAD gene therapy for advanced Parkinson's disease: a double-blind, sham-surgery controlled, randomised trial. The Lancet Neurology. 2011;10(4):309–19.

37. Hacein-Bey-Abina S, Hauer J, Lim A, Picard C, Wang G, Berry C, et al. Efficacy of gene therapy for X-linked severe combined immunodeficiency. N Engl J Med. 2010;363:355–64.

38. Scholler J, Brady TL, Binder-Scholl G, Hwang WT, Plesa G, Hege KM, et al. Decade-long safety and function of retroviral-modified chimeric antigen receptor T cells. Sci Transl Med. 2012;4(132):132ra53.

39. Kerkar SP. Model T cells: a time-tested vehicle for gene therapy. Front Immunol. 2013;4:304.

40. Naldini L. Medicine. A comeback for gene therapy. Science. 2009;326(5954):805–6.

41. Leboulch P. Gene therapy: primed for take-off. Nature. 2013;500:280–1.

42. Biffi A, Montini E, Lorioli L, Cesani M, Fumagalli F, Plati T, et al. Lentiviral hematopoietic stem cell gene therapy benefits metachromatic leukodystrophy. Science. 2013;341:1233158-1-11.

43. Trobridge GD, Horn PA, Beard BC, Kiem HP. Large animal models for foamy virus vector gene therapy. Viruses. 2012;4(12):3572–88.

44. Chen Z, Lin Y, Yang F, Jiang L, Ge S. Gene therapy for cardiovascular disease mediated by ultrasound and microbubbles. Cardiovasc Ultrasound. 2013;11(11). doi:10.1186/1476-7120-11-11.

45. Kazuki Y, Hoshiya H, Takiguchi M, Abe S, Iida Y, Osaki M, et al. Refined human artificial chromosome vectors for gene therapy and animal transgenesis. Gene Ther. 2011;18(4):384–93.

46. Nowakowski A, Andrzejewska A, Janowski M, Walczak P, Lukomska B. Genetic engineering of stem cells for enhanced therapy. Acta Neurobiol Exp. 2013;73:1–18.

47. Raftery R, O'Brien FJ, Cryan SA. Chitosan for gene delivery and orthopedic tissue engineering applications. Molecules. 2013;18(5):5611–47.

48. Jin L, Zeng X, Liu M, Deng Y, He N. Current progress in gene delivery technology based on chemical methods and nano-carriers. Theranostics. 2014;4(3):240–55.
49. Paul M, van Dolleweerd C, Drake PMW, Reljic R, Thangaraj H, Barbi T, et al. Molecular pharming: future targets and aspirations. Human Vaccines. 2011;7(3):375–82.
50. Melo E, Canavessi A, Franco M, Rumpf R. Animal transgenesis: state of the art and applications. J Appl Genet. 2007;48:47–61.
51. Gene Therapy Clinical Trials Worldwide. http://www.wiley.com/legacy/wiley-chi/genmed/clinical/. Accessed 6 May 2015.
52. Kaiser J. Gene therapists celebrate a decade of progress. Science. 2011;334:29–30.
53. Verma IM. Medicine. Gene therapy that works. Science. 2013;341(6148):853–5.
54. Walker MC, Schorge S, Kullmann DM, Wykes RC, Heeroma JH, Mantoan L. Gene therapy in status epilepticus. Epilepsia. 2013;54(Suppl 6):43–5.
55. Clinical Trials. 2014. http://www.asgct.org/general-public/educational-resources/clinical-trials-information. Accessed 6 May 2015.
56. Glybera. http://www.uniqure.com/pipeline/glybera/. Accessed 6 May 2015.
57. What price affordable access? Nat Biotechnol. 2013;31(6):467.
58. McCarthy J, McLeod H, Ginsburg G. Genomic medicine: a decade of successes, challenges, and opportunities. Sci Transl Med. 2013;5(189 sr4):1–17.
59. Plon S. Cancer genetics and molecular oncology. In: Runge M, Patterson C, editors. Principles of molecular medicine. 2nd ed. New York: Humana Press; 2006. pp. 27–33.
60. Huang Y, Davidson N. Breast Cancer. In: Runge M, Patterson C, editors. Principles of molecular medicine. 2nd ed. New York: Humana Press; 2006:728. p. 35.
61. BRCA1 and BRCA2. Cancer risk and genetic testing fact sheet. http://www.cancer.gov/cancertopics/factsheet/Risk/BRCA. Accessed 6 May 2015.
62. Davies G, Tenesa A, Payton A, Yang J, Harris SE, Liewald D, et al. Genome-wide association studies establish that human intelligence is highly heritable and polygenic. Mol Psychiatry. 2011;16(10):996–1005.
63. Research CoSMa. Sleep disorders and sleep deprivation: an unmet public health problem. Washington: National Academies Press; 2006.
64. Crocker A, Sehgal A. Genetic analysis of sleep. Genes Dev. 2010;24(12):1220–35.
65. Schenkein J, Montagna P. Self management of fatal familial insomnia. Part 1: What is FFI? MedGenMed. 2006;8:65.
66. Faraut B, Boudjeltia KZ, Vanhamme L, Kerkhofs M. Immune, inflammatory and cardiovascular consequences of sleep restriction and recovery. Sleep Med Rev. 2012;16(2):137–49.
67. Parish JM. Genetic and immunologic aspects of sleep and sleep disorders. Chest. 2013;143(5):1489–99.

68. Herculano-Houzel S. Neuroscience. Sleep it out. Science. 2013;342(6156): 316–7.
69. Underwood E. Sleep: the brain's housekeeper? Science. 2013;342:301.
70. Xie L, Kang H, Xu Q, Chen MJ, Liao Y, Thiyagarajan M, et al. Sleep drives metabolite clearance from the adult brain. Science. 2013;342(6156):373–7.
71. Edkins R, Cheek D. Pharmacogenetics. In: Runge M, Patterson C, editors. Principles of molecular medicine. 2nd ed. New York: Humana Press; 2006. pp. 34–8.
72. Table of Pharmacogenomic Biomarkers in Drug Labeling. http://www.fda.gov/Drugs/ScienceResearch/ResearchAreas/Pharmacogenetics/ucm083378.htm. Accessed 6 May 2015.
73. Ong FS, Deignan JL, Kuo JZ, Bernstein KE, Rotter JI, Grody WW, et al. Clinical utility of pharmacogenetic biomarkers in cardiovascular therapeutics: a challenge for clinical implementation. Pharmacogenomics. 2012;13(4):465–75.
74. Wells QS, Delaney JT, Roden DM. Genetic determinants of response to cardiovascular drugs. Curr Opin Cardiol. 2012;27(3):253–61.
75. Joly Y, Ngueng Feze I, Simard J. Genetic discrimination and life insurance: a systematic review of the evidence. BMC Med. 2013;11:25.
76. Kimelman Bleich N, Kallai I, Lieberman JR, Schwarz EM, Pelled G, Gazit D. Gene therapy approaches to regenerating bone. Adv Drug Deliv Rev. 2012;64(12):1320–30.
77. Zaneveld J, Wang F, Wang X, Chen R. Dawn of ocular gene therapy: implications for molecular diagnosis in retinal disease. Sci China Life Sci. 2013;56(2):125–33.
78. Simonato M, Bennett J, Boulis NM, Castro MG, Fink DJ, Goins WF, et al. Progress in gene therapy for neurological disorders. Nat Rev Neurol. 2013;9(5):277–91.
79. Willett K, Bennett J. Immunology of AAV-mediated gene transfer in the eye. Front Immunol. 2013;4:261.
80. Benedetti S, Hoshiya H, Tedesco FS. Repair or replace? Exploiting novel gene and cell therapy strategies for muscular dystrophies. FEBS J. 2013;280(17):4263–80.
81. Chen G, Twyman R, Manji H. p11 and gene therapy for severe psychiatric disorders: a practical goal? Sci Transl Med. 2010;2(54):54 ps51
82. Ozomaro U, Wahlestedt C, Nemeroff C. Personalized medicine in psychiatry: problems and promises. BMC Med. 2013;11:132.
83. Picker JD, Walsh CA. New innovations: therapeutic opportunities for intellectual disabilities. Ann Neurol. 2013;74(3):382–90.
84. Di Stasi A, Tey S, Dotti G, Fujita Y, Kennedy-Nasser A, Martinez C, et al. Inducible apoptosis as a safety switch for adoptive cell therapy. N Engl J Med. 2011;365(18):1673–83.

13

Stem Cells and Organ Transplantation: Resetting Our Biological Clocks

Backward, turn backward, O Time in your flight…

Elizabeth Akers Allen
"Rock Me to Sleep" (1860)

The human body has only a limited ability to repair itself. Illness, injury, and aging can overwhelm its built-in capability to replace dysfunctional, damaged, or destroyed tissues. We can at best only partly regenerate our organs and cannot grow back a whole limb.

This chapter focuses on two major approaches to renewing our bodies. In science fiction, futuristic medicine developed by humans or extraterrestrials might be able to regrow or replace body parts far beyond current abilities. One avenue of research that shows great promise for translating those imagined capabilities into actual practice involves the use of "stem cells." They are a particular class of cells found in the human body before and after birth.

There are many types of stem cells, each with its own particular capabilities. All share the common trait of being able to transform into one or more other types of cells, including the specialized, "differentiated" ones present in various tissues and organs (e.g. neurons in the brain). Stem cells could potentially be directed to change into different varieties needed to replace a person's own malfunctioning or destroyed cells.

The extensive role stem cells play in the human body's development and health, and particularly their potential therapeutic uses, have been recognized only relatively recently. Thus, while science fiction works have long employed the concept of regrowing body parts, their use of stem cells as the means for doing that is also more recent, paralleling growth of real-life knowledge about them. Stem cell research is a very active field, with major discoveries made over the past few decades and others likely to come in the near future. It is an area that shows great promise for eventually changing some science fiction medicine into everyday clinical practice.

A second method that is already used to replace some body parts is tissue and organ transplantation. This is a much more established technique that basically involves "swapping out" a damaged body part for a healthy one. Typically this involves using tissue or an organ obtained from a deceased donor. However, in some cases (e.g. bone marrow and kidney transplants) the donor may be alive. Occasionally a person's own tissue (such as bone marrow) may be obtained for later transplantation into his or her own body.

In science fiction organ transplantation be used for benign or sinister purposes. For example, standard medical practice requires that the latter meet strict criteria for brain death before donating an organ essential to life, such as a heart. It also requires that a person previously gave consent to become an organ donor or that permission to donate organs be given by an appropriate individual such as a spouse or family member.

However, both in medical thrillers and science fiction the ethical principles used for such "harvesting" may be considerably more lax. Robin Cook's 1977 novel *Coma* involves patients being deliberately rendered brain dead and used as spare parts for transplants. Larry Niven's "Known Space" works such as "The Organleggers" (1969, later republished as "Death by Ecstasy") depict wholesale "organlegging" and other methods such as extensive use of the death penalty for increasingly less violent crimes such as income tax evasion to keep a larger supply of organs available.

Both stem cells and transplantation techniques offer potential ways to improve health and increase life expectancy. As we will see, these two methods may be used together, such as by transplanting tissue (and, perhaps, someday whole organs) created using stem cells into a person. However, each one has its own particular uses and challenges.

13.1 Types of Stem Cells

Our bodies contain many "mature" cells that normally do not demonstrate the ability to change into or produce other types of cells. Once red blood cells, neurons, myocardiocytes (heart muscle cells, also known as cardiomyocytes), and many other varieties develop they remain as those same types of cells for the rest of their existence. Cells that have reached the end-stage of their development and are no longer able to divide are called "terminally differentiated."[1] For example, a red blood cell ultimately loses its nucleus and thus its ability to reproduce. Mature neurons and myocardiocytes, while retaining their nuclei, are also thought to not divide in adults. Myocardiocytes do, however, increase significantly in size from birth to adulthood, thus increasing the overall dimensions of the heart.

Stem cells are at the opposite end of the spectrum, with far greater abilities to reproduce and transform into other types of cells [2–4]. The human body uses them to develop, grow, and repair itself from the very beginning of life until its end. There are many varieties of stem cells, each with its own origin, location within the body, and range of capabilities. However, all of them share certain characteristics. Stem cells can divide and proliferate throughout most if not all of a person's entire lifetime. They can differentiate into some or all of the kinds of cells that create the various tissues in our bodies.

Stem cells are also self-renewing. They can replenish their numbers via either "symmetric" or "asymmetric" cell division [5, 6]. In symmetric cell division the stem cell can divide into either two "daughter" cells identical to the original, or into two cells that then differentiate into another kind of cell. In the former case the total number of stem cells increases, while in the latter the body gains new cells for a particular tissue. Asymmetric division produces one daughter stem cell and a second cell that then differentiates into another variety. This maintains the same total number of stem cells while also increasing that of a different type.

Embryonic stem cells (ESCs) are the class of stem cells with the greatest inherent ability to divide and form the widest range of tissues [7]. As the name implies they are present in a human (or non-human animal's) embryo, the stage of development in our species extending from fertilization (the union of a sperm cell and ovum to produce an initially single-celled "zygote") to about 8 weeks post-fertilization. In humans, the zygote and the cells created by its first two divisions are "totipotent," or "omnipotent" [7–9]. This means they have the ability to differentiate into any of the different classes of cells that the human body can have, as well as the tissue that will form the embryo's attachment to the mother's uterus after implantation. The "potent" part of these and subsequent terms refer to a cell's ability to change into another type.

An embryo between about 5–9 days after fertilization is called a "blastocyst." It consists of up to several hundred cells that have differentiated into two basic classes of cells. One class includes the cells that will ultimately help form the "extra-embryonic" tissues of the placenta. These "trophoblasts" form the outer layer of the blastocyst. The ESCs are located below that layer. The latter are "pluripotent" [10]. They can differentiate into any human cell except the ones that help make the placenta.

More technically, ESCs can produce cells arising from any of the three "germ" layers in our bodies—the endoderm, mesoderm, and ectoderm. The endoderm gives rise to the various kinds of cells that line nearly all of the gastrointestinal tract; the trachea, bronchi, and alveoli of the lungs; the urinary bladder; and parts of the urethra and thyroid gland. The mesoderm produces connective tissues such as cartilage and bone, all types of muscle cells, red and white blood cells, adipose (fat) cells, and germ cells (sperm and oocytes). The ectoderm differentiates into the cells of the brain and other parts of the nervous system, as well as into skin, hair, sweat glands, and the lining of the mouth.

ESCs can be obtained by extracting them from a blastocyst [11] A method for obtaining these cells from human embryos was first described in 1998 [12]. The standard technique for doing this results in the destruction of the embryo, raising ethical questions regarding "harvesting" such cells from human embryos. ESCs obtained in this way and maintained in cultures have the ability to divide indefinitely and develop into any type of human cell. This gives them the greatest potential flexibility for generating tissues or perhaps even entire organs that could be used to replace a person's damaged ones. However, as will be discussed shortly, ESCs have other characteristics that limit such use.

Fetal stem cells are present in the organs of a fetus, the stage of development between 8 weeks following fertilization to birth. These stems cells are "multipotent." They are limited to developing into closely related classes of cells within a single germ line. For example, those found in the liver of a fetus can develop into more than one kind of cell, but only those types found in the liver and not elsewhere in the body.

Following birth, our bodies retain reserves of "adult stem cells" in most and perhaps all tissues [6, 8, 13, 14] At least some of these too are thought to be multipotent. For example, neural stem cells can differentiate into neurons as well as various supporting cells such as oligodendrocytes and astrocytes. They are present in specific areas of the brain and also found within the spinal cord [15].

Mesenchymal stem cells are derived from the mesoderm germ line. They are located in multiple tissues in our bodies throughout life, including bone marrow, bone, adipose tissue, and blood [6, 16, 18] They are thought to assist in replacing mature, differentiated cells there and elsewhere by turning into the types needed within their specific range of transformation. Mesenchymal stem cells can differentiate into osteocytes (bone cells), chondrocytes (cells producing cartilage), fibroblasts, adipocytes (fat cells), myocardiocytes, and hepatic (liver) cells [16, 17] Recent reports also suggest that, under certain conditions, mesenchymal stem cells can even be induced to "transdifferentiate" into cells of a different germ line such as neural tissue, which derives from the ectoderm germ line [18]

Mesenchymal stem cells are also present in umbilical cord blood and especially in the cord tissue itself. By saving and preserving cord blood and tissue, these and other types of stem cells can be available for further study should the person develop a disease or need them for possible future tissue regeneration [13].

Other types of adult stem cells are "oligopotent." They produce two or more types of cells within a single, specific tissue. For example, hematopoietic stem cells were first identified over 50 years ago [5]. They are capable of differentiating into all types of myeloid (e.g. red blood cells and most kinds of white blood cells) and lymphoid (e.g. lymphocytes) cell lines. Hematopoietic stem cells are concentrated in blood obtained from the umbilical cord at birth and also reside in bone marrow throughout life [19]. Like mesenchymal stem cells, if they are preserved in stored umbilical blood they can potentially be used later in a person's life for study and therapeutic use should that individual develop a particular disease (e.g. one of certain kinds of leukemia) [20].

Hematopoietic stem cells constantly produce new blood cells to replace those continuously lost through "old age" or destruction. Red blood cells have an average lifespan of about 120 days, while most types of white blood cells have lifetimes ranging from a few days to weeks. Without hematopoietic stem cells to replenish their numbers our bodies would quickly "run out" of these essential blood cells.

Finally, some adult stem cells are "unipotent." They produce only a single type of mature cell and typically reside only in the tissue containing that variety of cell. Thus, muscle stem cells are present in skeletal muscles and can turn into mature muscle cells (myocytes) if needed. Some unipotent and oligopotent cells are called "progenitor cells." They may be quiescent—neither dividing nor differentiating—unless called upon to do that by injury to the tissue where they reside. These progenitor cells can represent an intermediate stage between stem cells and fully differentiated ones. Unlike stem cells they cannot usually self-renew, and they typically differentiate into more mature cells soon after they are produced [2].

The role of adult stem cells varies depending on their types and locations. As mentioned previously, hematopoietic stem cells are continuously replicating and differentiating due to the high turnover rate of blood cells. Stem cells in the lungs, liver, hair follicles, and gastrointestinal tract also do this to replace cells within them having short life cycles [5]

On the other hand, stem cells in organs and tissues such as the heart, skeletal muscles, pancreas, and nervous system are predominantly dormant. They form new, mature cells of only a specific type, and at a very slow pace that may increase due to an acute need such as damage to those mature cells. In some cases their rate of turnover is so slow that, in the case of the heart and nervous tissue, it was thought until recently that those organs had no ability to create new mature cells following birth. This is in contrast to early (e.g. embryonic) stages when stem cells divide and proliferate rapidly to ultimately produce all the differentiated cells that develop during the active growth process [21].

In short, small reserves of different types of adult stem cells are present in at least most tissues throughout life. They can help replace damaged, destroyed, or diseased cells with widely varying degrees of efficiency [22]. However, with aging the number of these stem cells and how well they function can decline [21, 23]. The latter may be associated with damage to their DNA, environmental changes involving the cells, alterations in their physiology, and reduction in their ability to suppress tumor formation [15, 21].

For example, as new "daughter" stem cells are produced they may not always be exact copies of their "parent" due to mutations. Some may also eventually lose the ability to proliferate due to shortening of telomeres (described in Chap. 9) and other causes [21]. After a certain point they can become senescent—still alive but with reduced function and no longer dividing. If they become sufficiently "defective" over time they may undergo "apoptosis," a process in which a cell deliberately self-destructs, thus decreasing the total number of stem cells if they are not replaced. And as with other cells, one or more mutations could also potentially make them turn cancerous (more about this soon.) This may be more likely to occur after the stem cell has undergone many divisions, perhaps due to each division having a certain risk of malignant transformation and the "odds" eventually catching up with the cell, or by simply accumulating enough deleterious mutations over time to ultimately turn them into a cancer.

Another way stem cells might deal with defects produced by their aging or when they are subjected to stress is by "autophagy." This means that the cells can "digest" their own defective organelles and other components, recycling them and synthesizing new ones. However, with further aging stem cells may lose the ability to prolong their "healthy" lives via autophagy, leading to them being destroyed or becoming senescent without being replaced by normal offspring or, once again, accumulating enough defects to make them turn malignant [21, 22].

A reduced pool of effective stem cells could partly explain the body's diminished ability to replace damaged mature cells in various tissues with aging. Even at best the number of stem cells in various parts of the body is small compared to other types. Also, while their life expectancy is typically much longer than other types of cells, it is still finite. Thus, if stem cells can in fact divide only a certain number of times over a person's life before they become senescent, die, or become dysfunctional (including turning cancerous), their ability to repair or regenerate tissue will also be limited [5].

Adult stem cells that are normally dormant occupy their own "microenvironment" within tissues. Chemical and other changes within that microenvironment act as stimuli for them to "spring into action" and respond to tissue injury by initiating various repair mechanisms. An important area of current research is to determine what factors, either intrinsic to the cells themselves or associated with their microenvironment, make some types of adult stem cells (e.g. hematopoietic stem cells) so efficient at replacing mature cells and regenerating the corresponding tissues while others (e.g. neural and cardiac stem cells) show considerably more limited capabilities [5].

For example, when I went to medical school in the mid-1970s it was thought that adults had a finite number of myocardiocytes that, if lost by damage or disease, were not replaced. However, it was first reported in 2003 that cardiac stem cells are indeed present in the heart, although in limited numbers [24, 25]. It is now thought that they actually replace about 1 % of myocardiocytes per year, and that about 40 % of those latter cells in an adult's heart were created after birth [26]. Put another way, it has been estimated that about half of the heart's mass has been renewed by age 50. However, the rate at which this "turnover" occurs is thought to decrease with age [13].

Why cardiac stem cells can perform this "routine maintenance" but do not seem to "pick up the pace" and perform a significant amount of repair work when the heart suffers major destruction of myocardiocytes (e.g. at the time of a myocardial infarction) is unknown. Finding some way to stimulate them to act more rapidly and extensively to repair large areas of a damaged heart would be highly desirable, but whether this is possible and, if so, how to do it have yet to be determined.

Some adult stem cells can be harvested relatively easily from a person's body. For example, hematopoietic stem cells can be obtained from a person's bone marrow, where they make up about 2–4 % of the cells there [27]. After administering a local anesthetic to the skin and surrounding tissue, a needle is inserted into a bone containing accessible marrow, usually the sternum (breastbone) or a part of the pelvic bone. A small amount of fluid marrow is then drawn up into a syringe. A more solid sample of marrow cells can be obtained from the pelvic bone as a biopsy using a hollow needle that goes deeper into the bone. In either case the blood cells obtained will be at various stages of development, and a small percentage will be stem cells.

Hematopoietic stem cells obtained from bone marrow and other sources are, in fact, currently used to treat a number of leukemias, lymphomas (a kind of cancer derived from lymphocytes, a type of white blood cell), other blood disorders, and some other conditions. One technique is to first partially or completely destroy a person's own "unhealthy" blood cell-producing system (including the immune system) via chemotherapy and radiation at doses that do not significantly injure other body tissues. The harvested normal bone marrow or stem cells can then be infused intravenously, where they migrate to the usual sites of bone marrow formation and establish themselves there, proliferating and restoring the person's hematopoietic and immune system.

The bone marrow and stem cells used for transplantation can be obtained from either the person being treated or from another individual who is a close match regarding shared antigens, particularly those of the "human leukocyte antigen" (HLA) system. When the transplant source is "autologous" (derived from the patient) the cells have a significantly lower risk of producing an immune response than when they are "allogenic" (derived from a donor). As described in Chap. 12, the latter can be associated with a "graft-versus-host" reaction in which the transplanted cells recognize the new host's cells as foreign and produce an immune response against them. Autologous transplants also tend to restore blood cell production and immune system function more quickly than allogenic ones. However, autologous transplants may be associated with a greater risk for recurrence of the patient's original disease [19].

Hematopoietic stem cells can also be derived from other sources. They are present in circulating blood, where they can be extracted via filtering techniques. As mentioned previously they can be obtained from umbilical cord blood, and even from amniotic fluid obtained at birth. In the latter cases, the blood and/or amniotic fluid must be collected at birth and cryogenically preserved (see Chap. 7) for long-term storage.

While hematopoietic stem cells can only differentiate into blood cells, mesenchymal stem cells are potentially more versatile due to their ability to generate a wider variety of useful tissues [28]. The particular types of cells they can differentiate into depend on the cellular environment into which they are introduced. Mesenchymal stem cells also have a natural ability to migrate to an area of injury within the body, attracted by chemicals such as cytokines released by damaged cells. After reaching their destination they can promote healing by helping to suppress inflammation and differentiating into cells that replace damaged or destroyed ones [16, 29].

Like hematopoietic stem cells, mesenchymal stem cells can be obtained from bone marrow.[1] Unfortunately they are present there in only very small numbers, comprising somewhere between 0.001–0.1 % of bone marrow cells [17, 27]. Once they are obtained and their numbers possibly augmented by culturing them, they can be used to treat a patient by giving them back via an intravenous or arterial infusion, or by a direct injection into the target tissue. However, the best ways to administer them for a particular application are still being assessed [29].

A point in favor of mesenchymal stem cells is that they have a relatively low tendency to provoke an immune response when transplanted from one person to another. This raises the possibility of storing them in "cell banks" where

---

[1] For completeness sake it should be noted that these cells are also sometimes referred to as "mesenchymal stromal cells."

their numbers can be increased via culture techniques for clinical use when needed [24]. Nonetheless, under certain circumstances they can increase their likelihood of being recognized as "foreign," so the actual feasibility of this approach is still uncertain [29].

So far more than 200 clinical trials involving mesenchymal stem cells have been conducted in the United States. However, the total number of patients studied is only a little over 2000, and these trials varied widely regarding the procedures used and the diseases being treated. How well mesenchymal stem cells can be used to help repair or replace tissues is a dynamic but still uncertain area of research, with overall results that are said to be currently both inconsistent and inconclusive [17].

Due to the type of tissues they can form, including bone, cartilage, and fat, mesenchymal stem cells may have particular applicability to medical fields such as orthopedics and plastic surgery [30]. For example, those derived from adipose tissue might be transplanted to stimulate creation of fat in areas of the face and limbs where it has been destroyed or is deficient, thus providing cosmetic and functional benefit. They might also be used for bone regeneration, producing new cartilage for damaged joints, assisting wound healing, rejuvenating skin, and perhaps even helping injured peripheral nerves to regenerate [30].

A practical issue for using adult stem cells or, for that matter, any other type is how many of them need to be introduced into a target tissue. Use too few and the treatment might be ineffective, while too high a dose might increase the odds of the cells having undesirable effects such as interfering with the function of normal cells or provoking an immune response.

A commonly used dose for mesenchymal stem cells is about $1–2 \times 10^6$ cells for each kilogram of body weight. However, as described previously the number of mesenchymal stem cells in bone marrow is very low, and obtaining enough to give that dose typically requires obtaining a small sample and culturing it. While these cells can replicate quickly, increasing their number a thousandfold within several weeks, there is the risk of them being damaged during that time or exhausting their extensive but finite ability to divide. Protocols for optimally processing them for therapeutic uses are still being studied [17].

Some areas, such as the brain, may not only require a large number of stem cells or differentiated cells derived from them for purposes of repair or rejuvenation but would also face other challenges. For example, the microenvironment in parts of the brain requiring treatment may not be hospitable for those types of cells or for initiating production of new neurons or supporting cells. Likewise, the newly created cells would also have to integrate into the neurons and other cells already there in a functionally meaningful way, e.g. forming synaptic connections. And if the conditions that produced the original injury or disease were still present, those "young," healthy cells may also be vulnerable to them too [15].

It is also important to determine at what point in a particular disease process introducing stem cells would do the most good. For example, studies involving infusions of mesenchymal stem cells after a myocardial infarction showed better improvement in heart function when the cells were administered 5 or more days after the injury rather than earlier [24, 31]. This might have been due to injured areas of the heart having a more hostile environment early on due to release of harmful substances, insufficient oxygen supply, etc. associated with the infarction.

Stem cells administered to a damaged tissue might help repair it by creating new, healthy cells that are then incorporated into the tissue. However, in some cases the predominant way they can help repair it is by their mere presence altering the microenvironment and stimulating the tissue to repair itself by various means, so-called "paracrine" effects [18, 25, 32]. This can include secreting substances that help promote healing [16].

For example, although stem cells derived from bone marrow have been introduced into damaged hearts either through infusion via a coronary artery or direct injection, the reported resulting improvements in heart function have ranged from none to at best mild [24, 27]. Even when improvement was found it might be only transient. In these particular instances any benefits may indeed have been due to temporary improvements caused by the presence of these stem cells rather than generation of a significant number of new, healthy myocardiocytes.

An alternative method first used in 2001 involved transplanting many hundreds of millions of myoblasts—undifferentiated cells derived from a person's skeletal muscle—into that same individual's heart [24] While a detectable improvement in heart function was sometimes reported as occurring, these cells had the decidedly undesirable effect in some patients of producing serious ventricular arrhythmias (abnormal heart rhythms). This was thought to be due to the skeletal muscle cells not integrating with the "native" myocardiocytes from an electrical standpoint and thus producing areas where those arrhythmias could start. This issue further demonstrates the all too common problem of a new treatment having unforeseen harmful effects.

Initial studies in small numbers of patients using actual cardiac stem cells and similar heart-derived cells have been somewhat more promising [26] However, this area of research will require much more study before definitive results can be obtained.

Other types of stem cells, such as those derived from adipose tissue, also represent potential venues for research [32] In particular neural stem cells and other types of pluripotent cells are being studied as possible means to treat a wide variety of common neurodegenerative diseases such as Alzheimer's disease and Parkinson's disease, as well as other neurological problems (e.g. spinal cord injuries) [33–35]. Specifically, it is hoped that neural stem cells could generate new neurons, myelin-producing cells, or the cells of the eye's retina based on a person's needs. Fetal and adult neural stem cells could also potentially be used for these purposes. An important issue regarding harvesting adult neural stem cells from a person is that those cells are located in discrete parts of the brain, which are obviously considerably less easy to safely access than other types of adult stem cells (e.g. the hematopoietic and mesenchymal varieties obtained from bone marrow) [36].

Another possible source for neural stem cells or more differentiated (e.g. progenitor) neurological cells is to create them from ESCs or mesenchymal stem cells. Fortunately, like their mesenchymal counterparts, neural stem cells have the ability to migrate to an area of injury or other site where they are needed. Thus, if an adequate number of them were introduced into the body, they would potentially "know" where to go to do the most good. However, also like mesenchymal stem cells, any neural stem or other cells obtained will likely need to be cultured to increase their numbers to a level high enough to potentially produce a beneficial clinical effect. Here again this poses the risks of cell death, the stem cells differentiating prematurely or into the wrong kind of cell, etc. Developing better culture methods to reduce such issues is a critical preliminary step in transplanting these cells for therapeutic uses.

13.2 Reprogramming Cells

As described in Chap. 10, experiments in the 1950s provided the first steps in demonstrating that somatic cells could be "reprogrammed" to an earlier stage of development [37, 38]. For example, a study published in 1958 reported the creation of cloned tadpoles after nuclei derived from the intestinal cells of a frog were introduced into enucleated egg cells [38] This was an important milestone in demonstrating that "mature" cells could be reprogrammed back to an earlier state of development. The creation of Dolly the sheep and subsequent mammalian clones was a further development of that somatic cell nuclear transfer method [7, 9, 39].

Induced pluripotent stem cells (iPSCs) are differentiated cells from an adult animal (e.g. a mouse or human) that, using one of a variety of methods, change into ones with properties similar to pluripotent ESCs [3, 7, 9, 40, 42]. This was first done in 2006 using fibroblasts derived from mice [43]. The fibroblasts had four genes encoding for "transcription factors" with the unpoetic names *Oct3/4, Sox2, Klf4*, and *c-Myc* introduced into them using a genetic engineering technique involving a retrovirus.

These four genes reprogrammed the fibroblasts so that they could then be induced to form cells from any of the three germ lines, just as ESCs can. A follow-up study from those same investigators published in 2007 used that technique on human fibroblasts and produced similar findings [44].[2] Subsequent work has shown that other transcription factors, with equally nonintuitive designations such as *Nanog*, can also be involved in maintaining pluripotency in ESCs and other cells [45, 46]. Other somatic cells besides fibroblasts, such as keratinocytes, neural cells, and lymphocytes have also been used to generate iPSCs [47].

The exact details of what changes occur at the molecular level when somatic cells are reprogrammed into iPSCs are not completely established. Basic research studies are, however, continuing to identify what happens during this process [48]. A greater understanding of the underlying mechanisms involved could lead to more efficient production of iPSCs as well as identifying what types of somatic cells might be most amenable to this process.

Recent studies indicate that iPSCs and ESCs are not entirely identical from either a genetic or epigenetic standpoint, but what those differences mean functionally is still being investigated [49–53]. One issue with producing iPSCs is that reprogramming is not an all-or-nothing process but involves a series of changes. The techniques used to create an iPSC may leave the original somatic cell (e.g. a fibroblast) only partially reprogrammed and never reaching the stage where it would be able to differentiate into the desired new cell type [7]. Even somatic cells that do become pluripotent may retain some of their original characteristics as well as have changes induced by the reprogramming process itself, with uncertain effects on the functional status and other attributes of differentiated cells produced by them [7].

One method used to demonstrate that iPSCs are truly pluripotent is to use "tetraploid complementation" in animal models (e.g. mice) [54, 55]. In this technique two embryos, each at an extremely early stage of development when it consists of only two cells, are fused together. The resulting single cell is "tetraploid"—that is, it has two pairs of each type of "somatic" chromosome (those that are not X or Y chromosomes) instead of the single pair in a normal "diploid" cell. When either ESCs or iPSCs are combined with the tetraploid cell at a stage no later than a blastocyst, the tetraploid components form the

---

[2] The 2012 Nobel Prize in Physiology or Medicine was awarded jointly to the lead investigator for these studies, Dr. Shinya Yamanaka, and to Sir John B. Gurdon, who pioneered the previously mentioned somatic cell nuclear transfer technique in frogs.

extra-embryonic tissues (e.g. the placenta) while the diploid ESC or iPSC provides the genetic material to create the animal.

Such "combined" cells have been implanted into mice and produced viable offspring. The latter is an approximate clone of the animal from which the ESC or iPSC was initially derived. However, production of the iPSCs produces enough genetic and epigenetic changes in the "donor" cells that the offspring will not be identical to the "parent." Nonetheless, this method represents a possible alternative to somatic cell nuclear transfer techniques for producing "clones."

It should be noted, however, that the efficiency of using this technique to create offspring surviving to birth has been reported to be low (0.3–13 %) [55] and that many of those created did not survive long after birth. Thus, any application of such techniques to humans would raise at least the same ethical issues as somatic cell nuclear transfer [56].

iPSCs are now being used for research purposes to create a "disease-in-a-dish" [13, 16, 25, 39, 52, 57–59]. For example, fibroblasts can be obtained from a person by removing a small sample of skin and culturing those cells. They can then be changed into iPSCs that in turn are induced to form cells of a tissue affected by a genetic or other disorder. Those cells can then be studied to determine what functional or physiological issues are associated with the disease. For example, myocardiocytes derived from iPSCs originating in patients with diseases that can produce life-threatening arrhythmias have been studied to see what changes in normal electrochemical processes or other abnormalities they have. A more direct way to make such assessments is to compare iPSCs produced from a person with a genetic disorder to "normal" control ones [57].

iPSCs have also been derived from cancer cells, including those from various types of leukemias and gastrointestinal cancers, to better understand their individual characteristics [40]. Interestingly, when cancer cells are made pluripotent they may lose at least some of their ability to produce tumors, and differentiated cells derived from them may also lack this and other hallmarks of cancer [39]. It is hoped that the knowledge obtained from these methods can be used to devise better treatment strategies, such as developing new drugs designed to correct or at least ameliorate the underlying defect(s) present in abnormal tissue or to destroy cancer cells.

ESCs can also be used to differentiate into particular types of cells used for disease modeling and drug testing [16]. However, iPSCs have the advantage of being "personalized" for a particular patient—derived from that individual's own cells—rather than providing the more general information ESCs give [40]. Thus, iPSCs might help identify very specific issues with genetic composition and cell function in an individual and thus help better direct potential therapies.

iPSCs could potentially be combined with gene therapy to treat some medical problems. One way of doing this could be to take cells such as skin fibroblasts and convert them outside the body ("in vitro") to iPSCs. The original genetic defect would then be corrected by gene therapy and the iPSCs then induced while still in culture to differentiate into the required type of cell—e.g. a blood or muscle cell, or one that is part of the nervous system [60]. This might be done by producing appropriate chemical and other conditions in the culture where the iPSCs are growing [27]. The "processed" cells could then be injected back in the person's body where they hopefully would then function properly. Alternatively the "repaired" iPSCs could be injected into the person at a location conducive to them differentiating into the "right" type of cells. Perhaps ideally some of these iPSCs would not initially differentiate but remain as an ongoing "pool" to potentially replace any mature tissue cells that are lost in the future [1].

Another potential approach might be to reprogram a fibroblast only partially so that, instead of becoming plurioptent and thus an iPSC, it travels only far enough back along its developmental path so that it can be changed directly into a myocardiocyte or other type of differentiated cell [27, 50]. This can, in fact, be accomplished by a process mentioned previously, "trandifferentiation," in which a fibroblast could be turned into a myoblast, a pancreatic cell into a hepatocyte, etc [50]. Because fibroblasts are present in the scar tissue that forms in the heart after a myocardial infarction, one possible application of this technique would be to reprogram the fibroblasts within the scar itself to turn into myocardiocytes. Whether this can actually be successfully accomplished remains to be determined, however.

One potential major advantage of using iPSCs as a source of replacement tissue is that, because they originate in a person's own body, they would not be expected to produce an immune response as would any type of material deemed by the body to be "foreign," such as ESCs. However, the new gene introduced into modified iPSCs or differentiated tissue derived from them might itself be considered foreign, and the possibility that it could therefore elicit an immune response must still be taken into account.

Moreover, as with any type of gene therapy a number of these cells large enough to be effective would need to be created; the method used to introduce a new gene must be sufficiently efficient to produce an adequate number of corrected cells; and the vector used to introduce that gene should not cause changes to the DNA of the cell that might cause delayed adverse effects, e.g. cancer. Once these cells are introduced into the body they must also do their intended "jobs" well enough to clinically benefit the patient. This might be an easier task for blood cells, which can be present in multiple areas of the body and do not need to organize into or interact in specific ways with "solid" tissues, such as nerves or the brain's synaptic networks.

Another factor in using adult stem cells is that at least some types of them may function differently in males versus females. In one report estrogen-related effects were found to significantly affect how hematopoietic stem cells divided and renewed in female mice compared to those same types of stem cells in males [61, 62]. How much or if such gender-related differences might occur in humans is uncertain.

Despite these potential limitations, iPSCs might be particularly suitable for treating single-point genetic disorders. This technique has been used with some success in a mouse model involving replacing the gene associated with sickle cell anemia with a "normal" one [63]. For this type of therapy to be successful the genetically corrected cells derived from iPSCs must survive and preferably, if it is not associated with other risks such as development of cancer, increase their numbers [42]. A limiting factor for this or other uses of iPSCs is that it typically takes weeks to develop them from a person's somatic cells—a delay that could become an issue depending on how quickly they are needed (e.g. to replace damaged tissues).

Recent reports suggest that techniques used to create iPSCs within a mouse rather than in vitro might actually produce cells that are totipotent rather than "merely" pluripotent [64, 65]. As described previously this means that they would not only be able to generate all the cell types that ESCs can but also the placental tissue that a blastocyst can. Whether these apparently totipotent cells could then be harvested and used to generate a complete animal as well as whether these findings could be applied to humans are questions that are still unanswered.

These results also raise the possibility of inducing a person's cells to become iPSCs and subsequently differentiating without first removing them from the body. The obvious issue, however, is to make sure only the appropriate cells do this and only in the way desired, without forming the wrong, dysfunctional, or otherwise unwanted tissue instead of a healthy replacement. As in many other instances, the road extending from a proof-of-principle to safe, effective, routine therapy can be very long and challenging.

Another intriguing observation is that some mature, differentiated cells in the body can display some capability for self-renewal on their own under certain circumstances. Macrophages, a type of white blood cell, develop from hematopoietic stem cells. However, recent studies suggest they may also proliferate by dividing on their own if needed [66]. Hepatocytes, a kind of cell specific for the liver, also have the ability to do this, consistent with the liver's much greater ability to regenerate itself compared to many other solid organs such as the heart. These findings raise the possibility that some other types of mature, differentiated human cells that appear to have very limited (if any) ability to replicate and replace damaged ones may have more "talent" in this area than we currently give them credit for.

Finally, both iPSCs and ESCs might also prove useful in other areas of research devoted to learning how our bodies develop and function. For example, due to their pluripotency they can be used to better understand how human tissues and organs form during embryonic development [59]. This can include uncovering the details of how particular genes act individually or interact with others to create specific cells and modulate their functions. Learning the genetic, epigenetic, and environmental factors that make an individual iPSC or ESC differentiate into one particular kind of different cell rather than another could be of great importance in creating new tissues for therapeutic use.

13.3 Risks of Stem Cells

Both ESCs and iPSCs have two major intrinsic risks that currently hold back their use in human clinical trials [30].Each one's ability to form tissues derived from the three germ lines actually represents a two-edged sword. If the tissues created can grow and function normally, they could indeed help replace similar, damaged tissues in a person. However, ESCs and iPSCs also have the potential to develop in undesired ways. One is that they can form a type of tumor called a "teratoma." In fact, the ability of a cell such as an ESC or iPSC to form a teratoma is a criterion used to verify that it is, in fact, pluripotent [64, 67].

Though uncommon, teratomas can occur spontaneously in a person, usually but not exclusively in an ovary or testis. They are thought to be typically present at birth, although their size at that time may range from very tiny to large [19]. Teratomas are characterized as containing tissue derived from at least two different germ layers. These tumors are typically "encapsulated" with a well-marked boundary of tissue, but they may contain strange mixtures of various ones, including hair and neurological tissue (ectodermal germ line) along with bone (mesodermal germ line). Teratomas are typically (although not always) "benign" in the medical sense of not spreading ("metastasizing") to other parts of the body, but only growing locally in one part of it. Nonetheless, if a "benign" teratoma grows large enough, ruptures, or otherwise interferes with normal surrounding tissues (e.g. a teratoma located in the brain) it can cause significant secondary health problems. Less commonly teratomas are "malignant" and can potentially spread to other parts of the body, similar to the behavior of many types of cancer.

Needless to say, developing a teratoma at a location where ESCs or iPSCs were used to "successfully" replace damaged tissue is highly undesirable. In some animal studies, transplanted tissue derived from ESCs did in fact produce teratomas, as well as other types of cancerous growths in the recipient [19]. This safety concern is one reason why there have been few human trials using ESCs, which have included one involving ESC-derived cells to treat a form of blindness and a since-discontinued one (based on financial rather than medical reasons) involving use of a certain kind of nervous system cell to treat spinal cord injuries [67, 68].

Another potential risk of iPSCs and ESCs (particularly the latter) is that, even if they do not form actual teratomas, they may be difficult to control when it comes to creating a particular type of desired cell or tissue rather than undesired ones. In this respect the pluripotency of iPSCs and ESCs might constitute a relative disadvantage. For them to be more useful for transplantation purposes the genetic and molecular mechanisms by which they maintain pluripotency might need to be identified and altered to restrict and direct their ability to differentiate.

In the case of iPSCs, the risk of developing tumors such as teratomas varies depending on the type of somatic cell used to generate them. This might be due to epigenetic factors and some genes in those cells being resistant to reprogramming [69]. It is possible, however, that if the genetic or other factors that lead to development of teratomas can be adequately identified, ESCs and iPSCs could be further modified to reduce this risk. However, more research is needed to address this issue.

Currently there have been no human clinical trials using iPSCs, although such studies have been proposed and may begin in the near future. Once again a primary consideration delaying such research is safety, based on the risk of iPSCs causing not only teratomas, but various types of cancer due to some of the methods used to create them. The original technique of using a retrovirus to introduce the four transcription factors needed to make an adult cell become pluripotent is associated with the retrovirus integrating itself into the host cell's DNA. As mentioned in Chap. 12, the virus could do this at a critical location that either interferes with an important gene or with sections of the DNA that suppress tumor formation [40]. The iPSCs themselves could also have significant chromosomal abnormalities [70]. The net effect is that a differentiated cell produced from iPSCs could either not function properly or become cancerous.

Newer techniques to create potentially "safer" iPSCs have been used with varying degrees of success. A major advantage of using retroviruses, particularly a lentivirus (see Chap. 12), to produce iPSCs is that they are considerably more efficient than some alternative methods for doing that, although their overall efficiency is still low. While some other types of retroviruses can introduce the "reprogramming" genes only into cells that are dividing, lentiviruses can also do this in cells that are not dividing [9]. However, lentiviruses too can integrate into a host cell's DNA, thus also having the potential risk of turning it cancerous.

Other methods for producing iPSCs include using adenoviruses and a small RNA virus called the Sendai virus as vectors, as well as chemical modification, plasmids (small DNA molecules), and modified RNA [40, 42, 47, 69, 71, 72]. Combinations of particular molecules and fewer than the original four transcription factors (e.g. using only two, *Oct4* and *Klf4*) have also been used to create iPSCs [50]. One such study reported that iPSCs could be produced from mouse fibroblasts by treating the latter with only a "cocktail" of seven different molecules selected to induce pluripotency [41, 71]. This method made up to 0.2% of the fibroblasts convert to iPSCs, an efficiency that was comparable to techniques using transcription factors. How useful iPSCs produced by such non-integrating methods would be to treat human diseases has yet to be determined, however.

In some cases these alternative techniques might involve activating genes already present in a cell to make it become pluripotent rather than introducing new genes into it. The hope is that at least some of these other methods will be safer regarding the risk of developing cancer. Methods that do not result in integration of new genetic material into the cell's DNA are considered safer than ones that do. However, it is still possible that they too might produce enough genetic and epigenetic changes to make the resulting iPSCs more susceptible to creating tumors rather than healthy differentiated cells. Moreover, most non-integrating methods are significantly less efficient at producing iPSCs than the original ones using retroviruses, although newer techniques may improve this [53]. In fact, recent reports indicate that a non-viral technique that modifies a single genetic factor could produce iPSCs with nearly 100% efficiency [73, 74].

For safety purposes, any method used to generate iPSCs would need to be determined to have a risk of teratoma formation that is as low as possible. Also, the differentiated cells produced from an individual patient's iPSCs created by such a method would need to be tested or otherwise ascertained to be safe before transplantation.

As mentioned previously, a particular limitation of ESCs and tissues derived from them is that, if used for transplantation, they represent cells foreign to a person's body. They are thus subject to being attacked and destroyed by the recipient's immune system.

This problem might be partially ameliorated by methods discussed in detail later in this chapter that are similar to those used for organ and other types of transplantation. These include matching the tissue to be transplanted as closely as possible with those in a person's body based on major "antigens" (substances that the body recognizes as either foreign or part of it) and use of medications that suppress immune function. However, the latter can also potentially increase the odds of developing serious infections and cancer by itself. This problem might also be addressed by processing differentiated cells prior to transplantation to reduce incompatibilities with the host, although this would be less feasible for using actual organs derived from ESCs [68, 75].

Moreover, as mentioned previously the process of creating the iPSCs might modify them enough so that they are no longer identical to the host but might have enough changes to activate the immune system [68, 69, 75]. In some cases even adult stem cells obtained from a person's own body might, when used for transplantation, potentially evoke an immune response due to changes in them caused by the culturing process used to increase their numbers. Fortunately, however, this does not seem to be a significant risk in the case of mesenchymal stem cells [19].

Tumors and cancers can also have their own pools of stem cells [16]. In fact, it is now thought that tumors and other forms of cancer can originate from adult stem cells and partially differentiated (progenitor) cells derived from them that have, due to genetic mutations or other factors, become cancerous [2, 14, 19]. While stem cells are already self-renewing—a characteristic that cancer cells (unfortunately) also possess—progenitor cells must typically acquire this ability as part of a process that turns them cancerous. In particular, cancer stem cells have been identified as part of various types of leukemia, isolated from tumors involving the central nervous system, and found in cancers involving areas such as the breast, lung, liver, pancreas, colon, and prostate [2, 14, 76–78]. Some of the same transcription factors mentioned previously such as *Nanog* that keep cells pluripotent also appear to contribute to the ability of cancer stem cells to resist chemotherapy as well as increase their ability to spread and invade normal tissue [14].

Another potential source of cancer stem cells is from "mature" cancer cells that have undergone "reprogramming" and reverted to a stem cell-like state [14, 19, 78, 79]. This mechanism could also help renew the number of cancer stem cells beyond that produced by their own replication and increase a tumor's ability to produce additional mature cancer cells.

Just as the "healthy" type of stem cell present in many tissues can differentiate and create replacement cells, cancer stem cells can replenish the very unhealthy ones in a tumor. As with normal stem cells, cancer stem cells may constitute only a very small fraction of total cancer cells. For example, in leukemia and multiple myeloma (cancers involving various types of blood cells), only 1–4 % and 0.001–1 % of malignant (cancerous) cells respectively were found to be cancer stem cells [4].

However, even a small number of cancer stem cells may act as a quiescent pool of cells that can produce new, mature cancer cells after the latter are killed during treatment. Most cells in a tumor are not involved at any given time in making it grow, and destroying them will cause the tumor to regress. However, the tumor can grow again due to division of cancer stem cells, which are typically resistant to chemotherapy, radiotherapy, and other treatments that destroy mature cancer cells [2, 80]. Such measures are generally most effective against dividing cells, and if the cancer stem cells are not active at a given time they would be expected to be less susceptible to those kinds of therapy. However, it is important to determine what other factors might contribute to this resistance to help improve treatment. These might include cancer stem cells possessing cellular systems designed to strongly resist apoptosis, efficiently repair their DNA, and/or remove toxic substances (including chemotherapeutic agents) from them [78].

Another concern is how sensitive normal stem cells are to injury from the chemotherapeutic agents and radiation doses used to treat cancer stem cells. If the former are more likely to be injured than the latter, those treatments could give the cancer stem cells a relative advantage for further growth. Also, while normal stem cells can protect themselves to a degree from cancerous mutations by becoming senescent or undergoing apoptosis when damaged by chemotherapy and/or radiation, cancer stem cells might do neither—again giving them a relative survival advantage. The challenge is to minimize damage to the "innocent bystander" healthy stem cells (as well as other normal cells) while targeting the cancerous variety [2, 14].

There is also a risk with obtaining stem cells from a person and increasing their numbers by culturing them outside the body before injecting them back into that individual for treatment. Stem cells themselves can become malignant after a sufficient number of divisions, and thus there is a chance that as they proliferate during culturing some might turn cancerous. It is possible that the "unnatural" environment of the culture itself could enhance this risk [19].

Even a person's own mesenchymal stem cells may be "fooled" into actually promoting the growth of breast and other types of cancers [16].They might also protect a tumor by reducing the body's immune response to it. This raises the possibility that, if someone with an undiagnosed tumor is treated with a dose of mesenchymal stem cells, they could paradoxically make the person worse rather than better.

However, in other contexts ESCs and either neural or mesenchymal stem cells might be used to treat certain types of cancers that are resistant to other methods. For example, they have been used as "carriers" to deliver genes, therapeutic chemicals, and other substances to gliomas, a type of brain tumor with an overall poor prognosis that can occur in either children or adults [81, 82]. The gene(s) introduced into the tumor by this method may trigger them to self-destruct via apoptosis or kill them in other ways, as would any cytotoxic chemicals carried by the stem cells.

Another issue with iPSCs or any other technique using adult somatic cells such as fibroblasts is that these cells may have relative "deficiencies" that could be passed on to any differentiated cells produced by them. As described in Chap. 9, telomere shortening in fibroblasts and other somatic cells that typically undergo many replications normally limits how many times they can divide before becoming senescent. Fibroblasts derived from an adult that are used to generate iPSCs already have some degree of telomere shortening. The amount of shortening may vary, however, since some fibroblasts in the body may have undergone fewer replications than others.

Such considerations raise the questions of whether the healthy longevity of at least some types of differentiated tissue derived from iPSCs or adult stem cells will be shorter than normal or at least relative to tissue obtained from genetically "younger" ESCs, and whether the risk of malignant transformation might also be increased. A study involving ESCs obtained from mice bred to have innately short telomeres suggested that such ESCs had a reduced ability to become stably differentiated [83]. Again, whether or how much this observation might apply to adult somatic cells with reduced telomere length in humans is uncertain but suggests the need for further investigation.

13.4 Organ Transplantation

The concept of replacing a damaged or diseased organ with a healthy one is a simple one. However, it took nearly the entire twentieth century and into the current one to develop reasonably successful techniques for harvesting various types of donor organs, performing surgery, and reducing the risks of postoperative complications [84, 85]. The first successful corneal transplant using tissue from a human donor was performed in 1905 [85]. Initial (though unsuccessful) attempts at kidney transplantation occurred a year later in 1906, using organs obtained from either a goat or pig [84, 85]. Despite other attempts over the ensuring decades, the first successful human kidney transplant was not done until 1956, with the donated kidney coming from the recipient's living, monozygotic ("identical") twin [84–86]. Current sources of kidneys for transplantation include both deceased and living donors.

Initial attempts at liver, lung, pancreas, and heart transplantations in the 1960s showed that the procedures could be successfully done from a surgical standpoint [87–90]. However, long-term survival for patients after these operations was initially poor, primarily due to rejection of the transplanted organs and infections associated with the "immunosuppressive" medications available at that time to reduce the risk of rejection. If the donated organ had sufficient differences in certain antigens from the recipient, particularly in key types such as the HLA ones described previously and the ABO blood groups,[3] the recipient's body would recognize it as containing "foreign" molecules such as peptides,[4] assume it was harmful (similar to the way it identifies "non-self" material in bacteria and other infectious agents), and activate its immune system to destroy the organ.

Unfortunately, on its own the human body is not "smart" enough to understand that, although the donated organ is not an original part of that individual's body, it is beneficial rather than dangerous like an infectious microorganism would be. The immune response that the body then generates, involving a variety of white blood cells and chemical pathways, is "non-selective"—a "shotgun" approach that does not distinguish between "innocent" targets like the transplanted organ and "guilty" ones such as harmful bacteria.

One way to work around this problem is via "tissue typing"—matching antigens as closely as possible between donor and recipient, particularly ones where differences would be most likely to trigger an immune response. These would include the HLA antigens, also known as the "major histocompatibility complex" (MHC) and, to a lesser extent, "minor histocompatibility antigens" (miH) [91]. This usually works best when the donors and recipient are close family members, who typically share more of the same antigens with each other than with non-related individuals. The degree of antigen "matching" is related to how close the family relationship is, with that in monozygotic

---

[3] A particular individual's red blood cells may have both the A and B antigens (type AB); only the A antigen, with antibodies against the B antigen in the bloodstream (type A); only the B antigen, with antibodies against the A antigen in the bloodstream (type B); or neither the A nor the B antigen, and antibodies against both the A and B antigens in the bloodstream (type O). An immune response that destroys red blood cells will occur if a person receives a blood type containing an antigen he or she does not have. Thus, an individual with type AB blood can receive any other type of blood; type A can only receive type A or O; type B can only receive type B or O; and type O can only receive type O. There should also be compatibility regarding another red blood cell antigen, the "Rh factor." Thus, a person with "O+" blood is type O and has the Rh factor, while one who is "O–" is type O but lacks the Rh factor and can have antibodies against it. Thus, an O + individual can safely receive blood from an O + or O– donor, while someone with O– blood can only receive it from another O– individual.

[4] Peptides are linear chains of amino acids that, if long enough, are called polypeptides. Proteins in turn consist of one or more chains of polypeptides that are biologically functional.

twins obviously being the closest. Depending on which antigens were inherited from each parent, siblings may or may not be close matches, a particular parent may share some but not all antigens with a child, and overall similarities decrease as relatives become more distant.

Another major method is to suppress the recipient's immune response so that it does not inappropriately "attack" the transplanted organ. Unfortunately, like the body's own immune response, all currently available immunosuppressive medications lack adequate selectivity in their effects. While they can successfully reduce the likelihood of rejecting a transplanted organ, they also decrease the ability of the recipient's immune system to appropriately recognize and deal with infections and malignant cells.

Infections are a common cause of complications following transplantation. This risk is generally highest soon after transplantation but remains significant thereafter, with the particular types of infections likely to occur varying depending on long it is after the transplant [92]. The overall risk of developing a malignancy is reported to be about 20 % at 10 years following transplantation and nearly 30 % after 20 years [93]. Skin and lip cancers are the most common type, representing about 36 % of post-transplantation malignancies. Most such cancers, including squamous and basal cell carcinomas, are amenable to treatment. However, though less common than those types, the risk of developing a considerably more serious skin cancer, malignant melanoma, is also increased in transplant patients.

Likewise the incidence of malignancies associated with infections involving particular viruses, such as Kaposi's sarcoma and ones associated with "post-transplant lymphoproliferative disorder" (PTLD), is also increased. In PTLD the body produces too much of a certain kind of white blood cell (B lymphocytes) [94]. It is typically associated with new or prior infection with the Epstein-Barr virus, which causes infectious mononucleosis. PTLD can also result in fever, shortness of breath, and development of cancers such as lymphoma, particularly non-Hodgkins lymphoma.

Besides their intrinsic risk of suppressing both the beneficial and harmful effects of the recipient's immune system, immunosuppressive medications can have other significant side effects. For example, the good effects of one such medicine, cyclosporine, comes at the particular cost of toxic effects on the kidneys [84, 95]. Another one, sirolimus, can also harm the kidneys and have other adverse effects, such as impaired wound healing.

A post-transplantation complication that can lead to failure of the donated organ and death is "transplant arteriopathy [96]". This involves the recipient's immune system mounting a response against the blood vessels of the transplanted organ. Transplant arteriopathy can affect both large and small arteries. It can cause enough loss of blood supply to damage or destroy parts of a transplanted organ, leading to deterioration of its overall function and perhaps ultimate failure.

When this vascular issue occurs in transplanted hearts it is called "cardiac allograft vasculopathy" (CAV) and represents a potentially fatal complication [96, 97]. CAV is associated with narrowing of the coronary arteries, reducing blood supply to the heart. However, it differs from "conventional" coronary artery disease (CAD) associated with atherosclerosis in several important ways. CAD is typically associated with localized blockages in one or more coronary arteries, and these lesions may be calcified. Conversely CAV can involve diffuse narrowing of very long lengths of all three coronary arteries, making those narrowed and typically non-calcified regions much less amenable to standard treatments such as coronary angioplasty and bypass surgery.

Moreover, the transplanted heart is at least partially "denervated"—that is, it does not have the usual connections with the body's nervous system. Thus, while a person with significant CAD might feel chest pain (angina) when the heart does not get enough blood, the patient with a heart transplant and CAV may not feel angina due to lack of nerve fibers carrying sensations of pain. Long-term surveillance of post-cardiac transplantation patients can thus involve use of coronary angiography, which (as noted in prior chapters) is invasive and carries a small amount of risk itself. Moreover, as noted above, even if CAV is identified treatment options may be limited. Overall, the incidence of CAV has been reported to be 52 % at 10 years following cardiac transplantation. It can be associated with myocardial infarction and arrhythmias that can cause "sudden death," with the individual unexpectedly developing a lethal arrhythmia.

Thus, given these and other risks, there can be a fine line between balancing the overall beneficial and harmful affects of immunosuppressive therapy and transplantation itself. Nonetheless, the long-term success of many types of organ transplants is directly related to development of immunosuppressant agents that can do this reasonably well. Widespread use in particular of cyclosporine beginning in the early 1980s significantly decreased the rate of failure/rejection of transplanted organs (e.g. kidneys and hearts) and increased the overall long-term survival of organ recipients. Along with older medications such as azathioprine and corticosteroids, the availability of newer agents such as tacolimus, mycophenolate mofetil, sirolimus (rapamycin), and many others to reduce the risk of acute rejection at the time of transplantation and for long-term use has led to increasing success in keeping transplant patients alive and healthy [84, 91, 95].

Transplantation procedures done mainly to improve quality rather than quantity of life raise other issues. For example, the first successful human face transplantation was performed in 2005. This technique involves using various types of facial tissues from a deceased donor that might include skin, muscles, tongue, nose, lips, eyelids, scalp, and bones such as the mandible ("jaw bone"). Reports published early in 2014 describe up to 28 individuals receiving facial transplants worldwide, with an overall mortality rate of 11.5 % in the first 26 patients [98, 99].

Individuals who receive face transplants must use immunosuppressive therapy for the rest of their lives. This increases their risk of developing the complications described previously, including various types of cancer and infections. Other risks of this procedure include those associated with the surgery itself and lifelong ones for acute skin rejection and graft-versus-host disease.

Overall, in select patients with very severe facial injuries, face transplantation potentially has cosmetic and often functional benefits for improved facial reconstruction compared to other methods. However, unlike a heart transplant that can, despite its risks, improve overall survival and quality of life, a face transplant is not only associated with its own increased morbidity but could decrease how long the recipient lives. Procedures such as a hand transplant, which can also have obvious functional benefits but is not essential for life, raise similar considerations. Such issues demonstrate the need to carefully weigh the risks and benefits of such a procedure, knowing that neither choice—to have the procedure or to decline it—is ideal and whichever one is made involves tradeoffs.

A major ongoing problem with organ transplantation is the scarcity of donors compared to prospective recipients. The "pool" of available organs has increased somewhat due to measures such as improved techniques for harvesting and preserving them for transplantation, as well as broadening the selection criteria to include donors who are older or had somewhat more "problematic" organs regarding the latter's function and other characteristics. However, the rate of increase in the number of potential recipients continues to outpace that of donors.

A number of factors limit the number of donors. One is the relatively small number of individuals who both volunteer to be donors (or, if unable to give consent, have it given by family members) and qualify as donors of at least some organs based on age and other criteria. The latter include that the organ must be reasonably healthy; the donor should not have known infections, malignancies, or other issues that could impact the health of the transplanted organ and ultimately the recipient; there must be a suitable "match" based on antigen profiles, sometimes size, and other factors between donor and recipient; and many organs (particularly the heart) have a very narrow window of time between when it must be harvested and transplanted (see Chap. 7).

Another factor is the obvious one that, with certain exceptions such as kidney transplants, donors must be deceased prior to donating their organs, with the heart being the most obvious example. This involves the critical issue of deciding when a donor actually *is* dead, which must be determined by strict medical and ethical criteria. It could also potentially involve heinous acts such as stealing organs from someone who is still alive or deliberately killing a person to obtain them.

As mentioned earlier, science fiction-related examples of using unwilling donors include Robin Cook's *Coma* (1977) and some of Larry Niven's "Known Space" stories. In the latter works the mismatch between supply and demand for organs is "solved" by either punishing individuals found guilty (with reduced regard for whether they are actually innocent instead) of what would now be considered trivial "crimes" with the death penalty, or by simply stealing them from victims ("organlegging"). However, the key practical problem with a society in which everyone (or at least the majority of the population) is at significant risk of being an involuntary donor/victim or recipient is that the net benefit/risk may be comparable for everyone, reducing if not eliminating the value of implementing of such an organ procurement system.[5] Also, simply following current standard medical procedures for organ tracking (e.g. it is not standard operating procedure to purchase organs via online auctions or other poorly documented sources) and ethical guidelines should curb (but unfortunately, given enough individuals lacking such ethical principles, certainly not entirely eliminate) widespread use of the organlegging technique.[6]

Besides informed consent for the procedure, organ transplantation for "vital" organs such as the heart requires that the patient be "brain dead." This can be defined as the "complete and irreversible cessation of all brain and brainstem function [100]". Using this as the "standard" for death replaces earlier concepts that death occurred when the heart stopped beating and breathing ceased. Current medical treatments include cardiac defibrillators, pacemakers, and medications that can restore what would otherwise be a fatal arrhythmia. These and other methods such as mechanical ventilators can support a person's cardiopulmonary system well enough to keep the rest of the body's remaining organs functioning even when brain death has occurred. An all too common scenario is that of a young adult who has suffered severe head trauma or another cause of isolated brain/brainstem injury that left the rest of the body reasonably intact and functional.[7]

---

[5] However, as a *reductio ad absurdum* thought experiment the concept as Niven portrays it is indeed very powerful.

[6] The overall less fatal procedure of "buying" an organ such as a kidney from a living donor also has its own serious ethical issues.

[7] Part of the evaluation to assess whether or not an individual who has been declared brain dead is a suitable donor for a heart transplant is to confirm that the heart is functioning adequately. During my career

A variety of guidelines have been developed to make the diagnosis of brain death [101]. Key points include ascertaining that no reversible reasons for loss of brain activity are present (e.g. medications or hypothermia—see Chap. 7), the presence of multiple specific findings on a neurological examination (including but not limited to criteria for coma and absence of many types of reflexes), and performance of tests designed to assess brain activity and/or blood flow. The challenge is to ensure that the best possible criteria for evaluating and confirming brain death, with complete and irreversible loss of brain and brainstem function, are used and applied properly. In particular, brain death must be distinguished from other diagnoses such as a persistent vegetative state (a condition associated with severe brain damage and reduced consciousness) that have at least the potential to improve. Given the critical ethical principle of not removing vital organs from an individual who is not truly brain dead, it is essential that current standards for that diagnosis continue to be discussed and refined as needed based on further reviews and findings.[8]

Another method to increase the supply of donor organs is to use ones from animals—"xenotransplantation." As noted previously, some early attempts at organ transplantation involved taking them from goats, pigs, etc. Many organs in pigs are similar to size to those of humans, and they can be obtained in great abundance. In fact, specially treated heart valves derived from pigs or from the pericardium (the thin sac of tissue enclosing the heart) of bovines are used to create certain kinds of artificial heart valves for humans.

However, there are at least three major biological issues for using animals as organ donors in humans. Some animal tissues (particularly those from pigs) can be genetically modified to produce certain types of human proteins that "fool" part of the recipient's immune system into "thinking" they are not foreign material. Unfortunately methods have not yet developed to protect those porcine or other organs from other components of the human immune systems and prevent rejection.

Also, although pig organs may be similar in size to the human versions, there are still major differences in their physiology, proteins and other constituents, cell function, etc. Cumulatively these factors could significantly reduce the ability of the transplanted organ to function and survive inside its new human host.

I have had to interpret echocardiograms (ultrasound studies of the heart) to make that determination. If I found that heart function was preserved, the organ could then be harvested. But even with the donor truly brain dead, with no hope of recovery, and knowing the heart may save someone else's life, the tragedy and finality of death in that situation is still inescapable.

[8] The plot of Robert J. Sawyer's *The Terminal Experiment* (1995) addresses this issue in a science fiction context, raising the question of when exactly death occurs.

A third, less well-established risk is that of "zoonotic infection." This means that a particular microorganism (e.g. a virus) that has infected the animal's organ but normally does not infect humans could, following transplantation, potentially cause an infection in the human recipient. For example, certain porcine retroviruses might be able to do this.[9]

Another approach to increase the supply of transplantable organs is to actually grow them in some way. An idea presented in works like the movies *Parts: The Clonus Horror* (1979) and *The Island* (2005) is that human clones would be created to serve specifically as organ donors for the "original" versions of the clones. The rationale for doing this is to ensure that tissue-matched organs would be available to possible recipients should they be needed.

Besides the obvious ethical issues involved, the practical difficulties involved with this method would be enormous. For example, it assumes that human clones could be created and develop normally (assumptions that, as described in Chap. 10, are by no means certain); that the clones would not become ill, die, or be at an inappropriate age (e.g. too young) before they were needed to become donors; that the potential recipient would eventually need an organ transplant and not die from other health problems for which an organ transplant would not be helpful; that someone would pay the great costs involved with creating, feeding, providing healthcare, etc. and otherwise supporting the clone before it might be needed to serve as a donor; that the clones and others will not learn of this arrangement and object to it (something that, as expected, the above movies do show happening); and so on. In short, this method to procure organs for transplantation is, besides being grossly unethical, tremendously complex and costly for the proposed "return on investment."

Fortunately considerably "simpler" methods are being investigated for creating new organs (and certain tissues) for transplantation. 3D printing techniques may be applied to create an inorganic structure in an appropriate shape (scaffold), such as to replace part of a bone in a person's body [102]. Osteoblasts and other bone cells could then be introduced to the scaffold where they could proliferate and help form part of a new bone. After implantation of this bone-like structure into a person's body, that individual's own cells could also be stimulated by its presence to promote further growth and healing. Similar concepts underlie methods used to create other artificial structures such as outer ears (auricles) [103, 104].

---

[9] Other viruses thought to have "crossed over" from non-human animals to humans include the HIV virus (a retrovirus), the H5N1 strain of avian influenza (very loosely termed the "bird flu"), and the MERS (Middle East respiratory syndrome) coronavirus.

While still an investigational technique, a small number of individuals have received transplants of bioengineered tracheas[10] that have been seeded using their own stem cells [105]. In this case the scaffold for the stem cells can be provided by an animal (e.g. pig) trachea that has been "decellularized," removing its own cells except for the tissue ("extracellular matrix") that gives it structure. In animal models other tissues and at least part of whole organs have been created using similar methods with various types of cells, including lungs, liver, and the heart [106–109]. Mammary gland cells and prostate cells have also been generated from individual stem cells and used in animal transplantation studies to help regrow those organs without requiring a scaffold.

Overall, such techniques for growing tissues and organs for transplantation are at varying stages of development. Tissues with a relatively simple structure and composition such as the cartilage in outer ears or parts of certain bones may be particularly amenable to scaffolding and similar (e.g. 3D printing) methods. Liver tissue, with its intrinsic ability to regenerate, might also prove amenable to production of grafts using a scaffold matrix.

However, the heart has very complex anatomic, physiological, mechanical, and electrical properties that leave little room for error. While animal studies have taken the first steps in creating at least a miniature version of a bioartificial heart, [108] this is a far cry from using a person's own cells to build a fully functional and healthy replacement should anything bad happen to the original. On the other hand, using stem cells to help repair a damaged heart might be easier. A recently described technique used yet another type of stem cell, "parthenogenetic stem cells," to create myocardiocytes [110]. In one study this type of stem cell was derived from mouse oocytes that were stimulated via chemicals to start dividing and developing. This produced pluripotent stem cells with some characteristics similar to ESCs, but unlike the latter they could not develop into a complete new organism.

The parthenogenetic stem cells differentiated into various cells types including myocardiocytes, which were then transplanted into the hearts of the mice that donated the oocytes and others that were compatible matches regarding important antigens. Those transplanted myocardiocytes were reported to both function well and not be associated with rejection. If similar techniques could be applied in humans, this variety of stem cell might serve as a source of (potentially banked) tissue for repairing hearts damaged by a myocardial infarction or other event.

As a science fiction connection to the history of heart transplantation, there actually have been and still are individuals with two hearts who were not from

---

[10] As described in Chap. 1, the trachea is the "tube" (located primarily in the neck) that connects the pharynx, the upper part of the "windpipe," to the bronchial tubes of the lungs.

Gallifrey. The first "heterotopic"heart transplant was done in 1974 by Dr. Christian Barnard, who performed the first "orthotopic" operation to replace one human heart with another in 1967. In a heterotopic operation a person's original, damaged heart is left in place and a new, donor heart is implanted into the right side of the chest. This procedure can involve connecting the new heart to the old one in such a way that it either assists only the left side of the original heart or both the right and left sided sets of chambers (see Chap. 1).

Though done far less frequently than orthotopic transplants, the heterotopic technique is still an option in individuals meeting certain criteria. These can include significantly increased pressures in the pulmonary (lung) blood vessels; if the available donor heart is considered marginal for being able to take over a person's entire circulation; or if it is not an ideal match to the recipient regarding HLA or other antigen typing and thus has an increased risk of rejection. An obvious advantage is that, if the donor heart is rejected, the patient still has the original heart present. A disadvantage is that the second heart can develop its own problems such as potentially serious arrhythmias and blood clots forming inside it. However, in selected patients it can be a lifesaving treatment [111].

13.5 The Bottom Line

The concepts of stem cell therapy/regenerative medicine and organ transplantation can serve as the primary focus of a science fiction work (e.g. *The Island*). Or it can be just element in a plot, such as in Lois McMaster Bujold's *Memory* (1996) which opens with the novel's protagonist recovering from major tissue and organ transplantation procedures. As will be discussed in the next chapter, two episodes of *Star Trek: The Next Generation* deal specifically with the artificial heart Captain Picard received after his original, biological one was damaged.

Current research in these fields is going in many different directions. The pace at which particular issues will be addressed—what types of stem cells will prove most useful and safe for tissue/organ repair or replacement, dealing with the complex immunological problems with transplantation, growing whole new organs—is uncertain. It is also unclear what the theoretical and practical limits of various avenues of research might be, such as whether the risks of using stem cells might (at least in some circumstances) outweigh their benefits or if complex organs such as the heart (much less the brain!) can actually be created at a level of function and safety as well as in numbers suitable for extensive clinical use.

Science fiction, unfettered by many real-world considerations of the scientific difficulties involved, can use ideas involving regenerative medicine and organ transplantation in many ways and settings. However, extending them too far for story purposes can reduce the work's realism regarding other considerations, as mentioned previously regarding *Parts: The Clonus Horror*. If such techniques do become feasible, all of their ramifications—including how practical they are to perform, how much they cost, how long it takes to perform them, what resources are needed to use them, etc.—should be taken into account if only implicitly.

Likewise, using organ transplantation or advanced techniques for tissue repair using stem cells or other means to help explain why a character is 200 years old is also suspect. While such methods can certainly help an individual (e.g. one requiring a heart transplant) live longer than would otherwise be possible, they might have at most little impact on extending maximal lifespan beyond the known (at least so far) upper limit of about 122 years. As noted in Chap. 9, aging is a process that involves the whole body and is in a fundamental sense different from "disease." While it certainly may act as a stopgap measure to prolong life, replacing a liver or kidney will do nothing to restore an elderly individual's immune system, clear that person's cells of accumulated DNA mutation and damage, prevent degeneration of the brain, etc. Other methods will have to be developed (if they can be) to address such issues. The net result is that many, many "medical miracles" will likely need to be developed (once again, if they actually can) in many, many different and distinct areas (genetic engineering, nanotechnology, use of stems cells and organ transplantation, etc.) to create a biologically youthful 200-year old human with intact cerebral function.

Many other medical issues might also be skirted over in science fiction with a similar loss of realism. For example, it takes a certain amount of time for cells to divide and grow. Thus, if a character receives a new liver grown from his or her cells, if those cells were obtained today there would simply not be enough time to have that organ be ready for transplantation tomorrow.

For example, H. L. Gold's short story "No Charge for Alterations" (1953) depicts a future physician literally remolding a "patient's" body like a sculpture as well as altering her personality. The physical alterations involve instantaneous plastic surgery that includes wholesale removal/reshaping of skin, bone, etc., and addition of new "cells" and "synthetic tissue." They are vividly portrayed by sentences such as "The flesh fled from the cathode and chased after the anode as he broadened the fine nose, thickened the mobile lips, squared the slender jaw and drew out carefully the delicately arched orbital ridges." However, the process described grossly underestimates the complexity of those parts of the body, the myriad types and organization of its cells and tissues, the body's ability to heal, etc., and overall has at best a tenuous relation with actual human biology.

Similarly F. L. Wallace's "Tangle Hold" (1953) opens with the main character recovering from what is essentially a total "skin transplant." He had suffered burns from scalding high-pressure water and other trauma so severe and extensive that, he is told, "…every square inch of your skin is now synthetic." His treatment is said to involve peeling off burned parts, fitting the synthetic skin, then spraying on a bandage, after which "New cells form with this synthetic substance as the matrix." However, as noted in Chaps. 1 and 2, human skin has multiple layers, many kinds of different cells, a complex pattern of nerves and blood vessels, and performs many more functions (e.g. heat regulation) than just protecting our internal anatomy. How the synthetic skin would allow something as simple but essential as sweating is not addressed. Likewise, burns sufficient to "require" replacing a person's entire skin would also presumably involve at least some fourth-degree burns to underlying tissues, including the most vulnerable ones such as those of the penis and scrotum. Likewise the eyelids would offer little protection to the eyes in that situation, but there is no description of the latter being "boiled."

However, the focus in both stories is clearly not on medical realism but exploring the ramifications *if* it were possible to do those essentially impossible things. That "if" *is* a valid idea to explore in a story, even at the cost of making it science fantasy instead of science fiction.

Moreover, in the real world major operations place great stress on the human body. They can also be associated with many potential operative and postoperative complications. Unless one invokes "magical" methods such as nanotechnology far in advance of current capabilities, a character will simply not be anywhere near fully recovered the day after a heart transplant. Our bodies also take a significant amount of time to heal and recuperate long after surgery. In fact, an individual may never recover completely in some ways and can have delayed complications associated with scarring, infections, etc.

In short, for procedures such as stem cell therapies or advanced transplantation techniques to become "routine" they must (among other things) have an overall high initial success rate as well as a sufficiently low rate of immediate and delayed complications and adverse effects. A futuristic science fiction story can take these things for granted, but ideally it should not ignore fundamental principles of biology (e.g. rate of cell division) or the basic limitations of the "standard" human body. And even if this idea is not directly reflected in a story, hopefully this chapter will help the writer understand how far medical science has yet to go to bring these techniques to their imagined (and possibly real) potential.

References

1. Doulatov S, Daley GQ. Development. A stem cell perspective on cellular engineering. Science. 2013;342(6159):700–2.
2. Jordan C, Guzman M, Noble M. Cancer stem cells. N Engl J Med. 2006;355(12):1253–61.
3. Kolios G, Moodley Y. Introduction to stem cells and regenerative medicine. Respiration. 2013;85(1):3–10.
4. Ma X, Zhang Q, Yang X, Tian J. Development of new technologies for stem cell research. J Biomed Biotechnol. 2012;2012:741416.
5. Fuchs E, Chen T. A matter of life and death: self-renewal in stem cells. EMBO Rep. 2013;14(1):39–48.
6. Sokolov M, Neumann R. Lessons learned about human stem cell responses to ionizing radiation exposures: a long road still ahead of us. Int J Mol Sci. 2013;14(8):15695–723.
7. Liang G, Zhang Y. Embryonic stem cell and induced pluripotent stem cell: an epigenetic perspective. Cell Res. 2013;23(1):49–69.
8. Shyh-Chang N, Daley GQ, Cantley LC. Stem cell metabolism in tissue development and aging. Development. 2013;140(12):2535–47.
9. Szabłowska-Gadomska I, Górska A, Małecki M. Induced pluripotent stem cells (iPSc) for gene therapy. Med Wieku Rozwoj. 2013;17(3):191–5.
10. Gao L, Thilakavathy K, Nordin N. A plethora of human pluripotent stem cells. Cell Biol Int. 2013;37(9):875–87.
11. Li W, Ding S. Human pluripotent stem cells: decoding the naive state. Sci Transl Med. 2011;3(76):76 ps10.
12. Thomson JA. Embryonic stem cell lines derived from human blastocysts. Science. 1998;282(5391):1145–7.
13. Terzic A, Nelson T. Regenerative medicine primer. Mayo Clinic Proc. 2013;88(7):766–75.
14. Wang ML, Chiou SH, Wu CW. Targeting cancer stem cells: emerging role of Nanog transcription factor. Onco Targets Ther. 2013;6:1207–20.
15. van Wijngaarden P, Franklin RJ. Ageing stem and progenitor cells: implications for rejuvenation of the central nervous system. Development. 2013;140(12):2562–75.
16. Bibber B, Sinha G, Lobba AR, Greco SJ, Rameshwar P. A review of stem cell translation and potential confounds by cancer stem cells. Stem Cells Int. 2013;2013:241048.
17. Ikebe C, Suzuki K. Mesenchymal stem cells for regenerative therapy: optimization of cell preparation protocols. Biomed Res Int. 2014;2014:951512.
18. Leatherman J. Stem cells supporting other stem cells. Front Genet. 2013;4:257.
19. de Sa Silva F, Almeida PN, Rettore JV, Maranduba CP, de Souza CM, de Souza GT, et al. Toward personalized cell therapies by using stem cells: seven relevant topics for safety and success in stem cell therapy. J Biomed Biotechnol. 2012;2012:758102.

20. Ballen KK, Barker JN. Has umbilical cord blood transplantation for AML become mainstream? Curr Opin Hematol. 2013;20(2):144–9.

21. Signer RA, Morrison SJ. Mechanisms that regulate stem cell aging and lifespan. Cell Stem Cell. 2013;12(2):152–65.

22. Bowman T, Zon L. Ageing: stem cells on a stress-busting diet. Nature. 2013; 494:317–8.

23. Li M, Belmonte JC. Ageing: genetic rejuvenation of old muscle. Nature. 2014; 506:304–5.

24. Hayashi E, Hosoda T. Therapeutic application of cardiac stem cells and other cell types. Biomed Res Int. 2013;2013:736815.

25. Wong W, Sayed N, Cooke J. Induced pluripotent stem cells: how they will change the practice of cardiovascular medicine. Methodist Debakey Cardiovasc J. 2013;9(4):206–9.

26. Michler R. Stem cell therapy for heart failure. Methodist Debakey Cardiovasc J. 2013;9(4):187–94.

27. Doppler SA, Deutsch MA, Lange R, Krane M. Cardiac regeneration: current therapies-future concepts. J Thorac Dis. 2013;5(5):683–97.

28. Wang J, Liao L, Wang S, Tan J. Cell therapy with autologous mesenchymal stem cells–how the disease process impacts clinical considerations. Cytotherapy. 2013;15(8):893–904.

29. Dimarino AM, Caplan AI, Bonfield TL. Mesenchymal stem cells in tissue repair. Front Immunol. 2013;4:201.

30. Salibian AA, Widgerow AD, Abrouk M, Evans GR. Stem cells in plastic surgery: a review of current clinical and translational applications. Arch Plast Surg. 2013;40(6):666–75.

31. Schächinger V, Erbs S, Elsässer A, Haberbosch W, Hambrecht R, Hölschermann H, et al. Intracoronary bone marrow-derived progenitor cells in acute myocardial infarction. N Engl J Med. 2006;355:1210–21.

32. Sanchez L. Use of stem cells in heart failure treatment: where we stand and where we are going. Methodist Debakey Cardiovasc J. 2013;9(4):195–200.

33. Gage FH, Temple S. Neural stem cells: generating and regenerating the brain. Neuron. 2013;80(3):588–601.

34. Hsu YC, Chen SL, Wang DY, Chiu IM. Stem cell-based therapy in neural repair. Biomed J. 2013;36(3):98–105.

35. Willyard C. Stem cells: a time to heal. Nature. 2013;503:S4–S6.

36. English D, Sharma NK, Sharma K, Anand A. Neural stem cells-trends and advances. J Cell Biochem. 2013;114(4):764–72.

37. Briggs R, King TJ. Transplantation of living nuclei from blastula cells into enucleated frogs' eggs. Proc Natl Acad Sci USA. 1952;38:455–63.

38. Gurdon JB, Elsdale TR, Fischberg M. Sexually mature individuals of *Xenopus laevis* from the transplantation of single somatic nuclei. Nature. 1958;182:64–65.

39. Siller R, Greenhough S, Park I, Sullivan G. Modelling human disease with pluripotent stem cells. Current Gene Therapy. 2013;13(2):99–110.

40. Fernandez Tde S, de Souza Fernandez C, Mencalha AL. Human induced pluripotent stem cells from basic research to potential clinical applications in cancer. Biomed Res Int. 2013;2013:430290.

41. Hou P, Li Y, Zhang X, Liu C, Guan J, Li H, et al. Pluripotent stem cells induced from mouse somatic cells by small-molecule compounds. Science. 2013;341(6146):651–4.

42. Sharkis S, Jones R, Civin C, Jang Y. Pluripotent stem cell-based cancer therapy: promise and challenges. Sci Transl Med. 2012;4(127):127ps9.

43. Takahashi K, Yamanaka S. Induction of pluripotent stem cells from mouse embryonic and adult fibroblast cultures by defined factors. Cell. 2006;126(4):663–76.

44. Takahashi K, Tanabe K, Ohnuki M, Narita M, Ichisaka T, Tomoda K, et al. Induction of pluripotent stem cells from adult human fibroblasts by defined factors. Cell. 2007;131(5):861–72.

45. Wang M, Chiou S, Wu C. Targeting cancer stem cells: emerging role of Nanog transcription factor. Oncology Targets and Therapy. 2013;6:1207–20.

46. Jauch R, Kolatkar P. What makes a pluripotency reprogramming factor? Current Molecular Medicine. 2013;13(5):806–14.

47. O'Doherty R, Greiser U, Wang W. Nonviral methods for inducing pluripotency to cells. Biomed Res Int. 2013;2013:705902.

48. Sancho-Martinez I, Belmonte J. Stem Cells: Surf the waves of reprogramming. Nature. 2013;493:310–1.

49. Bai Q, Desprat R, Klein B, Lemaitre J, De Vos J. Embryonic stem cells or induced pluripotent stem cells? A DNA integrity perspective. CurrGene Ther. 2013;13(2):93–8.

50. Ma T, Xie M, Laurent T, Ding S. Progress in the reprogramming of somatic cells. Circ Res. 2013;112(3):562–74.

51. Nguyen HT, Geens M, Spits C. Genetic and epigenetic instability in human pluripotent stem cells. Hum Reprod Update. 2013;19(2):187–205.

52. Wutz A. Epigenetic alterations in human pluripotent stem cells: a tale of two cultures. Cell Stem Cell. 2012;11(1):9–15.

53. Zhou H, Ding S. Evolution of induced pluripotent stem cell technology. Curr Opin Hematol. 2010;17(4):276–80.

54. Zhao XY, Li W, Lv Z, Liu L, Tong M, Hai T, et al. iPS cells produce viable mice through tetraploid complementation. Nature. 2009;461(7260):86–90.

55. Boland MJ, Hazen JL, Nazor KL, Rodriguez AR, Gifford W, Martin G, et al. Adult mice generated from induced pluripotent stem cells. Nature. 2009;461(7260):91–4.

56. Lo B, Parham L, Alvarez-Buylla A, Cedars M, Conklin B, Fisher S, et al. Cloning mice and men: prohibiting the use of iPS cells for human reproductive cloning. Cell Stem Cell. 2010;6(1):16–20.

57. Egashira T, Yuasa S, Fukuda K. Novel insights into disease modeling using induced pluripotent stem cells. Biol Pharm Bull. 2013;36(2):182–8.

58. Mercola M, Colas A, Willems E. Induced pluripotent stem cells in cardiovascular drug discovery. Circ Res. 2013;112(3):534–48.

59. Zhu Z, Huangfu D. Human pluripotent stem cells: an emerging model in developmental biology. Development. 2013;140(4):705–17.

60. Simara P, Motl J, Kaufman D. Pluripotent stem cells and gene therapy. Transl Res. 2013;161(4):284–92.

61. Leeman D, Brunet A. Stem cells: sex specificity in the blood. Nature. 2014;505: 488–90.

62. Nakada D, Oguro H, Levi BP, Ryan N, Kitano A, Saitoh Y, et al. Oestrogen increases haematopoietic stem-cell self-renewal in females and during pregnancy. Nature. 2014;505(7484):555–8.

63. Cherry AB, Daley GQ. Reprogrammed cells for disease modeling and regenerative medicine. Annu Rev Med. 2013;64:277–90.

64. de Los Angeles A, Daley G. Stem cells: reprogramming in situ. Nature. 2013;502:309–10.

65. Abad M, Mosteiro L, Pantoja C, Canamero M, Rayon T, Ors I, et al. Reprogramming in vivo produces teratomas and iPS cells with totipotency features. Nature. 2013;502(7471):340–5.

66. Sieweke MH, Allen JE. Beyond stem cells: self-renewal of differentiated macrophages. Science. 2013;342(6161):1242974.

67. Kuroda T, Ysauda S, Sato Y. Tumorigenicity studies for human pluripotent stem cell-derived products. Biol Pharm Bull. 2013;36(2):189–92.

68. Pearl JI, Kean LS, Davis MM, Wu JC. Pluripotent stem cells: immune to the immune system? Sci Transl Med. 2012;4(164):164ps25.

69. Okano H, Nakamura M, Yoshida K, Okada Y, Tsuji O, Nori S, et al. Steps toward safe cell therapy using induced pluripotent stem cells. Circ Res. 2013;112(3): 523–33.

70. Mayshar Y, Ben-David U, Lavon N, Biancotti JC, Yakir B, Clark AT, et al. Identification and classification of chromosomal aberrations in human induced pluripotent stem cells. Cell Stem Cell. 2010;7(4):521–31.

71. Cyranoski D. Stem cells reprogrammed using chemicals alone. 2013. http://www.nature.com/news/stem-cells-reprogrammed-using-chemicals-alone-1.13416. Accessed 15 April 2015.

72. Lin T, Ambasudhan R, Yuan X, Li W, Hilcove S, Abujarour R, et al. A chemical platform for improved induction of human iPSCs. Nat Methods. 2009;6(11):805–8.

73. Hanna JH, Saha K, Jaenisch R. Pluripotency and cellular reprogramming: facts, hypotheses, unresolved issues. Cell. 2010;143(4):508–25.

74. Rais Y, Zviran A, Geula S, Gafni O, Chomsky E, Viukov S, et al. Deterministic direct reprogramming of somatic cells to pluripotency. Nature. 2013;502 (7469):65–70.

75. de Almeida PE, Ransohoff JD, Nahid A, Wu JC. Immunogenicity of pluripotent stem cells and their derivatives. Circ Res. 2013;112(3):549–61.

76. Aguilar-Gallardo C, Simon C. Cells, stem cells, and cancer stem cells. Semin Reprod Med. 2013;31(1):5–13.

77. Shiozawa Y, Nie B, Pienta KJ, Morgan TM, Taichman RS. Cancer stem cells and their role in metastasis. Pharmacol Ther. 2013;138(2):285–93.
78. Wang K, Wu X, Wang J, Huang J. Cancer stem cell theory: therapeutic implications for nanomedicine. Int J Nanomedicine. 2013;8:899–908.
79. ElShamy WM, Duhé RJ. Overview: cellular plasticity, cancer stem cells and metastasis. Cancer Letters. 2013;341(1):2–8.
80. Nieto MA. Epithelial plasticity: a common theme in embryonic and cancer cells. Science. 2013;342(6159):1234850.
81. Binello E, Germano IM. Stem cells as therapeutic vehicles for the treatment of high-grade gliomas. Neuro Oncol. 2012;14(3):256–65.
82. Germano IM, Emdad L, Qadeer ZA, Binello E, Uzzaman M. Embryonic stem cell (ESC)-mediated transgene delivery induces growth suppression, apoptosis and radiosensitization, and overcomes temozolomide resistance in malignant gliomas. Cancer Gene Ther. 2010;17(9):664–74.
83. Pucci F, Gardano L, Harrington L. Short telomeres in ESCs lead to unstable differentiation. Cell Stem Cell. 2013;12(4):479–86.
84. Watson CJ, Dark JH. Organ transplantation: historical perspective and current practice. Br J Anaesth. 2012;108 Suppl 1:i29–42.
85. Dunning J, Calne R. Historical perspectives. In: Klein A, Lewis C, Madsen J, editors. Organ transplantation. A clinical guide. Cambridge: Cambridge University Press; 2011. pp. 1–8.
86. Merrill J, Murray J, Harrison J, Guild W. Successful homotransplantation of the human kidney between identical twins. J Am Med Assoc. 1956;160:277–82.
87. DiBardino D. The history and development of cardiac transplantation. Tex Heart Inst J. 1999;26:198–205.
88. Orens JB, Garrity ER, Jr. General overview of lung transplantation and review of organ allocation. Proc Am Thorac Soc. 2009;6(1):13–9.
89. Song AT, Avelino-Silva VI, Pecora RA, Pugliese V, D'Albuquerque LA, Abdala E. Liver transplantation: fifty years of experience. World J Gastroenterol. 2014;20(18):5363–74.
90. Han D, Sutherland D. Pancreas transplantation. Gut Liver. 2010;4:450–65.
91. Issa F, Goto R, Wood K. Immunological principles of acute rejection. In: Klein A, Lewis C, Madsen J, editors. Organ transplantation. A clinical guide. Cambridge: Cambridge University Press; 2011. pp. 9–18.
92. Kotton C. Major complications–infection. In: Klein A, Lewis C, Madsen J, editors. Organ transplantation. A clinical guide. Cambridge: Cambridge University Press; 2011. pp. 46–52.
93. Dey B, Spitzer T. Major Complications–Cancer. In: Klein A, Lewis C, Madsen J, editors. Organ transplantation. A clinical guide. Cambridge: Cambridge University Press; 2011. pp. 31–7.
94. Al-Mansour Z, Nelson B, Evens A. Post-transplant lymphoproliferative disease (PTLD): risk factors, diagnosis, and current treatment strategies. Curr Hematol Malig Rep. 2013;8(3):173–83.

95. Kumar V, Gaston R. Immunosuppression: past, present, and future. In: Klein A, Lewis C, Madsen J, editors. Organ transplantation. A clinical guide. Cambridge: Cambridge University Press; 2011. pp. 19–30.

96. Kushner Y, Colvin R. Major Complications–pathology of chronic rejection. In: Klein A, Lewis C, Madsen J, editors. Organ transplantation. A clinical guide. Cambridge: Cambridge University Press; 2011. pp. 38–45.

97. Parthasarathy H, Lewis C. Long-term management and outcomes. In: Klein A, Lewis C, Madsen J, editors. Organ transplantation. A clinical guide. Cambridge: Cambridge University Press; 2011. pp. 102–11.

98. Coffman K, Siemionow M. Ethics of facial transplantation revisited. Curr Opin Org Transpl. 2014;19:181–7.

99. Khalifian S, Brazio PS, Mohan R, Shaffer C, Brandacher G, Barth RN, et al. Facial transplantation: the first 9 years. The Lancet. 2014;384(9960):2153–63.

100. Cantu E III, Zaas D. Organ donor management and procurement. In: Klein A, Lewis C, Madsen J, editors. Organ transplantation. A clinical guide. Cambridge: Cambridge University Press; 2011. pp. 53–62.

101. Hwang DY, Gilmore EJ, Greer DM. Assessment of brain death in the neurocritical care unit. Neurosurg Clin N Am. 2013;24(3):469–82.

102. Bose S, Vahabzadeh S, Bandyopadhyay A. Bone tissue engineering using 3D printing. Mater Today. 2013;16(12):496–504.

103. Reiffel AJ, Kafka C, Hernandez KA, Popa S, Perez JL, Zhou S, et al. High-fidelity tissue engineering of patient-specific auricles for reconstruction of pediatric microtia and other auricular deformities. PLoS One. 2013;8(2):e56506.

104. Liu Y, Yang R, He Z, Gao W. Generation of functional organs from stem cells. Cell Regen. 2013;2:1–6.

105. Vogel G. Trachea transplants test the limits. Science. 2013;340:266–8.

106. Petersen TH, Calle EA, Zhao L, Lee EJ, Gui L, Raredon MB, et al. Tissue-engineered lungs for in vivo implantation. Science. 2010;329(5991):538–41.

107. Uygun BE, Soto-Gutierrez A, Yagi H, Izamis ML, Guzzardi MA, Shulman C, et al. Organ reengineering through development of a transplantable recellularized liver graft using decellularized liver matrix. Nat Med. 2010;16(7):814–20.

108. Ott HC, Matthiesen TS, Goh SK, Black LD, Kren SM, Netoff TI, et al. Perfusion-decellularized matrix: using nature's platform to engineer a bioartificial heart. Nat Med. 2008;14(2):213–21.

109. Badylak SF, Taylor D, Uygun K. Whole-organ tissue engineering: decellularization and recellularization of three-dimensional matrix scaffolds. Ann Rev Biomed Eng. 2011;13(1):27–53.

110. Didie M, Christalla P, Rubart M, Muppala V, Doker S, Unsold B, et al. Parthenogenetic stem cells for tissue-engineered heart repair. J Clin Invest. 2013;123(3):1285–98.

111. Newcomb AE, Esmore DS, Rosenfeldt FL, Richardson M, Marasco SF. Heterotopic heart transplantation: an expanding role in the twenty-first century? Ann Thorac Surg. 2004;78(4):1345–50; (discussion 50–1).

14

Bionics: Creating the Twenty-Four Million Dollar Man or Woman

Let us then conclude boldly that man is a machine…
Man a Machine (1748)
Julien Offray de La Mettrie

Gentlemen, we can rebuild him. We have the technology.
The Six Million Dollar Man
(Opening narration)

The unmodified human body has many physical limitations. Even the best Olympic athletes can only run so fast or jump so high. Our vision is limited to a tiny sliver of the electromagnetic spectrum and our hearing to a narrow range of frequencies. From an overall biological perspective healthy individuals differ over only a small range regarding what they can physically do and how easily their skin, bones, and other tissues can be damaged.

But while human biology has remained essentially unchanged for many thousands of years, there have been tremendous recent advances in development of new materials, electronics, and other branches of the physical sciences. "Bionics" is a term that, in a medical context, can be used to describe ways to combine electrical, mechanical, and/or electronic devices and materials with the human body. Current research focuses on the therapeutic uses of these methods, e.g. to help restore sight, hearing, mobility, and other "normal" functions to those who have lost them due to injury, illness, etc.

However, another potential reason for uniting (wo)man and machine—one at the cutting edge of real world research and a staple of science fiction—is to make us "more" than human. The question here is whether we can, through non-biological means, significantly improve the "design specs" for our bodies much more rapidly than the millions of years it took to achieve their current physical capabilities.

The current answer to that question is that there has been much more progress made on restoring functions than augmenting them. On the mid-1970s television series *The Six Million Dollar Man* and *The Bionic Woman* a check for 6 million dollars could rebuild Steve Austin and Jaime Sommers "better" than they were before their major, dismembering accidents. Unfortunately, even adjusting for inflation and spending 24 million dollars we really *don't* have the technology to do that even now. Nonetheless we are significantly closer to those goals than 40 years ago, and this field holds great potential for further advances.

14.1 Blending Machine and Flesh in Science Fiction

Characters or entire groups in science fiction can be mixtures of living and electromechanical parts. At one end of the spectrum one or more devices can be affixed to someone who clearly is biological in origin. After learning up close and personal how dangerous playing with light sabers can be, both Anakin Skywalker and his son Luke had lost limbs replaced by artificial ones. However, while Luke's new right hand is not shown as having strength or dexterity significantly different from his original biological one, his father's is depicted early in *Star Wars: A New Hope* (1977) as capable of lifting a Rebel Alliance officer with ease and inflicting bone-crushing terminal trauma to that unfortunate individual's neck.

Similarly the Six Million Dollar Man received a replacement pair of legs, a new left arm, and an artificial left eye while the Bionic Woman also gained two "improved" legs and right arm along with an electronically enhanced right inner ear. The titular character in the 1966 movie *Cyborg 2087* also is clearly human, but with augmented strength in his left arm, better-than-human legs, and a "panel" in his chest whose function is less clear. In comic books Lightning Lad of the Legion of Super-Heroes lost his right arm to the monstrous Moby Dick of Space (*Adventure Comics* #332, May 1965) and had it replaced with a robotic one. Tharok, a villain that group faced in *Adventure Comics* #352 (January 1967), had the entire left side of his body replaced by robotic parts following a terrible accident that occurred while he was committing a robbery.[1] Victor Stone, aka "Cyborg," is also a comic book character with roughly equal "original" and mechanical/electronic parts. The protago-

---

[1] The extensive operative procedure Tharok undergoes also raises questions about how internal organs located exclusively or partially in the left side of the chest (e.g. the heart) and abdomen (such as the spleen, part of the small and large intestines, spleen, etc.) were dealt with, since there would not normally be enough room to "squish" them and all their vascular, nerve, and other connections into his remaining biological right side. There would also be obvious issues with relatively symmetric parts in the back (e.g. the vertebral column and spinal cord), neck (such as the trachea and esophagus) and pelvis (including the anus, bladder, and external genitalia)—to say nothing of the left side of the brain itself. These are issues that did not occur to me when I first read that story many years ago but now seem a bit problematic. Perhaps the psychological trauma associated with having to live with such major alterations (including potential loss of the genitalia) might help explain why Tharok continued an antisocial life of crime after his accident...

nists in the movies *RoboCop* (1987) and the considerably more fanciful *Inspector Gadget* (1999) also undergo significant "scientifically" based physical and electronic alterations.

Likewise, the *Doctor Who* series has recurring groups (e.g. the Daleks and Cybermen) who began as biological beings before becoming overwhelmingly mechanical, along with one-shot individuals such as the Captain in the 1978 serial "The Pirate Planet." The Borg in the later *Star Trek* television series and movie *First Contact* (1996) are particularly aggressive regarding recruiting new members by "assimilating" them into their collective consciousness via methods that include technological modifications.

In written science fiction a human brain can be placed into a robotic body as in Neil R. Jones series of Professor Jameson stories beginning in 1931, or C. L. Moore's 1944 story "No Woman Born." Mixtures of flesh and spaceship feature prominently as "brainships" in Anne McCaffrey's 1969 novel *The Ship Who Sang*. Science fiction characters with less extreme replacements of the biological with metal and electronics are depicted in the Martin Caidin's novel *Cyborg*[2] (1972), and the opening pages of William Gibson's 1984 *Neuromancer* describe a bartender with a robotic prosthetic arm.

Rather than starting out as biological entities and adding mechanical parts, science fiction characters can also be created by reversing this order. The title character in the first *Terminator* movie (1984) is fundamentally a mechanical, artificial being covered with enough living tissue to make it resemble a future governor of California. The humanoid Cylons in the second *Battlestar Galactica* television series (2004–2009) follow a similar genesis. The heroic but tragic main character in the Harlan Ellison-written episode "Demon with a Glass Hand" for *The Outer Limits* television series (1964) also seems to fall into this category. Ray Bradbury's "The City" (aka "Purpose," 1950) describes another variation on this theme. In that short story human spacefarers are literally hollowed out and have their internal organs replaced by robotic parts and brains geared for a sinister purpose. The main character in Isaac Asimov's story "The Bicentennial Man" (1976) is a robot who yearns to become more human and succeeds in doing that.

14.2   How to Make a Cyborg

The term "cyborg"—generally taken as short for "cybernetic organism"—was proposed in 1960 [1] to describe individuals joined with mechanical systems to improve their ability to better survive space travel. The Anne McCaffrey

---

[2] This work formed the basis for the two television series *The Six Million Dollar Man* and *The Bionic Woman*.

work cited previously is a further extension of this idea. The term could also encompass anyone whose body is in intimate contact with a fabricated aid to restore or augment a "conventional" human function such as sight, or provide abilities beyond what a person could otherwise do.

However, using the definition too loosely would make even me a cyborg, in that I use glasses to improve nearsightedness. Likewise someone who has undergone total hip or knee replacement surgery, an individual with a pacemaker to manage heart rate, or a diabetic with a sophisticated insulin pump [2] could all be encompassed by a broad enough definition of the word. Even someone wearing a device such as the Google Glass now being tested, with its ability to keep a person continuously linked to the Internet and computer functions, could be called a cyborg.

Rather than play with semantics, the discussion here will involve more "cutting edge" technologies that are connected so closely to the body for extended times that they essentially become "one" with it. This will include various types of advanced "prosthetics"—an artificial item used to replace a missing body part, such as a limb—as well as devices to help provide function to biologically dysfunctional organs such as an eye or ear.

14.3 Bionic Limbs

Recent progress in development of "bionic" devices overlaps ones Steve Austin and Jaime Sommers received during their television careers. Prosthetic limbs have been used since ancient times to replace those lost via trauma or disease, as well as ones absent or deformed at birth [3, 4]. Losing limbs was apparently a common occupational hazard for fictional early eighteenth century-style pirates such as Captain Hook and Long John Silver. More seriously, limb losses due to war injuries, vehicular accidents, etc. have created a great need to restore function for many individuals.

Fortunately current limb prosthetics are much more sophisticated than a simple hook or pegleg. Upper extremity prostheses can have relatively simple claw-like pincers, or "wrists" and "fingers" with articulating metal joints that can give the user greater dexterity. The more complex varieties may have many joints with associated motors to move them, as well as sensors that allow the user to better control how much force the artificial hand exerts.

Modern replacement arms can include relatively lightweight metals such as aluminum and titanium as well as other strong, light materials like carbon fiber. In some cases they can be covered with a lifelike plastic to make them look closer to a biological arm. This fabricated "skin" and other parts of the artificial limb could include other sensors to detect temperatures and pressure.

"Touch" sensors can assess how much pressure is being exerted by measuring changes in electric resistance and corresponding current flow in the device. Another method is to use "piezotronics," based on the principle that changing the strain on a material can redistribute its positive and negative charges—a change that can produce a detectable electric signal [5]. Such methods might be used to produce arrays that could potentially be much more sensitive to pressure than human skin is.

The construction and energy needed to use lower extremity replacements varies depending on whether the residual part of a person's limb includes the knee or not. Overall, having an intact knee allows the individual to use less energy when walking than if the residual limb ends above the knee, as well as reducing the complexity of the prosthetic limb. The replacement(s) will also have to be strong enough to support some of a person's weight if only one is used, or all of that weight if parts of both lower extremities are missing.

The mechanics of walking are also different with a biological versus a prosthetic lower extremity. Walking or running with the former includes coordinated motions of the hip, knee, and ankle joints. Calf muscles generate almost 80 % of the mechanical work needed to perform a leg's work during walking. Current prostheses using carbon fiber springs and other means can store and release elastic energy when in contact with a surface, but they cannot generate positive work like those muscles can. The net effect is that a person walking with one of these prostheses must use about 10–30 % more metabolic energy to walk at the same speed as someone with two biological lower extremities. New leg prostheses incorporating improved ankle joints may help alleviate some of these issues and reduce energy expenditure during walking [6].

Besides improving the articulation of lower extremity prosthetic joints to better emulate biological ones, other approaches are also being taken. One prosthetic optimized for running, a J-shaped one using carbon fiber, has been used by athletes for competitive events [7]. It has both disadvantages and advantages compared to biological legs. This type of prosthetic is not as efficient for normal walking, the wearer has difficulty making turns when running in a straight line, and it is also more difficult to accelerate from rest and decelerate from running. However, a runner with this prosthetic may use less energy when actually running at a constant speed than someone with "normal" legs.

Prosthetic limbs are attached to an individual's body and controlled by various means. Straps, harnesses, and suction are all ways to fasten the prosthetic to the residual limb. The device typically contains a socket that the remaining biological part of the limb fits into, either directly or via an intermediary liner made of silicone or other protective materials. Certain harness arrangements allow the prosthetic arm to perform a function (e.g. opening and closing pincers at its end to grasp an object) when the opposite shoulder or arm moves.

An individual whose level of limb loss is above the elbow or knee could have a prosthetic limb surgically attached to the remaining long bone in the arm or thigh (humerus and femur, respectively). However, doing this via a rod or other conventional type of internal attachment poses the risk of fracturing the bone if too much stress is applied to the prosthesis.

The latter also means that the interface between Luke Skywalker's "bionic" hand and the end of his biological forearm would be particularly vulnerable due to the anatomy of the wrist. On the forearm side this joint consists of the ends of two parallel long bones, the radius on the thumb side and the ulna on the other. At the farther (distal) side of the joint there are two rows of four small "carpal" bones. Each of these 8 carpals has a different size and shape than the others. In reviewing the relevant scene in *The Empire Strikes Back* the level of Luke's injury seems to have most likely included the loss of his carpal bones, so they are (literally) out of the picture. Unfortunately, the remaining radius and ulna, either individually or together, do not represent a very good "anchor" or "attachment" sites for his new mechanical hand. Also, just how the synthetic skin of his prosthesis blends into his own flesh and how durable that connection would be is uncertain.

Meanwhile, in our own present-day and nearby Galaxy, an upper or lower extremity prosthesis capable of coordinated and possible fine motion using built-in motors, servos, and microprocessors needs to be controlled either via its own programming or by the user to emulate a biological limb as closely as possible. In a lower limb prosthesis with the equivalent of a knee joint, sensors could be used to detect the angle of the joint and adjust its movement to match the individual's walking speed. However, to change ambulation modes (e.g. walking on a flat surface, going up steps, or running) with some motorized lower limb prostheses, the user may have to use a key fob-type control or perform a series of programmed, exaggerated motions to change its pattern of action [8].

An alternative way to do this is to use the nerve signals present in the remaining muscles in the limb and their "electromyographic" (EMG) signals— the electrical activity they generate when those muscles contract or relax [8, 9]. One method of doing this in someone with loss of a lower limb above the knee involves placing a grid of electrodes on the end of the residual part of the limb. EMG signals are then detected and recorded as the individual tries to consciously flex and extend the nonexistent knee and ankle joints. The types and patterns of nerve signals that would have gone to the missing parts of the limb could then be used to help coordinate the actions of the sensors, motors, and other components of the robotic prosthesis. The microprocessor circuitry in the latter could decode EMG signals using a pattern recognition algorithm, combine it with sensor data regarding load, velocity, etc., and thus be better able to mimic "normal" walking [8].

Similar techniques can be used to help an individual control the separate "fingers" of a robotic hand. These can include "myoelectrodes" to detect contraction of muscles in the biological part of the arm or other types of sensors that recognize an exerted force. The signals produced are then sent to a processor in the unit that commands the digits to move in response to those stimuli. These methods typically require "training" both the user and device to make as sure as possible that a particular action and signal generates the desired response by the prosthesis.

The most sophisticated current limb prostheses can significantly improve the ability of someone with limb loss to do many activities of daily living, such as walking and picking up objects. The greatest progress in this area involves "motor" functions, in the medical sense of those involving initiation and performance of motion. Voluntary movement of a biological limb requires transmission of electrical impulses from the brain, through the spinal cord and finally to peripheral nerves innervating the limb's skeletal muscles. Robotic prosthetic limbs are becoming increasingly able to emulate those motor actions using systems that also include sensors to detect the position, velocity, pressure, etc. exerted on individual components (e.g. digits) and use this feedback to coordinate actions.

However, peripheral nerves in a biological arm carry more than motor instructions from the brain to move a limb. They also transmit sensory information back to the brain that we perceive as pain, temperature, pressure, stretch, vibration, etc. Both motor and sensory data are carried in peripheral nerves as bundles of "fibers." These consist of axons—the extensions of neurons in the central nervous system (brain and spinal cord)—with supporting cells and material such as myelin, which (very roughly speaking) acts like the insulation on a conventional wire. Some nerve fibers carry only sensory data, others only motor information, and some may transmit both types.

Our brains constantly monitor sensory and motor information from the limbs in a feedback loop. For example, when we shake hands with someone, sensory fibers in the nerves innervating the muscles of our hand send electrical signals back to our brains telling it both how much pressure we are exerting on the other person's hand and how much ours is experiencing. Our brains use this information to then give electrical instructions to the motor components of those nerves to make their corresponding muscles contract just enough to exert the desired pressure on the other individual's hand. And if that other person decides to squeeze too hard, sensory fibers send a "911" call back to the brain, which can route motor commands to the muscles of the rest of the affected arm to make it resist and pull back from that painful grip.

But while current robotic limb prostheses have made great progress in emulating the motor part of this process, they lag behind in providing the sensory component to our brains. To do this requires tapping directly into the person's nervous system, such as by stimulating nerves within the residual limb or by going even farther "upstream" into the areas of the brain itself responsible for processing sensory information [10–13]. One technique using a hand prosthesis involves implanting "transversal intra-fascicular multichannel electrodes" (TIMEs) into the remaining parts of the median and ulnar nerves in an individual's forearm [11]. These are connected to sensors in the hand prosthesis that provide those nerves with information corresponding to the force being exerted on an object by the artificial fingers. The person can in turn use this sensory information to provide motor commands to the prosthesis to adjust the force of the grip applied.

A recent report [14] involved a somewhat similar method employing "peripheral nerve cuff electrodes" to provide functional touch sensation in prosthetic hands. Each of these cuff electrodes was described as being "an electrically insulating silicone sheath with multiple electrical contacts spaced evenly around the outside of the peripheral nerve." An individual with the prosthetic perceived electrical pulses transmitted to residual peripheral nerves by these cuffs as sensations localized to different parts of the artificial hand. This stimulation helped the person to better control grasping strength and manipulate delicate objects. It also reduced "phantom pain,"[15] part of a spectrum of sensations that an amputee may experience, which includes feeling that an absent limb is present and painful. In its current form this method required that the implanted electrodes be connected to leads extending out from the upper extremity to an external nerve stimulator device. However, a completely self-contained prosthesis incorporating this technology could potentially be developed in the future.

Another technique involves directly stimulating the brain to allow "bidirectional" data exchange (sensory information goes to the brain and motor information comes from it, as in a fully biological limb) in a hand prosthesis [10]. This technique is currently being tested in animal models. It involves placing electrode arrays directly on the part of the brain that processes "somatosensory" data—a general term for the "input" the brain receives regarding touch, temperature, and other sensory information from skin, muscles, internal organs, etc. Initial results suggest that it could potentially emulate a biological limb's ability to detect pressure and thus allow a person to adjust the grip in a prosthetic hand accordingly.

These are all important steps for incorporating data from sensors into the process by which a person can control a prosthetic limb. However, the current "state of the art" is unable to provide anywhere near the full spectrum of pain and pleasure sensations, awareness of temperature, etc., and only some of the "feeling" of pressure and touch to such artificial limbs. There is also another major sensory deficit to overcome—the inability to judge position without visual or other cues. This normally relies on what has been called the "sixth sense"—proprioception [16]. It allows a person to "know" where limbs and the rest of the body are located in space and direct their movements accurately. An individual whose eyes are closed can still "feel" where the arms are and direct the fingers to touch an earlobe.

Thus, even if Luke had sensors in his right hand telling him how much pressure it was exerting on his light saber's handle, without having adequate tactile and proprioceptive feedback he could not be entirely sure his mechanical fingers were actually gripping it without directly seeing that he was. If it were to fly out of his grasp as he swung it backward during a heated battle with an evil Sith Lord, he might not realize it had been lost until he swung his prosthetic hand forward and saw it was empty. That delay could mean the difference between retreating from the enemy and retrieving the weapon using the Force or being on the receiving end of his foe's energy beam.

Likewise, while the best current prosthetic limbs have shown remarkable progress since the 1970s, their overall capabilities still fall far short of those possessed by Steve and Jaime. Their legs could reportedly propel them at speeds over 60 miles per hour as well as enable them to jump to great heights and land from ones that would break mere biological ones. The bionic arms they possessed were also much more powerful than the ones they lost.

However, to make prosthetic limbs better than the originals raises myriad questions concerning the physics involved, the durability and electronic reliability of their parts, how well they mimic the full spectrum of sensory and proprioceptive data that biological ones do, what to use for a power source to run them, how much force would be needed to unintentionally separate them from their interface with the biological limb, etc. For example, while the energy needs of their fictional bionic limbs were said to be supplied by an "atomic power unit," exactly how much electrical energy would actually be needed to use those limbs for even routine purposes such as walking or picking up light objects? How feasible would it be to build such a power unit small enough to provide that energy for long periods? How much electrical energy would be required and how efficiently could it be delivered to the mechanical parts of the legs to allow them to run faster and leap higher than the original biological ones? Could they do that without generating too much heat, exposing Steve or Jaime to dangerous levels of radiation, or have other potentially destructive effects? And might not the bionic limbs have multiple points of failure in their "joints", motors, electronics, etc. that could make them break down at an inopportune time such as while pursuing or being pursued by nefarious characters?

This list of questions could go on and on. The point is that, in the real world, there are practical limitations to what a bionic limb could do. However, we have not yet come close to reaching those limits. An artificial limb stronger than a normal one can be made with the right parts. The J-shaped lower extremity prostheses used for running, though not as functional as "normal" limbs for walking, accelerating, etc., can already let a person reach speeds comparable to athletes using natural ones. Mechanical wrists and fingers could perform motions that conventional ones cannot, such as rotating 360 degrees. Batteries and other power sources are becoming more compact and able to deliver more energy over time, potentially enabling advanced mechanical parts in artificial limbs to exceed flesh-and-blood capabilities.

In short, restoring and even "enhancing" humans with mechanical limbs falls well within the purview of "hard" science fiction, as long as the limitations described previously are taken into account (e.g. development of more durable materials, better power sources, etc.) via at least a passing reference, even if the exact (and perhaps dubious) science behind them is left unexplained. Even if some capabilities of the television versions of "bionic" individuals fall outside the range of what may be physically possible, they can still be enjoyed as science fantasy.

Science fiction works may also combine future advances in several fields to produce more impressive effects than might be possible to each alone. For example, Alan Dean Foster's *The Human Blend* (2010), the first novel in his "The Tipping Point Trilogy," opens with a description of one character fitted with "nanocarbonic" prosthetic legs that include modified muscle fibers and tendons as well as artificially elongated femurs. These modifications are said to allow the wearer to jump longer and higher than the best unmodified athlete, similar to Steve's or Jaime's abilities. A second character has his own modifications, including having half of his stomach and most of his intestines replaced by a "nutrient extractor and maximizer."

The point of using such body alterations in science fiction, which may involve a mixture of genetic engineering, advanced surgery, bionics, and advanced materials science, is not to show exactly how to do them or whether they would work in "real life." In fiction medical issues such as how good and durable the "attachments" between a person's artificial and biological limbs are, the strength of materials in the prosthetic legs, issues with possible development of pernicious anemia and other vitamin deficiencies, fluid balance, diarrhea, etc. in someone who has had much of his gastrointestinal tract replaced, etc. can be reasonably pushed aside in favor of presenting a story. As long as some nod is made to "real science"—done successfully in *The Human Blend* by alluding to how these (to the use the novel's term) "Meld" components are made or simply what they do—the writer can then proceed to go on to present what those modifications *mean* within the story and to its characters.

14.4 The Bionic Eye

Steve Austin's replacement left eye can do things that are closer to what Superman's can do than the ones we possess. It has a built-in zoom lens to provide "telescopic vision," can see much further into the infrared than a typical human eye (thus providing "night vision" and the ability to "see" heat), and tracks fast-moving objects better than a "normal" eye. Steve's new eye also lets him see in a more "conventional" manner.

The good news is that electronics and other "bionic" enhancements have been developed to help restore vision in those who have lost it. The bad news is that current devices can only partially provide only a little of the many functions a biological eye can do—much less make it "better" than it was.

Vision in humans requires an intact eye, optic nerve, and pathways through the brain to specific regions called the "visual cortex." Using our eyes to look in different directions, coordinating the movements of both eyes so that we do not "see double," and focusing our eyes and adjusting how much light enters them (via our pupils) requires that multiple muscles around and within our eyes work properly.

Current "bionic eyes" focus (so to speak) on trying to simulate an essential part of the eye, the retina. This is the tissue that lines the inner surface of the back of the eye. The retina plays a role similar to the film in a "traditional" camera or the light-sensitive detectors in a digital one. It is a complex structure that includes two types of photoreceptor cells, rods and cones. Rods far outnumber cones (about 120 million versus 6.5 million, respectively) [17] and are distributed roughly equally across the surface of the retina. They are more sensitive to light than cones and also help provide "peripheral" vision—the ability to see things we are not directly looking at.

However, rods see the world only in black-and-white. Cones are responsible for color vision and can detect fine detail and rapid motion better than rods. There are typically three types of cones. Each is sensitive to a particular range of wavelengths that we sense as "visible light." They are conventionally called "red," "green," or "blue" cones, although the range of wavelengths they are most sensitive to correspond only very roughly to the ones of those colors. Different shades of color are perceived by differences in how much each type of cone is stimulated by light. Congenital lack or dysfunction of one type of cone is a cause of color blindness, resulting in difficulties distinguishing certain colors (e.g. red and green or, less commonly, blue and yellow).

An important reason why cones have a far better ability to distinguish fine features than rods is that each cone has its own "interneuron" as part of its connection with the optic nerve, whereas many rods connect to a single one. This enables cones to have much greater "spatial resolution"—the minimum distance two separate objects must be for the eye and brain to distinguish them as separate—than rods do.

The eye's cones are heavily concentrated in a very small region, the "fovea centralis," near where the optic nerve exists the eye and approximately in line with the pupil. This location and the unique characteristics of the cones—their ability to detect color and fine detail, etc.—help explain why, under conditions of sufficiently bright light, we have our "best" vision when we look straight ahead. Conversely, because they are far more sensitive to light than cones, rods alone are responsible for what little we can see in the dark. Although they present only a colorless world with much less detail than cones do, the information rods provide still helps us to see large objects and movement to some degree. Their sensitivity also increases over a period of minutes in dim light, "adapting" our eyes to darkness and increasing what we can see.

Electrical signals produced by light striking the retina are processed by the brain to make us perceive the world as "upright" despite light rays being inverted by the lens of each eye. These signals are also needed for the brain to direct the strap-like "extraocular muscles" fastened to our eyeballs to contract in ways that allow each eye to move in the same direction. The overlapping images produced by each eye are combined in the brain into a single one that also provides "binocular vision" and depth perception.

Each eye also has its own visual field—the range over which it can see its surroundings to varying degrees of detail. Both eyes provide their own fields of view based on the way fibers in their respective optic nerves pass into the brain. At a location called the "optic chiasm" nerve fibers supplying the lateral ("ear side") of the retina go to the same-sided hemisphere of the brain. Thus, lateral fibers of the left optic nerve go to the left cerebral hemisphere, and lateral fibers of the right optic nerve pass to the right cerebral hemisphere. However, fibers from the medial ("nose side") of the retina cross at the optic chiasm to the opposite hemisphere. Therefore, medial optic fibers from the left eye go to the right cerebral hemisphere while the corresponding fibers from the right eye travel to the left cerebral hemisphere.

This "wiring" helps us preserve a uniform visual field by having the medial part of the retina in one eye work in conjunction with the lateral part of the other. One cerebral hemisphere processes a single side of our field of view (the medial part of one and the lateral of the other), while the other hemisphere processes the other side. Both hemispheres then act together to produce our complete (both halves) fields of view.

Current "bionic" eyes are just starting to scratch the surface for reproducing these abilities of a healthy eye [17–24]. Prostheses are now available for clinical use to treat blindness in individuals with degenerative diseases involving the retina, such as "retinitis pigmentosa." In the latter condition both rods and cones are destroyed, with vision deteriorating when enough of them are lost.

"Bionic eyes" make use of all the remaining "healthy" parts of the eye and act to restore the retina's lost function as much as possible. One method is to use an "epiretinal" implant—one positioned within the eye ("intraocular") just in front of the retina. The part of this artificial system inside the eye includes a microelectrode array in direct contact with cells connected to the optic nerve [17]. The individual wears a camera attached to an eyeglass-type frame that in turn includes a wireless transmitter and antenna. The rest of the system's exterior components include a small pack containing a processor and battery connected to the parts mounted to the eyeglass frame via a cable. Processed data from the signals produced by the camera are then wirelessly transmitted to a tiny receiver/antenna system within the eye connected directly to the microelectrode array. The latter then sends signals via the optic nerve to the brain for "natural" processing.

A second type of implant is the "subretinal" variety. The intraocular part of this device contains light-sensitive photodiodes, amplifiers, and a microelectrode array, with the latter inserted just beneath the retina itself. Light passing through the lens of the eye is detected, analyzed, amplified, and then sent through the retina's cells to the optic nerve for further processing. The implant system also includes a small receiver placed under the skin near one ear. The array within the eye is connected directly to that receiver via a thin cable. The latter device is in turn wirelessly connected to an external power supply and control unit that the user can adjust to increase the brightness and contrast of what is "seen" using the system. The receiver obtains power and control signals via a small antenna affixed to the scalp near it using a small magnet placed beneath the skin, with the antenna having a corresponding magnetized area to attract and attach to it.

A third method uses a "suprachoroidal" implant. In this technique the array is positioned between two tissues "behind" the retina, the "choroid" layer adjacent to the retina and the "sclera" (the "white" part of our eyes) that forms the outermost layer of the back our eyes. Otherwise this system is similar to the subretinal type.

Current implants allow an individual who would otherwise be blind to at least detect the presence of light and distinguish overall shapes. Visual acuity may at best be about 20/1400 (a person can see only at 20 ft what a "normal" eye could see at 1400 ft) for some types of implants [23]. A recently developed subretinal implant model [25] has reportedly bettered this to about 20/546 by increasing the number of independent "microphotodiode-amplifier-electrode elements" on the "chip" placed in the eye.[3]

However, none of these devices restores color vision—only shades of gray can be "seen." Having an implant in only one eye when both are blind also means the depth perception produced by binocular vision (using both eyes) remains absent. These implants currently have a total area (e.g. 9 mm^2) [24, 25] that is a small fraction of that of the retina itself, thus producing "tunnel vision" with a much narrower than usual field of view.

There are practical limits to the size and density of microelectrode arrays, as well as how strongly they can stimulate the cells that transmit signals to the brain. Those cells could be damaged by electrical impulses that are too powerful or if the array generates excessive heat in its operation. Such implants also require that the retina itself be otherwise healthy except for its loss of rods and cones, and that all the neurological pathways (e.g. the optic nerve) to the visual cortex and that part of the brain itself are intact. If there were any disruption of those other components, a retinal implant or any similar eye-emulating device would not work.

Overall, the weakest link in recreating or improving at least some abilities of the human eye is not in present-day, much less future microcircuitry and electronics. A more sophisticated version of eyeball-shaped webcams could be much more sensitive to light than our own eyes. It could also "see" over a much wider range of the electromagnetic spectrum, well into the infrared and ultraviolet, than its "natural" counterpart and provide a much wider range of "colors" to perceive. Electronic magnification could, like Steve Austin's bionic eye, give it telescopic vision. Appropriate components might also give it the ability to have a much wider field of detailed view than our eyes do.

On the other hand, how well the brain could process information far beyond its usual "design specs" from such a high-tech video camera is uncertain. However, the human brain possesses a significant degree of "plasticity"—the ability to change its functions to "make sense" of never-before encountered or otherwise different stimuli. What it would "feel" like to see deep into the infrared could only be subjectively experienced by someone with that ability, but there is at least a possibility that the brain could change enough to make it possible to do that.

Instead the greatest challenge for such electronically enhanced vision may lie at the junction between machine and brain. One could imagine high-tech artificial eyes, such as the ones Geordi La Forge had implanted to replace his VISOR sometime before the events of *Star Trek: First Contact* (1996), that

---

[3] This model also uses a wireless system with a small external coil connected to a battery pack that sends power and control signals to another, implanted coil. The latter then sends those signals to the subretinal implant itself via a cable.

could be plugged into the equivalent of Ethernet jacks in the back of our eye sockets ("orbits").[4] Better yet, they might be fitted with an "Eject" button and the futuristic version of a Bluetooth connection to the visual cortex. That way one could be popped out into your hand and turned as needed to see what is around and behind you as you walk facing forward. Such easy removal would also facilitate repairing or replacing such a bionic eye.

However, it is this interface between the electronic and biological components of a bionic visual system—whether the retinal implants described previously, or up to a fully artificial eye—that may represent the greatest challenge to success. The optic nerve and other components of the visual pathway deeper in the brain—the optic tract, the lateral geniculate nucleus, and optic radiation—must be intact and able to receive, encode, and transmit the data received from electronic signals to the visual cortex. In a healthy biological eye this requires many millions of microscopic nerve endings interacting with the retina's rods and cones.

Designing a connection, either direct or otherwise, between an artificial eye and all those myriad nerve endings that is able to transmit the huge volume of complex, constantly changing data representing "normal" or more than normal sight represents a huge technical challenge [26]. The microelectrode arrays in the retinal implants described previously show that such stimulation of the optic nerve can be done, but with severe limitations (e.g. no color vision, the size of the array being much smaller than the overall area of the retina itself, etc.) The amount of information they can transmit to the brain is far less than a healthy eye normally does. Moreover, here too the strength of the electrical stimulation delivered to the cells of the retina must be low enough to avoid damaging or destroying them, further limiting what can be done.

Another potential technique would be to bypass the optic nerve and either stimulate other components of the visual pathway leading from the eye—the optic tract, etc.—or go right to the visual cortex and interface directly with it instead [27]. Electrical stimulation of the visual cortex can produce a sensation of light. Thus, in theory a sophisticated enough means of providing that part of the brain with appropriate signals might work. One could imagine a high-tech equivalent of a fiber-optic cable connected to an artificial eye at one end and attached directly into the visual cortex of the brain at the other, with the cable having enough "bandwidth" to reliably transmit all the complex data needed for "sight" between them.

---

[4] The VISOR itself did, in fact, appear to interface with (presumably) Geordi's optic nerves or more directly with another anatomically appropriate part of his central nervous system via contacts at his temples.

Unfortunately this approach presents its own problems for reproducing all the sensations that we perceive as sight. The many millions of nerve endings connected to the retina are localized to one place, "bundled" within the optic nerve. In contrast, when these "connections" from the rest of the visual pathway reach the visual cortex they spread out within a much larger and complex three-dimensional structure composed of many hundreds of millions of neurons, making any sort of interface more difficult. Moreover, the visual cortex is not functionally uniform but has areas that are (at least relatively) specialized for different tasks, such as color perception, orientation, detection of motion, etc. Mapping exactly which neurons are responsible for receiving and processing all the complex information they receive from the eyes and then interfacing with them in a functional way would be very challenging.

In short, an artificial eye would have to not only acquire information corresponding to the light reaching it but also code that data in the form of analog or digital electric signals that the visual cortex could receive and decode—somewhat analogous to a computer network. But while the latter uses equipment and data processing protocols we humans have devised, our understanding of exactly how our brains process vision and ways we can successfully and safely "tap" into it are still very limited. Unfortunately Steve Austin will have to wait a while before he gets his new and improved bionic eye. On the other hand, hopefully enough progress will be made in the several centuries before Geordi La Forge is born to help him "see" by the time he is ready to join Starfleet.

Nancy Kress's story "Someone to Watch over Me" (2014) posits another variation on the concept of melding electronics with the human eye. In it a woman has the equivalent of a miniature video camera with wireless transceiver implanted beneath the cornea of one of her 6-month-old daughter's eyes. The purpose of that device, virtually invisible from the outside, is to spy on her estranged husband's activities with his new "significant other" when he has custody of the baby. The woman becomes obsessed with watching the video it transmits to her private monitor screen. Given adequate advances in electronics, this story—a successful blend of technology and psychology—might well be within the realm of possibility.

14.5 The Bionic Ear

In contrast to the cloudy future for artificial eyes, the prospects for Jaime Sommers's bionic ear sound much better. In fact, one type—the cochlear implant—has been in clinical use for several decades, with hundreds of thousands of people receiving it [28–32]. This device is used by individuals who are deaf or have profoundly impaired hearing. Like retinal implants, cochlear implant systems generally have both internal and external components. The user typically wears a small outside unit containing a microphone, speech processor, and transmitter hooked over the ear. A small receiver is implanted beneath the skin near the ear. It receives processed signals from the transmitter and then, via a stimulator, sends those signals to an electrode array just behind the eardrum. This array then stimulates the auditory nerve, which in turns sends those signals to the brain.

Cochlear implants can enable the hearing impaired to both hear and understand speech. However, the way these devices stimulate the auditory nerve is different from how the human body's own internal structures do it. Our outer ears consist of the "pinna," or "auricle"—the fleshly flap that we conventionally call our ear—and the ear canal that leads to the eardrum, or "tympanic membrane." The latter acts as the outer boundary of the "middle ear." The eardrum is connected on its inner surface to three tiny bones—the smallest in our bodies—called the stapes, malleus, and incus. These transmit the vibrations that sound waves produce on the eardrum to a membrane that divides the middle ear from the fluid-filled inner ear. The major structure there is the cochlea, shaped in a spiral form similar to a snail's shell. Inside it is the "organ of Corti," which acts as a transducer converting mechanical energy (e.g. vibrations produced by sound) into electrical signals. The auditory nerve is connected to the cochlea and sends these impulses to the appropriate regions of the brain.

The cochlea has a complex method for detecting different frequencies or "tones." A membrane within it vibrates in resonance with different frequencies, and this information modulates how "hair cells" within the cochlea send signals to the auditory nerve. As the name suggests, the electrode array of a cochlear implant works at the level of the cochlea, emulating some of its functions. Unlike the "natural" way this is done, the signals sent to the auditory nerve that represent lower and higher tones are related to the electrode array's position on and the depth of electrode contacts within the cochlea—e.g. contacts inserted deeper inside it correspond to lower tones and more superficial ones to higher tones.

The electrical stimulation the auditory nerve receives from a cochlear implant can provide a person with a useful model of external sounds and speech. However, because the type of stimulation is different from the "standard" biological method, the individual's brain has to "learn" what those signals mean and decode them into words and other information. The fact that the auditory cortex and other areas of the brain can do this well enough to greatly increase understanding of spoken words and other types of sounds is another strong reflection of the brain's plasticity [33].

But a cochlear implant also has limitations compared to a healthy biological auditory system. For example, complex patterns of sounds such as music present difficulties for it [34]. However, the ability to distinguish and "understand" melodies can improve with practice due, presumably, to modifications in the brain's ability to process that information. Also, the range of frequencies a cochlear implant could detect is restricted by the intrinsic limitations of the cochlea and its components. While dogs, bats, and other animals have inner ears "constructed" to hear higher frequencies, ours are made to operate within, at best, a range of between roughly 20 and 20,000 Hz (aka cycles per second). Thus, any hope of "hearing" far into the ultrasonic range will require other types of modifications to the human body and different techniques.[5]

Near-future advances could significantly improve cochlear implants by further miniaturizing and "internalizing" more of their components [35]. The auditory nerve is also not the only site that can be stimulated to produce "hearing." Individuals who have experienced damage to that nerve itself can receive "auditory brainstem implants." These stimulate a deeper part of the auditory pathway within the brainstem, the "ventral cochlear nucleus," using electrodes placed within it. Current devices and techniques provide some improvement in ability to "hear," but speech recognition and similar functions are generally not as good as with cochlear implants [36].

Recent "proof of principle" reports suggest other possible avenues for improvement [37]. An artificial ear made using 3D printing techniques and cartilage can have electronic circuitry integrated into it. Such a device could have "built-in" enhancements such as the ability to receive radio frequency (radio) signals.

Overall, it seems likely that advances in restoring and, eventually, perhaps improving hearing above "normal" standards will come sooner than ones for vision. As complex as they are, hearing and the processes associated with it are not as complicated as the visual system is. And even if current "bionic ears" are still not as good as the "original" version, they are getting closer.

14.6 Other "Bionic" Body Parts

The human heart can be aided or replaced by various devices. Pacemakers can either partially or entirely take over for the heart in producing and regulating its rhythm. They are used in patients whose own hearts either intermittently or always beat too slowly to pump an adequate supply of blood to their

---

[5] Or, as in J. G. Ballard's 1960 short story "The Sound-Sweep," we could just learn to better experience and "appreciate" ultrasonic music without actually hearing it.

bodies. Depending on what kind of problems the heart has "mechanically" and/or with its electrical system, pacemakers can stimulate that organ to contract in a manner up to a level roughly equivalent to its "normal" operation. In some patients whose left ventricles pump too weakly and out of synchronization with their right ventricles, pacemakers can also be used to better coordinate how those two ventricles work together and improve how well their hearts function overall.

Some early pacemaker models, first developed in the late 1950s, were at least partly "external," with electrodes attached to the outer surface of the heart and connected via wires through the skin to an external control device. However, contrary to scenes in *The Outer Limits* episode "It Crawled Out of the Woodwork" (1963) these types of "wearable" devices were not used to restore anyone to life after being literally scared to death by a malevolent energy being.[6] On the other hand, an individual totally dependent on such a device to maintain heart rhythm could indeed die if it were suddenly disabled.

Fortunately modern pacemakers are both fully implantable—typically placed under the skin beneath one clavicle (collarbone)—and much more reliable. Depending on the medical problem(s) being treated, up to three leads (plastic coated electrode wires) are passed through a vein into the right side of the heart and connected to a (typically) disk-shaped pacemaker generator. Current models commonly function 5–10 years before their built-in generators/battery systems require replacement. Pacemakers also have sophisticated electronic circuitry/computers that enable them to be wirelessly reprogrammed by an external transceiver, autonomously adjust when and how fast they fire based on the heart's own rhythm and a person's activities, and record information regarding the heart's rhythm for transmission and review by medical personnel. They are sometimes combined with the additional capabilities of an "implantable cardioverter-defibrillator" (ICD) that can sense when an abnormal rapid heart rhythm (tachyarrhythmia) in a ventricle (e.g. ventricular tachycardia or ventricular fibrillation) occurs and take steps to treat it (e.g. by delivering one or more "shocks" to the heart, or briefly pacing the latter more rapidly to interrupt the tachyarrhythmia).

Mechanical devices that help or replace the heart itself are also in clinical use. In some cases the left ventricle, right ventricle, or both are aided by an implanted "ventricular assist device" that acts as a mechanical "pump." Earlier total artificial hearts such as the Jarvik 7, used in the 1980s, required large

---

[6] Or at least the professors who taught me in medical school never mentioned this risk. Fortunately I have not yet encountered any patient who has suffered a cardiac arrest due to such a creature. However, despite my professors' silence on the subject, if such an unfortunate individual ever does come to my hospital I will be able to make the diagnosis and render appropriate care based on what I learned from that science fiction show. Who says television isn't educational…

external machines to run them that, of necessity, severely limited a patient's mobility. One current model approved for use in the United States is connected via tubing passing through the individual's skin to external battery and driver systems. These outside components can, however, be light enough to be worn in a backpack or shoulder bag, allowing the patient increased mobility [38, 39]. While this is not as convenient as the completely implanted model Captain Jean-Luc Picard has (as described in the *Star Trek: The Next Generation* episodes "Samaritan Snare" and "Tapestry" from 1989 and 1993, respectively), that model can be used for at least several years pending transplantation of a biological heart.[7] Completely implanted artificial heart systems, with the power supply and controller all inside the body, are also being used on a limited basis [40]. Their batteries need to be recharged periodically using an outside charger that sends energy through the skin to the power supply. However, all current types of total artificial hearts require that patients take blood thinners, with their attendant risk of bleeding problems.

There is wide variability in the need to replace or take over the functions of other body organs. Work is currently being done to develop artificial, implantable lungs, but these are not yet available. Dialysis machines, although far from an implantable solution for individuals with severe kidney failure, can effectively replace that organ's primary filtering and waste removal functions. Efforts to create replacements for the liver and pancreas focus on biological means such as using stem cells to grow parts of new ones, although the use of non-biological devices such as insulin pumps can at least partially replace that particular function of the pancreas. There does not appear to be a great need for a bionic gallbladder or spleen, since their functions are helpful but not absolutely essential for health. Devices to stimulate the nerves of the bladder to help control its function are already in use, and others to improve motility of the stomach and intestines are also in development.

"Advanced" forms of such "artificial" organs also find a place in science fiction. In Kevin J. Anderson's novel *Resurrection, Inc.* (1988) a company employs synHearts, synEyes, synLivers, and synLungs as needed to reanimate the dead. While doing the latter is not yet part of routine medical practice, presumably such synthetic organs could also be used to heal the living. However, the details of how they are constructed and work are not described in the

---

[7] It is puzzling, however, why Captain Picard still has an artificial heart instead of a transplanted biological one. As noted in Chap. 13, heart transplantation has been a well-established procedure since the late 1960s. Presumably by the twenty-fourth century current remaining issues such as potential rejection of that organ, the need to take immunosuppressive medications, and development of cardiac allograft vasculopathy would have been solved. Likewise techniques for using stem cells or other means to grow a new biological heart might be standard practice by that time. This may be a situation in which drama trumps strictly medical considerations.

novel, and as noted previously it will at best take some time for clinical reality to catch up with imagination.

14.7 Connecting Brains to Computers and Machines

As important as all these other developments are, from both a "real world" and science fiction standpoint a topic of particular interest is fusing electronics with our central nervous systems, particularly our brains. Along with all its complex biochemistry and myriad functions, the human brain and the rest of the nervous system "run" on electricity. Beyond that, however, there are many differences in details between how those parts of our bodies work and their electrical and electronic analogs.

As noted previously, in the simplest terms our peripheral nerves and spinal cord act as "wires" carrying electrical impulses to and from the brain. However, those parts of the nervous system are far more complicated in how they are constructed and work than the copper and other metallic wiring in a house or power cord. Though not as complex as those in the eye, their connections with muscles, internal organs, skin, etc. are much more intricate than those needed to transfer power from a wall outlet to a refrigerator or desktop computer. The axons and their associated cells that provide the body's "wiring" are both microscopic and numerous. They are capable of modulating their activity in a variety of ways based on many types of chemical and physical stimuli, including those that we "will" them to do.

Those characteristics of peripheral nerves and the spinal cord are magnified and expanded by the most complex part of our nervous system, the brain. Although our ability to either emulate or "eavesdrop" on its most complex functions is still very limited, we can detect, record, and use the electrical energy it produces at very basic levels—and employ those "cerebral signals" to interface with computers and machines.

For example, there are now clinically available devices used to stimulate particular parts of the brain to help control abnormal electrical activity and systems within it [41]. Thus, implanted "deep brain stimulators"—similar in some ways to pacemakers—can improve motor function (movement) in individuals with Parkinson's disease, significant tremors, and other problems that impair their ability to move. Such devices work by partially restoring control functions that certain parts of the brain have lost. Likewise, in some people with severe seizure disorders stimulating appropriate areas of the brain interrupts the abnormal activity of the neurons responsible for seizures and either suppresses or terminates the latter.

The term "neuroprosthetic" describes a device that is electrically "coupled" to the brain or other parts of the nervous system, sharing information and control between them in either a one-way or bidirectional flow. Electrical impulses from the brain can be detected and interfaced with external devices by various means. These methods include some type of electrode system, a signal processor to "interpret" and direct signals generated or received by the brain, and an "effector"—an electronic device (e.g. a computer), mechanical device (such as a robotic limb), etc. with which the brain ultimately interacts. The result is a "brain-computer interface" or "brain-machine interface" system [41–45].

As demonstrations of what can be accomplished with current equipment, monkeys have been trained to use one or two robotic arms at a time, [46, 47] and move a computer cursor by "thought" alone [42]. Humans have also operated prosthetic limbs and performed computer-related tasks such as moving a cursor, opening e-mail, operating a television, and even used a computer screen to accurately spell words [48]. Such interfaces could also be used to help an individual with motor impairments due to a cerebrovascular accident ("stroke"), spinal cord injury, a motor neuron disease[8] (e.g. amyotrophic lateral sclerosis), or other causes to operate a motorized wheelchair or other machine by "thinking" alone [41].

One of the least "invasive" ways of harnessing the brain's electric signals is with an electroencephalography (EEG, also "electroencephalogram") machine [41, 45, 48, 49]. Electrodes are positioned along a person's head and used to detect electrical activity from different parts of the brain through the overlying scalp and skull. While this technique does not require placing anything within the body, it has significant limitations. EEG detects firing from multiple groups of neurons from relatively large volumes of the brain, making it much more difficult to separate the desired signals from other ones that, in this context, represent "noise."

The net effect is that an individual trying to use EEG signals to control a computer or machine requires extensive training and mental effort to do it. Even with practice the response time between when a person "thinks" a command and how long it takes a device to respond is delayed due to the time required for appropriate signals from neurons to pass through the skull and scalp to the electrodes, then be processed by the external EEG apparatus. Moreover, the electrodes and other parts of the EEG system can also respond to electric fields in the environment produced by wiring, other equipment, etc. instead of those produced by the person's brain, thus reducing control of an "effector." Even the user's eye and head movements can produce electrical artifacts that interfere with the actual brain signals detected via the EEG.

---

[8] Physicist Stephen Hawking is a well-known individual with this type of disease.

Some of these issues can be addressed by getting the electrodes closer to the brain. "Electrocorticography" (ECoG) involves placing an electrode array inside the skull, against a part of the brain involved with a certain function. The array is positioned beneath one layer, the dura mater, of the tissue (meninges) lining the brain without actually penetrating the latter. ECoG thus allows "finer" control of external devices than can be accomplished by an EEG. Cerebral signals are also no longer attenuated and delayed by passing through the skin and scalp, and more of the signals received by the electrode array originate from the area of interest rather than surrounding ones.

However, while the signals received are better than those from an EEG, they are not as high a quality as those obtained by going deeper into the brain (more on this shortly). Also, placement of the array requires an operation going through the skull, with all its attendant risks of anesthesia, infection, etc. Even after a successful operation the presence of that foreign body could trigger the body's immune system against it, stimulating development of inflammation and other effects [41, 45].

Microelectrodes can also be placed adjacent to single neurons in the brain to record each one's activity, or that of electrical "spikes" generated by a small number of neurons near the electrode ("local field potentials"). As expected, getting closer to the source of electric "action" in the cerebrum can potentially deliver the highest quality neural signals delivered to the equipment controlling a neuroprosthetic. However, besides the greater invasiveness of this approach there are significant issues with long-term stability of this "connection" to neurons. The presence of a microelectrode also provokes an inflammatory response from surrounding cells that can result in its "encapsulation," effectively decreasing its "connection" with one or more neurons and degrading the signal it sends to external electronic parts of the control system. The position of microelectrodes can even be altered significantly by the normal slight motion of the brain within the skull caused by an individual simply moving his or her head.

Other potential issues with implantable electrodes and arrays is that the leads that connect them to the outside may fracture (break), they may malfunction in other ways, and they present obvious difficulties with upgrading. There are good reasons why the human skull does not have the equivalent of a zipper to obtain easy access to the brain. Unfortunately this absence does make placing and replacing implants considerably harder.

Overall, as with ocular and auditory implants, the interface and characteristics of human anatomy present intrinsic limitations to being coupled with electronics. The materials used for electrodes also play a part in how well they "connect" with the brain. Electrodes can be fashioned from thin metal wires or use nonmetal substrates (e.g. silicon, ceramic, etc.) coated with a metal. In an overlap with nanotechnology, single electrodes and electrode arrays made of tungsten, stainless steel, or other materials that are then coated with carbon nanotubes may prove useful for creating an effective brain interface with outside machines or computers [50].

Besides these other issues, the living brain itself has a consistency that can be described as "spongy," making it more difficult to achieve a physically and electrically stable interface with implanted wires, arrays, etc. Also, being a three-dimensional structure, some areas of interest in the brain do not lie on its surface but deeper within it. Going into those areas from the outside at the macroscopic level presents obvious risks of injury. Any heat generated by or electrical activity associated with an electrode should also obviously fall below the threshold for injuring neurons.

Likewise, reaching the desired areas internally from blood vessels (e.g. arteries) greatly limits the size of the devices that can be used, including any of the potential nanobots or similar nanoscale machines etc. described in Chap. 11. Such an approach would also face challenges in crossing the blood-brain barrier and making sure that even an appropriately tiny device traveled through the appropriate branches of a complex maze of capillaries to reach the "right" part of the brain.

Another approach to get deep into the brain would be via the spaces called "ventricles" within it. These are part of the system through which cerebrospinal fluid circulates around the brain and spinal cord. However, this approach also presents great difficulties getting electrodes and devices where they "belong" without damaging nearby structures.

A further problem is that, although certain areas of the brain are known to be associated with motor and sensory functions, for more complex activities multiple scattered areas of it might need to be activated. This raises the question of just how much "hardware" must be put inside the head and how complex it has to be. For certain "simple" actions, such as moving a robotic limb via thought, an area such as the "primary motor cortex" is localized enough that interfacing with it could produce effective motions of an artificial limb. Learning how to use a brain-machine interface for more complicated tasks, however, may involve multiple and specific areas within a region of the brain [51, 52].

Science fiction stories may go far beyond these (by comparison) "straightforward" functions of the brain to ones that are considerably less well understood. For example, devices capable of acting synergistically with or "taking over" the brain's far more complex activities, such as those described in terms such as "consciousness" or as "memories," are extrapolating far beyond current capabilities.

Further miniaturization of internal devices used on or inside the brain as well as the development of materials less likely to invoke an inflammatory response may reduce some of these issues. This is an area where advances in medical nanotechnology in particular might prove helpful. This type of technology could involve creation of new materials for microelectrodes to increase their "biocompatibility" and efficiency, make their delivery to the appropriate parts of the brain easier, and perhaps reduce the size of sensors, microprocessors, etc. to the point where they could more effectively interface with the brain.

The leap that some science fiction makes to self-contained brain implants (or perhaps their nantotech equivalents) that require no external machinery or obvious power source to work is also well ahead of current capabilities. Implants with built-in telemetry circuits and a capacitor to help provide power have been suggested [53]. Also, while wireless communication may be a near-future possibility, for now the individual microelectrodes and arrays located inside the head must be literally "wired" to signal processors and effector equipment outside it to communicate with them.

One way this is currently done for long-term applications is to connect an internal array via a thin wire cable to a metal stud. This stud, located on the outside of a person's head and anchored to the skull, [41] can then act as a connection point for external equipment. Such an arrangement presents risks such as infection, local skin reactions, dislodgment, bleeding, etc [53]. However, its potential benefits may, especially as associated technologies improve, ameliorate these issues.

Any future implants small and complex enough to work as independent units or wirelessly with external equipment will also need some way to supply power to them. This might require further advances in battery technology, storage of sufficient power in sufficiently small capacitors, systems capable of inductive charging from outside the head, or other methods—all of which, once again, should not generate enough heat to or otherwise damage nearby brain tissue. Another potential way to power tiny implants in or near the brain is to have them "live off the land." For example, one proposal envisions implanting a fuel cell that uses glucose within the cerebrospinal fluid surrounding the brain to power very small neuroprosthetics located there [54]. At present, however, the feasibility of using such techniques in humans has not yet been demonstrated.

For now, clinical use of implants within the brain focus on uses described previously, such as managing otherwise intractable seizures, controlling motion in patients with Parkinson's disease and other movement disorders, and helping individuals with severe neurological deficits (e.g. spinal cord injuries resulting in quadriplegia) to better interact with their environment. Other possible uses include helping individuals who have disorders of consciousness in which they are in a "locked-in" or in an apparent "vegetative" state—conditions in which they cannot effectively speak or otherwise physically interact with the external environment—to communicate using a noninvasive EEG brain-computer interface or more invasive methods [55–57].

Brad Aiken's story "Locked In" (2010) is a realistic exploration of how near-future technology might allow an individual with locked-in syndrome to communicate and regain a measure of mobility.[9] This syndrome is typically associated with a blood clot (thrombus) occluding part of the basilar artery[10] in the brain, resulting in damage to an area of the brainstem called the "pons." As depicted in the story, initial treatment can include removal of the clot using a special catheter ("endovascular thrombectomy"), with infusing medications (e.g. "thrombolytic agents") into a vein or artery to "dissolve" the thrombus being an additional method.

Unfortunately permanent damage to the pons can occur if its blood supply is not restored soon enough. As shown in "Locked In," affected individuals can be fully conscious. However, they cannot speak and are unable to move any of their limbs ("quadriplegia") and either all or nearly all of their other "voluntary" muscles (those under conscious control) [58]. The muscles of the eyes are typically (but not always) spared, however, allowing those patients to blink. The story shows a character with locked-in syndrome reestablishing contract with the world by means of a brain implant coupled to a voice synthesizer and capable of interfacing with external computer networks. He also receives a high-tech wheelchair operated by his thoughts.

In short, this work is an example of science fiction that does an excellent job of using "real" medicine and knowledge of the current state of research in brain-computer interfaces to describe advances that are not only plausible but that may well reach clinical practice in the near future. With such a firm foundation in "actual" rather than "speculative" science it can then proceed to show how that technology affects its characters and the world they live in.

14.8 Uploading Our Minds

As mentioned in Chap. 9, one potential way that what we think of as our "minds" might outlive our "birth bodies" is to transfer them to a new biological one or "upload" them to a synthetic analog. In simple conceptual terms,

---

[9] The fact this story is so accurate in its medical details and scientifically "solid" in its speculations is not surprising given that the author is a physician specializing in physical medicine and rehabilitation.
[10] This artery and the vertebral arteries that it derives from are described in Chap. 2.

the latter might be thought of as transferring the processes and encoded information that encompass our thoughts, personalities, memories, reasoning and artistic abilities, etc.—all the things we subjectively subsume under the pronoun "I"—to another medium with a similar (or perhaps better) way to store and use them.

An early example of this idea is described in Frederick Pohl's story "The Tunnel Under the World" (1955), which envisions transferring the patterns of human minds directly into miniature robot bodies. The concept also figures in various ways in a wide range of written and media-based science fiction. For example, Walter Gibson's *Neuromancer* (1984) describes memory chips holding the "saved" consciousness of a person. The plot of Gregory Benford's *Eater* also involves transferring a person's mind into a computer. Robert Silverberg's *To Live Again* (1969) posits the ability to "record" a person's personality and even download it into another individual. Greg Egan's *Diaspora* (1997) presents a future society dominated by the non-corporeal "software" analogs of previously living biological humans. Robert J. Sawyer's *The Terminal Experiment* (1995) includes the creation of variants of a character's personality in computer simulations, while his *Mindscan* (2005) deals with uploading one's consciousness into an artificial body.

In the movie *Avatar* (2009) a person's mind can be incorporated into a non-human biological body, while the *Doctor Who* episode "Forest of the Dead" (2008) portrays deceased individuals each of whose consciousness has been "saved" into a computer. This concept could also be employed to achieve "immortality" by transferring one's thoughts and memories into a cloned younger body, as the alien Asgard do in the television series *Stargate SG-1*. A related idea is that of mind transfer, with two characters temporarily "exchanging" minds, as shown in *The Outer Limits* episode "The Human Factor" (1963) and the final first-run story of the original *Star Trek* series, "Turnabout Intruder" (1969).

The concept that a person's mind can be effectively simulated in, duplicated, and/or transferred into a different medium (biological or otherwise) makes some intrinsic assumptions about what a "mind" is and how it works. First and foremost is that we know enough about what a "mind" is, at least from a functional standpoint, to emulate or manipulate it as if it were (in a very crude analogy) like a software program that can be copied intact to another computer and even have its "code" changed or "improved." This can also be thought of as a "mind as information and information processing" model that, in some ways, makes it seem to be something that can be effectively "disembodied"—that is, to exist, though perhaps be unable to function independently, of a physical medium such as a body or computer.

However, while the idea of a mind existing independent of a "thing" it is associated with, if only as information, can be imagined, the question is how much or even if that concept has any functional meaning. While not necessarily the same in details, thinking about what mind uploading and transfer might entail could evoke the image of the "ghost in the machine." Whether that "ghost" is thought of as a self-contained entity, complex data and information patterns alone, fits one of the definitions of a "soul," etc. the common assumption in these concepts is that a "mind" can, at least in some sense, be separated from the original substrate of our bodies. Subjectively each of us feels as if as the "I" that is considered our "self" is in charge of running a particular, finite conglomeration of muscles, bones, organs, etc.[11] That relationship is a bidirectional one, with both external and internal changes in that body affecting how "I" think and feel (e.g. stubbing a toe or becoming hungry).

This concept raises many more questions. For example, what exactly does constitute a "human mind" in general and an individual one in particular? How much of what it is and does do we really understand? What actual methods would be needed to extract, duplicate, or at least model all the processes that go on our in brains to effectively create, separate, and/or transfer an individual consciousness? Even if a mind were "only information," how can we take a "snapshot" of what is a fundamentally dynamic and ever-changing process based on the interactions of billions of neurons, trillions of synapses, myriad neurotransmitters and other chemicals, etc.? Even if we had all that information, how could it be translated into the much different "language" and processing methods a computer would use? Or, if the goal were to "imprint" that information onto another human brain that was "tabula rasa," how would one make that brain's neurons both physically form new synaptic connections and function exactly the same as the original's based on that data, recreate the latter's memories and precisely how his or her limbic and other individual brain systems worked, make the new brain's cells respond to neurotransmitters and hormones identically to how the first one's did, etc.?

Thinking about potential answers to these and other questions extends beyond the medical/scientific realm into that of philosophy and theology. Whatever the human mind "is," its activities are centered in the brain. A purely mechanistic view would be that the brain and the various biochemical, electrical, and other actions within it *is* the mind. Assuming such methods are technically and not just theoretically feasible, reproducing or simulating a particular brain's activities would effectively create a duplicate mind. On the

---

[11] Actually that is the way it (correctly or incorrectly) feels to me. I can only infer you feel that way too. There is a reason why solipsism and other philosophical conundrums are perhaps best left to professional philosophers. Thinking about them too much might confuse us non-professionals and distract us from dealing with the real world of everyday life—assuming, of course, that there actually *is* a real world...

other hand, could the brain also work in essential ways that we are not yet aware of, or do our minds have characteristics that go beyond brain function alone and what we call "natural?"

The "ultimate" answers to such questions must, for now, remain speculative. Nonetheless, regardless of whether the human brain and mind are synonymous or the latter is, in some way, "more" than the former, as a pragmatic point the scientific study of the "mind" is limited to the matter and energy that compose the brain and how that organ works. Even within those limited material parameters our current knowledge of what the brain is and exactly how it performs its functions is very incomplete, particularly regarding its "thought" processes. Moreover, far from being a relatively homogeneous organ like the liver or kidney, what we think of holistically as the "brain" is actually a compendium of many different components with a large number of diverse activities. It is more analogous to a whole "city" composed of many sundry and diverse "citizens," each performing various tasks (hopefully) in harmony with the others, rather than a solitary one.

Large books can and have been written giving the details of what is currently known about brain anatomy and the functions of its various regions. This section can only scratch the surface of this subject and give a general idea of how complex that organ is. As described in Chap. 2 the brain can be divided into different gross anatomic structures. These include but are not limited to the brainstem (with its own various divisions such as into midbrain and hindbrain), the two cerebellar hemispheres with their lobes, the cerebrum (composed mainly of the cerebral cortex) with its own pair of hemispheres, and systems made up of interacting "subcortical" (beneath the cerebral cortex) structures such as those that comprise the limbic system.

Conscious and particularly abstract thought, language, and other "higher" functions involve the frontal lobes of our cerebral hemispheres but also other structures throughout the brain with their own particular activities/roles. For example, the limbic system includes the amygdala, hippocampus, mammillary bodies, hypothalamus, anterior thalamic nuclei, parahippocampal cortex, and many other substructures associated with emotions, memories, reproductive behavior, and other at least partially "non-thinking" functions. The basal ganglia include many other anatomic areas (e.g. the striatum, substantia nigra, pallidum, and nucleus accumbens) involved with behavior, initiating movement, motivation, learning new skills, etc.

Thus, instead of only being similar to the single or multi-core CPU of a conventional computer or even standard parallel processing systems, the human brain has many very different, specialized processing units that interact among each other in ever-changing and complex ways. These parts of the brain share some common characteristics, such as the presence of neurons

connected in intricate patterns via synapses. However, there are wide differences among the types of neurons and supporting cells in various regions,
many different ways to make synaptic connections, and a wide variety of
neurotransmitters (e.g. dopamine, serotonin, GABA, glutamate, to name a
few) that are present in different areas and have differing effects. The "analog"
aspects of synapses, with responses and electrical activity modulated by how
much of a neurotransmitter is released and taken up at synapses, also greatly
increases the complexity of how our brains function and their capabilities.

Modern day supercomputers already exceed in at least some ways the raw
computational capability and memory storage of a human brain, although the
latter (at least for now) remains more efficient in its energy needs and much
more compact pound-for-pound than its silicon competition. An inexpensive
calculator far outstrips a typical person's ability to do mathematical calculations, and more sophisticated devices and software can perform algorithmic
assessments, combine inputs in different ways, and "judge" different alternatives well enough to write music and other "creative" functions. However,
making calculations and both formulating and following algorithms are only
part and by no means the whole of what the human brain does.

Far more is going on "under the hood" of both our brains and bodies besides our conscious thoughts and memories. This includes not only what is
loosely called the "subconscious" but the activities of the aforementioned limbic system, the autonomic nervous system that helps regulate actions that
are at least partly involuntary (e.g. heart rate and digestion), maintenance
of balance and position (including proprioception), the effects of hormones
directly on our brains or indirectly by their effects in other parts of the body,
etc. For example, our brains and the rest of us need appropriate blood levels
of thyroid hormones (T_4 and T_3) to maintain health, including our ability to
think clearly. The effects of sex hormones such as estrogen, testosterone, and
others have both obvious and subtle effects on how an individual's brain functions. Our emotions and sensations are also intimately connected in feedback
loops with our physiology as well as specialized parts of the brain such as the
limbic system. For example, fear makes us secrete more catecholamines such
as epinephrine (adrenaline) that in turn increases heart rate and breathing
rate, which in turn have their own effects on how (for good or ill) our brains
function.

Overall, any transfer, duplication, or accurate simulation of a human
"mind" would have to take into account how the many different parts of the
brain and the rest of the body work separately and together. Besides their
similarities there are also profound, fundamental differences between how
standard modern computing devices process information and our brains
do, as well as the types of data each uses. Our brains make extensive use

of "neural networks," with many individual processing elements working together through a large number of interconnections, rather than only the digital methods typical of computers. While artificial neural networks can be created, their complexity compared to how the brain works is limited both by their hardware and, more fundamentally, by our lack of complete, detailed knowledge of the ways "higher" functions of the brain, particularly what we experience as "self-awareness" or "sense of self," actually work.

The only method we currently have to create a fully functioning human brain and "mind" is strictly biological and described in Chap. 10. Present-day techniques to "scan" the anatomy, physiology, and electrical activity of the brain holistically, such as functional magnetic resonance imaging (fMRI, see Chap. 8) and EEG, provide far too meager information to make a functioning "copy" of an individual's brain. At this point in our knowledge we cannot be entirely certain it is possible to ever do this. Perhaps the "mind" is too intricately bound to the brain and rest of the body to duplicate. But even if it is not, the sheer scale of reproducing or emulating all the processes, relationships, and information going on in the many diverse parts and cells of a human brain is staggering.

Thus, works that describe the human mind as the equivalent of software at the very least greatly underestimate the complexity of that software and the processes by which so much of its "code" is being constantly rewritten by interactions with the external environment, internal ones with the rest of the body, and whenever "I" is reflecting on itself and initiating its own thoughts. The question of what actually constitutes self-awareness rather than simulated action and algorithms is within our ability to speculate about but not to answer yet.

Similarly, for the events of Robert J. Sawyer's *Mindscan* to unfold, the issues and challenges to actually scan a human brain so thoroughly that its "mind" can be transferred to a fully functional, enhanced artificial human-like body (a tremendously challenging and significant achievement in itself!) must be greatly oversimplified. On the other hand, making the concept plausible (which, at the level of a work of fiction rather than the considerably more exacting standards of a scientific research paper, this novel does well) requires introducing some (but not all) relevant concepts and tacitly underestimating the difficulties involved.

For example, early chapters in that novel describe a process to completely scan the structure and interconnections of all of the neurons, neurotransmitter concentrations, and other neural components of a person's brain, comparing it to an MRI but with much finer resolution. However, for the story's plot to proceed, just how much finer that resolution must be—many orders of magnitudes beyond current technology—must be left unsaid. Moreover, the novel also presents the reasonable idea that such a scan must be done very quickly due to the ever-changing functional and physiological state of the brain. Here too, however, just how dynamic processes involving "thought," "personality," etc. actually are—even at the relatively crude levels of neurotransmitter effects, membrane electrical potentials, electrolyte and hormone levels, physiological parameters such as pH and oxygen concentrations, etc.—must be downplayed to make the scanning process seem plausible.

Even more speculatively, the procedure by which a brain is scanned is described as using "quantum fog," with a person's head reportedly being permeated with subatomic particles that are "quantally entangled with identical particles" to be injected into the "artificial braincase" of that individual's new body. However, as suggested by Chap. 8's description of quantum entanglement, the leap between doing this at the level of individual particles or even diamonds and the far greater complexity of the human brain and its many individual atoms, molecules, cells, etc. is very great indeed.

Science fiction typically mixes various amounts of established and speculative science. In some works, like *Mindscan*, there can be a particularly wide gap between present-day knowledge/capabilities and those their creators depict. This could make it more challenging for writers to figure out ways to convince their audience that what are currently "imaginary" medical or other scientific advances might conceivably become real someday. Authors can add at least a semblance of realism to their works by including true, factual information that supports the plausibility or at least possibility of such advances occurring. Creating that sense of verisimilitude may, however, require greatly deemphasizing or ignoring equally or more valid information stressing the difficulties with developing them. If "necessary," simple ad hoc pronouncements that what is now uncertain ("The human mind *might* be only the equivalent of software.") has been confirmed as being true ("The human mind *is* only the equivalent of software.") could also be used for story purposes.

Such methods would not be appropriate for a balanced, objective, scientific critique of what might be learned or become feasible in the future. However, they could be considered a valid means to "jumpstart" the plot of a successful fictional work by making it possible for the events shown in it to unfold and thereby explore their ramifications. Rather than concentrate only on how accurate and plausible the medicine/science in a successful work like *Mindscan* is, after a certain point it might be better to concentrate on the "big picture" of its overall plot rather than the perhaps overly optimistic assumptions it makes about future advances.

Looking at the Mona Lisa too closely would show that, at a strictly material level, it is "only" some oil-based pigments smeared on wood, which in turn are composed of particular arrangements of atoms and molecules. It is only by stepping back and examining it more holistically that one can appreciate its other dimensions and meanings. Similarly, even if some of the medicine and other science in them are not nearly as certain to be true as they claim, imaginatively speculative science fiction creations like *Mindscan* can also be enjoyed as successful and effective works of art.

14.9 The Bottom Line

Bionic limbs already exist and will continue to improve mechanically and electronically. Interfacing them with peripheral nerves or directly to the brain to restore functions similar to or, potentially, in some ways superior to biological limbs represents difficult but not insurmountable challenges. A robotic arm might indeed be stronger than an average biological one, and so the death grip that Darth Vader can deliver could be feasible. One might also, for example, have removable modular attachments for the hand to temporarily replace it with specialized tools like a drill or, in a science fiction setting, perhaps a weapon.

On the other hand, a robotic arm or limb can be no stronger than its materials or move with more force than the power that can be delivered to it. Although Steve Austin's bionic left arm was said to have the strength of a bulldozer, in the real world you actually need a bulldozer (or its equivalent), with all its associated metal, power, mass, etc. to have that strength. And even if Steve's arm is mechanically stronger and able to lift more than a standard biological one can, he must be very careful not to overload its "weakest link" connection to his flesh-and-blood part lest the bionic arm get ripped off or his body be otherwise injured, even if the artificial arm itself is durable enough to remain undamaged.

Bionic eyes will also certainly improve but face formidable challenges to approach, much less exceed the healthy characteristics of the "normal" biological variety. Current implants are foreign bodies in the eye, requiring surgery to place or replace, and carry risks of infection, scarring of adjacent tissues, bleeding, dislodgment, etc. Increasing the area of the array used and the density and types of electrodes in it might increase the risk of producing too much heat and damaging the cells of the retina or other problems. Also, current implants partially replace the functions of damaged rods and cone alone. They require that the other cells of the retina be healthy and use them as the "middle man" to interface indirectly with the optic nerve. If the damage to the retina includes those other cells too, current implants cannot work.

Nonetheless, despite its challenges, this is an area in which further research could make great progress. The electronic components of a "bionic eye" would certainly be expected to improve. And if better techniques are found to interface with the optic nerve or directly with the visual cortex itself, eventually the capabilities of such "artificial" vision could conceivably approach or perhaps surpass that of "normal" vision.

From a practical standpoint current bionic ears, in the form of cochlear implants, significantly improve the ability to hear but are still not as good as a healthy inner ear. Here too the issue with "improving" the sense of hearing lies in the interface with and intrinsic limitations of the biological part of the system.

Fortunately bionic replacements are not the only game in town for restoring or enhancing the human body. Methods discussed in previous chapters—stem cell therapy, organ and limb transplantation, genetic engineering to improve the performance of biological parts, and medical nanotechnology represent other alternatives either individually or as part of a possible "multidisciplinary" approach. For example, hand transplantations are already a reality, with current limitations including needing to take immunosuppressive medications with their attendant risk of infections and how much neurological function can be restored to the "new" hand. Growing back lost limbs is considerably more speculative. While doing this with stem cell therapy, genetic engineering, or nanotechnology used alone or in some combination is conceivable, these methods both individually and collectively represent many challenges. The complete replacement of a biological eye with another depicted in the 2002 movie *Minority Report* far exceeds current capabilities. However, if issues regarding transplant rejection and the serious surgical challenges to connecting the new eye to the optic nerve (again, perhaps aided by stem cells, nanotech, etc.?) could be addressed, it might be feasible to do this one day for someone with a damaged eye.

Besides actual artificial limbs, individuals who still have their lower extremities but cannot move them due to a spinal cord injury could use a mechanical "exoskeleton" to move [59]. A person with paraplegia can be strapped into the device and employ crutch-like supports or simply bars manipulated with both arms to walk. Some models include jointed metal lower extremities strapped to the nonfunctional biological ones. These artificial limbs include motors, sensors, and computer control to enable the individuals to move them and their own lower extremities in such a way that they can climb up steps.

Current exoskeletons use manual controls such as joysticks, but later ones might use EEG and other methods to "tap" electrical signals from a person's brain to move the device. Another limitation of available models is their size and mass, with one model weighing 38 kg. Further developments in the electronics and materials involved in creating an exoskeleton will hopefully improve these limitations.

Brain-machine and brain-computer interfaces can already translate electrical activity produced by areas of the brain and even single neurons into a form that can be used to control external devices. However, these capabilities can utilize only (comparatively speaking) simple electric impulses. Trying to decipher the electric "language" underlying far more complex functions such as speech, "thought," retrieval of memories, etc. and using that knowledge to connect our brains directly to computers or other electronic devices will be much more complicated and its feasibility, reliability, and risks uncertain. Because we simply do not have enough knowledge to determine whether it is possible or not, it is certainly not improper for a science fiction work to take the "optimistic" approach that it will be possible someday to effectively blend our native, biological information processing capabilities with those of artificial processors. Based on the current state of the art, however, the chance of this happening to a meaningful degree anytime soon seems (pending further research and discoveries) unlikely.

The prospects are even more nebulous for uploading, copying, etc. a human "mind." We have reasonably solid knowledge regarding basic principles of how the brain and its individual cells work, what areas of it are involved in specific functions, etc. However, the exact details of how the myriad individual neurons and other cells within a particular human brain create a unique sentient being remain a mystery. Likewise, at least for the foreseeable future the idea of a "synaptic scanner" or doing a "mind transplant" to an artificial or different biological "receptacle" is almost the equivalent of using a magic wand.

However, because of that "almost" the idea of uploading a mind and similar concepts still remains within the realm of science fiction rather than crossing indelibly into science fantasy. At worst, for story purposes one can introduce plot elements such as technologically advanced aliens[12] or time travelers from the far future to whom these techniques are commonplace. Setting the story far enough in humanity's future, long after all those pesky details about how to upload a mind etc. have been solved, would also obviously "work." If it is used, however, I suggest the approach of merely presenting it as a "matter of fact" without addressing the challenges that had to be solved to make it so, or touching on them only lightly. Otherwise more "real life" questions may be raised than answered, such as exactly how one goes about making a "synaptic scanner" or similar device.

In summary, it is too early to tell whether and to what degree specific advances in bionics will one day enter clinical medical practice or if they will forever

---

[12] This is a plot twist that, despite all these reservations about its feasibility, I have used in one of my own published stories.

remain in the realm of fantasy. Until then, they can (with varying degrees of realism) play a useful role healing and enhancing characters in science fiction.

References

1. Clynes M, Kline N. Cyborgs and space. Astronautics. 1960(September):25–6, 74–6.
2. Dolgin E. Managed by machine. Nature. 2012;485:S6–8.
3. Thurston A. Paré and prosthetics: the early history of artificial limbs. ANZ J Surg. 2007;77(12):1114–9.
4. Brumfiel G. The insane and exciting future of the bionic body. 2013. http://www.smithsonianmag.com/innovation/the-insane-and-exciting-future-of-the-bionic-body-918868/?page=2. Accessed 15 April 2015.
5. Service R. Bioelectronics: the cyborg era begins. Science. 2013;340:1162–5.
6. Herr HM, Grabowski AM. Bionic ankle-foot prosthesis normalizes walking gait for persons with leg amputation. Proc Biol Sci. 2012;279(1728):457–64.
7. McCarthy B. Flex-Foot Cheetah. 2011. http://www.ele.uri.edu/courses/bme281/F11/BrookeM_2.pdf. Accessed 15 April 2015.
8. Hargrove LJ, Simon AM, Young AJ, Lipschutz RD, Finucane SB, Smith DG, et al. Robotic leg control with EMG decoding in an amputee with nerve transfers. N Engl J Med. 2013;369(13):1237–42.
9. Dawley J, Fite K, Fulk G. EMG control of a bionic knee prosthesis: exploiting muscle co-contractions for improved locomotor function. IEEE Int Conf Rehabil Robot. 2013;(June):1–6.
10. Tabot G, Dammann J, Berg J, Tenore F, Boback J, Vogelstein R, et al. Restoring the sense of touch with a prosthetic hand through a brain interface. Proc Natl Acad Sci. 2013;110(45):18279–84.
11. Raspopovic S, Capogrosso M, Petrini FM, Bonizzato M, Rigosa J, Di Pino G, et al. Restoring natural sensory feedback in real-time bidirectional hand prostheses. Sci Transl Med. 2014;6(222):222ra19.
12. Kwok R. Once more, with feeling. Nature. 2013;497:176–8.
13. del Valle J, Navarro X. Interfaces with the peripheral nerve for the control of neuroprostheses. Int Rev Neurobiol. 2013;109:63–83.
14. Tan DW, Schiefer MA, Keith MW, Anderson JR, Tyler J, Tyler DJ. A neural interface provides long-term stable natural touch perception. Sci Transl Med. 2014;6(257):257ra138.
15. Pirowska A, Wloch T, Nowobilski R, Plaszewski M, Hocini A, Ménager D. Phantom phenomena and body scheme after limb amputation: a literature review. Neurologia i Neurochirurgia Polska. 2014;48(1):52–9.
16. Abbott A. In search of the sixth sense. Nature. 2006;442:125–7.
17. Zrenner E. Fighting blinding with microelectronics. Sci Transl Med. 2013; 5(210):210ps16.

18. Chuang A, Margo C, Greensberg P. Retinal implants: a systematic review. 2013. http://bjo.bmj.com/content/early/2014/01/08/bjophthalmol-2013-303708.abstract?sid=3e338ecd-584a-4346-bc29-6a8b16951f55. Accessed 15 April 2015.
19. Luo Y, da Cruz L. A review and update on the current status of retinal prostheses (bionic eye). 2014. http://bmb.oxfordjournals.org/conte-nt/early/2014/02/12/bmb.ldu002.abstract?sid=fff951f7-9033-40d0-951f-6371bccdb700. Accessed 15 April 2015.
20. Guenther T, Lovell N, Suaning G. Bionic vision: system architectures: a review. Expert Rev Med Devices. 2012;9(1):33–48.
21. Ong J, da Cruz L. The bionic eye: a review. Clin Exp Ophthalmol. 2012;40(1):6–17.
22. Merabet LB. Building the bionic eye: an emerging reality and opportunity. Prog Brain Res. 2011;192:3–15.
23. Nirenberg S, Pandarinath C. Retinal prosthetic strategy with the capacity to restore normal vision. Proc Natl Acad Sci. 2012;109(37):15012–7.
24. Greenemeier L. FDA approves first retinal implant. 2013. http://www.nature.com/news/fda-approves-first-retinal-implant-1.12439. Accessed 15 April 2015.
25. Stingl K, Bartz-Schmidt KU, Besch D, Braun A, Bruckmann A, Gekeler F, et al. Artificial vision with wirelessly powered subretinal electronic implant alpha-IMS. Proc Biol Sci. 2013;280(1757):20130077.
26. Ghovanloo M. An overview of the recent wideband transcutaneous wireless communication techniques. Conf Proc IEEE Eng Med Biol Soc. 2011;2011:5864–7.
27. Pezaris JS, Eskandar EN. Getting signals into the brain: visual prosthetics through thalamic microstimulation. Neurosurg Focus. 2009;27(1):E6.
28. Eshraghi AA, Gupta C, Ozdamar O, Balkany TJ, Truy E, Nazarian R. Biomedical engineering principles of modern cochlear implants and recent surgical innovations. Anat Rec (Hoboken). 2012;295(11):1957–66.
29. Eshraghi AA, Nazarian R, Telischi FF, Rajguru SM, Truy E, Gupta C. The cochlear implant: historical aspects and future prospects. Anat Rec (Hoboken). 2012;295(11):1967–80.
30. Lenarz T, Pau H, Paasche G. Cochlear implants. Curr Pharm Biotechnol. 2013;14:112–23.
31. Wilson BS, Dorman MF, Woldorff MG, Tucci DL. Cochlear implants matching the prosthesis to the brain and facilitating desired plastic changes in brain function. Prog Brain Res. 2011;194:117–29.
32. NIDCD Fact Sheet. Cochlear Implants. http://www.nidcd.nih.gov/staticresources/health/hearing/FactSheetCochlearImplant.pdf. Accessed 15 April 2015.
33. Fallon JB, Irvine DRF, Shepherd RK. Neural prostheses and brain plasticity. J Neural Eng. 2009;6(6):065008.
34. Marozeau J, Innes-Brown H, Blamey PJ. The acoustic and perceptual cues affecting melody segregation for listeners with a cochlear implant. Front Psychol. 2013;4:790.

35. Lawand N, Ngamkham W, Nazarian G, French P, Serdijn W, Gaydadjiev G, et al. An improved system approach towards future cochlear implants. Conf Proc IEEE Eng Med Biol Soc. 2013;2013:5163–6.

36. Vincent C. Auditory brainstem implants: how do they work? Anat Rec (Hoboken). 2012;295(11):1981–6.

37. Mannoor M, Jiang ZS, James T, Kong Y, Malatesta K, Soboyejo W, et al. 3D printed bionic ears. Nano Letters. 2013;13(6):2634–9.

38. Parker MS, Fahrner LJ, Deuell BP, Olsen KM, Kasirajan V, Shah KB, et al. Total artificial heart implantation: clinical indications, expected postoperative imaging findings, and recognition of complications. AJR Am J Roentgenol. 2014;202(3):W191–201.

39. Copeland J. SynCardia total artificial heart. Tex Heart Inst J. 2013;40(5):587–8.

40. What Is a Total Artificial Heart? 2014. http://www.nhlbi.nih.gov/health/health-topics/topics/tah/. Accessed 15 April 2015.

41. Lee B, Liu CY, Apuzzo ML. A primer on brain-machine interfaces, concepts, and technology: a key element in the future of functional neurorestoration. World Neurosurg. 2013;79(3–4):457–71.

42. Lebedev M, Tate A, Hanson TL, Li Z, O'Doherty JE, Winans J, et al. Future developments in brain-machine interface. Clinics. 2011;66(S1):25–32.

43. Leuthardt EC, Schalk G, Roland J, Rouse A, Moran DW. Evolution of brain-computer interfaces: going beyond classic motor physiology. Neurosurg Focus. 2009;27(1):E4.

44. Nair P. Brain-machine interface. Proc Natl Acad Sci U S A. 2013;110(46):18343.

45. Nicolas-Alonso LF, Gomez-Gil J. Brain computer interfaces, a review. Sensors (Basel). 2012;12(2):1211–79.

46. Ifft PJ, Shokur S, Li Z, Lebedev MA, Nicolelis MA. A brain-machine interface enables bimanual arm movements in monkeys. Sci Transl Med. 2013; 5(210):210ra154.

47. Hochberg L. Turning thought into action. N Engl J Med. 2008;359(11): 1175–7.

48. Shih JJ, Krusienski DJ, Wolpaw JR. Brain-computer interfaces in medicine. Mayo Clin Proc. 2012;87(3):268–79.

49. Thakor N. Translating the brain-machine interface. Sci Transl Med. 2013; 5(210):210ps17.

50. Keefer EW, Botterman BR, Romero MI, Rossi AF, Gross GW. Carbon nanotube coating improves neuronal recordings. Nat Nanotechnol. 2008;3(7):434–9.

51. Blake D. How brains learn to control machines. Nature. 2012;483:284–5.

52. Koralek AC, Jin X, Long JD, 2nd, Costa RM, Carmena JM. Corticostriatal plasticity is necessary for learning intentional neuroprosthetic skills. Nature. 2012;483(7389):331–5.

53. Ryu SI, Shenoy KV. Human cortical prostheses: lost in translation? Neurosurg Focus. 2009;27(1):E5.

54. Rapoport B, Kedzierski J, Sarpeshkar R. A glucose fuel cell for implantable brain-machine interfaces. 2012. http://www.plosone.org/article/info%3Adoi%2F10.1371%2Fjournal.pone.0038436. Accessed 15 April 2015.
55. Lule D, Noirhomme Q, Kleih S, Chatelle C, Halder S, Demertzi A, et al. Probing command following in patients with disorders of consciousness using a brain-computer interface. Clin Neurophysiol. 2013;124(1):101–6.
56. Chatele C, Chennu S, Noirhomme Q, Cruse D, Owen A, Laureys S. Brain-computer interfacing in disorders of consciousness. Brain Inj. 2012;26(12):1510–22.
57. Naci L, Monti M, Cruse D, Kubler A, Sorger B, Goebel R, et al. Brain-computer interfaces for communication with unresponsive patients. Ann Neurol. 2012;72(3):312–23.
58. Schoen J, Boysen M, Warren C, Chakravarthy B, Lotfipour S. Vertebrobasilar artery occlusion. West J Emerg Med. 2011;12(May):233–9.
59. Gwynne P. Technology: mobility machines. Nature. 2013;503:S16–7.

15

Summing Up: Using Medicine in Science Fiction

The desire to take medicine is perhaps the greatest feature which distinguishes man from animals.

Sir William Osler
Science and Immortality (1904)

The best of all physicians
Is apple-pie and cheese!

Eugene Field
"Apple-Pie and Cheese" (1889)

The preceding chapters provide a broad overview of how science fiction can include medical care and concepts. Medicine's contributions to the genre may involve depicting the ministrations of human (or otherwise) practitioners of the healing arts, a cornucopia of methods to threaten or inflict physical and psychological harm on characters, and using futuristic technologies to heal, rejuvenate, or even remake the human body. This final chapter will summarize and expand on key ideas presented previously as well as introduce several new ones.

Real-life or speculative medicine can be incorporated into science fiction in a variety of ways. It can be an essential part of the primary plot or subplot of a work, serve as a background for the events in it, help define characters and explain why they act as they do and what happens to them, form the basis of entire human or alien societies, or play an essentially subliminal role in the overall action. Weaving medical elements into a work can be as obvious as including a physician as a character or positing tremendous advances in genetic engineering and nanotechnology that redefine what it means to be human. Or it might merely involve characters being injured in the course of their adventures, receiving medical care, and then resuming their participation in the main plot.

To be most effective, medicine in science fiction should have at least a veneer of realism. Even if it includes extremely imaginative, implausible, or even technically incorrect components whose use can only be justified by dramatic license and that require suspension of disbelief, at least some of the depiction should be grounded (however tenuously) in actual medicine and human biology. Balancing all those considerations in a way that does not impede and hopefully contributes to the overall plot requires careful weighing of what to include and exclude, as well as how to blend those medical elements into the overall narrative.

The suggestions in this chapter on how to use medicine in science fiction are by no means exhaustive. They represent general guidelines on how to smoothly combine the two, whether the medicine involved is an overt component of the story or only an unobtrusive subtext. Following these suggestions should at least increase the odds that a science fiction work will not stretch medical credibility beyond its breaking point.

15.1 "As You Know, Bob…"

To be effective a story must provide at least a modicum of information about where and when it occurs, the character(s) in it, background information about its settings, and other facts the reader requires to understand and appreciate what is going on in it. This is especially important in science fiction, which can take place anywhere in time, space, parallel universes, alternate realities, etc. The classic recommendation "show, don't tell" applies here as in other writing contexts. This melodramatic description of a character—"He was a villain dark and deep who did bad things to people."—certainly conveys information about him, but in a nonspecific and strictly "telling" way. A more detailed sentence—"Beakers bubbled and sparks flashed from the massive equipment lining the secret laboratory's walls as the wizened cackling white-coated doctor adjusted his thick goggles and prepared to pull the heavy lever that would commence his weird experiment on the gorgeous young redhead who screamed and struggled to free herself from the wide leather straps securing her voluptuous scantily clad body to the examining table."—presents that male character's evil intentions in a much more vivid and picturesque way.[1]

A writing technique commonly used in early science fiction but generally frowned upon in current styles is the "expository lump," also known as the "info dump." This "tell, don't show" technique typically involves matter-of-factly unloading a great deal of information regarding the background,

---

[1] However, while it might have been at home in a 1930s pulp magazine with a title such as *Spicy Science Stories*, for a variety of reasons I do *not* encourage anyone to use this particular sentence or anything similar in a modern, serious SF work.

characters, important plot elements, etc. of a science fiction work in one place—perhaps even doing that multiple times in a longer work such as a novel. This might be done in a purely first- or third-person narrative form, or by a character delivering a monologue to one or more other characters concerning what has already happened or is about to happen.

For example, two seminal science fiction novels by H. G. Wells—*The Time Machine* (1895) and *The Invisible Man* (1897)—use this writing method to different degrees. In the former work the bulk of the novel is devoted to the Time Traveler telling his companions about his past/future experiences. In the latter an overall more real-time, action-oriented style is interrupted by a lengthy section in which the title character gives a speech explaining how he achieved his invisible state and how he plans to use it.

Instead, generally speaking, providing only the most essential expository material and giving it in small doses during a story is more reasonable. Thus, just as in their antecedents in old-time movie serials, the crawling credits at the beginning of the *Star Wars* movies succinctly provide just enough "telling" background information to lay the foundation for subsequent dynamic "showing" of other essential plot points. The challenge for science fiction writers is to avoid overwhelming the reader with too much information at one time. Giving dollops of it instead, such as through brief descriptions of a setting or scene, what a character does, or via dialogue rather than monologue can be more effective.

However, providing information via dialogue whose style has little if anything to do with the way real people speak is also less than ideal. In his excellent and admirably terse book *Writing to the Point: A Complete Guide to Selling Fiction* (1999) science fiction writer Algis Budrys describes the issue of what he calls "maid-and-butler" dialogue. It refers to characters talking in a way whose blatant purpose is only to provide information relevant to the story, spoken in a way that no two people would actually talk to each other. Here is my own, "original" example of this somewhat stilted technique:

Space Doctor No. 1: "Perhaps it is a blessing in disguise that the Saturnian ambassador became ill at the Solar System Peace Conference here at Moon-Base 5. If we can successfully operate on the inflamed green scales near its dorsal excretory orifice perhaps the ambassador will be grateful and strike a more conciliatory stance towards our Commonwealth of Inner Planets, thus bringing an end to humanity's decades-long interplanetary war with the hydrogen-breathing denizens of the Outer Worlds."

Space Doctor No. 2: "Yes. And even more importantly, I understand the ambassador has an excellent health insurance plan that pays physicians promptly and generously."[2]

An even more jarring variation of this style is for one character to tell another at great length what the second already knows. Consider this example (which also involves much dubious science): "As you know, Bob, the transcalcinator ray you developed during the 5 years you worked as a research scientist on Deneb IV for the Trans-Galactic Federation founded in Terrestrial Year 2491 is the only weapon known to be effective against the aliens you named the Blue Extragalactic Marauders when you encountered them during your hyperlight exploratory expedition to the Andromeda Galaxy 2 years ago." This sentence presumably includes a large amount of salient information regarding further plot developments. However, if I were Bob I would be inclined to think, "How can someone say all that in one breath? And why on Deneb IV would he?"

15.2 Too Much or Too Little Medicine?

Another important issue is how much medical detail to include in a science fiction work. This can involve a delicate balancing act involving many and perhaps conflicting considerations. However they are used, medical elements should not "get in the way" of the story itself. For example, if a character is injured there is no need to give a highly detailed account, no matter how accurate, of how he or she is treated unless those details are germane to the work as a whole. Thus, if the only thing needed is to get injured or ill characters back on their feet, tentacles, pseudopods, etc. there may be no need to do more than give a minimally sketched account of their medical care.

On the other hand, my novelette "Hearts in Darkness"[3] includes a real-time account of a character being treated for a myocardial infarction (heart attack). In this case, however, providing a great deal of accurate detail regarding the medications, tests, and procedures used to address that medical issue was, in my opinion, not only justified but necessary. As mentioned back in Chap. 5, the story's protagonist is a cardiologist on the International Space Station (ISS). This setting requires that he must "make do" with the limited medical supplies and other resources there, as well as being ultimately respon-

---

[2] I hasten to add that this particular dialogue never has nor ever will appear in any of my own science fiction stories…although I have been guilty of occasionally using a similar style for purposes of parody and satire.

[3] *Analog Science Fiction and Fact*, March 2002.

sible for the difficult choice of whether to keep his patient on the ISS or return him to Earth for definitive care.

Presenting this scene realistically meant that I had to use only medications and equipment that the ISS either currently has or could realistically have in the story's near-future time frame. That included naming the specific medications used, including chewable aspirin, sublingual nitroglyerin, and intravenous amiodarone; describing a little of the pathophysiology of a myocardial infarction; and the role of an electrocardiogram in its diagnosis. Such medical details were *not* included to "show off" the medical knowledge of the author, but because they were in fact integral to the plot and leaving them out would have weakened the story.

Realistic depictions of medical details in science fiction also serve another purpose. They can be didactic, providing at least some readers with new information about medical science and how it is applied. Thus, after reading "Hearts in Darkness" someone not familiar with the challenges of treating a patient in microgravity, what causes a myocardial infarction, and how it is treated will know more about them than before reading the story. The challenge for the author is to add this "teaching" function into a work well enough that it neither calls attention to itself nor impedes the flow of the plot.

This idea is illustrated by the marked contrast in tone and priorities between a medical textbook or research paper and an SF work that includes medical details. The (hopefully!) strict nonfiction of the former is interested in presenting only true information and the state-of-the-art of present-day medical knowledge and care—certainly noble goals in their own right. On the other hand, a good science fiction story is not limited to only bare facts but can use them selectively, "fudge" them to a degree when needed, and extrapolate far beyond what is currently known and can be done. Its nonfiction, "educational" function—however beneficial—remains subservient to the need to create an entertaining, thought-provoking, poetic, etc. work of fiction.

The only caveat I would give here is that hopefully the science fiction story, television show, movie, etc. will not "teach" readers and viewers things about medicine and medical care that are really incorrect and thus leave their audience thinking they are actually right. Hopefully it should be easy for adults to recognize that, as noted in Chap. 6, being bitten by a radioactive spider will not confer superpowers on the bitee. On the other hand, unless they read Chap. 3 some individuals might not recognize a depiction of an astronaut immediately "freezing" into a large chunk of ice when exposed to the vacuum of space as wrong.

Medical details that are overtly explanatory and at least partly didactic carry an increased risk of slowing down the flow of a story or even bringing it to a crashing halt. In general they are best presented using the "divvy it out in small bits" suggestion of the previous section rather than turning into a lecture or including information that, however accurate or interesting it might be by itself, can slow the pace of story to a crawl. On the other hand, I intentionally violated this principle several times in one of my stories, "Primum Non Nocere."[4] I used this "exception to the rule" to satirize the pomposity of the very unpleasant and unprincipled physician delivering long medical lectures (which, for the most part, actually contained sound information and advice!) to a literally captive audience of patients. I counterbalanced the deadening effects of his info dumps by injecting tension into those scenes. In one of them the doctor acts in a clearly menacing bedside manner to the helpless protagonist as he delivers his speech, while in another his prolonged droning leads to growing anger from his patients and ultimately a riot against him. This combining of "static" medical information with "dynamic" action can, like a spoonful of sugar, indeed help the medicine go down.

Overall, in my own science fiction stories there is a roughly inverse relationship between how "speculative" the medicine in them is and how many details I give about how it works. Thus, when I use "futuristic" concepts such as suspended animation and rejuvenation treatments I typically limit myself to focusing on what they do and describing their effects rather than trying to "prove" they can actually work and the precise methods they employ. My prime rationale for this minimalist approach is that, based on our current limited knowledge, we really *don't* know if they are actually feasible or exactly how to do them.

Because of this uncertainty, trying too hard to explain how particular problems with developing advanced techniques in medical nanotechnology, genetic engineering, etc. were "solved" or being too specific about the procedures employed to create "medical miracles" with them could be counterproductive. Every particular "explanation" given to support their feasibility could be countered by "But what about…" objections from anyone (including me as the author!) who recognizes the many complex issues and challenges involved with making them actually work. Rather than unnecessarily slow down the story's plot by exhaustively describing all the known issues/problems with developing these techniques and how they were addressed/solved, I typically try to keep things brief and simple. This may include presenting just a few core concepts (e.g. what biological processes would need to be altered to rejuvenate the elderly), or even just showing techniques such as suspended animation in action with little explanation of what goes on "under the hood" to make them work.

---

[4] *Analog Science Fiction and Fact*, December 2010.

On the other hand, if the medicine involved is well established or likely to be feasible in the near future, I am far more willing to provide a much greater amount of and more specific details about how medical care is delivered and the concepts underlying it. My goal in doing that is to add enough realism to a story so that the reader can see how medicine is actually practiced and the ideas behind it, while not adding so many facts and minutiae that they bog down the flow of the story.

Note, however, that this is method of being laconic when describing speculative medicine and more loquacious with the "real" variety is only *one* valid writing technique. To improve its level of realism science fiction certainly can include descriptions and "explanations" of how particular advanced medical technologies work that are not comprehensive or address all of the actual issues with developing them. Even using a dash of technobabble or very shaky medicine (although hopefully not too much of either!) to move the story along may also work, although that could run the risk of it crossing the sometimes blurry boundary between "science fiction" and "science fantasy." Ultimately, short of using medicine that is unnecessarily wrong, the range of reasonable ways to include medical concepts in science fiction is wide.

15.3 Say or Suggest?

In the same vein, integrating medicine into science fiction can use the more general writing principle of "less is more." Reinforcing what was mentioned in the previous section, rather than provide readers or viewers with every detail of what is going on, it may be more effective to provide just enough information so that they can then "fill in the blanks" on their own. Thus, when Dr. McCoy takes out his medical tricorder and it makes a funny sound over a patient, we do not need to know how it is constructed or what exact diagnostic information it provides in order to follow the story. In fact, having the good doctor give a detailed explanation on those points would, while definitely of great interest to a few of us in the audience, destroy the effectiveness of the scene for the overwhelming majority of viewers. Instead, for story purposes all we really need to hear is his diagnosis, prognosis, and/or treatment plan.

This principle of minimalism also overlaps with that of "show, don't tell." The first few pages of William Gibson's *Neuromancer* (1984) quickly and matter-of-factly introduce a time and place where robot arms, organ transplantation, and a variety of sophisticated electronic enhancements to the human body are routine. For the purpose of that particular story there is no need to explain what scientific breakthroughs and developments led to their creation. Instead they are simply presented in passing as "givens" in the novel and their use woven intricately into the novel's overarching plot. Such a bare minimum of medical "facts" is enough to help paint the picture of a whole society and how it functions. This method lets readers use their own impressions and imagination to "connect the dots" and see how all these elements fit together rather than overtly telling its audience how they do.

Another successful approach, this time involving medical (and other) nanotechnology, is used in Edward M. Lerner's novel *Small Miracles* (2009). In its opening scene a character wearing an experimental nanotechnology-based protective suit explains its capabilities to another individual (and the reader). Instead of merely "dumping" this important information, however, the narrative integrates the conversation they have into other action and in a context in which it would be reasonable for those explanations to be given. Moreover, the character receiving that information has a need to learn it, would not have known it beforehand—and, perhaps most importantly, is not named "Bob." The net effect is that the information is not artificially grafted into the story but provided in a realistic scenario and manner.

15.4 The Limits of Extrapolation

Many concepts and technologies first shown in science fiction have come "true" (if not always in their details) in the real world. Less than two decades separate the 1950 movie *Destination Moon* from Apollo 11, or the use of communication relay stations in space depicted in George O. Smith's *Venus Equilateral* stories from the 1940s with Telstar 1's launch in 1962. Despite the weakness of its fictional aspects, parts of Hugo Gernsback's early science fiction work *Ralph 124 C 41 +* (1911) foretell many later inventions such as television, tape recorders, and radar.[5] The high-tech submarine *Nautilus* depicted in Jules Verne's *Twenty Thousand Leagues Under the Sea* (1870) lent its name to the world's first nuclear-power submarine, launched in 1954. And although many of the reproductive methods described in Aldous Huxley's *Brave New World* have not yet been realized, as described in Chap. 10 we are considerably closer to developing them than when that novel was first published in 1932.

Along with such "prophesies" of future scientific developments, in some cases science fiction has actually inspired the real world development of new technologies. Flip-style cell phones owe much of their form and function to the communicators first shown on the original *Star Trek* series. Modern day devices such as pulse oximeters used to noninvasively measure and display

---

[5] We are still waiting, however, for some of its other predictions to come true, such as floating cities and weather control stations.

heart rate and oxygen saturation in blood fulfill some of the functions of Dr. McCoy's medical tricorder. Current robotic systems used to help surgeons perform operations are at least conceptually descendants of the remote manipulators Robert A. Heinlein described in his 1942 story "Waldo."

But science fiction's role of "predicting" or even "shaping" the future is only a tiny subset of what it does. When it does deal with the "future" science fiction is not limited to the single, solitary future[6] that continues to unfold with every passing second. Overall, whether or not any of those possible futures become "true" is less important than that we have an opportunity to vicariously live in them for a time. That experience may prove entertaining or thought-provoking, enjoyable or alarming, inspiring or depressing, etc.

Those and other effects can be produced by works that have essentially no chance of being real (e.g. the odds of there being sentient extraterrestrial life elsewhere in the Universe seems very high, but there are no actual Vulcans with a specific history and culture); ones whose events once described the future but have now become part of the past, such as the original timelines of Ray Bradbury's *The Martian Chronicles* stories and so many other science fiction works from the early and mid-twentieth century; as well as those that still have the "If this goes on…" character of dealing with developments that really might occur in the future, e.g. routine human cloning.

In short, science fiction has value even if it is "wrong" about the future or what is possible based on later scientific discoveries. While it is still too early to tell if the Earth of A.D. 802,701 will be populated by Eloi and Morlocks, the kind of time machine H. G. Wells describes in his 1895 novel *The Time Machine* is, like the TARDIS, now confined to the realm of science fantasy rather than "real" science. Likewise, unless used for satiric/nostalgic purposes such as my story "Neighborhood Watch,"[7] the many "Golden Age" science fiction works as disparate in tone as Stanley G. Weinbaum's "A Martian Odyssey" (1934) or Isaac Asimov's "Victory Unintentional" (1942) that posit intelligent aliens native to our Solar System's other worlds are "obsolete" by current scientific knowledge. However, they and others can still be enjoyed and admired for their plots and ideas.

These observations regarding science and the "future" are certainly not original, but basic truisms well known to readers and writers of the genre. However, I restate them here to serve as background to highlight a particular style of science fiction. It is one that covers a broad range extending between

---

[6] There are also a variety of theoretical "many-worlds" interpretations involving quantum mechanics in which every possible alternative event/future does indeed become "real." However, until that idea is confirmed (assuming it can be, even if true), from a practical standpoint I will refer here to the sole future that we humans collectively seem to experience.

[7] *Analog Science Fiction and Fact*, January/February 2013.

works successfully predicting the "real" future and those depicting ones that did not or cannot happen. This is the realm of "plausible" science fiction—stories, movies, etc. that show how current political, social, economic, etc. issues might actually play out in coming years, and especially how technology might advance and change the human condition.

The use of or at least acknowledgment of *real* science to do this is part of what is described as "hard" science fiction. Dr. Stanley Schmidt, whose credentials as a physicist, science fiction writer, and the editor of *Analog Science Fiction and Fact* for over three decades give him a unique perspective on this topic, includes two key concepts in his definition of a hard science fiction story.[8] They are: "A story in which (1) some element of scientific or technological speculation is so deeply integrated that it can't be removed without making the whole story collapse, and (2) the author makes a creditable attempt to make the speculation plausible."

Dr. Schmidt uses Daniel Keyes's *Flowers for Algernon* to illustrate the first point.[9] In it a mentally disabled man has his intelligence temporarily increased to genius level by an experimental medical technique, with profound effects on his life and the people around him. If that latter scientific element were absent not only would the story fail to be science fiction but the dramatic events it describes would simply not occur.

According to Dr. Schmidt the second point can include either "extrapolation"—speculation that "makes correct use of well-established science"—or "innovation," which "posits new science that hasn't been discovered (and maybe never will), but conceivably could be without contradicting what's already known." Furthermore, "Successful innovation requires knowing enough established science to know how to invent something new that could be consistent with it. Successfully using it in a story requires working out the logical consequences of the imagined new science, and being consistent with those. And if something in a story looks impossible in terms of known science, the story should at least hint at how it might be possible anyway."

Dr. Schmidt's observations are also germane to the use of medicine in science fiction. The basic medical facts and current capabilities regarding such science fiction concepts as suspended animation, life extension, genetic engineering, etc. described in earlier chapters give only a "snapshot" of what we currently know and can do. Trying to predict what our knowledge and abilities in those areas will be decades from now, much less centuries or millennia, is difficult for many reasons. It involves extrapolating well beyond our

---

[8] Personal communication.
[9] This work was originally published as a short story in 1958, expanded into a novel first published in 1966, and was the basis for the 1968 movie *Charly*.

present-day, limited data points—and we simply cannot be sure what lies beyond them.

Using an example borrowed from physics, a graph plotting an object's velocity (y-axis) versus its mass (x-axis) will appear to be a straight vertical line as the velocity increases over a wide range starting from zero, with its mass (seemingly) remaining constant. However, this "Newtonian" or "classical" relationship—thought for centuries to be the only one between them —no longer applies when the object reaches velocities that are a significant fraction of that of light in a vacuum. At that point "Einsteinian" or "relativistic" effects become prominent and the plot on that graph deviates more and more from a straight line, with the object's mass increasing (see Chap. 7) and that line curving increasingly to the right as its velocity comes closer and closer to but never reaches that of light. But until this latter effect was confirmed by acquiring more data, a physicist c. 1800 would have drawn that line as straight up to and even beyond the speed of light based on the information available at that time.

This can be applied as an analogy to the pace of medical research and what it might accomplish in the future. Any "straight line" or other extrapolations beyond its current status in a particular field are by their nature uncertain. True "game changing" events, such as the widespread use of vaccines, development of antibiotics, successful organ transplantation, and the advanced electronics present in modern diagnostic equipment are, in the "big picture," rare and inherently unpredictable as to when or if they occur. The time between when a medical "breakthrough" is made in the laboratory and when it enters clinical practice is often measured in decades or even longer. New drugs and procedures are tested first in animals for safety and efficacy before—if they pass those tests—being evaluated in humans. Much research that at first seems promising or to be a breakthrough turns out to be a dead end or worse.[10] Other findings based on animal studies that initially suggest that a great advance in medical care may soon be forthcoming might turn out to have limited applicability to humans, be hard to implement, have unanticipated adverse effects, etc.

For the medicine in a science fiction work to be "plausible" rather than just "possible" (even if only theoretically so, such as an alien landing tomorrow and bestowing a cure for cancer on us)[11] it must take into account where we are right now in our understanding. Looked at in this light, the chances of most of the topics covered previously—long-term suspended animation, dra-

---

[10] See my article "Bad Medicine: When Medical Research Goes Wrong" (*Analog Science Fiction and Fact*, September 2010) for more about this.

[11] This scenario was depicted in *The Twilight Zone* episode "The Gift" (1962).

matically increasing the maximal human lifespan while maintaining health, regenerating complex human organs via stem cells or other means, nanotechnology that improves on our own cells' abilities for repair and functioning, using genetic engineering to make us "more" than human, uploading our minds into machines—seem, at least for now, decades away at best. This is true for no other reason than that, even if methods to do these things were discovered tomorrow, it would takes years of testing to confirm that they work and are not associated with significant risks of their own.

Worse, for some of these ideas we may hit a roadblock imposed by Nature, such as the speed of light for material objects. Thus, perhaps it will prove to be essentially impossible to safely place a whole human body in its present form into suspended animation indefinitely and revive it without extensive brain damage. Or it may prove theoretically possible to do something but, at a given time, "impossible" from a practical standpoint due to the enormous difficulties, cost, risks, etc. it entails. For example, the laws of physics allow us to build a crewed starship that could travel at a significant fraction of the speed of light to the Alpha Centauri star system. However, there are obviously great engineering and other scientific difficulties involved with building such a craft and keeping its crew alive long enough to reach their destination, to say nothing of the political, economic, and other impediments involved. Hopefully that starship will be built someday—but it will definitely not be tomorrow or the day after.

In science fiction it is perfectly acceptable to go from point A—what we know and can do now—to point Z, all those wondrous new things we might learn and do sometime in the future—and gloss over the details of all the letters in-between that explain exactly how we get between those two points. But in real life, whether in medicine or any of the other sciences, it does *not* work that way. Experts may have to perform intensive research for years to get to point B in a particular medical or other scientific area, more decades to reach point C—and then perhaps discover that, no matter how hard they try, they cannot get to point D. Even worse, researchers may not know whether that is because point D does not exist or because they are not looking for it in the right way. Ultimately the only proof that it really is possible to get from point A to point Z is to actually do it. Until then, we humans simply cannot be certain what is possible and what is not.

Based on the current status of the different topics discussed in previous chapters, it is possible to estimate how plausible they are overall and when they might become "real." The two main medical issues with long-term living in space (Chaps. 3 to 6)—the effects of microgravity/reduced gravity and radiation exposure—could be dealt with adequately if not completely in the relatively near future by either engineering means (e.g. generation of artificial gravity while in space or better physical shielding against radiation) or biological ones (such as better medications to maintain musculoskeletal mass and protect against radiation).

Solutions to the engineering issues could theoretically be done now, but their implementation is limited more by practical issues (e.g. available launch vehicles, the mass of shielding on spacecraft, the need to construct large habitats in space to generate close to "normal" terrestrial gravity, etc.) than by lack of knowledge. Biological methods for protecting the human body and helping it stay anatomically and physiologically healthy in space will remain a challenge to varying degrees for some time to come. At present the medications used to do this have limited effectiveness, and the issues described in Chap. 10 regarding babies and children in space will be even more difficult to manage. Altering the human body itself via genetic engineering, nanotechnology, or other advanced means to better tolerate being in space has a considerably more uncertain future. Barring radical breakthroughs in these areas they seem to be most optimistically decades away at best, with a greater possibility but no certainty that they will prove feasible even a century or more from now.

The outlook for placing a whole person into suspended animation and successfully reviving that individual also appears pessimistic. As noted in Chap. 7, the sheer size of the human body and the number, complexity, and vulnerability of its many cells make any proposed methods—cooling (even with vitrification), drugs, etc.—very difficult if not nearly impossible to implement. Using suspended animation for space travel or other purposes in science fiction can, based on current capabilities and projections, be considered "possible but highly implausible."

On the other hand, further developments such as improved methods for temporarily replacing human blood and cooling a person could, though still not producing complete suspended animation, reduce metabolic activity longer than can currently be done. Being able to do the latter for even a few hours at a time could be critical for treating a severely injured person, even if the method is inadequate for helping reach Tau Ceti. Inducing a mildly hypothermic state that in some ways mimics "hibernation" might be feasible, but its role as a practical and reasonably safe technique for extended (e.g. weeks or more) periods of time is still very uncertain. Conversely, significantly better techniques for preserving whole organs for transplantation could well be developed in the near future.

The likelihood of either discovering or developing the paranormal abilities described in Chap. 8 is dismal. Technological analogs for a few of these abilities might perhaps be possible, such as an implanted or wireless "telepathic transceiver" being used to detect electrical signals in the brain corresponding to thought and send them to another similarly equipped individual. However, overall and most optimistically those concepts are firmly entrenched in the "highly implausible" to "impossible" categories.

The prospects for significantly increasing *maximal* human lifespan are also pessimistic at present. As noted in Chap. 9, the sheer complexity of the task, our still very limited knowledge of the aging process throughout the human body, and the simple need to wait many years to see how well a particular method actually works in humans make any major advances in the near future highly unlikely and, for science fiction story purposes, implausible. Any significant impacts on this may require breakthroughs in multiple fields such as genetic engineering and nanotechnology that, relatively speaking, are still themselves in their infancy. Medical advances over the next hundred years will with near certainty include significant ways to improve "healthspan"—how long and well we stay healthy during our lifetimes—and allow more individuals to live for a decade to two beyond a century in (hopefully) reasonable health. Likewise average life expectancy could and hopefully will come closer to what currently appears to be maximal lifespan. However, unless perhaps if they are set far into the future, science fiction depictions of characters routinely living for centuries fall into the "possible but highly implausible" category.

On the other hand, successful human cloning and development of an artificial uterus may lie not too far in the future, at least as proofs of principles. As noted in Chap. 10, the overall efficiency of current cloning techniques for mammals is low, with relatively few healthy live births compared to the number of attempts. If enough such attempts were made with humans, however, it is possible that one of them might be (at least relatively) successful. Thus, looked at strictly from the standpoint of what is biologically feasible, human cloning as a near-future capability can indeed be considered "plausible." As will be discussed in the next section, however, this is an area where there may be a wide gap between what "can" be done and, from an ethical standpoint, "should" be done, particularly given the risks involved with the methods currently available.

The likelihood of a safe, effective artificial uterus being developed in the next few decades seems low, primarily due to the difficulty of providing and maintaining all the complex life-sustaining functions of the natural maternal circulation to a human embryo or fetus. However, given enough time and effort, it is plausible that an artificial uterus could become a reality sometime over the next century. Here too, however, if such a device were too complex, resource intensive, costly, and/or had a low probability of bringing a healthy fetus to term, its use might be limited to being a "last resort" to save the life of an embryo or fetus rather than a routine replacement for "standard" pregnancy.

As described in Chap. 11, medical nanotechnology is advancing at a pace far slower and in much less "spectacular" ways than envisioned in a work such as K. Eric Drexler's 1986 book *Engines of Creation*. In this case the path from "concept" to "application" has been a long one, with even the most advanced current "proof of principle" experiments in the laboratory being far away from clinical use in humans to heal, rejuvenate, or improve our bodies. More promising approaches such as the use of nanotechnology to assist in treatment of cancers or to help grow new organs via stem cells or other means still have to go through a long period of animal and human testing before their overall benefits and risks compared to more "conventional" methods can be assessed.

Based on the current status of medical nanotechnology, science fiction depictions of nanobots repairing cells or making dramatic changes to the inside or outside of the human body, will be solidly in the "implausible" category for many years to come, if they happen at all. It is certainly conceivable that one day the equivalent of a nanobot might help repair a cell in the laboratory. However, translating that ability to the much more complex environment of the human body, with all the issues of getting a sufficient number of nanobots (either directly or by their own replication) to a target organ, shielding them from the body's immune system, keeping them within the body long enough to do their intended job without damaging it, testing such techniques in enough individuals to assess their overall safety and efficacy, etc. will take much longer—and with no guarantee of clinically relevant success. In short, many of the "magical" properties of medical nanotechnology depicted in science fiction must—at least for now—be described as "possible but highly implausible."

Like nanotechnology, genetic engineering is imbued in some science fiction with astounding capabilities. However, as shown in Chap. 12, in real life such techniques are still very limited. Such manipulations are essentially confined to single genes, particularly in human studies, and both the techniques to do this and ways of assessing their effectiveness are still in their infancy. Note too that current efforts are confined to therapeutic "repairs" of congenital genetic defects or damaged/diseased tissues. This is in contrast to much science fiction, in which genetic engineering is used to make wholesale changes to the human body (for good or ill), even ones such as "intelligence" that are intrinsically complex, polygenic, and associated with many epigenetic and environmental factors.

It is perhaps not too much of an exaggeration to compare the current status of genetic engineering and the amazing advances in that field depicted in some science fiction to the difference between the bone thrown into the air and the spaceship it "changes" into early in *2001: A Space Odyssey*. That analogy illustrates that, although it might indeed be possible to go from the "technology" of bone-as-weapon to that of a spaceship, the figurative road between them can be long and complicated.

As described in Chap. 13, stem cell research is a very active field at the laboratory and animal studies levels, and starting (however tentatively in some areas) to cross over into more extensive clinical trials. Its path and pace are more difficult to evaluate. Each type of stem cell has its own potential risks and benefits, and which ones will prove most useful in different situations is still uncertain. Within the next few decades stem cells may well play a relatively routine role in repairing, regenerating, or even possibly rejuvenating tissues, as well as helping grow new organs, etc. These uses have at least some claim to being "plausible" over that time frame. How "easy" such treatments will be, their cost, how long it would take to implement them, and other details are less certain.

Organ transplantation is already a reality in many areas, with head/brain transplantation being a major exception. The challenges in this field are significant but primarily involve improving current techniques in particular areas, such as preserving organs, increasing the supply of donor organs, avoiding long-term complications such as rejection, etc. rather than requiring any radical innovations. Significant near-future improvements in these areas are quite plausible. However, it should be emphasized once again that, while organ transplantation can improve the recipient's quality of life and increase that individual's lifespan, this technique has—at least by itself—considerably less value as a way for a person to live beyond the overall natural human lifespan of a little over 120 years. Having the option of swapping out a damaged or "old" liver or heart for a new one will be of limited value for this purpose if cancer, the general loss of critical functions (including those of our brains), and other effects of aging kill a person first or render quality of life too poor.

The outlook for some areas of bionics and body-machine interface are, as noted in Chap. 14, very good. These mainly involve the rapid rate of improvements in electronics, computer technology, and materials science and engineering. Interfacing these technologies with body parts, particularly the components of the nervous system that act as biological "wiring"—including peripheral nerves and parts of the brain involved with motor/sensory functions—has been increasingly successful. Although not without further challenges (such as improving "bionic eyes" and the like), these areas should continue to advance and are indeed "plausible."

However, no matter how small electronic circuitry becomes or computer processing power increases, coupling them with the "higher" functions of our brains or "uploading our minds" represent dramatically different, difficult, and uncertain issues. The critical one is our very incomplete knowledge regarding how the parts of the human brain associated with memory, personality, "thinking," and the like actually do this except in the most general details of anatomy, physiology, cell biology, etc. For example, creating a "neural implant" in someone's brain to provide that person with information presupposes that it actually can communicate in some way with the "organic" part of its host and that the latter can process and (either consciously or perhaps unconsciously) become aware of and/or use that data.

Thus, even if an implant has a great deal of information stored on its "chip" or can access the Internet or other external database for updates, there is no known "cerebral shortcut" to get all that data into our "awareness" via real-time transfer, storage as "memories" in organic or electronic media, etc. Until we learn a great many more details about what the human "mind," "consciousness," "personality," etc. are (at least from a purely functional standpoint) and ways to interface electronics with "them," neural implants that do this and techniques to "upload the mind" or even emulate them to create a "person-on-a-chip" must fall into the "highly implausible" realm.

Of course, all these preceding assessments of the pace and probability of science fiction-related medical advances are by their nature tentative, with a wide margin of uncertainty. Any of a number of unexpected breakthroughs could radically shorten the time frame for achieving them, or unforeseen difficulties either push them farther into the future or render them (at least from a practical standpoint) "impossible." As a general statement, knowledge of medical science and how to apply it will (barring the Earth being struck by a large enough asteroid or other catastrophe any time soon) clearly continue to grow, with some advances being sudden, unexpected, dramatic, and true "game changers."

However, it is a logical fallacy to argue from this general idea to a specific one that *particular* advances and discoveries will be made, or to make all but the broadest assessments of what it will be possible to do within a certain number of years. Any opinion on this topic must be based on actual facts, a reasonable assessment of current capabilities and the problems that must be solved to make a major advancement—and be flexible enough to change if circumstances/knowledge (e.g. a breakthrough occurs or a promising area of research fails) also change.

Of course, it bears repeating that science fiction always has the option of throwing "plausibility" to the wind for story purposes and going instead with "possible though highly unlikely." It can even include the "impossible," with the latter justified particularly as a "thought experiment" with only partial grounding in reality, such as an exploration of human nature, social or philosophical commentary, etc. However, if a writer's goal is to make a work more realistic or write "hard" science fiction, the issues regarding medical plausibility I have just described should at least be taken into account.

15.5 What Price Progress?

There is a subtle but significant medical issue with science fiction depicting some futuristic medical technologies as being routine. Such methods as human cloning, an artificial uterus, genetic engineering and nanotechnology to "enhance" humans, advanced stem cell therapies, brain/neural implants, or uploading the mind into a computer, etc. will inevitably require a great deal of testing before they become "mature" techniques and potentially part of standard clinical practice. The medical research that will be needed to do that, however, raises a variety of ethical issues.

The Declaration of Helsinki, first adopted in 1964 by the World Medical Association, includes core ethical principles for conducting human medical research. It follows other formal statements regarding the physician-patient relationship such as the Nuremburg Code (1947) and the Declaration of Geneva (1948), with the latter two created in response to atrocities committed by some physicians during World War II. Such documents include important statements regarding the duties of physicians and the rights of patients. These include that the well-being of the patient is the first and most important consideration; that voluntary, informed consent must be obtained from a patient/research participant or, if the latter cannot give it due to being too young or because of mental incapacity, by a legal guardian; and that every effort be made to minimize risk to the patient/participant.

Thus, based on these principles and the general need to try to maximize benefits and keep risks as low as possible in delivering medical care, using genetic engineering for therapeutic purposes on young children with severe genetic disorders that will inevitably disable, cause great suffering, and eventually kill them is reasonable. In such cases the potential benefits of gene therapy are great and its expected overall risks, though perhaps significant, still less than not doing it. These issues would need to be discussed with the child's parent(s) or legal guardian, who can then decide whether or not to give informed consent based on their own assessment of what is in the child's best interest.

Such issues become murkier in other situations. In the case of a "healthy" adult who consents to genetic engineering, being injected with nanobots, or some other treatment whose goal is to make him or her "healthier" or "enhanced" in some way, the ethical question raised is whether the possibility of *any* potential serious risks, no matter how remote, would outweigh these possible benefits as considerations. In science fiction the writer can decide whether the treatment is a resounding success, as it was for Steve Rogers, aka "Captain America." In real life, however, such a procedure might seriously injure or kill a previously healthy individual—hence the ethical issue. And even if the person is willing to take that risk, the physician involved also has the responsibility to make an ethical decision regarding what is in the best interests of that individual.[12]

Similar considerations of risk versus benefit also enter into decisions regarding using an electrode array surgically implanted in the brain. This type of operation would carry both the short-term risks of the operation and long-term risks of potentially causing an infection in the brain or other damage. However, if this method could help a paraplegic walk by using "thought" to control electronically enhanced leg braces—a "therapeutic" intervention designed to improve quality of life—its benefits could reasonably be assessed as outweighing its risks. On the other hand, a hypothetical neural implant designed to increase a person's intelligence might carry significantly higher risks, particularly when the device was first tested in humans and its full range of possible adverse effects was still unknown. Along with those risks there would also be no absolute guarantee the implant would work and benefit the individual. These issues would render such a procedure more problematic from an ethical standpoint.

Likewise, a person expected to die soon from advanced age or a terminal disease would have little to lose and much to gain if his or her "mind" could be uploaded to a computer, even if that process required (as an extreme and somewhat facetious example) removing the brain (thus effectively killing the person's body) and running it through a high-tech blender to scan its "contents." On the other hand, testing such a procedure on young, healthy volunteers would be more ethically problematic, considering that it could potentially cause their very untimely demise if the experimental process did not work rather than conferring a type of "immortality" on them.

These issues become more acute when conducting "experiments," no matter how well intended, on human subjects/patients who innately cannot give informed consent. This includes stages in the development of human life ranging from embryos to anytime during childhood, before a person can give informed consent as an adult. Thus, the potential use of gene therapy in utero before a severe genetic disorder begins to cause damage prior to birth could be justified on a risk/benefit ratio. Conversely, using genetic engineering as an experimental procedure—once again, a necessary stage before it might eventually be "perfected" and become "routine" to create "designer babies" or other reasons—on an otherwise normally developing embryo/fetus prior to birth, even if the goal is to "improve" it, is associated with serious ethical questions.

---

[12] As illustrated by Robert Louis Stevenson's *Strange Case of Dr. Jekyll and Mr. Hyde* (1866), it is also prudent for the physician him- or herself to not become the first human test subject for evaluating a new medication.

Similarly, using an investigational artificial uterus to try to save very premature fetuses that would otherwise die would have a reasonable risk/benefit ratio. Testing that artificial uterus on an embryo specifically created to assess the device's safety and efficacy, with the attendant risks of destroying the embryo or bringing it to term with severe damage caused by the process, is considerably more ethically suspect. These considerations also apply to any attempt to successfully bring a human "clone" to term. To produce one "Dolly" might well result in many embryos and fetuses dying before birth or, even worse, surviving to birth with problems associated with the cloning process that cause severe disabilities, suffering, and/or early death.

Let me emphasize once again that all these procedures—human cloning, an artificial uterus, advanced genetic engineering, etc.—will not suddenly and magically appear in full, perfected form followed by a physician or other scientist shouting, "Eureka!" They will all require extensive testing on human subjects (after, presumably, "passing" the stages of laboratory and animal evaluation) to become practical realities—and there is no guarantee that any particular series of tests will prove successful. Thus, the net "benefits" in knowledge and capabilities gained after inflicting a tremendous amount of human suffering and death from such testing might still be minimal.

Moreover, another fundamental principle of medical research is that, although it can be done to help the health of others besides the patient—to provide a larger "social good"—that consideration should *not* outweigh the need to keep the patient's welfare as the highest value, and to conduct the research only with the patient/test subject's informed consent. Put another way, the "need to break some eggs to make an omelet" idea cannot, based on standard medical ethics, be used to justify deliberately injuring or even putting at unnecessary risk a small number of individuals solely or even primarily to benefit a larger number.

Of course, individuals, groups (e.g. corporations), and governments might choose to ignore all these ethical issues and simply test the limits of what it is physically possible to do. Science fiction and medical thrillers are replete with such situations, and unfortunately they could certainly occur in the "real world" too. However, "standard" (although not necessarily universal) procedure is for a research protocol to undergo a formal review process that includes taking the ethical principles described previously into account as primary considerations before it is approved. Such reviews act to at least curb (although obviously cannot by themselves eliminate) flagrant abuses involving medical research. Moreover, in some cases there can be reasonable disagreements on how to interpret and apply ethical principles. Nonetheless, the primary goal is to make sure they are applied as best as can be determined.

In short, science fiction does not have to depict a profit-first multiplanetary corporation or tyrannical government using techniques such as advanced genetic engineering, an artificial uterus, etc. for flagrantly harmful purposes to raise ethical issues. Even when those methods are shown as being employed for (at least to some degree) benign and beneficial reasons, the research that led up to them may have unavoidably involved actions that did in fact violate the principles described in this section. That is something for the science fiction writer and reader to consider if they choose to ponder a plot and what it really means beneath its surface, even if the latter has other elements that are appealing.

15.6 The Bottom Line

Modern-day, futuristic, and alien medicine can all play a primary, pivotal, or supporting role in science fiction. Its depictions of how medicine is practiced and the medical information presented can reasonably include widely varying proportions of "science" and "fiction," ranging from the strictest realism to elements whose plausibility or even possibility are extremely suspect. At one end of the spectrum, too much realism can actually detract from a story and impede its flow, such as by adding unnecessary details that do nothing to advance the plot, provide setting or characterization, etc. At the other end, flagrant medical errors—especially ones that could have been corrected without harming the story—are best avoided.

Science fiction writers can choose to use any or none of the facts, analyses, and other information contained in this book for a particular work dealing with medical issues. Science fiction readers are free to suspend disbelief as much or little as they wish when reading a work that contain depictions of medicine that, after perusing this and preceding chapters, they will (hopefully) recognize as being not entirely accurate. Hopefully the material I have presented will give a better appreciation for the complexity of the human body, both the science and art of medicine, the challenges that must be overcome to improve health and longevity, to establish a permanent human presence away from Earth, and perhaps even "improve" our biology.

Ultimately no one can make more than an "educated guess" about what the medicine of the future will actually be like. However, science fiction can grant us visions far beyond the visible horizon of mere reality to whole new worlds where medical science may or may not one day go. I hope this book will help provide a background and points of comparison to better appreciate, understand, and perhaps even create the medicine in science fiction.

Appendix

Discussions of Specific Medical Issues in Science Fiction

This is a chapter-by-chapter summary of detailed discussions regarding specific medical issues in science fiction. The page number where each discussion begins is given in parentheses.

Chapter 1. How the Human Body Works: From Quarks to Cells

1. Why the subatomic worlds depicted in early works such as Ray Cummings's 1922 novel *The Girl in the Golden Atom* are not realistic. (Page 3)
2. A general comparison of humans and sentient aliens in science fiction. Examples cited include the Kzinti and Puppeteers of Larry Niven's "Known Space" works and the Mesklinites of Hal Clement's 1954 novel *Mission of Gravity*. (Page 7)
3. Green blood in *Star Trek*'s Vulcans and its relation to terrestrial biology. (Page 33)
4. "Shapeshifters" in science fiction and their relation to actual biology. Examples cited include John W. Campbell's 1938 story "Who Goes There?" and Ray Bradbury's 1949 work "The Martian." (Page 35)

Chapter 2. Hurting and Healing Characters

1. Depictions of major infections in science fiction. Examples cited include H.G. Wells's novel *The War of the Worlds* (1898), C. L. Moore's and Henry Kuttner's story "Vintage Season" (1946), and the 1971 movie *The Andromeda Strain*. (Page 47)
2. Use of sentient (and otherwise) alien parasites and symbiotes. Examples cited include Robert A. Heinlein's 1951 novel *The Puppet Masters* and the Goa'uld from the *Stargate SG-1* television series. (Page 50)
3. Effective depictions of neurological and musculoskeletal disorders in science fiction. Examples cited include "Bendii Syndrome" in the 1990 *Star Trek: The Next Generation* episode "Sarek" and congenital myasthenic syndrome in Robert A. Heinlein's 1942 story "Waldo." (Page 53)

4. The "manual cardiotomy" performed in the 1984 movie *Indiana Jones and the Temple of Doom*. (Page 62)
5. Nature of the fatal injuries inflicted on Qui-Gon Jinn in the 1999 movie *Star Wars: The Phantom Menace*. (Page 66)
6. Depictions of fatal head and neck injuries in science fiction. Examples cited include the 1962 movie *The Brain That Wouldn't Die* and 2013's *Man of Steel*. (Page 67)
7. Potential difficulties Dracula-style vampires might encounter when attempting to drain blood from the necks of victims. (Page 68)
8. Rapid recovery from head injuries in fictional works such as many Republic Pictures serials compared to their actual effects. (Page 70)
9. Fictional versus actual effects of explosions on the human body. (Page 73)
10. How physicians are presented and used in science fiction. (Page 76)
11. Feasibility of the *Star Trek* hypospray. (Page 80)
12. Depictions of death in science fiction. (Page 82)
13. Fictional presentations of cardiopulmonary resuscitation. (Page 82)

Chapter 3. Space Is a Dangerous Place
1. Comparison of early fictional depictions of space travel to actual ones. (Page 89)
2. Real versus imagined effects of walking on Mars using supplemental oxygen alone, without a spacesuit. Example cited is the 1964 movie *Robinson Crusoe on Mars*. (Page 93)
3. Realistic versus erroneous depictions of exposure to vacuum in space. Examples cited include the movies *2001: A Space Odyssey* (1968), *Mission to Mars* (2000), and *Total Recall* (1990). (Page 94)
4. Comparison of preparations for extravehicular activity in fiction compared to actual procedures. (Page 98)
5. Risk of collision between spacecraft and orbiting debris. Example cited is the 2013 movie *Gravity*. (Page 109)
6. Collisions by asteroids, comets, and rogue planets with the Earth. Examples cited include Larry Niven and Jerry Pournelle's *Lucifer's Hammer* (1977), the 1998 movie *Armageddon*, and 1933 novel *When Worlds Collide* by Philip Wylie and Edwin Balmer. (Page 112)
7. Modifying humans to live in extraterrestrial environments. Examples cited include Kevin J. Anderson's *Climbing Olympus* (1994) and Anne McCaffrey's *The Ship Who Sang* (1969). (Page 114)
8. "Freezing" in space. Example cited is Edward M. Lerner's 2012 novel *Energized*. (Page 115)

Chapter 4. Microgravity and the Human Body

1. A functional biological modification to better adapt the human body for living in microgravity. Example cited is the "Quaddies" from Lois McMaster Bujold's 1988 novel *Falling Free*. (Page 136)
2. Early concepts for producing artificial gravity in space. Examples cited include the television series *Men into Space* (1959) and *2001: A Space Odyssey* (1968). (Page 138)
3. Using whole worlds as "spaceships" to provide roughly Earth-level artificial gravity. Examples cited are Stanley Schmidt's *Lifeboat Earth* (1978) and the *Fleet of Worlds* series (2007–2012) by Larry Niven and Edward M. Lerner. (Page 141)
4. Thought experiment on using white dwarf or neutron star material as a means of artificial gravity on a spacecraft. (Page 141)
5. Limitations of using "magnetic boots" as a partial simulation of artificial gravity. Example cited is the movie *Destination Moon* (1950). (Page 144)

Chapter 5. Space Medicine: Paging Dr. McCoy

1. General science fiction depictions of medical care in space and extraterrestrial environments. Examples cited include the movie *The Empire Strikes Back* (1980) and Larry Niven's novel *Ringworld* (1970). (Page 151)
2. Clinical decisions regarding management of a spacefarer who has an acute myocardial infarction ("heart attack") on the International Space Station. Example cited is my story "Hearts in Darkness" (2002). (Page 159)
3. Difficulties in treating severely ill or injured astronauts far from Earth and the possibility they cannot be saved. Example cited is my story "The Last Temptation of Katerina Savitskaya" (2008). (Page 160)
4. Use of futuristic technology in science fiction to provide improved medical care. Examples cited include *Star Trek*'s medical tricorder and the instruments described in C. M. Kornbluth's story "The Little Black Bag" (1950). (Page 178)
5. Extended discussion of the feasibility of the "autodoc" described in Larry Niven's *The Patchwork Girl* (1980) and other works. (Page 178)

Chapter 6. Danger! Radiation!

1. Descriptions of how radiation affects living beings in science fiction. Examples cited include Walter M. Miller, Jr.'s novel *A Canticle for Leibowitz* (1960) and the movie *Beginning of the End* (1957). (Page 187)
2. Depictions of lethal radiation exposure in science fiction. Examples citied include the 1998 novel *Lethal Exposure* by Kevin J. Anderson and Doug Beason, and the movie *Star Trek II: The Wrath of Khan* (1982). (Page 197)

3. Early science fiction depictions of nuclear power plant accidents. Examples cited are Robert A. Heinlein's "Blowups Happen" (1940) and Lester del Rey's "Nerves" (1942). (Page 199)
4. Effects of space radiation on astronauts. Example cited is James A. Michener's 1982 novel *Space*. (Page 202)
5. Real-life radiation hazards near Jupiter ignored by science fiction works. Examples cited include Isaac Asimov's "Christmas on Ganymede" (1942) and the 1956 movie *Fire Maidens of Outer Space*. (Page 203)
6. Using radiation exposure realistically in science fiction. Example cited is Robert A. Heinlein's short story "The Long Watch" (1949). (Page 207)

Chapter 7. Suspended Animation: Putting Characters on Ice
1. Uses of suspended animation in science fiction. Examples cited include Philip Francis Nowlan's *Armageddon 2419 A.D.* (1928) and A. E. van Vogt's "Far Centaurus" (1944). (Page 211)
2. Long-term suspended animation induced by freezing in science fiction compared to the human body's actual response to cold. Examples cited include Edgar Rice Burroughs's 1937 story "The Resurrection of Jimber-Jaw" (1937) and Sir Arthur C. Clarke's 1997 novel *3001: The Final Odyssey*. (Page 217)
3. Potential brain damage caused by induction of cryogenic suspended animation described in the 1987 novel *The Legacy of Heorot* by Larry Niven, Jerry Pournelle, and Steven Barnes. (Page 232)
4. Possible reference to the potential use of hydrogen sulfide gas to help induce suspended animation mentioned in the novel *Bowl of Heaven* by Gregory Benford and Larry Niven (2012). (Page 236)
5. Issues with science fiction depictions of head/brain transplantation and the "brain in a jar." Examples cited include the 1962 movie *The Brain That Wouldn't Die*, Robert A. Heinlein's novel *I Will Fear No Evil* (1970), and Curt Siodmak's 1942 novel *Donovan's Brain*. (Page 238)
6. Alternatives to biological means of producing suspended animation based on real or imagined physics. Examples cited include Poul Anderson's novel *Tau Zero* (1970), Vernor Vinge's *The Peace War* (1984), and my own short story "To Him Who Waits" (1999). (Page 239)
7. Science fiction depictions of biological methods for inducing suspended animation that do not primarily involve lowering body temperature. Examples cited include the movie *The Empire Strikes Back* (1980) and Allen Steele's 2002 novel *Coyote*. (Page 242)
8. Specific use of suspended animation for space travel. Examples cited are *2001: A Space Odyssey* (1968) and Charles E. Gannon's 2013 novel *Fire with Fire*. (Page 243)

Chapter 8. Telepathy, Using the Force, and Other Paranormal Abilities

1. Depictions of a variety of "psychic powers" in science fiction. Examples cited include the 1968 movie *The Power*, Jerome Bixby's story "It's a *Good* Life" (1953), and Alfred Bester's 1956 work *The Stars My Destination*. (Page 249)

2. Some uses of telepathy in science fiction. Examples cited include Robert A. Heinlein's novel *Time for the Stars* (1956) and A. E. van Vogt's 1946 novel *Slan*. (Page 250)

3. Use of telekinesis in science fiction. Examples cited include the 1961 *Twilight Zone* episode "The Prime Mover" and Robert A. Heinlein's 1961 novel *Stranger in a Strange Land*. (Page 259)

4. The physics of "using the Force" to move objects via telekinesis. Example cited is *The Empire Strikes Back* (1980). (Page 262)

5. Issues with use of precognition in science fiction. Examples cited include Phillip K. Dick's short story "A World of Talent" (1954) and his 1956 novel *The World Jones Made*. (Page 264)

6. Basic principles of physics and their relation to possible ways of effectively teleporting objects and people. Examples cited include Alfred Bester's 1956 work *The Stars My Destination* and A. E. van Vogt's novel *The World of Null-A* (1948). (Page 270)

7. Practical issues with individuals being correctly or incorrectly teleported. Examples cited include the *Star Trek: The Next Generation* episode "Second Chances" (1993) and the movie *Star Trek: The Motion Picture* (1979). (Page 274)

Chapter 9. The Biology of Immortality

1. Fictional depictions of extreme longevity. Examples cited include Robert A. Heinlein's novel *Methuselah's Children* (1958) and the *Twilight Zone* television series episodes "Escape Clause" and "Long Live Walter Jameson." (Page 281)

2. A detailed description of rejuvenation treatments in Bruce Sterling's novel *Holy Fire* (1996). (Page 309)

3. Social and psychological issues with extreme longevity. Examples citied include "The Trade-Ins" episode (1962) of *The Twilight Zone* and my own story "The Best is Yet to Be" (1996) (Page 310)

Chapter 10. Sex in Science Fiction

1. Some ways of incorporating sex into science fiction. Examples cited include Isaac Asimov's "What Is This Thing Called Love?" (1961) and Ursula K. Le Guin's *The Left Hand of Darkness* (1969). (Page 321)

2. Use of the artificial uterus in science fiction. Examples cited include the novel *Brave New World* (1931) by Aldous Huxley and the movie *Attack of the Clones* (2002). (Page 329)

3. Issues with children living in space as portrayed in science fiction. Examples cited include Robert A. Heinlein's novel *Farmer in the Sky* (1950). (Page 349)

4. Potential factors limiting sexual attraction/intercourse between humans and aliens as well as production of offspring. Examples cited include Edgar Rice Burroughs's characters John Carter and Dejah Thoris. (Page 351)

Chapter 11. The Promises and Perils of Medical Nanotechnology

1. Some depictions of nanotechnology in science fiction. Examples cited include Neal Stephenson's 1995 novel *The Diamond Age* and the novel *Small Miracles* by Edward M. Lerner (2010). (Page 362)

2. Possible ways to use medical nanotechnology in science fiction. Examples cited include Arlan Andrews, Sr.'s stories "Hail, Columbia!" and "Other Heads" (both 1993). (Page 383)

Chapter 12. Genetic Engineering: Tinkering with the Human Body

1. Some portrayals of genetic engineering in science fiction. Examples cited include Edmond Hamilton's "The Man Who Evolved" (1931) and "The Sixth Finger" (1963) episode of *The Outer Limits* television series. (Page 389)

2. Reasons for performing genetic engineering of humans as depicted in science fiction. Examples cited include Charles Sheffield's *Sight of Proteus* (1978) and James Blish's *The Seedling Stars* (1957). (Page 401)

3. Discussion of a method for genetically enhancing intelligence described in the 1991 novella "Beggars in Spain" by Nancy Kress. (Page 410)

4. Technical challenges to making major genetic changes to the human body. Examples cited include Lois McMaster Bujold's novel *Falling Free* (1988) and Edmond Hamilton's 1938 story "He That Hath Wings." (Page 418)

Chapter 13. Stem Cells and Organ Transplantation: Resetting Our Biological Clocks

1. Fictional depictions of organ harvesting for later transplantation. Examples cited include Robin Cook's 1977 novel *Coma* and Larry Niven's 1969 story "The Organleggers."(Page 430)

2. Discussion of criteria for brain death as a necessary condition prior to harvesting of organs essential for life. Robert J. Sawyer's novel *The Terminal Experiment* (1995) is cited as a depiction of this issue in science fiction. (Page 454)

3. Use of human clones as sources of organs for transplantation. Examples cited are the movies *Parts: The Clonus Horror* (1979) and *The Island* (2005). (Page 456)
4. Science fiction depictions of using cellular alteration, transplantation, and other techniques to make major changes in a person's body. Examples cited are H. L. Gold's short story "No Charge for Alterations" (1953) and F. L. Wallace's "Tangle Hold" (1953). (Page 459)

Chapter 14. Bionics: Creating the Twenty-Four Million Dollar Man or Woman
1. Depictions of interfacing electronic and mechanical enhancements directly to the human body. Examples cited include the mid-1970s television series *The Six Million Dollar Man* and *The Bionic Woman*. (Page 467)
2. Description of different ways of blending human tissue with artificial parts. Examples cited include the 1966 movie *Cyborg 2087* and C. L. Moore's 1944 story "No Woman Born." (Page 468)
3. An analysis of how Luke Skywalker's robotic right hand might interface with the remainder of his biological upper extremity and its potential limitations. The 1980 movie *The Empire Strikes Back* is cited. (Page 472)
4. Use of future advances in several fields to create artificial enhancements for the human body. Example cited is Alan Dean Foster's novel *The Human Blend* (2010). (Page 476)
5. Discussion of Geordi La Forge's VISOR as a means of providing vision in the *Star Trek: The Next Generation* television series and movies. (Page 480)
6. Description of a novel way to "enhance" vision described in Nancy Kress's story "Someone to Watch over Me" (2014). (Page 482)
7. A comparison of modern-day partial or complete mechanical replacements for a heart compared to type implanted into Captain Jean-Luc Picard in the *Star Trek: The Next Generation* television series. (Page 485)
8. Description of various types of "synthetic" organs in Kevin J. Anderson's novel *Resurrection, Inc.* (1988). (Page 486)
9. Discussion of a realistically depicted brain-machine interface in Brad Aiken's story "Locked In" (2010). (Page 492)
10. Ways of depicting transfer of human consciousness to electronic or other artificial media in science fiction. Examples cited include Frederick Pohl's story "The Tunnel Under the World" (1955) and Robert Silverberg's novel *To Live Again* (1969). (Page 492)
11. Specific issues potentially limiting the ability to transfer a human "mind" into a computer as depicted in Robert J. Sawyer's novel *Mindscan* (2005). (Page 497)

Chapter 15. Summing Up: Using Medicine in Science Fiction

1. Examples of "tell, don't show" style of writing in two novels by H. G. Wells, *The Time Machine* (1895) and *The Invisible Man* (1897). (Page 509)

2. A discussion of when using detailed medical information in science fiction can be reasonable, using my story "Hearts in Darkness" (2002) as an example. (Page 510)

3. Satiric use of "tell, don't show" style of writing, using my story "Primum Non Nocere" (2010) as an example. (Page 512)

4. Employing a "minimalist" style of writing to quickly introduce a science fiction setting and situation. Examples cited include William Gibson's *Neuromancer* (1984). (Page 513)

5. Successfully integrating medical information into a science fiction work without slowing the story down. Example cited is Edward M. Lerner's novel *Small Miracles* (2009). (Page 514)

6. Limits to extrapolating development of future medical and other advances in science fiction. Examples cited include Hugo Gernsback's novel *Ralph 124C 41 +* (1911) and Jules Verne's *Twenty Thousand Leagues Under the Sea* (1870). (Page 514)

7. The value of science fiction even when it is unsuccessful in "predicting" the future or development of imagined technology. Examples cited include H. G. Wells's *The Time Machine* (1895) and Stanley G. Weinbaum's "A Martian Odyssey" (1934). (Page 515)

8. Defining "hard" science fiction. Example cited is Daniel Keyes's *Flowers for Algernon* (published in 1966 as a novel). (Page 516)

Index

A

Abbott, E.A., 275
Acceleration and deceleration, 113, 114
  during landing on Mars, 113
  gravitational acceleration, 142, 143
  injuries from, 55, 66, 152
    neck, 66
  linear, 136, 137
  spacecraft, 113, 136, 137
    Apollo capsule, 113
    Mercury capsule, 113
    Soyuz capsule, 113
    Space Shuttle and, 113
Adenosine triphosphate (ATP), 15, 215,
    216, 292
Advanced resistive exercise device
    (ARED), 131, 229
Aging, 44, 284, 291, 292
  as natural process, 284
  cognitive decline, 308
  delaying, 288
  Hayflick limit, 302
  healthspan, 296, 301, 520
  immortality, 281, 308, 493, 525
  life expectancy, 214, 295, 297, 301,
    308, 397, 430, 520
  longevity, 282, 288, 300, 301
    genes and, 287
  mice
    effects of dietary restrictions, 300
  rate, 214
  senescence, 292, 302, 304
  stem cells and, 434

  target of rapamycin (TOR), 289, 294
  telomeres, 303–307, 434
  theories of
    antagonistic pleiotropy, 294
    death genes, 294
    disposable soma, 294
    mutation accumulation, 295
    role of accumulated injuries, 292
    role of hyperfunction, 293
    role of oxygen, 292
    wear-and-tear, 44, 292
Algae, 8, 13
Aliens, 7, 76, 91, 151
  biochemistry, 7, 48
  in science fiction, 77
    examples based on nonhuman ter-
      restrial life, 7
    humanoid, 7
    interactions with humans, 42, 351
    longevity, 282
    medical care, 77, 151
    shapeshifters, 35
    teleportation, 249
  microorganisms, 48, 49
  sex and, 322, 351–355
  telepathy and, 250, 257, 278
Alleles (*See also* Genes), 391, 392
  aging and, 291, 295
  beneficial variants, 400
  co-dominant, 394
  definition, 17
  disease and, 291
  dominant, 392, 393, 395

heterozygous, 392–394, 396, 409
homozygous, 392–395, 409
recessive, 392–394
X-linked, 395, 403
Alzheimer's disease, 54, 284, 291, 296
genes associated with, 290
heritability of, 291
Amino acids, 9
aliens and, 49, 354
nutrition, 300
proteins and, 10, 12, 13, 48, 50, 354,
364, 391
right- and left-handed, 10
and terrestrial life, 10
Ammonia
alien life and, 8
as coolant, 100, 102
as part of comets, 111
generation by humans, 102
Anderson, K.J., 197, 486
Andrews, A., 383
Anesthesia, 175, 176
in microgravity, 175, 176, 348
suspended animation, 236
Antibodies, 34, 128, 331, 374, 381
Armstrong limit, 92
Arrhythmias, 25, 83, 127, 229, 347
atrial fibrillation, 219
definition, 25
in microgravity, 127
sudden death, 46, 452
ventricular fibrillation, 46, 219
commotio cordis, 69
ventricular tachycardia (VT), 46, 83,
485
with decompression sickness, 347
with heterotopic heart transplant, 458
with hypothermia, 218
with myoblast transplantation, 438
Artificial intelligence, 362
Asexual reproduction, 323, 324, 352
Asimov, I., 2, 54, 75, 203, 251, 259, 322,
362, 390, 469, 515
Aspirin, 412, 511

Atherosclerosis
and reactive oxygen species, 293
definition, 44
fibrous plaque, 45
in the elderly, 284, 291, 296
possible use of gene therapy for, 414
Atoms
binding with carbon, 6
comparison to the Solar System, 3
conversion to energy, 271
definition, 3
effects of a magnetic field, 260
free radicals and, 191
in a human body, 273
isotopes, 3
nucleus, 3
size and mass, 3
Austin, S., 467, 470, 482
Autodoc, 77, 151, 179
challenges to creating, 180, 182
comparison with needs for suspended animation, 244
definition, 178
needed capabilities, 183
nutrition and waste disposal, 180
preventing sores and ulcerations,
180
providing nutrients, 180
value in space, 178

B

Bacteria, 13
as causes of disease, 47, 167, 168
blood-brain barrier, 381
following high-level radiation exposure, 197
in spacecraft, 166, 168, 337
treatment with silver nanoparticles,
371
asexual reproduction, 323
beneficial, 28, 301
cell wall, 14, 382
cilia, 366
cyanobacteria, 13

effects of hypothermia, 229
extraterrestrial, 48
extremophiles, 168
   Deinococcus radiodurans, 168
genetic modification
   Escherichia coli, 408
in food, 166
in spacecraft, 166
number in the human body, 167
sexually transmitted diseases (STDs)
   and, 337
Streptococcus mitis, 168
Big Bang, 5, 241, 247
energy and, 5
matter and, 5
Bionics
cyborg, 468, 469
ears, 482
   cochlear implants, 482–484
eyes, 477, 479, 499, 522
limbs, 470–475, 476, 499
Blake, A., 390
Bones
appendicular skeleton, 30
bone marrow, 32, 130
   hematopoietic stem cells and, 433,
   435
   transplant, 421
cartilage and, 31, 61, 62
comparison of humans and birds, 418
cryopreservation of, 231
development, 6
   calcium and, 6, 30
   phosphorus and, 6, 30
effects of aging, 296
effects of dehydroepiandrosterone
   (DHEA), 298
effects of growth hormone, 31, 297
effects of microgravity
   animals, 349
   calcium loss, 130
   children, 349
   intervertebral disks, 128

loss of bone mass, countermeasures,
   130, 135, 229, 349, 350
loss of bone mass, in children, 349
effects of strontium-90, 198, 199
fibroblasts, 302
fractures, 58, 70, 130
   healing, 170
   traction for, 172
genetic modification, 413, 418, 424
growing replacements, 457
in ear, 483
injuries to, 55
   facial bones, 59
   in space, healing, 153, 170, 171
   joints, 57, 170
   limbs, 56
   neck and back, 104
   ribs and sternum, 62
measurement of bone mineral density,
   131, 298
modification by genetic engineering,
   400
number and range of sizes, 30
osteoporosis
   use of bisphosphonates, 131, 223
transplantation of facial bones, 452
vertebral column, 30, 60
Bradbury, R., 1, 35, 47, 89, 91, 469
Brain
aging and, 307, 309
   cognitive reserve, 309
anatomy, 21, 495
   meninges, 51, 60, 489
   ventricles, 490
blood supply, 66, 68, 69, 96, 347, 492
   effects of hypothermia, 218
blood-brain barrier, 11, 381, 384, 490
brainstem, 20, 59, 454, 492, 495
cerebellum, 21, 22
cerebral hemorrhage, 59, 348
cerebrum, 21, 22, 293
comparison to computers, 487, 495
damage, 46, 59, 219, 237, 346

cerebral hemorrhage, 59, 82, 348
concussion, 70, 74, 103
contrecoup injury, 70, 74
decompression sickness, 96, 347
due to hypothermia, 221
explosions and, 73
prions and, 49
radiation, 197, 198
stroke, 44, 68, 96, 164, 221
suspended animation, 216, 218,
    219, 237
development of, 350
effects of teleportation, 273
epilepsy, 416
imaging, 253, 497
interface with machines and comput-
    ers, 487–492
microelectrodes, 489, 491
isolated, 238
locked-in syndrome, 492
neuroplasticity, 309
neurovestibular system, 123, 139
number of neurons, 251
plasticity, 480, 483
power consumption, 255
relation to mind, 495, 497
specialization of structures in, 495
surgery, 221
synapses, 252, 308, 494, 496
transfer of memories and conscious-
    ness, 492, 494, 497
transplantation, 77, 238
Brown, F., 7
Bujold, L.M., 54, 136, 418, 458
Burns
hydrazines and, 100
sunburns, 172, 191
treatment of, 172
    silver nanoparticles, 371
types of, 72
    first-degree, 72, 172
    fourth-degree, 72, 460
    internal, 72, 73
    second-degree, 72, 172
    third-degree, 72, 172
Burroughs, E.R., 217, 282, 351

C

Calcium
abundance in the human body, 7, 274
atherosclerosis and, 45, 163
blood levels, 31, 32
    with cancers, 51
bones and, 19, 30
loss in microgravity, 130, 345
relation to kidney and ureteral stones,
    131
roles in the human body, 130
supplemental, 297
    in pregnancy, 345
total parenteral nutrition and, 227
urinary excretion of, 130
Calment, J.L., 282
Campbell, J.W., 35, 250
Cancer
aging and, 284, 289, 294
brain, 51, 444, 449
breast, 52, 297, 409
    genetics and, 394, 409
    radiation exposure and, 197
    treatment of, 372
definition, 191, 304
diagnosis of
    using nanoparticles, 369, 380
dwarfism and, 289
genetic screening for
    in astronauts, 162
lung, 52, 299, 370, 411
    beta-carotene and, 299
ovarian, 162, 394, 409
    genetics and, 394, 409
paraneoplastic syndrome, 51, 52
prostate, 294
radiation and, 197, 350
risk of developing, 43, 193, 292, 294
    in different organs, 52
role of lymphocytes against, 34
skin, 52, 162, 405, 451

melanoma, 52, 162, 405
stem cells, 434
   origination from, 447
teratomas, 444
thyroid
   radiation exposure and, 197
treatment of
   using gene therapy, 377, 379
   using hypothermia, 220
   using nanoparticles, 369, 371
   using nanotechnology, 367, 379
   using radiation, 376
   using telomerase inhibitors, 306
tumors and, 51
Carbohydrates (*See also* Glucose)
cellulose, 9
fructose, 9
galactose, 9
glycogen, 9, 181
lactose, 9
monosaccharides, 9
polysaccharides, 9, 407
sucrose, 9
   use as cryoprotectant, 231
Carbon
abundance in the human body, 274
abundance in the universe, 7
amino acids and, 9
carbohydrates and, 9
fullerenes
   use for nanotubes, 373
graphene
   use for nanotubes, 372
involvement in life processes, 6, 7
lipids and, 8, 10
Carbon dioxide
as part of comets, 111
atmospheric, 25
   photosynthesis and, 205
effects of increased levels on vision,
   132
hypercapnia, 116
in atmosphere of Mars, 93
levels on spacecraft, 132, 165

respiration and, 25, 44
Carbon monoxide, 346
fires and, 100
Cardiopulmonary resuscitation, 46, 82
in microgravity, 174
use in treatment of hypothermia, 220
Cardiovascular system (*See also* Heart)
aorta
   anatomy of, 23
   atherosclerosis and, 44
   effects of aging on, 44
   injury to, 63, 66
   Marfan syndrome and, 396
arteries
   atherosclerosis and, 44, 365
   basilar artery, 68, 492
   carotid, 23, 67
   decompression sickness and, 96
   definition of, 22
   effects of aging on, 44, 292
   penile, 327
   pulmonary, 23
   radial, 57
   spurting of, 58, 174
   subclavian, 23, 68, 180
   systemic, 58, 96, 229
   vertebral, 68, 132
atria
   anatomy of, 22, 62, 96
blood pressure, 21
   circadian rhythms and, 107
   definition, 25
   diastolic, 25
   during coitus, 336
   effects of anaphylactic reaction on,
     173
   effects of resistive exercise on, 132
   exercise tests and, 162
   exposure to vacuum and, 94
   hemorrhagic shock and, 58
   hypertension, during pregnancy,
     44, 284, 347
   hypothermia and, 218
   measuring, 25, 157

orthostatic reflexes and, in micro-
    gravity, 126, 159
pericardial tamponade and, 62
regulation of, 21, 29, 363
septic shock and, 65, 134
systolic, 25, 58, 336
telemedicine and, 158
vasovagal reflex and, 96, 116
capillaries, 23, 232, 365, 490
cardiac cycle
systole and diastole, 24
components of, 22
deep vein thrombosis, 229
ductus arteriosus, 330, 347
foramen ovale, 96, 229, 330
heart rate, 21
    atrioventricular node and, 25
    circadian rhythms and, 107
    during coitus, 336
    effect of exercise on, 25
    exercise tests and, 162
    hypothermia and, 218
    monitoring during extravehicular
        activity, 127
    orthostatic reflexes and, in micro-
        gravity, 126
    regulation of, 21
    sinus node and, 25
    sinus tachycardia, 84
    telemedicine and, 158
    vasovagal reflex and, 96
    ventricular tachycardia and fibrilla-
        tion, 46
thrombus, 44, 79, 224
valves
    aortic valve, 23
    mitral valve, 23
    number of leaflets, 24
    pulmonic valve, 23
    tricuspid valve, 23
veins
    bleeding in microgravity, 174
    decompression sickness and, 96
    definition of, 22

effects of microgravity on, 125
inferior vena cava, 23, 66
jugular, 67, 132, 180, 224
oozing of, 174
placement of intravenous line, 224
pulmonary, 23, 58
varicose in pregnancy, 345
ventricles
    anatomy of, 22, 24
    echocardiography of, 162
    myocardial infarction and, 45
Cartilage, 19
    ears and, 457, 484
    gene therapy and, 414
    intervertebral disks and, 60
    knees and, 43, 57
    production by chondrocytes, 432
    ribs and, 61
Catecholamines, 32, 127, 135, 496
Cells
    cell membrane, 11, 14, 215, 254, 255,
        367, 372, 376
    cytoplasm, 14, 215
    cytoskeleton, 14
    definition of, 14
    division of (See Mitosis), 16
    effects of freezing, 214
    epithelial, 18
    eukaryotes, 15
    fibroblasts, 19
    germ, 192, 193, 294, 304, 332, 350,
        352, 353, 398, 400, 401
    number in the human body, 233
    organelles
        aging and, 309
        definition of, 14
        endoplasmic reticulum, 14
        Golgi apparatus, 14
        mitochondria, 15
        vacuoles, 14
        vesicles, 14
    prokaryotes, 15
    somatic, 192
    specialized, 13

stages of development, 18
tetraploid, 440
tissues and, 19
transdifferentiation of, 432, 442
Chekov, A., 54
Chromosomes
    artificial, 407
    autosomal, 392
        number of, 392
    chromatin, 16
    definition of, 16
    diploid, 16
    genes and, 17
    haploid, 16
    histones, 16
    number of, 16
    sex chromosomes
        X chromosome, 16, 325, 395, 396
        Y chromosome, 16, 325, 395
    telomeres and, 303
Circadian rhythms, 107
    aging and, 284
Clarke, A.C., 218, 233, 272
Cloning
    planarians and, 286, 323
    somatic cell nuclear transfer (SCNT),
        332, 439, 441
    teleportation and, 274
    types of animals cloned, 333
    use of clones as organ donors, 456
Clopidogrel, 412
Computed tomography (CT), 134, 370
    for brain imaging, 253
    for diagnosis of coronary artery dis-
        ease, 163
    for diagnosis of ureteral stone, 134
    spacecraft and, 79, 152
    use with gold nanoparticles, 370
Cook, R., 390, 430, 454
Coronary angiography, 163, 452
Coronary artery disease
    comparison with cardiac allograft
        vasculopathy (CAV), 452
    coronary angioplasty, 412, 452

definition of, 45
evaluation by exercise tests, 162
genetic variants and, 284, 291
screening for in astronauts, 162
Crew Medical Officers (CMOs), 158,
        159, 171
    definition of, 158
Crichton, M., 47, 362, 380
Cummings, R., 3

D

Danner, H., 390
Decompression sickness, 96, 97, 99, 347
Dehydroepiandrosterone (DHEA), 298
Deoxyribonucleic acid (DNA)
    definition of, 11
    nucleic acids, 11
    sequencing, 17
Diabetes mellitus, 53, 135, 164, 295, 410
    risk of developing, 295
Disseminated intravascular coagulation,
        48
Diverticulitis, 155
Down syndrome (Trisomy 21), 284, 395,
        416
Drexler, K.E., 361, 364, 368

E

Echocardiography, 162, 407
Egg cells (ova), 16
    cloning and, 332
    cryopreservation of, 231
    effects of radiation on, 192
    fertilization of, 231, 324, 326, 327
        in vitro, 328
    genetic issues and, 398
    meiosis and, 16
    menstrual cycle and, 326
    monozygotic twins and, 397
    natural loss of, 327
    parthenogenesis and, 324
    sex chromosomes and, 325
    telomerase and, 306
Einstein, A., 5, 190, 241, 313

Electrocorticography (ECoG), 253, 489
Electroencephalography (EEG), 252,
   253, 488, 489, 492, 500
Electrons, 3
   annihilation by positrons, 272
   atom and, 3, 4
   particulate radiation and, 188
   quantum mechanics and, 256
   quantum teleportation and, 270
   size and mass, 3
   Sun and, 3
Electroreception, 257, 267
Elements
   compounds and, 6
   definition of, 3
   radioactive, 190, 194, 198, 254
      in fallout, 199
      positron emission tomography
         (PET) and, 254
   trace
      total parenteral nutrition and, 227
Endocrine system
   definition of, 31
   glands, 31
      adrenal, 32
      mammary, 332, 457
      parathyroid, 32
      pineal, melatonin and, 293, 298
      pituitary, 31, 293, 297, 307, 326
      prostate, 30, 169, 325
      thyroid, Graves' disease, 363
      thyroid, radioactive fallout and,
         199
Energy, 2
   basal metabolic rate and, 115
   dark, 5
   definition of, 4
   electrical, 4
   electromagnetic, 5, 143, 189
      active galactic nuclei and, 204
      telepathy and, 254, 255
   forms of, 4
   gravitational, 5
   in space, 75

   kinetic, 4
      meteoroids, 108
   magnetic, 4
   nuclear, 4
   relation to radiation, 188
   thermal, 5
Entry Motion Sickness
   definition of, 125
Enzymes
   definition of, 9
Epigenetics, 18, 396
Eukaryotes, 15, 323, 324
   mitochondria and, 15, 16
Exercise
   aerobic, 125, 129, 131, 165
   astronauts and, 125, 129, 131
   resistive, 129, 165, 229, 349
Explosions
   injuries caused by, 73
      blunt trauma, 74
      flash burns, 74
      primary, 74
      psychological effects, 74
      secondary, 74
      tertiary, 74
   shock wave, 73
   supernovae, 201, 204
Extracorporeal membrane oxygenation
   (ECMO), 220, 331
Extrasensory perception (ESP), 266, 268,
   269
Extravehicular activity
   definition of, 98
   monitoring during, 127
   physical effort during, 129
   preparation for, 98
   risks of, 123
   spacesuits and, 98

F
Frankenstein, V., 78
Frogs
   levitation of by a magnetic field, 144
Fruit fly (*Drosophila melanogaster*), 16

lifespan extension in, 288
  effects of dietary restriction, 299
number of chromosomes in, 16
sexual reproduction in space, 341
typical lifespan of, 290
Functional near-infrared spectroscopy
    (fNIRS), 253
Fungi, 14, 169, 399
  in spacecraft, 166, 169

G

Gannon, C.E., 244
Gastrointestinal system
  components of, 27
  esophagus, 18, 26, 27, 133
    injuries to, 63
  gallbladder, 28
  intestines, 18, 34, 43, 65
    large, 28
    small, 28
  liver, 29
  hepatocytes, 304
  pancreas and, 28, 29, 32, 65, 71, 125
    stem cells in, 433
  stomach, 18–20, 26, 27, 29, 65
Gene therapy, 397
  adeno-associated viruses, 404
  adenoviruses, 402
  aging and, 297
  effectiveness of, 408
  foamy virus, 406
  for cardiovascular disease, 402
  herpes virus, 404
  human research trials, 408
  lentiviruses, 406, 414, 446
  murine leukemia virus, 405
  naked DNA, 407
  nanoparticles and, 377
  of fetus, 399
  of germ cells and embryo, 400
  poxviruses, 407
  retina, 401
  retroviruses, 405
  risks of, 403

causing cancer, 406
  leukemia, 406
treatment of burns, 414
treatment of cancer, 377
treatment of skeletal muscle degenera-
    tion, 414
using induced pluripotent stem cells,
    443
vectors for, 402
Genes
  beneficial, 400
  BRCA1 and *BRCA2* genes, 162, 394,
    409
  codon, 13, 391
  DNA and, 17, 391
  inheritance of, 393
    Mendelian laws, 393
  in humans, 17
  intelligence and, 401
  number of, 396
    in humans, 17
  oncolytic, 405
  polygenic, 397, 410
  prenatal screening, 328
  sirtuin, 290
Genetic diseases, 396
  alpha-thalassemia, 399
  beta-thalassemia, 399
  Fragile X syndrome, 416
  hemophilia A, 284
  Huntington's disease, 284, 400
  Leber congenital amaurosis, 399, 404
  Lesch-Nyhan syndrome, 399
  lipoprotein lipase deficiency, 408
    treatment with gene therapy, 408
  ornithine transcarbamylase deficiency,
    403
  progeroid syndromes, 291
  Tay-Sachs disease, 399, 400, 410
Genetic engineering
  brain and, 415, 416
  dangers of, 420
  for prevention and treatment of can-
    cer, 401

nonhumans and, 390
Genome, 391, 396, 398, 413
    definition of, 17
Genomic medicine, 409
Genotype, 394, 397, 412
    definition of, 392
Gibson, W., 469, 493, 513
Glial cells, 20, 252, 405
Glucose
    adenosine triphosphate (ATP) and
        aerobic metabolism, 215, 292
        anaerobic metabolism, 215, 292
    as cryoprotectant, 215
    as energy source in the human body,
        9, 216
    as fuel for neuroprosthetics, 491
    as fuel for respirocytes, 365
    brain imaging and, 254
    glucagon and, 32
    glycogen and, 9, 29, 216
    hypothermia and, 216, 227
    insulin and, 32, 227
    oxidative phosphorylation and, 15
    relation to lactose, 9
    relation to sucrose, 9
    sirtuin gene and, 290
    total parenteral nutrition and, 226
Gold, H.L., 459
Gravity
    artificial
        Coriolis force, 139, 140
        gradients and, 140, 141
        gradients involving, 140
        limitations to, 139
        rotation and, degree of, 137, 138
        technical issues, 136
    as curvature in spacetime, 122
    at Earth's surface, 121
    center of, 138
    comparison to telekinesis, 259
    low-Earth orbit and, 121
    lunar, 127
    Newtonian description of, 122

on Mars, 127
    relative strength of, 260
Growth hormone, 31, 297, 408
    circadian rhythms and, 107
    effects on aging in mice, 289

H

Hasse, H., 3
Hayflick, L., 302
Heart (*See also* Cardiovascular System)
    artificial, 485
    cardiac cycle
        systole and diastole, 24
    congestive heart failure, 171, 296
    echocardiography and, 162, 407
    effects of teleportation, 273
    exposure to vacuum and, 94
    injury to, 62
    myocardiocytes, 19, 193, 304, 402,
        430, 435, 438, 441
    puncture of, 62
    risk of developing tumors, 52
    stem cells in, 433, 435
    transplantation, 450, 457, 459, 486
Heinlein, R.A., 7, 50, 54, 89, 187, 199,
        207, 238, 251, 281
Hematopoietic system (*See also* Red blood
        cells and White blood cells)
    components of, 32
Hemoglobin, 24, 32, 58, 220
    beta chain, 391
    beta-thalassemia, 399
    color of
        arteries, 24
        veins, 24
    comparison to hemocyanin, 33
    effects of hypothermia on, 220
    genetic variants, 32, 391
Hormesis, 293
Hormones
    definition of, 18
    endocrine system and, 31
    insulin, 32

parathyroid hormone, 32
  replacement, 297
  sex, 11, 32, 294, 297, 325
  thyroid stimulating hormone, 31
Hoyle, F., 7
Hubbard, L.R., 76
Human immunodeficiency virus (HIV), 47, 400
  AIDS, 47, 400, 406
  treatment of, 377, 400
Huxley, A., 321, 514
Hydrogen
  abundance in the human body, 7, 274
  abundance in the universe, 7
  amino acids and, 9
  atom
    size of, 3
  bomb, 272
  Bussard ramjet and, 137
  carbohydrates and, 9
  deuterium, 3
  fusion of in stars, 142
  in atmosphere of outer planets, 93
  interstellar gas and, 204
  involvement in life processes, 3
  lipids and, 10
  peroxide, 293
  sulfide, 235, 236, 293, 352
  tritium, 3
Hypospray, 80, 81
  issues with using, 80
  jet injectors and, 80
Hypothermia
  definition of, 218
  mild, 218, 222, 223, 228
    induced for space travel, 222, 519
  moderate, 218, 221
  profound, 219, 221
    in dogs, 222
  severe, 219
  therapeutic, 221
  treatment of, 220

I
Immune system
  aging and, 284, 459
  autoimmune diseases, 43, 284, 363
  components of
    cells of, 33
    tissues and organs, 34
  failure to recognize cancer cells, 363
  genetic engineering and, 401
  immunosuppressive medications and, 451
  in children, 350
  low-level stimulation by intestinal bacteria, 167
  microgravity and, 128, 162
  nanobots and, 367
  nanoparticles and, 374
  rejection of transplanted organs and tissue, 50, 450, 452
  sleep and, 411
  suppression of
    risk of cancer from, 447
Infectious diseases
  aging and, 284
  alien microorganisms and, 48, 49
  cellulitis, 47
  diagnosing using genomic sequencing, 413
  post-operative wound healing in space and, 170
  preflight quarantine of astronauts, 169
  prostatitis, 169
  radiation exposure and, 196
  urinary tract infection, 225
  zoonotic, 456
Injuries
  blunt trauma
    abdomen and pelvis, 71
    bones and, 70
    definition of, 69
    face and, 71
    skull and, 70, 74

subdural hematoma and, 60
penetrating, 54
effects of, 55
to eye, 172
Integumentary system
definition of, 31
Intermittent resistive exercise device
(IRED), 131
International Space Station
altitude of, 201
atmosphere on, 98
foods on, 166
gravity and, 121
habitable volume of, 90
measuring time on, 108
medical problems on
classification, 158
medical supplies and equipment on,
157
medications stocked on, 155
noise levels on, 338
radiation exposure on, 201
rotating to provide artificial gravity,
140
sound levels on, 155
space debris and, 109
toilets on, 133
Ions, 3, 4, 14, 23
ionization, 3
Isotopes
definition of, 3
radioactive, 198

K

Kalpana space habitat, 139
Kidney and ureteral stones
history of in astronaut screening, 165
risk of from acetazolamide, 99
risk of in microgravity, 130, 134
salt intake and, 166
Kornbluth, C.M., 77, 178
Kuttner, H., 47, 187, 250

L

Lerner, E.M., 115, 141, 329, 362
Leukemia, 51, 402, 433
definition of, 197
radiation exposure and, 197
Life
amino acids and, 10, 48
definition of, 6, 8
extraterrestrial, 7, 10
life processes, 3, 4, 7, 13
origin of, 49
processes, 3, 4, 7, 13, 49, 364
terrestrial, 6–8, 11, 12
carbohydrates and, 9
elements used by, 21, 274
lipids and, 8, 10
nucleotides and, 8, 12, 303
proteins and, 9
Light
infrared, 189, 191, 267
and vision in animals, 267
near-infrared, 370, 372, 377
ultraviolet
and vision in animals, 267
burns caused by, 172
effects on production of amino
acids in space, 10
keratitis and, 173
production by stars, 204
tanning caused by, 191
wavelengths of, 190
visible, 189–191, 205, 370
wavelength range of, 190, 477
Lipids
adipose tissue, 11
cholesterol, 11, 44, 284, 301, 400,
412, 414, 415, 421
definition of, 10, 11
fatty acids, 11
beta oxidation of, 15
nanoparticles and, 372
phospholipids, 11
triglycerides, 11, 227

Lungs
   alveoli and, 26
   injuries to, 63
      exposure to vacuum, 94
      hemothorax, 63
      pneumothorax, 63
   pressure in, 92
   pulmonary embolism, 44, 181

M

Magnetic resonance imaging (MRI), 152,
   260, 370, 497
   functional magnetic resonance imag-
     ing (fMRI), 253
Magnetoencephalography (MEG), 253
Magnetoreception, 267
Marfan syndrome, 393, 394, 396
Mars
   asteroid belt and, 49, 110
   atmospheric pressure, 93
   flight to, 127, 162, 230
      artificial gravity, 138
      bone loss during, 130
      hibernation during, 222
      radiation exposure, 203, 346
      radiation protection, 201
      restrictions on medical equipment,
        152
      sleep disturbances during, 108
   length of day on, 107
   possible evolution of life, 49
Matter
   baryonic, 5
   conversion into energy, 136, 272
   dark, 5
   definition of, 4
Meiosis, 16
Melatonin, 107, 298
Meteoroids, 108–110
Methridge, S.T., 77
Mice
   hibernation in, 213, 235
   levitation of by a magnetic field, 144,
     260

lifespan extension in
   effects of dietary restriction, 288
   typical lifespan of, 290
Microgravity
   anemia and, 127
   bleeding in, 174
   cardiovascular system and
     effects on the heart, 127
   children and, 349
   eating and drinking in, 166
   effect on vision
     increased intracranial pressure, 132
     papilledema, 132
   effects on the endocrine system, 135
   effects on the lungs, 135
   effects on use of medications, 157
   genetic engineering, 136
   immune system and, 22
   infections in, 128, 168
   neurovestibular system and
     inversion illusion, 124
   personal hygiene in, 168
   surgery in
     anesthesia and, 169
   use of intravenous fluids in, 174
   waste disposal in, 134
Mitochondria
   DNA in, 15, 293
Mitosis, 16
Molecules
   definition of, 3
   effects of a magnetic field, 260
   hydrophilic versus hydrophobic, 11
   polymers, 371
Moore, C.L., 47, 187, 250, 469
Moorhead, P., 302
Muscles
   effects of microgravity
     loss of muscle mass, countermea-
       sures, 128, 129, 140
     loss of muscle mass, in children,
       349
   extraoccular, 21
   relation to bones, 19

rhabdomyolysis, 412
skeletal, 19
smooth, 19
Musculoskeletal system
components of, 30
Myasthenia gravis, 54
Myocardial infarction, 45
cardiac allograft vasculopathy and, 452
definition of, 45
increased risk of with sleep deprivation, 410
risk of during coitus, 336
stem cells and, 442
Vitamin E and, 299
Werner's syndrome and, 291

N

Nanoparticles
methods for administering, 373
Nanotechnology, 34
carbon nanotubes
definition of, 372
interaction with the immune system, 374
quantum dots and, 377
risks of, 381
use to help grow new tissues, 376
definition of, 361
for gene therapy, 377
for growing new tissues, 376
for organ transplantation, 377
gold nanoparticles, 370
iron oxide nanoparticles, 370, 382
liposomes, 372
microarrays, 379
microbivore, 365
nanobots
motion of, 365
need for an energy source, 366
nanomachines, 364
nanofactory, 366
nanomotors, 376
natural, 480
noocytes, 362

photothermal therapy, 370, 373, 377
quantum dots, 371, 376
respirocyte, 365
silver nanoparticles, 371
stem cells and, 377
theranostics, 377
Nervous system (*See also* Brain)
autonomic, 21, 124
central, 20
components of, 20
cranial nerves, 20
brainstem and, 20
neurons, 21
peripheral, 20, 61
spinal cord, 20, 61
cauda equina, 61
vertebral column and, 61
stem cells in, 433
Neutrons
atom and, 3
particulate radiation and, 188, 198
radiation quality factor and, 194
size and mass, 3
Nitrogen
abundance in the human body, 3
amino acids and, 9
atmospheric
on Earth, 25
level in spacecraft, 98
decompression sickness, 96, 347
bubble formation in blood, 96
in atmosphere of Mars, 93
in atmosphere of Titan, 93
in blood, 96
involvement in life processes, 364
liquid, 231, 236
Niven, L., 42, 77, 139, 151, 178
Nourse, A.E., 77
Nucleotides, 8, 12, 303, 391
adenine, 12, 391
Hemoglobin A and, 391
cytosine, 12, 391
definition of, 12
guanine, 12, 391

single-nucleotide polymorphism
  definition of, 396
thymine, 12, 391
  Hemoglobin S and, 391
triplet and, 13
uracil, 12
Nucleus
  atom, 3
  cell
    lack of in red blood cells, 32
    nucleoskeleton, 15
  comet, 111
Nutrients
  antioxidants, 166, 293, 299
Nutrition
  total parenteral, 181, 226

O

Organ and tissue transplantation, 232
  adult stem cells, 432
    mesenchymal, 447
  allogenic, 436
  antirejection medications, 452
  autologous, 436
  brain death, 221, 454
  brain, 238
  complications
    cardiac allograft vasculopathy
      (CAV), 452
    development of cancer, 451
    infections, 451
    post-transplant lymphoproliferative
      disorder (PTLD), 451
    transplant arteriopathy, 451
  cryopreservation of organs, 232
  embryonic stem cells (ESCs)
    rejection and, 446
  face, 453
  general use of, 429
  graft-versus-host disease, 421
  head, 238
  heart, 450, 452
    artificial, 485
    heterotopic, 458

human leukocyte antigen (HLA) sys-
    tem for tissue matching, 436
immunosuppressive medications, 451
kidney, 449
liver, 450
lung, 450
pancreas, 450
scarcity of donors, 453
timing after harvesting, 232, 453
tissue typing, 450
use of animals, 455
  risk of infection, 456
use of bone marrow and stem cells, 435
vitrification of large organs, 234
Organ systems
  definition of, 20
Orthostatic hypotension, 126, 127
Oxygen
  abundance in the human body, 6, 10,
    274
  aerobic metabolism, 30
  amino acids and, 9
  atmospheric, 25
    level in spacecraft, 98
  carbohydrates and, 9
  hyperbaric, 97, 347
  in atmosphere of Mars, 93
  in atmospheric of Titan, 93
  involvement in life processes, 3
  lipids and, 10
  respiration and, 23
  saturation, 23
    assessment by functional magnetic,
      resonance imaging (fMRI), 253
  supplemental, 92
  suspended animation and, 212

P

Pacemaker, 454, 484
  cells, 25
Parkinson's disease, 293, 382, 439
  deep brain stimulators and, 487
Phenotype, 325, 392–394, 397, 398
  definition of, 392

Phosphorus
  abundance in the human body, 274
Photodynamic therapy of cancer, 376
Plants, 7, 9, 15, 390, 392, 408
  as sources of microorganisms in space,
    167
  genetic modification of, 408
  uptake of strontium-90 by, 199
Poisons, 55, 75
  hydrogen cyanide, 75
  potassium cyanide, 75
Positron emission tomography (PET),
    253, 254
Positrons, 188, 200, 254
  annihilation by electrons, 272
Potassium, 14, 29
  cell functions and, 6
  total parenteral nutrition (TPN) and,
    226
Pournelle, J., 42, 112, 232
Precognition, 264, 266
Prokaryotes, 15, 323
  bacteria, 323
  mitochondria and, 15
Proteins, 9, 15, 17
  albumin
    nanoparticles and, 372
  amino acids and, 9
  coding by DNA, 12
  polypeptides, 17, 450
  size of, 361
Protons
  atom and, 3
  isotopes and, 3
  particulate radiation and, 188, 200
  quantum teleportation and, 270
  radiation quality factor and, 194
  size and mass, 3
  Sun and, 203
Protozoa, 13, 323, 337
Psychological stress in space, 105, 107,
    116
Pulmonary system
  components of, 25

Q

Quaddies, 136
  comparative anatomy of upper and
    lower extremities, 418
  genetic modifications in, 419
Quarks, 3

R

Radiation, 187–189, 204
  as plot element in science fiction,
    188, 197
  background, 91, 194, 277
  cosmic rays, 194, 200, 201, 205–208
  definition of, 188
  effects during pregnancy, 346
  effects in children, 346, 350
  effects on bone marrow cells, 194
  effects on the brain, 197
  effects on the gastrointestinal system,
    194
  electromagnetic, 5, 143, 189, 190,
    194, 198, 200, 201, 205, 206
  from nuclear explosion, 198
  gamma rays, 190, 191, 199–201, 203,
    205
    active galactic nuclei and, 204
  gamma-ray bursts and, 205
  gray
    definition of, 194
  in space, 200
  ionizing, 191, 192, 205, 207, 352,
    376
    definition of, 190
    for treating cancer, 376
  mutations produced by, 193
  near neutrons stars and black holes,
    204
  non-ionizing
    definition of, 191
  particulate, 188, 191, 198, 200, 201,
    204–206
    alpha particles, 188, 194, 198
    beta particles, 188, 194, 198

rad
definition of, 194
rem
definition of, 194
risk of developing cancer, 194, 197,
202, 350
in astronauts, 202
shielding, 199, 201, 206
use of regolith, 206
sievert
definition of, 194
survival of microorganisms in, 166,
337
therapy, 52
timing of hair loss, 196
Type Ia supernova and, 205
white dwarf and, 143
x-rays, 161, 162, 170, 189, 191, 198,
200, 201, 203
active galactic nuclei and, 204
Rapamycin, 289, 452
as treatment for cancer, 289
Reactive oxygen species (ROS), 293,
298, 300
Red blood cells (RBCs), 32, 433
ABO blood group system, 394
average lifespan of, 433
definition of, 32
effects of hypothermia on, 220
recycling by the spleen, 34
reduced number in microgravity, 418
Resveratrol, 290
Rhesus monkeys, 299, 300
Ribonucleic acid (RNA), 12, 48, 49, 303,
307, 350, 363
definition of, 11
messenger RNA, 13
primer, 303
transcription, 12, 307
translation, 13
Ribosomes, 15
size of, 361
Roundworm (*Caenorhabditis elegans*)
lifespan extension in, 288–290, 307

effects of dietary restriction, 299
sexual reproduction in space, 341
transgenerational inheritance, 307
typical lifespan of, 290

S
Sawyer, R.J., 455, 493, 497
Schmidt, S., 141, 241, 516
Septic shock, 64, 65, 197, 230
Severe combined immunodeficiency
(SCID), 398, 405
Sexual reproduction
artificial uterus, 329, 330, 332, 355
520, 524, 526, 527
coitus, 327, 328, 336–340, 356
microgravity and, 338
penile fracture caused by, 339
comparison to asexual reproduction,
323, 324
definition of, 325
estrous cycle in non-primates, 328
function of, 323
genetic diversity and, 323
genetic engineering and, 353
germ cells and, 192
in non-primate mammals, 328
in space, 335
nonhuman animals and, 341
menstrual cycle
effects of microgravity on, 343
fertility and, 326
hormones and, 326
oral contraceptives and, 342
possible dysfunction of in space, 343
Turner syndrome and, 325
placenta
abruptio placentae, 348
formation of, 329, 332
functions of, 330
production of hormones by, 331
pregnancy
blood pressure during, 347
eclampsia, 348
ectopic, 155, 348

human chorionic gonadotropin (hCG) and, 331, 380
  initiation of, 326
  morning sickness and, 345, 347
  preeclampsia, 347, 348
  radiation exposure in space, 342
  risks in space, 345
  role of amniotic fluid, 329
Shelley, M., 78
Simak, C.D., 93
Siodmak, C., 238
Skin
  cryopreservation of, 231
  fibroblasts and, 302
Skywalker, L., 77, 262, 472
Sleep
  circadian rhythms and, 107, 108
    disruption of, in space, 107
  disorders of, 411
  melatonin and, 298
  need for, 410, 411
Smith, E., 1, 250, 282, 390
Sodium
  cell functions and, 14
Solar flares, 200, 203
Solar wind, 200
  coronal mass ejections, 200
  solar particle events, 200
Sommers, J., 467, 470, 482
Space Adaptation Syndrome (SAS)
  definition of, 122
Space debris, 108, 109
Space Motion Sickness (SMS)
  comparison to entry motion sickness (EMS), 125
  definition of, 122
  frequency of, 123
  medications for, 124
  pregnancy and, 345, 347
  severity of, 123
  size of spacecraft and, 124
  susceptibility to, 123
Space Shuttle
  space debris and, 109

Sperm cells
  anatomy of, 327
  cryopreservation of, 231
  formation of, 325
  genetic issues and, 398
  intracytoplasmic injection, 328
  meiosis and, 16
  number produced, 328
  radiation and, 192, 194, 196
  rate of production, 192
  sex chromosomes and, 325
Star
  neutron, 141–143, 204
  white dwarf, 141, 142
Stem cells
  adult
    definition and types of, 429, 432, 433
    harvesting of, 435
    life expectancy of, 434
    possible difference in males and females, 443
    role of, 433
  cryopreservation of, 231
  definition of, 429
  division of, 431
  embryonic
    as source of neural stem cells, 439
    comparison to induced pluripotent stem cells (iPSCs), 439
    definition of, 431
    for treatment of cancer, 449
    immune response and, 442
    risks of, 444–449
    source of, 432
    tissues produced by, 431
  fetal
    definition of, 432
  hematopoietic, 433, 435, 436
    cells derived from, 32, 433
  in muscles, 433
  induced pluripotent cells (IPSCs), 439
    comparison to embryonic stem cells (ESCs), 440, 442

current uses, 441
definition of, 439
human clinical trials and, 445
immune response and, 442
longevity of, 449
potential use with gene therapy,
442
production using viruses, 445
risk of cancer, 445
risks of, 444, 445
tetraploid complementation and,
440
types of cells created from, 439
mesenchymal
as a possible source of neural stem
cells, 439
characteristics of, 436
definition of, 432
dose used for treatment, 437
for treatment following a myocardial infraction, 438
human clinical trials and, 437
numbers in bone marrow, 436
possible use with gene therapy, 413
risk of immune response and, 436
transdifferentiation and, 432
telomerase and, 306
teratomas and, 444
Stewart, G.R., 47
Sulfur
amino acids and, 9
Surgery
in space, 169
Suspended animation
comparison to therapeutic hypothermia, 222
comparison to travel at relativistic
speeds, 240
cryoprotectants, 215, 234
cryostat, 237
definition of, 211
for cells, 230
hibernation
in animals, 213–215

rewarming, 214
hydrogen sulfide and, 235
in animals, 212
neuropreservation, 237
of whole human body, 214–217
space travel and, 212, 222, 239
torpor, 213, 217, 222, 223, 236
vitrification, 234, 235, 237, 242, 243
Symbiotes, 50, 51

T

Telekinesis, 259–264
Telemedicine, 158, 160, 170
Telepathy
aliens and, 250, 259
definition of, 259
theoretical mechanisms for, 252
Teleportation
cloning and, 274
definition of, 270
matter transference and, 271
quantum entanglement, 256, 502
quantum teleportation, 270, 271
qubits, 270
Temperature
in space, 99
Teratomas
definition of, 444
induced pluripotent stem cells (iPSCs)
and, 445
relation to stem cells, 444
Tissue
adipose, 11, 19
hibernation and, 214
connective, 19
abnormalities in Marfan syndrome,
393
cryopreservation of, 231
definition of, 19
Toxins in spacecraft
ethylene glycol, 102
formaldehyde, 101
halocarbons, 102
hydrazine, 100

nitrogen tetroxide, 100
  risks during pregnancy, 346
Tsiolkovsky, K., 89, 138
Twins
  dizygotic, 291
  monozygotic, 291, 334, 397
    antigen matching, 450
    definition, 397

U

Urinary system
  bladder, 29
  components of, 29
  kidneys, 29
  ureters, 29

V

Vacuum
  explosive decompression with, 97
  exposure to, 94, 95
    decompression sickness, 96
    survivability of, 94
    time of useful consciousness, 97
  speed of light in, 5
  survival of microorganisms in, 168
Van Allen belts, 201, 203, 204
  electrons and, 201
  protons and, 201
Verne, J., 89, 514
Viruses
  animal, 455
  as causes of disease, 128, 337, 378
  Epstein-Barr, 451
  extraterrestrial, 49
  Sendai virus, 446

sexually transmitted diseases and, 337
  use in gene therapy, 402
Vision
  binocular, 478, 480
  cataracts, 173, 197, 291, 346
  color, 477, 480
  restoring lost, 477
  retina
    cones, 190, 477
    rods, 477

W

Wallace, F.L., 460
Warfarin, 411
Water
  abundance in the universe, 7
  as part of comets, 111
  blood-brain barrier and, 382
  electroreception and, 257
  glass transition temperature, 234, 237
  importance of for life, 6
  involvement in life processes, 8
Wells, H.G., 7, 47, 78, 89, 211, 260,
    321, 390, 509
White blood cells (WBCs)
  average lifespan of, 433
  definition of, 32
  effects of radiation on, 194
  telomere length, 307
  types of, 33
Williamson, J., 42

Z

Zelazny, R., 264

Printed in the United States
By Bookmasters